Environmental Noise Control

Naveen Garg

Environmental Noise Control

The Indian Perspective in an International Context

Springer

Naveen Garg
Senior Principal Scientist and Head
Acoustics and Vibration Standards
CSIR-National Physical Laboratory
New Delhi, Delhi, India

ISBN 978-3-030-87827-6 ISBN 978-3-030-87828-3 (eBook)
https://doi.org/10.1007/978-3-030-87828-3

This Springer imprint is published by the registered company Springer Nature Switzerland AG
The registered company address is: Gewerbestrasse 11, 6330 Cham, Switzerland

Foreword

Globally, environmental noise pollution has become a staid predicament. In Indian scenario, the authorities and related establishments have been very serious to tackle this grave issue, effectively working in synergy to chalk out new plans and strategies as well as smoothening the existing activities to control the accentuated traffic noise levels in the cities. This book is an excellent example and initiative encompassing various sources of noise and related issues in a very precise, concise and a single-window reference source. Also elaborated are the possible suggestions and recommendations on noise polices and noise action plans from national and international perspectives. One of the main features of the book is an elaborative discussion on the various noise sources, solutions and recommendations for formulating the noise polices and action plans. Being a noted researcher from the National Metrology Institute of India, the author has rightly included and discussed the dissemination of an unbroken chain of measurement traceability, starting from national primary sound standards of CSIR-National Physical Laboratory, New Delhi, India, to the various stakeholders across the country through the national legal metrology framework. I strongly feel that the present book would prove to be an outstanding and useful contribution for practising engineers, researchers, university faculty, students, town planners and pollution control boards as a reference book for environmental noise control.

The book is comprised of ten chapters covering the basics of environmental noise, noise policies, ordinances across the globe, strategies adopted for noise mapping, planning for control of the environmental noise levels and effects of noise on health. The implications and conclusions derived from the analysis of environmental noise levels of various cities are brownie points of the book. The comprehensive and illustrative analysis carried out and reported would be highly useful to the administrative authorities and policy makers in devising and formulating noise standards in future. The discussion on noise mapping of cities, its control, and impact on health and society is extremely impressive and motivational. I am sure that this book will be well taken and greatly appreciated by science and engineering graduates, research scholars, pollution control authorities, policymakers and researchers in this field. I congratulate Dr. Naveen Garg for his untiring efforts and timely endeavour to bring his long experience and expertise in the form of a source book and extend my best wishes for the grand success and wide circulation of the book.

Dr. Sanjay Yadav
Chief Scientist and Head, Physico-Mechanical Metrology Division,
Professor-AcSIR, Faculty of Physical Sciences and
Editor-in-Chief, MAPAN-Journal of Metrology Society of India
CSIR-National Physical Laboratory, New Delhi, India

Foreword

Dr. Naveen Garg has been a scientist at the CSIR-National Physical Laboratory, New Delhi, for about 18 years. As a former chairman of the National Committee for Noise Pollution Control (for 18 years) and president of the Acoustical Society of India, I have known Dr. Garg all along as an active researcher in the field of environmental noise control. He has been deeply involved in international key comparison exercises for sound and vibration parameters with other national metrology institutes of the world that has led to the publication of Calibration and Measurement Capabilities (CMCs) of CSIR-NPL, India, in the BIPM Paris Key Comparison Database (KCDB). He has been internationally recognized for his notable contribution to acoustic and vibration metrology. In summary, Dr. Garg is well qualified to write this book on environmental noise control.

This book comprehensively focuses on basics of environmental noise, various noise sources, legal noise regulations and standards followed in national and international perspectives, measurement facilities, methodologies and their traceability to the national standards, noise control measures, methodologies for conducting noise impact assessment studies, noise mapping studies, and health issues associated with environmental noise.

Dr. Garg presents a novel approach of multiple-criteria decision-making (*MCDM*), known as the Fuzzy TOPSIS (Technique for Order Preference by Similarity to an Ideal Solution) method, for prioritizing noise action plans when the opinions of experts vary, and so this will be very indispensable for the constitutional committees set up for recommending noise control strategies.

The comprehensive discussion on the evidence-based linking of noise exposure with health aspects in national and international perspectives; concise summary of various empirical formulations correlating the noise exposure with health effects and the presentation on the concept of disability-adjusted life years (DALYs) and a case study for estimation of DALYs; recommendations on sampling strategies for noise mapping of metropolitan cities of India; noise labeling of appliances; plan for a national planning policy framework for environmental noise management and control in Indian perspectives, acoustics-induced vibration in building elements;

proposal of a noise abatement program and noise abatement goal; cost-benefit analysis of various noise control strategies; accuracy of long- term and short-term and temporal sampling strategies for noise monitoring; urban morphology and planning for traffic noise control; concept of soundscape approach in noise management and control; and soundscape mapping make this book distinctive and inimitable.

This book will prove to be an outstanding and valuable contribution for practicing engineers, researchers, university faculty, students, town planners, pollution control boards, and policy makers, and a reference book for environmental noise control.

AICTE Distinguished Chair Professor
Indian Institute of Science
Bengaluru, Karnataka, India
August 2021

Professor M. L. Munjal

Preface

Environmental noise has emerged as a serious problem globally. The alarming increase in vehicular population and excessive use of horns in some countries have become a nuisance for society, causing serious annoyance within. The complex and pervasive problem of noise pollution has no single solution, rather it requires a combination of short-, medium- and long-term approaches and careful consideration of the nature of the noise source. There are many sources of noise pollution, each requiring tailored abatement measures. Roadway noise and aircraft noise are the most pervasive sources of environmental noise. The main areas of environmental noise mitigation or abatement are: transportation noise control, architectural design, use of passive control measures and urban planning. The present book is written with an objective of presenting the various aspects of noise control particularly in Indian perspectives. A long list of references, elaborate discussion of various noise sources, recommendations on noise polices and noise action plans are the strong features of this book. It is envisaged that the present book shall be very helpful for the practising engineers, researchers, university faculty, students, town planners and pollution control boards as a reference book for environmental noise control.

The book is divided into ten chapters covering the basics of environmental noise, noise policies and ordinances across the globe, strategies adopted for noise mapping and control of ambient noise levels, and health effects of noise. Chapter 1 deals with the basics of environmental noise and the fundamental theory of noise including the noise measurement scale, outdoor sound propagation and legal framework in India for controlling noise pollution. An overview of various sources of noise pollution and their characteristics is provided in Chap. 2. Chapter 3 presents an overview of acoustics metrology and the primary and secondary sound standards realized for the establishment of traceability chain in free-field sound measurements. Chapter 4 presents a retrospective view of noise policies and ordinances across the globe. The Noise Pollution (Regulation and Control) Rules of India and those followed in other countries are discussed. Chapter 5 presents an overview of measurement strategies carried out across the globe for evaluation and analysis of ambient noise levels in urban cities. It discusses the ambient noise levels measured in past 10 years from 2011 to 2020 and presents the implications and conclusions derived from the analysis of noise monitored data available from the 70 sites. Chapter 6 presents an overview of noise impact assessment studies for road, railways and aircraft noise. The various traffic noise models developed in recent years and noise impact assessment criteria are discussed in detail. The various aspects of noise control and active and passive noise control strategies are discussed in Chap. 7. Chapter 8 presents a review of noise mapping procedures for conducting noise impact assessment studies. The noise mapping strategy, benefits of noise mapping and concerned noise control action plans are discussed in elaborate. Chapter 9 reviews the various health effects associated with environmental noise and empirical formulations proposed in recent years for correlating annoyance and sleep disturbances with noise exposure. Finally, Chap. 10 presents a retrospective study for selection of regional road traffic noise management strategies using an *MCDM* approach known as Fuzzy TOPSIS (Technique for Order Preference by Similarity to an Ideal Solution).

The author wishes to acknowledge the inspiration and support of Professor Venu Gopal Achanta, Director, CSIR-National Physical Laboratory, New Delhi. The support of Mr. Vishal Gandhi and Mr. A. K. Sinha of Central Pollution Control Board; Dr. Sanjay Yadav, Divisional Head, Physico-Mechanical Standards; Dr. Chitra Gautam, Senior Scientist, Acoustics and Vibration Standards and Mr. Abhishek Sharma of CSIR-National Physical Laboratory is sincerely acknowledged. The author would also like to specially thank Dr. Birgit Rasmussen, Senior researcher at SBi, Danish Building Research Institute, Aalborg University Copenhagen (AAU-CPH), and Dr. María Machimbarrena of Applied Physics Department, E.T.S. Arquitectura, University Valladolid, Spain, for their helpful comments and suggestions for Chap. 7. The author is extremely thankful for the noise-monitored database published in CPCB, Delhi; DPCC, Delhi; and MPCB, Mumbai websites, which has been presented in Chap. 5 of this book. Author is extremely thankful to all the eminent researchers for their kind permission to utilize their research data, text, figures and tables in this book.

The author would like to thank his father, Mr. B. R. Garg; his mother, late Mrs. Neelam Garg; sister, Pooja; and brother, Vivek for all the love and encouragement. The author expresses his thanks to his wife Mrs. Vishesh Garg and daughters Rhythm and little angel Arshita (Chahat) for their wholehearted support. The ***author wishes to dedicate this work to his parents, late Mrs. Neelam Garg and Mr. B. R. Garg, for their love, endless support, encouragement and sacrifices.***

Above all, the author wishes to remember and pay his obeisance and reverence to "The Almighty" for his graciousness. ***Thanks for everything!***

New Delhi, India

Naveen Garg

Contents

About the Author

Dr. Naveen Garg is presently working as Senior Principal Scientist and Head, Acoustics and Vibration Standards, CSIR-National Physical Laboratory, New Delhi. He had been working in the field of acoustics and vibrations since past 18 years. He is a mechanical engineer specializing in machine design, vibrations and acoustics, measurement science, and working in development and up-gradation of primary standards of sound pressure and vibration amplitude, testing of the acoustical materials for sound insulation and absorption characteristics, machinery noise diagnostics and control, *in-situ* noise and vibration measurements and control and R&D in applied acoustics. He has done his doctorate in mechanical engineering from Delhi Technological University, Delhi and M. Tech in Machine Design from IIT Delhi. He has been involved in many consultancy and sponsored projects pertaining to environmental impact assessment studies in respect of noise and vibration for govt. bodies like the Central Pollution Control Board (CPCB), Archaeological Survey of India (ASI), Delhi Metro Rail Corporation (DMRC), and Bangalore Metro Rail Corporation Limited (BMRCL), India etc. He has published many papers in national and international journals in the field of acoustics and vibrations and has been actively involved in the key comparison exercises in acoustics and vibration metrology with other National Metrology Institutes (NMIs) of the world that has led to publishing of Calibration & Measurement Capabilities (CMCs) of CSIR-NPL, India, in BIPM, Paris Key Comparison Database (KCDB). He has been Managing Editor of *MAPAN-Journal of*

Metrology Society of India published by Springer, and reviewer of many international journals, and has enormously contributed towards evaluation, analysis and control of noise pollution. He was awarded **APMP Iizuka Young Metrologist Prize** for Developing Economies by Asia Pacific Metrology Programme (APMP) Secretariat, Japan, in November 2017 for his notable contributions to acoustics and vibration metrology. Dr. Garg is actively involved in many national and international committees on acoustics and vibration metrology, noise and vibration monitoring and control and building acoustics.

Chapter 1
Environmental Noise Control: A Review

Abstract This chapter presents an overview of aspects of environmental noise. Environmental noise is an accumulation of noise pollution that occurs outside due to transport, commercial, industrial, and recreational activities. Sound is a pressure wave that propagates through an elastic medium at some typical speed. For this wave motion to exist, the medium has to possess inertia and elasticity. While vibration relates to the wave motion in structural elements, sound relates to such wave motion in fluids (gases and liquids). This chapter highlights the introductory theory of sound, the decibel scale of sound measurement, loudness and weighing networks, directivity, hearing mechanism, outdoor sound propagation, and highlights the challenges ahead in Indian scenario for noise pollution abatement and control.

1.1 Introduction

Sound can be of different types, coming from different sources such as human speech, birds chirping, musical instruments, machines, transportation sources or due to religious ceremonies. Noise is generally considered as an unwanted sound. The word "unwanted" indicates that it is subjective as it may appear music to one young person, but may appear as unwanted sound or noise to another person who is trying to sleep or study. However, in general, noise is random in character and is sensed as unpleasant sound by most of the people, for example, the construction noise or traffic noise. The sound or noise heard by most people is generally the *immission* that may come from different sources and is dependent upon the environment in which the source is located. The source position, size of room, and amount of absorption effects the noise *immission* [1]. The outdoor noise levels vary with varied distance from the source owing to the changes in weather conditions and are influenced by topographical features, for example, ground cover, hills, and other obstacles between the source and the receiver. The environmental noise can have serious effects on human health such as annoyance, sleep disturbance, cognitive impairment, cardiovascular effects, and even hearing impairment. Noise pollution has been cited as an epidemic cause of psychological and physiological disorders internationally. There had been various studies carried out in many countries focused

N. Garg, *Environmental Noise Control*,
https://doi.org/10.1007/978-3-030-87828-3_1

on establishing a definite relationship between the health effects and traffic noise [2]. However, despite several fragmented policies, legislation and global efforts have been addressed; the noise pollution has been still a severe problem for many countries in the world. The harmful effects of noise pollution are evident from the World Health Organization analysis that around 250 million population worldwide are exposed daily to the dangerous noise levels. In the USA, 10 million of adults and 5.2 million of children approximately are suffering from noise-induced hearing disorders [3]. Thus, the community noise ordinances and global noise policies are being revised across the globe to enforce more stringent noise limits and criteria. In Indian perspectives also, the National Green Tribunal (NGT), New Delhi in its orders dated March 15, 2019 has directed Central Pollution Control Board (CPCB) of India to develop noise maps and identify the hotspots and propose a remedial action plan for controlling the noise pollution in cities. According to it, the state pollution control boards in India may undertake the noise level monitoring in conjunction with the police department and take remedial actions. Thus, it is the need of hour to monitor and assess the ambient noise levels and devise suitable noise control action plans for control of noise pollution in Indian cities.

1.2 Basic Acoustical Quantities

Sound or noise levels are measured in decibels (dB) which is a logarithmic unit for expressing the ratios of two values of physical quantities—sound pressure, intensity or power. Decibel is one-tenth of bel which is named after Alexender Graham Bell. The sound pressure level (*SPL*) is defined as:

$$SPL = 20 \log_{10}\left(\frac{p}{p_0}\right) \tag{1.1}$$

where p is the measured sound pressure and p_0 is the reference sound pressure of 20 μPa. Pascal (Pa) is the unit of pressure (N/m^2). A reference value of 20 μPa is chosen as it is the minimum sound pressure that human ear can detect. Exposure to higher noise levels may cause pain whereby the threshold of pain occurs at 140 dB. The sound intensity level is defined as:

$$SIL\,\text{or}\,L_I = 10 \log_{10}\left(\frac{I}{I_0}\right) \tag{1.2}$$

where I_0 is 10^{-12} W/m^2. The sound power level is defined as:

$$SWL \text{ or } L_w = 10 \log_{10}\left(\frac{W}{W_0}\right) \tag{1.3}$$

where W_0 is 10^{-12} W. Sound power is the amount of energy per unit time radiated from the source and measured in Watts. The sound power level of a source is independent of the environment. As a general rule, every increase of 3 dB represents a doubling of sound intensity or acoustic power. Decibel is a dimensionless unit and is a logarithmic value, so it is not proper to add them by normal algebraic addition. So, if 90 dB sound pressure level is added to 90 dB, the result will be 93 dB and not 180 dB. For a free-field environment, where there are no reflections, the correlation between the sound intensity and the root mean square acoustic pressure is:

$$I = \left(\frac{p_{rms}^2}{\rho c}\right) \tag{1.4}$$

where ρ is the density of air (kg/m^3) and c is the speed of sound (m/sec). The quantity ρc is called the "acoustic impedance" and is equal to 414 Ns/m^3 at 20 °C and one atmosphere. The sound intensity and sound pressure are correlated in far field of source as:

$$L_I = L_p + 10 \log_{10}\left(\frac{400}{\rho c}\right) \tag{1.5}$$

$$L_I = L_p + 26 - 10 \log_{10}(\rho c) \tag{1.6}$$

Thus, at the sea level and at 20 °C, the characteristic impedance ρc is 414 kg m^{-2} s^{-1}, so $L_I = L_p - 0.2$ (dB). The addition of two or more levels of sound pressure has a physical significance only if the levels to be added were obtained in the same measuring point. Consider a case wherein three machines are generating sound pressure levels of 90 dB, 88 dB, and 85 dB. The combined sound pressure level is measured as:

$$L_p = 10 \log_{10}\left(10^{(90/10)} + 10^{(88/10)} + 10^{(85/10)}\right) = 92.9 \text{ dB} \tag{1.7}$$

Similarly, the subtraction of sound level is performed. For instance, the sound level measured in an industrial setup is measured as 92 dB with a noisy machine in operation and 88 dB with noisy machine turned off, the sound level due to machine alone is calculated as:

$$L_p = 10 \log_{10}\left(10^{(92/10)} - 10^{(88/10)}\right) = 89.8 \text{ dB} \tag{1.8}$$

In cases wherein the difference in two sound levels exceeds 10 dB, the less noisy source has the least contribution. Also, for noise measurements, the total noise should exceed the background noise by 3 dB or more [4].

1.3 Loudness and Weighting Networks

Loudness is the subjective human response to sound that depends primarily on sound pressure levels and also on frequency of incident sound wave. Loudness level is expressed in phons, equivalent to the sound pressure level at 1000 Hz. In general, the human ear is least sensitive to extremely high and low frequencies. Table 1.1 summarizes the subjective perception of sound level changes [4].

The weighting networks are implemented by using electronic filters so as to provide a sound level meter response approximating the way the ear responds to loudness of pure tones. The weighting curves correspond to the Fletcher/Munson equal loudness contours. The commonly used weightings are: A-weighting (approximates 40 phon line); B-weighting (approximates 70 phon line); C-weighting (100 phone line); and D-weighting developed for aircraft flyover noise. However, A-weighting is most commonly used as it correlates reasonably well with how human ear responds to sound. These weightings are assigned in a way to reflect the higher sensitivity of human ear to sound in the mid- and high-frequency band. Table 1.2 shows the A-weighting correction values to the frequency range to mimic the way that the ear perceives the relative loudness of sound. The sound pressure level with the A-weighting "dB(A)" is thus a representative of the subjective response of the human ear. These weightings are assigned in a way to imitate the higher sensitivity of human ear to sound in the mid- and high-frequency band. The B-weighting is rarely used. The response of the human ear varies with the sound level. At higher sound pressure levels, the ear's response is flatter. C-weighting can

Table 1.1 Subjective effect of changes in sound pressure level

	Change in power		
Change in sound level (dB)	Decrease	Increase	Change in apparent loudness
3	1/2	2	Just perceptible
5	1/3	3	Clearly noticeable
10	1/10	10	Half or twice as loud
20	1/100	100	Much quieter or louder

Source: Hansen (2010, [4])

Table 1.2 A-, C-, and Z-weighting correction values

Frequency (Hz)	31.5	63	125	250	500	1k	2k	4k	8k	16k
A-weighting	−39.4	−26.2	−16.1	−8.6	−3.2	0	1.2	1	−1.1	−6.6
C-weighting	−3.0	−0.8	−0.2	0	0	0	−0.2	−0.8	−3.0	−8.5
Z-weighting	0	0	0	0	0	0	0	0	0	0

be used for peak measurements and low-frequency noise. It is commonly used for sounds with impulsive characteristics such as firearms, shooting ranges, pile driving, etc. C-weighting can be more informative as compared to A-weighting for certain noise sources with considerable sound energy below about 250 Hz. Most of the sound level meters have A- and C-weighting networks and flat or linear frequency response (i.e., no weighting). Z-weighting is a flat frequency response of 0.5 Hz to 20 kHz expressed as dB (Z) and is the actual noise with no weighting (Z for zero). Z-weighting has recently been used to measure explosive sounds and in the assessment of low-frequency noise [4–6].

1.3.1 Implications of A-Weighting

A-weighting is not a reliable measure of loudness for complex sounds consisting of multiple pure tones and/or broadband noise. The A-weighted sound level provides no sign of the frequency content of a complex noise, so it is ineffective for identifying or separating noise sources or for designing noise control measure [7]. Two sounds with the same A-weighted level can have quite different levels of annoyance [8]. The annoyance from the low-frequency noise was observed to be greater than that from the higher frequency noise at the same A-weighted level in a study [9]. The dB(A) underestimates annoyance for frequencies below about 200 Hz [10, 11]. At equal dB(A) levels, the noise dominated by the low-frequency component was perceived as 4–7 dB louder and 5–8 dB more annoying than the other noise [12].

1.4 Hearing Mechanism

Human ear is a complex mechanism consisting of several subsystems. The ear consists of outer, middle, and inner parts. The outer ear receives sound waves and sends them through auditory canal to eardrum causing it to vibrate. The ear drum which is a thin diaphragm completely seals the auditory canal. The middle ear has three small bones that work as a set of levers. They enhance the vibration energy from the eardrum and transfer it to the inner ear which has semicircular canals and "cochlea" that is concerned with hearing. The cochlea is the sensory system that converts the vibration energy of sound into electrical signals that are sent to brain for detecting the sounds. The cochlea is a coil of tube wound like a snail and has liquid-filled cavity divided into two canals by the "basilar membrane." The three stages involved for the ear to congregate, transmit, and perceive sounds from the environment are [5, 13]:

- Stage 1: The outer ear receives the incident sound waves and sends them to the eardrum through auditory canal causing it to vibrate.

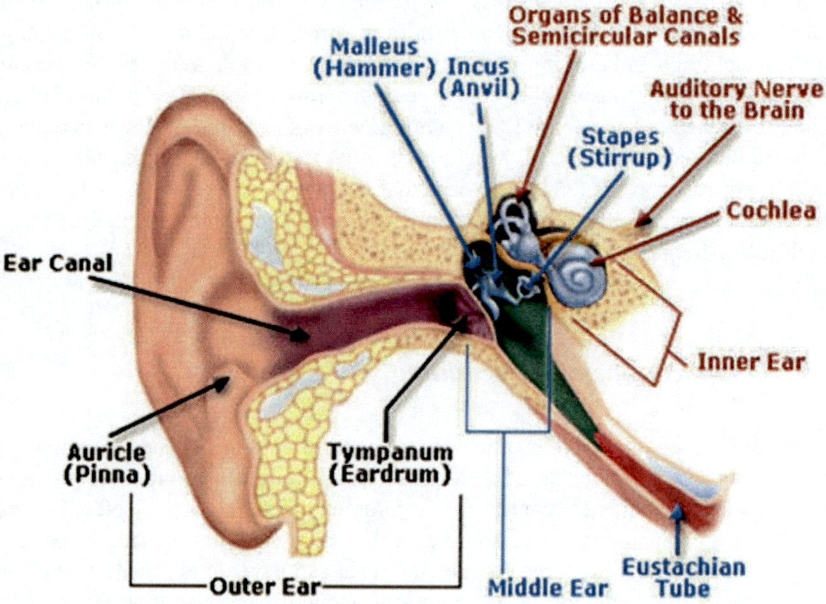

Fig. 1.1 Anatomy of the human ear (OSHA Technical Manual, [13])

- Stage 2: The middle ear has three bones that amplify the vibration and transfer it to the inner ear (the cochlea). The cochlea converts the vibration energy into electrical signals which are sent to brain.
- Stage 3: The cochlea has a basilar membrane that has extremely sensitive hair cells with neurons connected to them that convert the wave motion into nerve impulses. These impulses thus activate the auditory nerve to transmit information that the brain can interpret as pitch and loudness (Fig. 1.1).

The human ear can only hear sound which is in the frequency range of 20 Hz to 20 kHz. The noise perception is both frequency and pressure dependent. The subjective response to frequency is called pitch. Low frequencies are identified as low pitched, while the higher frequencies as high pitched [5, 13].

1.5 Noise Rating Curves

Noise curves are widespread ways to measure and specify background noise in unoccupied buildings and spaces with an idea of producing a single-value representation of a sound spectrum. The process of obtaining the single value is to measure the background sound frequency spectrum in the unoccupied room, which is then placed over onto a template graph containing the noise curves for assessing the

Table 1.3 Numerical value for the NCB curves

Octave-band center frequency (Hz)	NCB 50	NCB 45	NCB 40	NCB 35	NCB 30	NCB 25	NCB 20	NCB 15	NCB 10
16	89	87	85	83	82	81	81	81	81
31	80	77	74	71	68	66	64	62	61
63	68	65	61	58	54	51	48	45	43
125	63	59	55	51	47	43	38	35	31
250	58	53	49	44	39	35	30	26	21
500	55	50	45	40	35	30	25	20	15
1000	52	47	42	37	32	27	22	17	12
2000	48	43	38	33	28	23	18	13	8
4000	45	40	35	30	25	20	15	10	5
8000	42	37	32	27	22	17	12	7	2

Source: Schomer and Bradley (2000, [17])

applicability of noise criteria. There are varied indoor noise criteria used to quantify the level of background noise in rooms, including Noise Criteria (NC), Balanced Noise Criteria (NCB), Room Criteria (RC), Room Criteria Mark II (RC Mark II), and others. A fifth set of curves, called RNC Curves reported by Schomer, 2000 [14], attempts to address cyclic variation of low-frequency sound produced by large air ventilation system. Noise Criteria (NC) is one of the earliest methods designed in 1957 to be extensively used [15]. This system provides a one-number rating, determined by comparing the background sound levels to a set of defined NC curves. The curves were developed following measurements made in several commercial spaces and include octave bands from 63 Hz to 8000 Hz. The shape of the curves approximates the equal loudness contours developed in the early 1930s by Fletcher and Munson (1933). The NC procedure does not adequately distinguish between different sound spectra. Accordingly, two newer methods were developed and are currently recommended by ANSI Standard S12.2-1995: Balanced Noise Criteria (NCB) and Room Criteria (RC). These curves extend from the 16 Hz to the 8000 Hz. The standard defines the values for each individual curve from NCB-10 to NCB-65. In addition, the NCB-0 curve is defined as the threshold of audibility for continuous sound in a diffuse field. These two systems provide not only a sound level rating but also a sign of the frequency content of the noise. This method involves the estimation of the speech interference level (*SIL*) from the sound pressure level (*SPL*) at various frequency bands as follows [16]:

$$SIL = \left(\frac{SPL_{500} + SPL_{1000} + SPL_{2000} + SPL_{4000}}{4} \right) \tag{1.9}$$

The NCB rating of a spectrum is equal to the *SIL* rounded to the nearest decibel. Table 1.3 shows the numerical value for the NCB curves [17]. The second step involves a determination of the perceived balance between low- and high-frequency sound. A spectrum rich in low-frequency sound (16 Hz to 500 Hz) is defined as "rumbly." A spectrum rich in high frequency sound (1000 Hz to 8000 Hz) is defined

Table 1.4 Proposed room noise criteria curve values

Octave-band center frequency (Hz)	RNC 50	RNC 45	RNC 40	RNC 35	RNC 30	RNC 25	RNC 20	RNC 15	RNC 10
16	81	79.3	77.7	76	74.3	72.7	71	69.3	67.7
31	76	73.5	71	68.5	66	63.5	61	58.5	56
63	71	67.7	64.3	61	57.7	54.3	51	47.7	44.3
125	66	61.8	57.7	53.5	49.3	45.2	41	36.8	32.7
250	61	56	51	46	41	36	31	26	21
500	56	51	46	41	36	31	26	21	16
1000	52	47	42	37	32	27	22	17	12
2000	48	43	38	33	28	23	18	13	8
4000	44	39	34	29	24	19	14	9	4
8000	40	35	30	25	20	15	10	5	-

Source: Schomer (2000, [14])

as “hissy.” The rumble criterion is defined as the NCB curve with a value 3 dB higher than the curve determined on the basis of *SIL*. The NCB method requires that a specified speech interference level *SIL* is not be exceeded, and that the deviations from an NCB spectrum with a rating number equal to the measured *SIL* should not exceed +3 dB.

The original RC method was developed from measurements made in unoccupied office buildings by Blazier (1981) conventionally used in assessment of spaces with mechanical system as the dominant noise source. The rating is found by calculating L_{MF}, which is the average of the mid-frequency sound pressure levels in the 500, 1000, and 2000 Hz octave bands [16]:

$$L_{MF} = \left(\frac{SPL_{500} + SPL_{1000} + SPL_{2000}}{3}\right) \tag{1.10}$$

There had been further improvements in RC methodology with an introduction of the RC Mark II curves that are identical to the RC curves. The RC Mark II also includes a Quality Assessment Index (*QAI*) that provides an estimate of occupant evaluation, ranging from “acceptable” to “objectionable.” The RC curves lay down criteria levels that are below the threshold of hearing so as to avoid modern HVAC systems generating turbulent low-frequency fluctuations and fan surging. The NCB curves lay down criteria levels that are based on “well-behaved” HVAC systems, wherein the turbulence generation is diminished and fan surging does not prevail. As such, these criteria are inappropriate, alone, for a standard. Schomer (2000) recommended a set of criteria known as RNC curves with a purpose of avoiding penalty to a well-designed system and also to avoid turbulence-producing fan-surging HVAC system from meeting the acceptability criteria. The proposed RNC curves are such that in the 31 Hz octave band, a 10-unit change in the criterion level corresponds to a 5-dB change in the sound level [14–16]. Table 1.4 shows the proposed Room Noise Criteria curve values [14].

Table 1.5 Recommended specifications of HVAC systems for rooms (unoccupied) with various activities

Occupancy	NCB criterion curve
Concert halls, opera houses, and recital halls	15–18
Small auditoriums	25–30
Large auditoriums, large drama theaters, and large churches	20–25
Private residences (bedrooms)	25–30
Private residences (apartments, family rooms, and living rooms)	30–40
Schools (lecture and classrooms with area < 70 m^2)	35–40
Schools (lecture and classrooms with area > 70 m^2)	30–35
Hotels (meeting rooms)	25–35
Conference rooms (large)	25–30
Conference rooms (small)	30–35
Hospitals (private rooms)	25–30
Hospitals (wards)	30–35
Hospitals (operating rooms)	25–35
Hospitals (laboratories)	35–45
Libraries	35–40
Courtrooms	30–35
Churches	30–35
Movie theaters	30–40
Light maintenance shops, industrial plant control rooms, kitchens, and laundries	45–55
Offices (larger area with conference tables)	30–35

Source: Beranek (1997, [19])

There is currently no agreement among researchers on the best available criteria due to the lack of correlation with human perception. However, appropriate background noise criteria are essential for diagnosis of problems in existing buildings and prevented in newly constructed or renovated buildings [18]. Some of these curves are widely used in building codes as recommended limits for acoustical quality. For instance, a representative purpose of use of recommended specifications of HVAC systems for rooms (unoccupied) with various activities pertaining to NCB criterion curves is shown in Table 1.5 [19].

1.6 Directivity and Directivity Index

A point source radiates sound equally in all directions, but the sound from practical sources which are large in comparison with the wavelength of the sound they generate is directional, that is, not equal in all directions. In the far field, "directivity factor" describes the angular dependence of the sound intensity. The mean averaged sound intensity is calculated as $I = W/4\pi r^2$. If I is the direction dependent intensity, then directivity factor Q is given as:

$$Q = \frac{I_0}{I} \tag{1.11}$$

The directivity index (DI) is defined as:

$$DI = 10 \log_{10} Q \tag{1.12}$$

$$\text{or}\, DI = 10 \log_{10} I_o \text{-} 10 \log_{10} W + 10 \log_{10}(4\pi r^2) \tag{1.13}$$

where W is the sound power of the source and r is the distance from the source. The directivity factor is unity for an ideal point source which generates sound evenly in space. It is equal to 2 when such a source is above an acoustically reflective surface in an otherwise free field [5].

1.7 Octave Bands

The detailed analysis of the complex sound wave is interpreted by the frequency bands such as 1/1 octave and 1/3 octaves. The octave band analysis is indispensable to characterize a noise source, identify the dominant frequencies, and devise suitable strategies/methodologies for reducing the noise levels at dominant frequencies and in the overall noise level radiated by a particular source. Also, it is very important from audiometric point of view for diagnosing the hearing thresholds of patients at various frequencies. The frequency band analysis in terms of 1/1 octave or 1/3 octave band filters is present in almost all the sound level meters and analyzers. A frequency band is said to be an octave in width when its upper band-edge frequency f_u is twice the lower band-edge frequency f_l. Each *octave band* is named for its center frequency (geometric mean), calculated as follows: $f_c = (f_l f_u)^{1/2}$, where f_c = center frequency and f_l and f_u are the lower and upper frequency band limits, respectively.

$$f_c = \sqrt{2} f_l, \tag{1.14}$$

$$f_u = \sqrt{2} f_c \tag{1.15}$$

For instance, an octave filter with a center frequency of 1 kHz has a lower frequency of 707 Hz and an upper frequency of 1.414 kHz. In case of 1/1 octave, f_u is twice the f_l. Each octave band has a band number and central frequency. For one-third octave bands, $f_u = 2^{1/3} f_l$. The band center frequency can be calculated from the band number n as $10^{(n/10)}$. Table 1.6 shows the band number and center frequency for 1/1 and 1/3rd octave bands [20]. For 1/3rd octave filters, the center frequency is calculated as:

Table 1.6 Band number and center frequency for octave bands

Band no.	Nominal center frequency (Hz)	1/3 Octave pass bands (Hz)	Octave passbands (Hz)	Band no.	Nominal center frequency (Hz)	1/3 Octave pass bands (Hz)	Octave passbands (Hz)
1	1.25	1.12–1.41		11	12.5	11.2–14.1	
2	1.6	1.41–1.78		12	16	14.1–17.8	11.2–22.4
3	2	1.78–2.24	1.41–2.82	13	20	17.8–22.4	
4	2.5	2.24–2.82		14	25	22.4–28.2	
5	3.15	2.82–3.55		15	31.5	28.2–35.5	22.4–44.7
6	4	3.55–4.47	2.82–5.62	16	40	35.5–44.7	
7	5	4.47–5.62		17	50	44.7–56.2	
8	6.3	5.62–7.08		18	63	56.2–70.8	44.7–89.1
9	8	7.08–8.91	5.62–11.2	19	80	70.8–89.1	
10	10	8.91–11.2		20	100	89.1–112	

$$f_c = 2^{1/6} f_l, \tag{1.16}$$

$$f_u = 2^{1/6} f_c \tag{1.17}$$

1.8 Sound Fields

Sound fields are of three types: near field, far field, and diffuse field. Near field of a sound source is the region where the source cannot be treated as a point source and in which the inverse square law does not hold good. Sound pressure measurements are normally avoided in this region. Far field is the region wherein the inverse square law is obeyed. "Free-field" is the term used for sound field where there are no reflections present like in open space or in specially made rooms called "anechoic rooms." On the other hand, a diffuse field has multiple reflections from all directions. In this case, reflections combine so that the average acoustic energy per unit volume is the same throughout the field. Such fields are present in enclosed spaces. A reverberant field having reflection from all sides can be made with nonparallel reflective walls and ceiling. The following considerations are there for the sound fields as follows:

- In a free field, a 6 dB decrement in the measured sound pressure level is observed when doubling the distance from a sound emitting object. This behavior is explained by the inverse square law.
- In a diffuse field, like in a reverberant chamber, the spatial variation of sound level is minimal.

The typical variation of the net sound pressure level with distance from the source is shown in Fig. 1.2. Normally, sound pressure level measurements are avoided in

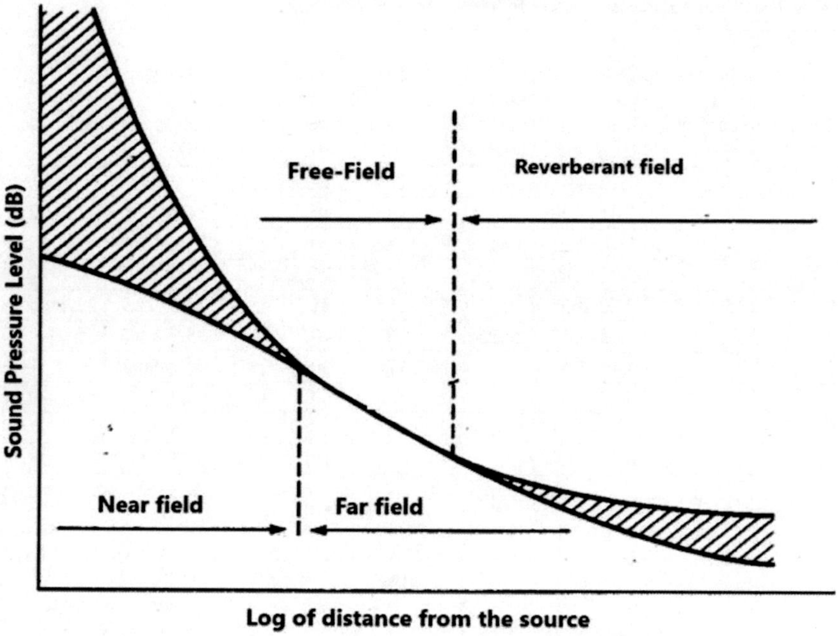

Fig. 1.2 Sound pressure variation with distance in a room (Tandon and Nakra, 2017, [5])

near field that extends a few wavelengths from the source. After this, free field exists, where direct sound pressure level dominates. The region in which reflected sound dominates is called the reverberant field, wherein the sound pressure level remains constant [5].

1.9 Noise Measurements and Analysis

1.9.1 Microphones

The most commonly measured acoustic quantity is sound pressure level. The basic sound measuring system consists of a transducer which is called a microphone, a preamplifier, and a meter to display the levels. The microphone can be condenser, electrets, piezoelectric (ceramic), or dynamic (moving coil) types. Condenser microphones (Fig. 1.3) are the most commonly used in precision sound level meters followed by electrets microphones. A condenser microphone basically acts as capacitors with a constant charge and have metallic diaphragm which is set in motion by oscillating sound pressure. The deflection of the diaphragm changes the capacitance between the diaphragm and a flat electrode backplate that causes the variation in capacitance and is used to develop an electric signal when a high

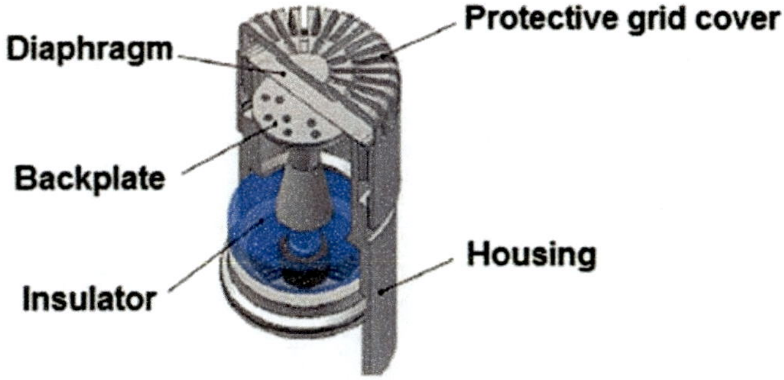

Fig. 1.3 Components of a condenser microphone (Microphone Handbook, [21])

polarization voltage is applied across to the capacitor. Air leak holes (vent) is provided in the microphones for equalization of ambient pressure so that static pressure inside and outside the microphone cavity is the same as the static pressure variations can be larger than the sound pressure usually encountered. The backplate has damping holes so that air flows through these holes with the movement of diaphragm. The damping effect is used to control resonant peaks of the diaphragm response. The changes in sound pressure leads to the deflection of diaphragm that generates a voltage proportional to the incident pressure oscillation corresponding to the individual microphone's sensitivity [5]. Some of the microphone designs include Carbon Microphones, Dynamic, Magnetic Microphones, Fiber Optic Microphones, Laser, Ribbon and Micro Electro-Mechanical Systems (MEMS). The most commonly used diameters of the condenser microphones are 1/4" (6 mm), 1/2" (12 mm), and 1" (25 mm). The response of a precision condenser microphone can be of three types: free-field, pressure, and random incidence. A free-field microphone is designed to measure the sound pressure at the diaphragm, as it would appear if the microphone were not present. These are suitable for working in open areas with no reflective surfaces and anechoic chambers. A pressure field is a sound field where the sound pressure has the same magnitude and phase at any position in the sound field. A pressure microphone is used commonly in an enclosure which is small in size when compared to wavelength. Testing of sound pressure exerted on walls, on air plane wings, or inside structures such as tubes, housings, or cavities are examples of pressure type microphone applications. The random incidence microphone is designed to be omnidirectional and measures sound pressure incident from multiple directions and reflections. These are suitable for taking measurements in reverberation chambers, areas with hard reflective walls, and sound from multiple sources [21, 22].

Based on the charge on the backplate of microphone, they are divided as externally polarized and prepolarized microphones. Externally polarized microphones need a DC voltage across the diaphragm-to-backplate air-gap to fix the charge and thus a power supply is required. Prepolarized microphones do not need external polarization voltage—the charge is fixed by a thin layer of charge-holding

Table 1.7 Sound pressure levels and their sources

Acoustic pressure (RMS)	Sound pressure level	Equivalent source
0.00002 Pa	0 dB	Threshold of hearing
0.02 Pa	60 dB	Business office
0.2 Pa	80 dB	Shop noise
1 Pa	94 dB	Large truck
2 Pa	100 dB	Jackhammer
20 Pa	120 dB	Chain saw
200 Pa	140 dB	Airplane takeoff

Source: Microphone Handbook [21]

electrets material on the backplate. Microphones are classified as LS1P or LS2P based on dimensions and field characteristics. For instance, LS2P refers to a laboratory standard microphone of mechanical configuration 2 having nearly constant pressure sensitivity as a function of frequency. Measurement microphones may be referred to by their IEC 61094 Classifications, for example: WS1F denotes the working standard, 1.0" microphone for free-field response, WS2P denotes the working standard, 1/2" microphone for pressure response, WS3D denotes the working standard, 1/4" microphone for diffuse field response. The condenser microphones have good acoustic sensitivity, low self-noise and low sensitivity to external vibrations but are susceptible to high humidity, and in such environment, they produce excessive background noise due to electrical leakages. Half inch diameter microphones are the most commonly used, while quarter inch microphones are used for the measurement of high noise levels. The sensitivity of a microphone is defined as the output voltage generated (in mV) for an incident sound pressure of 1 Pa (94 dB). It can also be expressed in decibels by the following expression [22]:

$$\text{Sensitivity} = 20 \log_{10}\left(\frac{Vp_0}{V_0 p}\right) \tag{1.18}$$

Thus, a microphone giving an output signal of 10 mV for an incident pressure of 94 dB has a sensitivity of 10 mV/Pa or -40 dB re 1 V/Pa. The sensitivity is measured in mV/Pa or dB re 1 V/Pa. Here $p_0 = 1\text{Pa}$ and $V_0 = 1$ volt. The open circuit sensitivity is defined as the pressure-field sensitivity valid with an idealized preamplifier that does not load the microphone. The higher the sensitivity, the lower the sound pressure level that may be measured, and conversely, the lower the sensitivity, the higher the sound pressure levels that may be measured. The selection of the microphone is based on sound pressure levels and frequencies to be measured and the acoustic field. For measuring the sound pressure level of a machine, it should ideally be placed in a free-field environment or at least in a very absorbing room and a free-field microphone should be selected [23]. The dynamic range refers to the difference between the highest sound pressure level a microphone can measure to the lowest. The lowest amplitude that a healthy human ear can detect is 20 millionths of a Pascal (20 μPa) (Table 1.7).

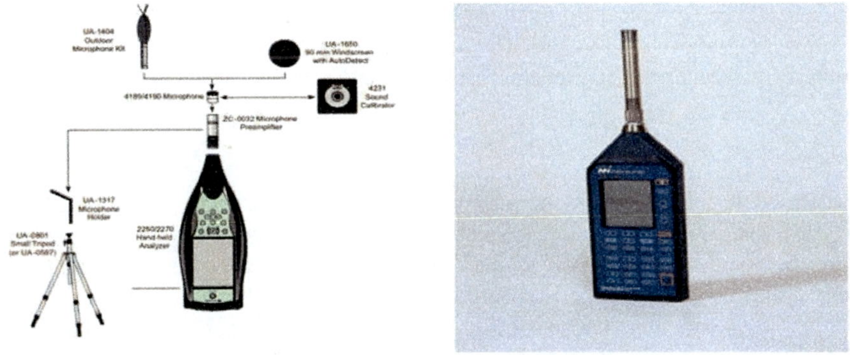

Fig. 1.4 Pictorial view of a B&K Sound Level Meter, Model No. 2250 (LHS) and Norsonic Sound Level Meter (Model No. Nor118)

Usually, the manufacturers specify the maximum sound pressure level based on the design and physical characteristics of the microphone. The maximum sound pressure level refers to the point where the diaphragm deforms in a nonlinear fashion, or where total harmonic distortion (THD) goes more than a specified value, typically 3% THD [21].

1.9.2 Sound Level Meter

A sound level meter consists of a microphone, preamplifier, weighting networks, amplifier with adjustable gain, rectifier, and rms detector with a meter to display the measured level. The sound level meters are classified as: Type 0: precision laboratory use, Type 1: general purpose, Type 2: low price, and Type 3: not used in practice (too wide tolerances). The Type 0 sound level meter is intended as a laboratory reference standard as shown in Fig. 1.4.

Fast and slow measurement settings are provided in the meter, wherein the response time in "fast" setting is around 200 ms and between 500 ms and 1 second in "slow" setting. Impulse mode can be of very short response time of the order of 50 μs for the measurement of impulse noise in dB(AI).

1.10 Outdoor Sound Propagation

The sound propagation is influenced due to a number of factors including absorption of sound in air, meteorological effects such as refraction and turbulence, and interaction with an absorbing or reflecting ground surface and diffraction by noise barriers. ISO 9613-2 specifies an engineering method for calculating the attenuation of sound during propagation outdoors so as to predict the noise levels at a distance from variety of sources [24, 25]. The equivalent continuous downwind octave band sound pressure level at a receiver location L_{ft} (DW) can be calculated as [24]:

$$L_{\mathrm{ft}}(\mathrm{DW}) = L_{\mathrm{W}} + D_{\mathrm{C}} - A \tag{1.19}$$

where L_{W} is octave band sound power level in dB(A) radiated by sound source in reference to sound power of 1 pW, D_{C} is directivity correction in dB in direction from center of source to receiver, and A is combined attenuation due to geometrical divergence (A_{div}), atmospheric absorption (A_{atm}), ground effect (A_{gr}), barrier attenuation (A_{bar}), and miscellaneous effects, viz., foliage, industrial sites, housing, etc. The attenuation term A is given by:

$$A = A_{\mathrm{div}} + A_{\mathrm{atm}} + A_{\mathrm{gr}} + A_{\mathrm{bar}} + A_{\mathrm{misc}} \tag{1.20}$$

New strategies are being devised in recent times pertaining to outdoor sound propagation with an objective of improving the accuracy of traffic noise models in recent times. Numerical propagation methods, viz., parabolic equation (PE), boundary element method (BEM), and straight ray approach (RAY), have been employed in various models for computing the effects of atmosphere, ground surface, and obstacles on sound waves. In the case of spherical spreading from a point source, the sound level is reduced by 6 dB for each doubling of distance from the source. However, for a line source, the attenuation is 3 dB for each doubling of the distance from the source. Reflection from the ground can also reduce the sound levels. When the source and receiver are both close to the ground, the sound wave reflected from the ground may interfere destructively with the direct wave. This effect (called the ground effect) is normally noticed over distances of several meters and more, and in the frequency range of 200–600 Hz. Depending on their relative phases and amplitudes, the reflected and direct waves may constructively add or destructively interfere, and the sound pressure level at the receiver may increase up to 6 dB [26]. The method of calculating the ground effect specified by ISO 9613-2 is applicable only to ground which is approximately flat, either horizontally or with a constant slope.

The most universal meteorological effects on sound propagation are refraction, scattering by turbulence, and atmospheric absorption. Refraction of sound rays occurs if the sound velocity and/or wind speed changes, that is, wind and temperature gradients are encountered. In downwind conditions and/or temperature inversion, the sound rays are bent downward, and in upwind conditions and/or lapse, they are bent upward. In the upwind direction, the sound speed decreases with altitude, sound waves are directed upward, away from the ground, forming a "shadow zone" into which no direct sound penetrates. Sound constantly refracts toward the lower sound speed. The effect is to refract sound waves in the direction of lower sound velocity in the incidence of a temperature gradient. Thus, in sunny afternoon, sound waves will bend upward in all directions from the source, forming a circular shadow zone due to the high surface temperature known as positive lapse rate. The reverse state often occurs at night due to the rapid cooling of air at the surface called as inversion or negative lapse and the sound waves are bent downward. This fact explains why sound sometimes travels much better at night, as it is focused along the ground instead of radiating upward [26]. Some recent models such as ASJ RTN Model-2008 recommended changes in ± 2.5 dB L_{Aeq} due to vector wind speed of 5 m/s at a distance of 50 m from road center and ± 3.5 dB L_{Aeq} at a distance of 100 m from road center [27]. The recent noise models include meteorological effects that can be seen if the normalized sound pressure level from road surface is plotted versus distance source-receiver [28, 29]. A number of 25 classes have been recommended that causes a determination of long-term average sound levels with an error of almost 2 dB at 1000 m range. However, the calculation for different conditions with a weighted average at end makes it complicated to retrace all probable results and transparency will be condensed. Some recent models such as French method for road traffic noise prediction (NMPB-Routes-2008 [30]) have introduced new replacements in ISO 9613-2 based attenuation model so as to account the mean curvature of rays (refraction) and its scattering (turbulence).

1.11 Noise-Control Engineering: Concept and Methodology

Noise can be controlled by treating the source, the sound transmission path, and the receiver of any combination of these options. The details of these control measures are as follows [13].

1.11.1 Source Treatment

Identification of the source noise characteristics is primary consideration for the source noise control. The methods utilized for source treatment are: modification, retrofit, substitution, and relocation. Modifications such as reducing excessive driving forces, reduced or optimized speed, minimizing distance between impacting parts, dynamically balancing the rotating equipment, use of vibration isolation table, etc. are the various methods for source noise control. The retrofit applications include vibration damping, vibration isolation, and installation of silencers. Vibration damping is an effective retrofit for controlling the resonant tones radiated by vibrating metal panels or surface areas. The two basic damping methods are free-layer damping and constrained layer damping. The single-layer damping treatments are viscoelastic materials that include filled bitumens, silicone sealant, and elastomeric polymers, available in the form of adhesive sheets or thick liquids for spraying or troweling on to the thin surface to be treated. The damping material thickness must be between one and three times the thickness of the surface being damped for effective results. Constrained layer damping treatments consist of a layer of viscoelastic material sandwiched between the structure and a layer of steel or aluminum

[31]. Vibration isolators such as metal springs, elastomeric mounts, resilient pads, etc. serve to decouple the "solid" connection between the source and the receiver. Use of dissipative or reactive silencers, or combination of both, is also recommended especially for automotive applications. Another source treatment, called the substitution of the source, involves using alternative equipment or materials that are inherently quieter and performs the same function at lower noise levels. It includes: using the belt drives over gears, using belt conveyors instead of rollers, replacing the omnidirectional fans or electric motors with unidirectional aerodynamic fans, using perforated panels instead of solid panels, hydraulic rather than mechanical presses, replace gear drives with belt drives, modification of gear teeth, by replacing spur gears with helical ones, etc. Controlling the noise level by locating or relocating the source shall be indispensable to minimize the overall background noise level [31].

1.11.2 Path Treatment

The path treatment includes adding sound-absorptive materials to the room or equipment surfaces, installing sound insulative materials between the source and receiver, use of acoustical enclosures or barriers, or combination of these treatments. Path treatment should be explored when all the available options of controlling the noise at source are considered, and there is no further noise reduction that can be accomplished at the source itself.

1.11.3 Receiver Treatment

The final option involves reducing the noise at the receiver. The receiver protection is of utmost importance especially when the measures implemented at the source and transmission path cannot effectively control the accentuated noise levels. This can be achieved by enhancing the sound insulation of dwellings occupied by people, and protecting an individual by means of ear-protecting devices such as ear plugs, semi-inserts, ear muffs, and helmets.

1.12 Anechoic and Reverberation Chamber

An anechoic chamber is a specially designed room for complete absorption of sound waves, thus absorbing energy in all directions. Full anechoic chambers are designed for absorbing the energy in all directions, while semi-anechoic chambers have a solid floor acting as work surface for supporting items for testing of emitted sound levels. The floor is damped and floated on absorbent buffers for vibration isolation. These chambers are used for calibration of electro-acoustic equipment like sound level

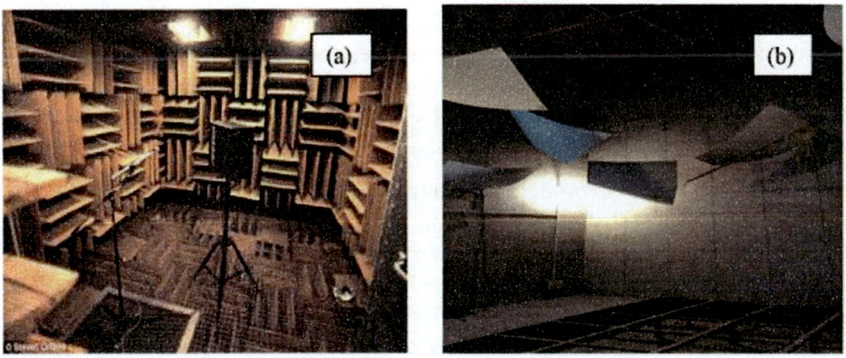

Fig. 1.5 (**a**) Pictorial view of Anechoic chamber; (**b**) Reverberation chamber [33]

meter, loudspeakers, hearing aids, etc., sound power level measurements, and directional characteristics of sound sources like sirens. There are few anechoic chambers in India in some of the institutes like one in CSIR-National Physical Laboratory of India that provides a free space of $3.5 \times 3.5 \times 3.5\ m^3$ with a lower cut-off frequency of 70 Hz [32] and others at Automotive Research Association of India, Pune; Fluid Control Research Institute, Palakkad, Kerala, etc. (Fig. 1.5).

A reverberation chamber or room is a room designed to create a diffuse or random incidence sound field (i.e., there is uniform distribution of sound energy in room). Reverberation chambers tend to be large rooms and have very hard exposed surfaces and the resulting diffuse sound field exists in the room [33]. These are used in: measurement of the sound power of a source; measurement of the sound absorption coefficient and sound transmission loss of a material, and in diffuse field microphone calibration.

1.13 Noise Pollution Control Policy

A national policy framework is the basis of noise management, wherein a legal framework is a must. The objectives of noise policy are principally based on estimated health impacts, such as the number of citizens annoyed by various noise sources. A three-pillar approach, that is, a stringent noise sources policy, insulation of dwellings, and zoning (i.e., spatial separation of noise intrusive activities, such as transport and industries, and other noise sensitive activities), is characteristic of noise policy [34]. Noise policy has both global and local aspects or hybrid of two depending upon the nature of noise source. Both are complementary to each as a noise policy that covers only the local aspects, or one that only considers the global aspect, cannot be considered inclusive and is unable to accomplish the desired noise reduction goals [35]. The European Parliament and Council adopted Directive 2002/49/EC, called "END," pertaining to the assessment and management of environmental noise on June 25, 2002, with an objective to "*define a common approach intended to avoid, prevent or reduce on a prioritised basis the harmful effects, including annoyance, due to the exposure to environmental noise.*" The END directive is an excellent illustration of a harmonized noise policy followed among the Member States with underlying principles [36]:

- **Monitoring the environmental problem** by developing "strategic noise maps" for major roads, railways, airports, and agglomerations, using harmonized noise indicators L_{den} (day-evening-night equivalent level) and L_{night} (night equivalent level) so as to assess the number of people annoyed and sleep-disturbed respectively throughout Europe
- **Informing and consulting the public** about noise exposure, its effects, and the measures considered to address noise
- **Addressing local noise issues** by requiring competent authorities to draw up action plans for noise control
- **Developing a long-term EU strategy** with an objective of reducing the number of people affected by noise, and framework for developing existing Community policy on noise reduction from source

Article 6.2 of END recommends in establishing the common assessment methods for determination of L_{den} and L_{night}. The Commission Directive (EU) 2015/996 of May 2015 stipulates a common methodological framework for the assessment of environmental noise (CNOSSOS-EU) for consistent and accurate reporting of strategic noise maps (including exposure of populations) by Member States as required by END 2002/49/EC starting from December, 2018 onwards [37]. Furthermore, sharing-based common data-related practices for harmonizing and accessing geospatial data by the Infrastructure for Spatial Information in the European Community (INSPIRE) for the assessment and development of strategic noise maps will be implemented [38]. The administrative costs are low (€0.15 for noise maps and €0.03 for action plans per citizen, every 5 years). Cost-benefit analysis showed that END was proficient with a favorable cost-benefit ratio of 1:29. However, the Directive does not fully utilize its potential attributed to the delays in implementation and requires a planned and careful execution [39].

Wise policy decisions entail information about the social and economic considerations of noise-control policies. A recent American study estimated that reducing environmental noise by 5 dB L_{dn} (day-night average sound level) would reduce hypertension cases by an estimated 1.4% and coronary heart disease cases by 1.8%. The associated cost savings and productivity gains were estimated as \$3.9 billion annually [40]. A Spanish study for Valladolid city on socioeconomic considerations revealed that authorities should spend an average of €9.54 per person per year to reduce noise contamination which corresponds to 1.38% of the city annual budget and 12.66% of the environmental budget of city. Fifty percent of the population would be willing to pay an average quantity of €7.22 per person per year to reduce noise contamination [41]. Istamto et al. (2014) assessed the willingness-to-pay (WTP) for noise pollution as €90–€320 per person per year to avoid an increase in the noise level from 50 to 65 dB [42]. A recent study by Huh and Shin (2018) for

South Korea calculated that the monthly mean WTP of respondents was KRW 3392.3 (USD 2.97) for residential noise reduction and KRW 3507.1 (USD 3.07) for transport noise reduction [43]. The Noise Policy Statement for England (NPSE) serves as a guiding document [44] that includes noise policy vision as: "*Promote good health and a good quality of life through the effective management of noise within the context of government policy on sustainable development*" with an objective of mitigating and minimizing the adverse impacts of noise on health and quality of life. It applies to all types of noise including environmental noise, neighbor noise, and neighborhood noise except occupational noise. Noise policies including neighbor noise are not addressed in some countries, but could certainly be meaningful for discussion, enforcement, and execution [45].

1.14 Challenges Ahead in Indian Perspective for Noise Pollution Control

A noise survey conducted by Delhi Pollution Control Committee (DPCC) in Delhi between August and October 1996 highlighted that increasing traffic was an environmental concern and traffic noise is a major nuisance. The main sources of noise pollution are automobiles followed by construction equipment, loudspeakers, bursting of crackers, etc. India currently has about 15 million cars, which is equivalent to 13 cars per 1000 population. Some cities like Delhi, Chennai, and Coimbatore have more than 100 cars per 1000 population. Delhi city is endowed with one of the most complex road network systems in the country which is used by more than $1/10^{th}$ of the country's vehicular population. A key observation Master Plan Delhi (MPD)-2021 mentioned that compared to the period of the last Master Plan, there is a phenomenal growth of automobiles which has resulted in congestion, pollution, safety of travel and parking, etc. As on March 31, 2016, the total number of motor vehicles on road in Delhi touched to 97.05 Lakhs, showing an increase of 9.94% over previous year. Thus, the Ministry of Urban Development has implemented a National Transit Oriented Development (TOD) Policy for Delhi with an objective of integrating the land use with the transport infrastructure to make the cities livable, healthy, and smart [46]. In recent times, the National Ambient Noise Monitoring Network comprising 70 Noise Monitoring Stations spread over 7 metropolitan cities which include Bangalore, Chennai, Delhi, Hyderabad, Kolkata, Lucknow, and Mumbai (10 stations in each city) has been established by CPCB, India. The network was developed in two phases. In first phase, 35 stations were installed in above said seven cities (05 each) during 2011, and another 35 stations were added in same seven cities (5 each) in the second phase during 2015. The Noise Monitoring Network so established is a unique and one of the largest noise monitoring networks of its kind across the globe. An analysis of the median of ambient noise levels obtained in 2017

[47] for the 7 cities correlated with the population density in (people per square km) yields a relationship similar to that proposed by Schomer et al. (2011) [48] as:

$$L_{\text{Aeq,24h}} = 26.9 + 10\ \log\ (\rho) \tag{1.21}$$

$$L_{\text{dn}} = 30.7 + 10\ \log\ (\rho) \tag{1.22}$$

where ρ is population density in people per square km, $L_{\text{Aeq,24h}}$ is equivalent sound pressure level for 24 hours, and L_{dn} is day-night average sound level. It may be noted that these equations have very high uncertainty due to lack of precise information about the population density for the areas wherein noise monitoring networks have been established and especially for areas having low population density. As a result, the coefficient of determination R^2 for both the equations (1.21) and (1.22) is less than 0.44. Also, the knowledge of the vehicular density shall further provide a more comprehensive relationship as developed by Tang et al. (2003) [49] expressed as $L_{\text{Aeq,24 h}} = a\log_{10} Q + b\log_{10} P + c$, where Q is daily traffic volume and P is local population density within each km^2 of area and constants a, b, and c for $L_{\text{Aeq,24h}}$ have been found to a=2.59, b=2.33, and c=47.7. Substituting the value of Q as 32,000 and P as 10,000 for a metropolitan city of the country, the value of $L_{\text{Aeq, 24 h}}$ comes out to be 70 dB(A), which is rather very high value when compared to the ambient noise standards. Thus, with growing population and vehicular population, the ambient levels will be very high as compared to the recommended limits in next 5 years, which necessitates the implementation of noise action plans for environmental noise control. The major factors responsible for the increase in noise pollution in Indian cities are listed as:

- Indiscriminate honking.
- Encroachment of road due to illegal construction and parking.
- Retrofitting of silencer and horns of vehicles.
- Use of music system during religious gatherings, marriage ceremonies and other ceremonies on open public places and without any mitigation measures.
- Use of diesel generator sets without any acoustic measures.
- Carrying out construction activity without taking mitigation measures.
- Lack of awareness among general public to approach correct authority for noncompliance of noise rules and also about noise pollution effects and control.
- Absence of penalty provisions in terms of monetary ground.
- Multiplicity of zones or mixed-use land pattern.
- Multiplicity of authorities and lack of information on execution of powers and responsibilities vested under Noise Pollution (Regulation and Control) rules among the authorities.
- Nonavailability of noise maps or noise inventories that show the existing noise situation or scenario.
- Lack of a comprehensive Traffic Management Plan including effective implementation and extension of traffic restrictions and construction of subways and

flyovers. Also, there is absence of a synchronized traffic signaling in some traffic corridors including National Highways within the city limits.
- No consideration on erection of noise barriers around the hospitals, schools, and other locations in silence zones.
- Lack of a comprehensive policy for the redevelopment of congested areas.

Recently, the *TomTom Traffic Index* created to help the cities around the world for dissemination of information pertaining to the congestion levels in urban areas estimates 57% of countries included witnessed increase in traffic congestion and global average congestion level of 29%. Bengaluru city of India tops the list of most congested cities, and Mumbai and Pune cities makes up in top five most congested cities of the world. New Delhi comes in the top ten cities among 416 cities across 57 countries in 2019 [50]. Thus, there is need of National Planning Policy Framework (NPPF) for environmental noise management and control in coordination with the Master Plan 2021 for land-use patterns and Urban Transport policy for balanced regional sustainable urban transport system with an objective of reducing the environmental noise in Indian perspectives. In recent years, the Government of India is very much serious about these issues and is taking many suitable steps like establishment of a diversified ambient noise monitoring network in seven major cities of India, development of a draft for execution of noise mapping of cities, installation of noise barriers for elevated metro corridors, and devising new transport policy (Transit Oriented Development (TOD) Policy) that will inculcate synergy between land-use and transport, foster integrated multi-modal public transport system to reduce dependence on personalized vehicles, advocate restructuring of existing network through expressways, elevated roads, arterial roads, distributor roads, and relief roads, and consider provision for introducing cycle tracks, pedestrian features in arterial and sub-arterial roads. However, The Dutch National Traffic and Transportation plan for the next two decades (NVVP 2001–2020) with a goal of [51]: *decreasing the number of houses exposed to a noise level of >70 dB(A) by 100%, the number > 65 dB(A) by 90% and the number > 60 dB(A) by 50% to be realised in 2030 can be a great inspiration for the developing countries like India to effectively deal with noise pollution in next two decades.*

References

1. Technology for a Quieter America (2010) The National Academies Press, Washington DC
2. Yuen FK (2014) A vision of the environmental and occupational noise pollution in Malaysia. Noise Health 16(73):427–436
3. Seidman MD, Standring RT (2010) Noise and quality of life. Int J Environ Res Public Health 7(10):3730–3738
4. Hansen C H (2010) Fundamentals of Acoustics, University of Adelaide, Australia. Handbook 2010, e-Book. http://www.who.int/occupational_health/publications/noise1.pdf
5. Tandon N, Nakra BC (2017) Vibration and noise engineering. I. K International Pvt. Ltd., New Delhi

6. Singal SP (2000) Noise Pollution and Control. Narosa Publishing House. March, 2000, 279 pages
7. Lamancusa JS (2000) Noise metrics and regulations, Penn State, 12/4/2000, https://www.mne.psu.edu/lamancusa/me458/4_metrics.pdf
8. Zwicker E, Fastl H, Dallmayr C (1984) Basic program for calculating the loudness of sounds from their 1/3 octave band spectra according to ISO 532B. Acust 55:63–67
9. Persson K, Björkman M, Rylander R (1985) An experimental evaluation of annoyance due to low frequency noise. J Low Freq Noise Vib 4:145–153
10. Persson K, Björkman M, Rylander R (1990) Loudness, annoyance and the dBA in evaluating low frequency sounds. J Low Freq Noise Vib 9:32–45
11. Leventhall HG (2004) Low frequency noise and annoyance. Noise Health 6(23):59–72
12. Kjellberg A, Goldstein M, Gamberale F (1984) An assessment of dB(A) for predicting loudness and annoyance of noise containing low frequency components. J Low Freq Noise Vib Act Control 3:10–16
13. Occupational Safety and Health Administration OSHA Technical Manual (OTM), Section III, Chapter 5-Noise, Department of Labour, United States.
14. Schomer PD (2000) Proposed revisions to room noise criteria. Noise Control Eng J 48(3):85–96
15. Beranek L, Ver I (1992) Noise and vibration control engineering: principles and applications. Wiley-Intersci 1992
16. Tocci GC (2000) Room noise criteria – the State of the Art in the Year 2000. Noise News Internation 8(3):106–119
17. Schomer PD, Bradley JS (2000) A Test of proposed revisions to room noise criteria curves. Noise Control Eng J 48(4):124–129
18. Wang LM, Bowden EE (2003) Performance review of indoor noise criteria. Archit Eng-Faculty Publication 19
19. Beranek LL (1997) Applications of NCB and RC noise criterion curves for specification and evaluation of noise in building. Noise Control Eng J 45(5):209–216
20. National Programme on Technology Enhanced Learning (NPLTEL), Mechanical engineering, Fundamentals of Acoustics, Octave band analysis, https://nptel.ac.in/courses/112104212/68. (Accessed on 10.02.2021)
21. Microphone Handbook, PCB Piezotronics, USA, 20 pages, http://www.pcb.com/Linked_Documents/Vibration/Microphone_handbook.pdf
22. Brüel and Kj□r (B&K) Microphone Handbook (2019) Technical Documentation, Vol. 1, March, 2019, BE 1447-1.
23. Malchaire J, Sound measuring instruments, 125–140, https://www.who.int/occupational_health/publications/noise6.pdf
24. ISO 9613-2 (1996) Acoustics-Attenuation of sound during propagation outdoors-Part 2: General method of calculation.
25. ISO 9613-1 (1993) Acoustics-Attenuation of sound during propagation outdoors-Part 1: Calculation of absorption of sound by the atmosphere.
26. Lamancusa J S (2009) Outdoor Sound Propagation, Penn State 7/20/2009, ME 458:Engineering Noise Control, https://www.mne.psu.edu/lamancusa/me458/10_osp.pdf
27. Yamamoto K (2010) Road traffic noise prediction model ASJ RTN-Model 2008: Report of research committee on road traffic noise. Acoust Sci and Tech 31:1–55
28. Probst W (2010) Accuracy and precision in traffic noise prediction, Proceedings Internoise, Lisbon, Portugal, 15–16, June, 2010, 1–6.
29. Probst W (2010) New Techniques in noise prediction, Proceedings of 20th International Congress on Acoustics, ICA, 23–27th August, 2010, Sydney.
30. Dutilleux G, Defrance J, Ecotière D, Gauvreau B, Bèrengier M, Besnard F, Duc EL (2010) NMPB-Routes 2008: the revision of the French method for road traffic noise prediction. Acta Acust United Ac 96:1–11
31. Hansen C H, Goelzer B I F, Engineering Noise Control, https://www.who.int/occupational_health/publications/noise10.pdf

32. Pancholy M, Chhapgar AF, Mohanan V (1981) Design and construction of an anechoic chamber at the National Physical Laboratory of India. Appl Acoust 14:101–111
33. Yadav S, Shivagan DD, Bapna K, Jaiswal VK, Sharma P, Saha S, Singh M, Garg N, Soni K, Titus SSK, Aswal DK (2020) Physico-mechanical metrology, Part III: Thermal, optical radiation and acoustic metrology. Metrology for Inclusive Growth of India. Springer, Singapore, pp 307–376
34. Weber M (2013) Noise policy: sound policy? A meta level analysis and evaluation of noise policy in the Netherlands. Utrecht University Repository.
35. Lang WW (2003) Global versus Local Issues in Noise Control Policy. Noise Vib Worldwide 34(2):17–22
36. European Noise Directive (2002) 2002/49/EC of the European Parliament and of the Council of 25 June 2002 relating to the assessment and management of environmental noise. Off J Europ Commun
37. Kephalopoulos S, Paviotti M, Anfosso-Lédée F, Van Maercke D, Shilton S, Jones N (2014) Advances in the development of common noise assessment methods in Europe: the CNOSSOS-EU framework for strategic environmental noise mapping. Sci. Total Environ 482-483:400–410
38. Abramic A, Kotsev A, Cetl V, Kephalopoulos S, Paviotti M (2017) A spatial data infrastructure for environmental noise data in Europe. Int J Environ Res Public Health 7:726. https://doi.org/10.3390/ijerph14070726
39. Licitra G, Ascari E (2018) Noise Mapping in the EU: State of Art and 2018 Challenges. Proceedings Internoise, 26–29 August, 2018, Chicago, Illinois.
40. Swinburn TK, Hammer MS, Neitzel RL (2015) Valuing quiet: an economic assessment of US environmental noise as a cardiovascular health hazard. Am J Prev Med 49(3):345–353
41. Martín MA, Tarrero A, González J, Machimbarrena M (2006) Exposure–effect relationships between road traffic noise annoyance and noise cost valuations in Valladolid, Spain. Appl Acoust 67:945–958
42. Istamto T, Houthuijs D, Lebret E (2014) Willingness to pay to avoid health risks from road-traffic-related air pollution and noise across five countries. Sci Total Environ 497–498:420–429
43. Huh SY, Shin J (2018) Economic valuation of noise pollution control policy: does the type of noise matter? Environ Sci Pollut Res 25:30647–30658
44. Noise Policy Statement for England (NPSE), DEFRA, 2010. www.defra.gov.uk; https://www.gov.uk/government/uploads/system/uploads/attachment_data/file/69533/pb13750-noise-policy.pdf (Accessed on 10.02.2021)
45. Rasmussen, B. (2018) Building acoustic regulations in Europe – Brief history and actual situation. In Proceedings of the Baltic-Nordic Acoustics Meeting 2018 (2018 ed., pp. 1–16). Reykjavik: Nordic Acoustic Association. Joint Baltic-Nordic Acoustics Meeting (BNAM) Proceedings.
46. Central Pollution Control Board, India (2017) Delhi's ambient noise levels influenced by traffic flow – Case studies, Control of Urban Pollution Series, CUPS/87/2017–18, September, 2017.
47. Central Pollution Control Board (2018) Status of Ambient noise level in Indian in 2017, CPCB report, *NANMN*/05/2017, August, 2018.
48. Schomer P, Freytag J, Machesky A, Luo C, Dossin C, Nookala N, Pamdighantam A (2011) A re-analysis of Day-Night Sound Level (DNL) as a function of population density in the United States. Noise Control Eng J 59(3):290–301
49. Tang SK, Chan WY (2003) Predictability of noise indices in a high-rise residential environment (L). J Acoust Soc Am 114:1222–1225
50. The Tom Tom Traffic Index., https://www.tomtom.com/en_gb/traffic-index/. (Accessed on 10.02.2021)
51. Nijland R, Vos E, Hooghwerff J (2003) The Dutch Noise Innovation Program Road Traffic (IPG). In Proceedings Internoise 2003, Seogwipo, Korea, 25-28 August, 2003. http://citeseerx.ist.psu.edu/viewdoc/download?doi=10.1.1.537.4305&rep=rep1&type=pdf (Accessed on 10.02.2021)

Chapter 2
Noise Sources: Characteristics and Control

Abstract This chapter presents an overview of various sources of noise pollution and their characteristics. There are two major noise pollution sources, viz., community noise and industrial noise. Major sources of community noise are transportation, automobiles, construction work, loudspeakers, recreational activities, fireworks, etc. The impact of noise pollution is associated with the mental, physical, emotional, and psychological well-being of an individual. The damaging effects from various noise sources are the potential health hazards that must be controlled through planning, executive, legal and scientific framework. The planning and execution of Noise Action Plans (NAPs) and inculcating awareness amongst the community towards the associated health hazards is thus a fundamental requisite for environmental noise management and control.

2.1 Introduction

Noise pollution in recent years has emerged as one of the important pollutants of environment. The indiscriminate use of loudspeakers from religious places and performance of religious ceremonies sometimes cause annoyance to the nearby community. The sources of noise pollution in an industrial society are manifold. In urban cities, these are associated with transportation, automobiles, motorcycles, metro trains, buses, fire brigade, police cars, ambulances, airplanes, industries, musical instruments, loudspeakers, etc. During religious or domestic functions and in political meetings particularly during the elections, noise pollution often increases due to the use of electronic medium of sound generation. The main source of noise, however, comes from transport. The number of vehicles plying on the roads has increased the noise pollution levels enormously. This problem is more aggravated in metropolitan cities where a large number of vehicles ply simultaneously and their noise is unbearable. The horns and sirens used by these vehicles, especially the pressure horns, are very noisy. Community noise (also called environmental noise) is the noise emitted from all sources, except noise at the industrial workplace, whereby the major sources are motor vehicles, construction activities, loudspeakers, fireworks, marriage processions, or religious ceremonies or other recreational activities,

N. Garg, *Environmental Noise Control*,
https://doi.org/10.1007/978-3-030-87828-3_2

etc. [1]. Thus, it is imperative to understand the characteristics of these noise sources and devise suitable mitigation measures and legislations for noise pollution control. In Indian scenario, noise is regarded as a pollutant under the air (Prevention and Control of Pollution Act, 1981 in India). Motor vehicles, which are a very significant part of this urban environment, are the main source of urban noise emission, contributing about 55% to the total noise [2–4]. It may be noted that the type of zone, geographic features, landscape, and topography are the vital factors on which noise emission and transmission depends [4]. Thus, devising the noise action plans (NAPs) and legislations should consider all these aspects exclusively for controlling noise pollution in urban cities.

2.2 Community Noise Pollution

Community noise, or environmental noise, involves the main sources of road, rail, and air traffic, industries, and other public works. The indoor noise sources are ventilation systems, domestic machines, home appliances, recreation systems like television, music systems, etc., and neighbor noise. In metropolitan areas, road traffic is a major contributor to ambient air pollution and the dominant source of community noise. Gan et al. investigations pertaining to a population-based cohort study in Vancouver, Canada revealed that the population in the highest noise decile had a 22% increase in coronary heart disease (CHD) mortality compared with persons in the lowest decile. These findings suggest that there are independent effects of traffic-related noise and air pollution on CHD mortality. A 10-dB(A) elevation in residential noise levels was associated with a 9% increase in the risk of death from CHD [5]. The European Union estimated about 40% of the population exposed to road traffic noise with noise level exceeding 55 dB(A) daytime, and 20% exposed to levels exceeding 65 dB(A). In totality, more than half of all European Union citizens is estimated to live in zones that do not ensure acoustical comfort to residents. The effects of noise in dwellings, typically, are sleep disturbance, annoyance, and speech interference. Indoor guideline values for bedrooms are 30 dB L_{Aeq} for continuous noise and 45 dB L_{Amax} for single sound events. The background sound level should not exceed 35 dB L_{Aeq} during teaching sessions in classrooms. For ward rooms in hospitals, the guideline values indoors are 30 dB L_{Aeq}, together with 40 dB L_{Amax} during night. During the day and evening, the guideline value indoors is 30 dB L_{Aeq} [6]. WHO published community noise guidelines (CNG) and night noise guidelines (NNG) for Europe in 1999 and 2009, respectively. Recently, the World Health Organization (WHO) Regional Office for Europe has developed environmental noise guidelines for the European Region, proposing a rationalized set of public health recommendations pertaining to the exposure to environmental noise due to various sources: transportation (road traffic, railway, and aircraft) noise, wind turbine noise, and leisure noise [7]. The Environment Action Programme of European Union of 2020, "Living well, within the limits of our planet" is committed to safeguarding the EU's citizens from environment-related risks to health by ensuring that by 2020 "Noise pollution in the European Union has significantly decreased, moving closer to WHO-recommended levels".

2.3 Noise Sources

2.3.1 *Road Traffic Noise*

Road traffic noise depends on the speed of vehicles/traffic flow rate, percentage of heavy vehicles in the traffic, road surface, and the change in engine speed. The two main sources are the propulsion noise and the rolling noise consisting of tire-road interaction and aerodynamic noise. The sources of propulsion noise are: power unit consisting of engine, transmission (gearbox), air inlet and exhaust, cooling fan, brakes, and vehicle body rattle. The performance depends largely on the engine speed. Sandberg studies indicated that the tire-pavement interaction noise dominates for passenger vehicles with the speed of above 40 km/h and for trucks with the speed of 70 km/h [8]. Engine noise is dominant at a low speed in low frequencies [9], while the aerodynamic noise is important only at a very high speed [10] and only at low frequencies (50–400 Hz [11]). The intake and exhaust noise also centers around low frequencies. The difference ranging from 9 to 14 dB(A) depending upon a single pavement or different types of pavement for same driving conditions can be there [12].

2.3.2 *Railway Noise*

Railways are considered to be the second most dominant source of environmental noise in Europe, with nearly seven million people exposed to noise levels higher than 55 dB L_{den} in 2012. On average, at European level, 4.65% of people living inside urban areas are exposed to railways traffic noise above 55 dB L_{den} [13]. The noise from road and rail traffic is estimated to cost the EU EUR 40 billion per year [14]. The three different sources of railway noise are identified: engine noise, rolling noise, and the aerodynamic noise. The most important source of noise is rolling noise, as this is relevant for both freight and passenger trains, while the aerodynamic noise is very important for high-speed trains. The rolling noise, equipment noise, and noise from the motor fan are the main contributors to the overall noise level when the train is traveling at a speed of less than 300 km/h [15]. The rolling noise and wheel noise can be controlled by composite brake blocks (freight wagons), resilient wheels, or wheel dampers. Rail noise can be controlled by rail dampers, resilient track pads, and installation of noise barriers of different heights. A majority of studies in European continent had reported that railway noise causes less annoyance than other transportation noise sources [16–19]. This is a so-called "railway bonus" in noise regulations of some European countries. However, some of the Japanese studies have produced very different results and have shown that railway noise annoyance was nearly the same as or even a little higher than road traffic noise annoyance [20, 21]. Field test results proved that the noise of the ballastless track is approximately 5 dB higher than that of ballasted track for the receiver at the boundary of the railway [22]. Ballastless track emit higher sound levels owing to the reduced connecting impedance for the rail, reduced vibration decay rates along the rail, and reduced absorption [23].

2.3.3 Aircraft Noise

Aircraft noise is one of the most damaging aviation environmental effect. It may lead to community annoyance and sleep disturbance, affect academic performance of children, and may increase the risk for cardiovascular disease of people living in the vicinity of airports [24]. The main sources of aircraft noise are: fan and compressor noise; landing gears; fan, turbine, and combustion noise; jet mixing noise; and airframe noise. All of these sources thus contribute to the noise signature of the aircraft as a whole for a given set of flight conditions [25]. Although the exposure–response relationships have been recommended for assessing the annoyance response due to aircraft noise, they are not applicable to assess the short-term effects. In Indian perspectives, exposure–response relationships are required for health impact assessment and annoyance. For instance, in case of a new runway 29/11 commissioned sometimes ago for Indira Gandhi International Airport, Delhi, there were some complaints filed by the local resident welfare association and a nearby hospital pertaining to the aircraft noise and its annoyance especially while landing and undergoing the reversed thrust. Subsequently, an expert committee was set up for suggesting the preventive measures for aircraft noise abatement and control. Some of the noise control measures like the installation of a 3.5 m tall barrier had been suggested for providing relief to the nearby community. Thrust reversal, also known as reverse thrust, is a momentary process wherein the direction of the exhaust from aircraft engine is directed forward after landing to slow down the aircraft. The process leads to the reduced wear on the brakes and enables shorter landing distance. It is typically applied immediately after aircraft touchdown to recover deceleration during landing. However, the application of reverse thrust causes an abrupt increase in noise in lateral direction. Figure 2.1 shows the superimposed octave band spectrum of 12 different measurements of aircraft while landing and undergoing reversed thrust at 150 m away from the runway with an average emitted noise level of 85 $\pm$ 3.5 dB(A).

The Aviation Environment Circular 3 of 2011 (File No. 04–01/2010- AED dated August 10, 2011) recommended on restricted use of thrust reverser after landing already existed for Indira Gandhi International (IGI) Airport. The noise exposure due to the aircraft noise is a fundamental issue for assessing the aviation impact and correlating with health effects and annoyance. A fundamental issue with conventional noise metrics or descriptors is that they may not represent noise exposure

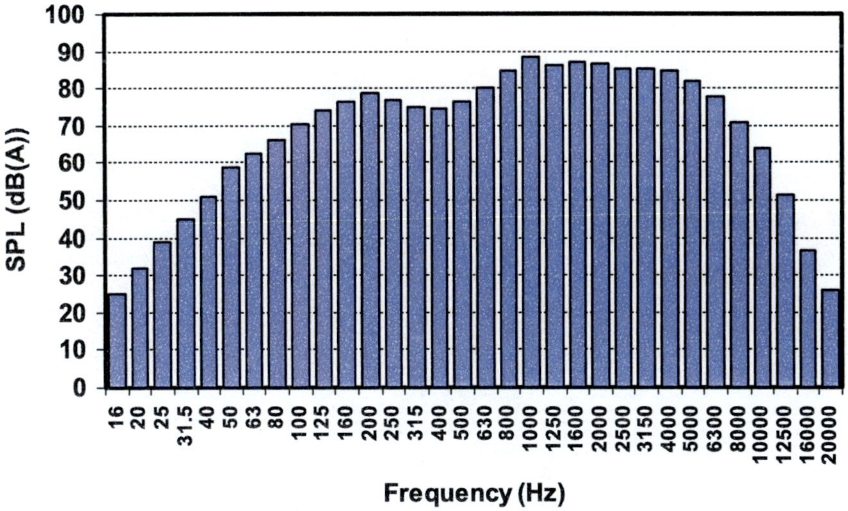

Fig. 2.1 Superimposed octave band spectrum of 12 different measurements of aircraft while landing and undergoing reversed thrust at 150 m away from the runway (Darpe et al. [26])

Table 2.1 Aircraft noise metric classifications and supplementary metrics (Hooper et al. [27])

Classification	Example metric
Single event maximum sound levels	L_{AMax} (dB)—maximum sound level
Single event energy dose	SEL (dB)—sound exposure level
Cumulative energy average metrics	L_{eq} (dB)—equivalent sound level
Cumulative time metrics	TA (minutes)—24-h time above
Flight path movements	Show individual movements or numbers of movements over a given time period
Respite charts	Number of hours with no aircraft movements; expressed as a % of the total number of hours during the period of interest
N70 contours	Combine single event noise level information with aircraft movement data. Maps show number of events louder than 70 dB(A)
Person Event Index	Allows total noise load generated by an airport to be computed by summing, over the exposed population, the total number of instances where an individual is exposed to an aircraft event above a specified noise level over a given time period

patterns in ways that a member of the general public can relate to or understand. However, there is no consensus as to the best means of describing the aircraft noise exposure. The common queries pertaining to this information requires the knowledge of: number of aircraft, noise levels, and the time of day [27]. Table 2.1 shows the aircraft noise metric classifications and supplementary metrics used widely.

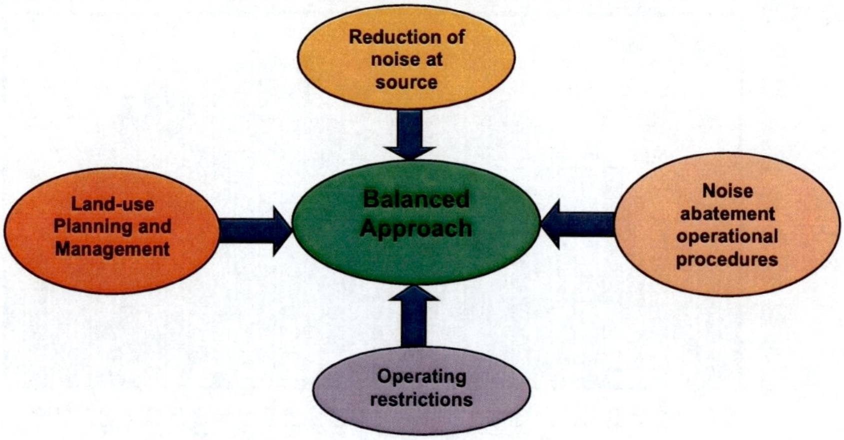

Fig. 2.2 Principal elements of ICAO balanced approach to noise [28]

The balanced approach to noise management recommended by International Civil Aviation Organization (ICAO) is an approach to manage noise at an airport through the examination of four principal elements, namely, reduction at source, land-use planning and management, operating restrictions, and noise abatement operational procedures, as shown in Fig. 2.2. The selection of a particular measure or combination of measures over others and their execution depends upon specific local and national circumstances. It is always imperative that noise mitigation measures for residential development should take into account while considering the airport development. The effectiveness of the noise control measures for both existing and new airports should be considered for each particular situation. The noise abatement may include one or more of the following as suitable: the use of noise preferential runways to direct the initial and final flight paths of aircraft away from noise-sensitive areas; the use of noise preferential routes to assist aircraft in avoiding noise-sensitive areas on departure and arrival; and the use of noise abatement takeoff or approach procedures designed to reduce the noise emissions, while maintaining the required level of safety. Reduced power/reduced drag techniques; Continuous Descent Approach (CDA); use of reversed thrust; and use of ground-based operational procedures such as limiting aircraft engine ground running are some of the noise abatement operational procedures suggested by ICAO [28].

Figure 2.3 shows the detailed aspects of principal elements of ICAO balanced approach to aircraft noise control [28]

2.3.4 *Industrial Noise*

Noise is a common occupational health hazard in industrial units also. The dominant noise frequencies from the machinery are observed to be in the low (>22–500 Hz)

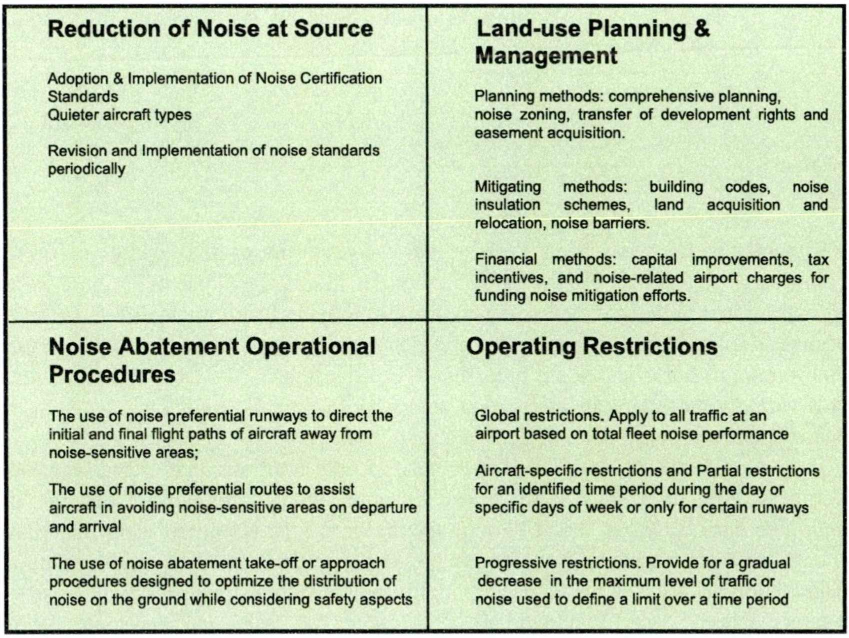

Fig. 2.3 Detailed aspects of principal elements of ICAO balanced approach to aircraft noise control [28]

and mid (>500–2 kHz) octave band center frequency characteristics in most of industrial setup [29]. Mid- and high-frequency noise is attenuated by propagation through the atmosphere and also by attenuation due to its passage over acoustically soft ground. The low-frequency noise emitted by various sources such as pumps, boilers, heavy industry, ventilation units, fans, etc. is sometimes confused with vibrations as it may cause certain parts of the human body to vibrate due to resonance. For example, the chest wall can resonate at frequencies of about 50–100 Hz and the head at 20–30 Hz [30]. Emmerich et al. [31] showed that exposure to industrial noise produces frequency-specific changes reflected in morphological changes of the cochlear hair cells, thus indicating specific damage to outer hair cells in animals (e.g., guinea pig). The review study reported by Prashanth and Venugopalachar showed that the auditory damage risk and physiological system response of human system are related to noise frequency [29]. This study provided strong evidence for the claims that noise with a major frequency characteristic of around 4 kHz has auditory effects. Worldwide, 16% of the disabling hearing loss in adults (over four million DALYs) is attributed to occupational noise, ranging from 7 to 21% in the various subregions. The effects of the exposure to occupational noise are larger for males than females in all subregions and higher in the developing region [32]. High noise level exposure may affect the auditory system. A maximum permissible level of occupational noise limit by the international standards organization is 85–90 dB(A) for 8 h/day. A noise level of 90 dB(A) for 8 h a day and 5 days a week is the maximum recommended safe limit by the Ministry of Labour, Government of India under Factories Act [33].

2.3.5 *Vehicular Horn Noise*

The vehicular horn has been a major noise source among the sources of noise pollution particularly in the developing countries like India. Noise pollution arising from the horns in the vehicles is a major nuisance; the indiscriminate use of pressure horns is a source of serious noise pollution. In case of vehicle horns, both maximum and minimum noise levels are prescribed. The minimum level is specified so that horn attracts the attention and serves the purpose and the maximum level from hearing damage point of view.

Figures 2.4 and 2.5 show the spectrum of a bike horn and that of two wheeler horn, car horn, and SUV horn. It can be observed that frequency spectrum is dominated in octaves of 2.5–4 kHz for all the horns. Thus, the building materials should have a better sound insulation in this region and control the sound insulation degradation due to the coincidence dip in this range to attenuate the honking noise to a larger extent. The Indian standard, IS 15796: 2008, Automotive vehicles-Horn installation requirements, recommends the maximum sound pressure level when determined in accordance with method of test given shall meet the following:

- Between 83 dB(A) and 112 dB(A) for horns fitted on two wheelers or three wheelers of a power less than or equal to 7 kW.
- Between 93 dB (A) and 112 dB (A) for the horn fitted on all other motor vehicles including tractor, two wheelers and three wheelers of a power greater than 7 kW.

The National Ambulance Code for road ambulance for Type L and M—National Ambulance Code (AIS 125) formulated by Ministry of Road Transport and Highways (MoRTH), India prescribed noise limits for sirens fitted on ambulance in the

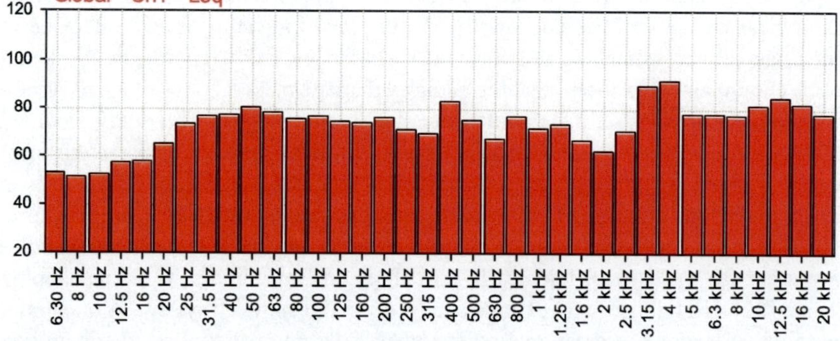

Fig. 2.4 Spectrum of a bike horn (L_{eq} = 95.6 dB (A), 100.5 dB (C))

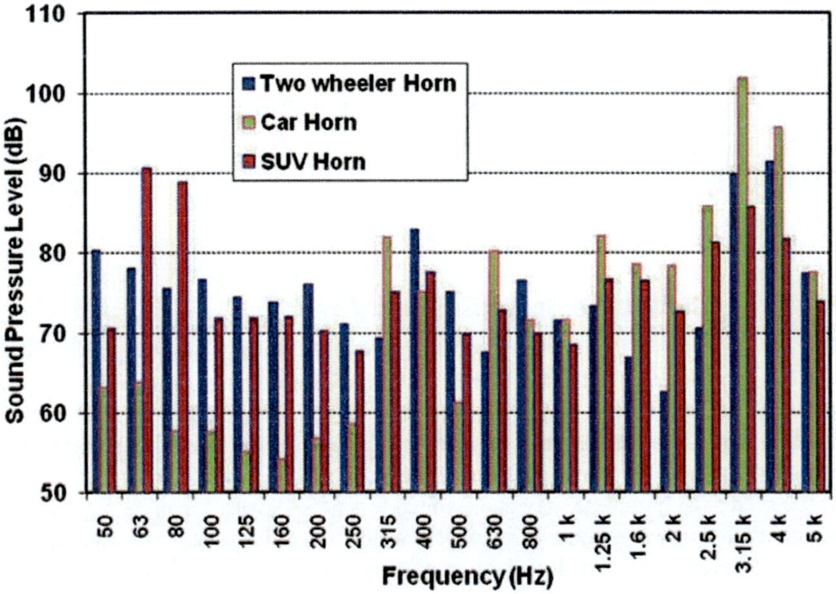

Fig. 2.5 Spectrum of a two wheeler horn (L_{eq} = 95.6 dB(A), 101.3 dB (C)); car horn (L_{eq} = 101.4 dB(A), 102.4 dB(C)); and SUV horn (L_{eq} = 96.3 dB(A), 100.5 dB (C))

range of 110 dB(A) to 120 dB(A). In a 2013 ruling, Supreme Court of India had banned pressure horns, multiple sound emitting horns, and musical horns in vehicles. The National Green Tribunal (NGT) of India imposed ban on use of pressure horns in all vehicles in the national capital particularly trucks and buses in 2016. MoRTH proposed for levying the penalty on vehicle owners installing multitoned and air horns and issuing challan to defaulters for fitting pressure horns. In 2017, the Delhi Police fined more than ten thousand persons for noise-related offences in national capital region of India.

2.4 Assessment of Community Noise

Community noise is an environmental problem to a community due to development process, transportation noise, community activities, etc. and can be detrimental to the quality of life of individuals in case the ambient noise levels are high. There had been various noise descriptors used for evaluation of the community noise. One elementary issue that must be accounted in the choice of an environmental noise metric(s) is the purpose for which the metric will be used [34]:

- To employ public policy on noise immission from one or more sources.
- To disseminate information about noise exposures easily understandable to the public.
- To assess a noise scenario for noise management and control.

The various descriptors used for assessment of community noise are as follows [35].

2.4.1 Equivalent Continuous Sound Level (L_{eq})

Machinery noise may generate a consistent sound pressure level, but there are cases where noise levels may vary with time over a wide range such as road traffic/ environment noise. In the case of road traffic, passage of trucks or motorcycles can cause high noise levels whereas that of a two-wheeler scooter or lean mixed traffic such as at nighttime may cause much less noise levels. It can be regarded as the continuous steady noise level which would have the same total A-weighted acoustic energy as the real fluctuating noise measured over the same period and is defined as:

$$L_{\text{eq}} = 10 \log_{10} \frac{1}{T} \int_{0}^{T} \left(\frac{\rho_A(t)}{\rho_0} \right)^2 dt = 10 \log_{10} \left(\frac{1}{N} \sum_{n=1}^{N} 10^{L_n/10} \right) \tag{2.1}$$

where T is the total measurement time, $\rho_A(t)$ is the A-weighted instantaneous sound pressure, ρ_0 is the reference sound pressure (20 μPa), and L_n is the noise level for each of N equal intervals. The duration of measurement in hours is to be specified with these levels.

2.4.2 Day-Night Average Sound Level

Day-night average sound level L_{dn} is widely used to evaluate the community noise problems. These are modifications of L_{eq} with the night time measured levels increased by 10 dB to take into account the increased annoyance caused by noise at night time (10 p.m. to 7 a.m.). So, basically these are L_{eq} levels with a night time penalty of 10 dB on measurements for the night hour.

$$L_{\text{dn}} = \log_{10} \left[\frac{1}{24} \left(15 \times 10^{(L_{\text{day}}/10)} + 9 \times 10^{(L_{\text{night}}+10)/10} \right) \right] \tag{2.2}$$

where L_{day} and L_{night} are equivalent continuous noise levels for the day and night times, respectively. Similarly, the day-evening-night sound level L_{den} is calculated as:

$$L_{den} = \log_{10}\left[\frac{1}{24}\left(12 \times 10^{(L_{day}/10)} + 4 \times 10^{(L_{evening}+5)/10} + 8 \times 10^{(L_{night}+10)/10}\right)\right] \quad (2.3)$$

where $L_{evening}$ is A-weighted equivalent noise level over the 4-hour evening period from 19:00 to 23:00 h. A night is normally the 8 h between 23 and 7 h or the 9 h between 22 and 7 h. However, individual countries define night differently, for example, 22 h to 6 h in some countries. $L_{evening}$ and L_{night} have a 5 and 10 dB weighting applied to each, respectively, in order to account for the difference in annoyance due to the different time during a day.

2.4.3 *Statistical Indicators*

Percentile noise levels L_n, where n may be anything from 1 to 99, is that noise level exceeded for n% of the measurement time. By definition of percentiles, L_1 must be greater than or equal to L_2, which must be greater than or equal to L_3, etc. Statistical indicators such as L_{10} and L_{90} are used widely for the environmental noise. L_{10} is the A-weighted noise level that is exceeded 10% of the time and L_{90} is the level exceeded 90% of the time. L_{10} is regarded as a measure of the more serious intruding noise levels and L_{90} is considered the average ambient or background level. L_{50} is the level exceeded for 50% of the time and represents the median of the fluctuating noise levels. For a varying sound, L_{10} is greater than L_{50}, which in turn is greater than L_{90} [36]. The L_{Max} or maximum sound level descriptor is the highest sound level measured during a single noise event (such as a vehicle pass by), in which the sound level changes value as time goes on. The maximum sound level is important in judging the interference caused by a noise event with common activities. According to Robinson (1969), the L_{eq} is related to percentile levels as:

$$L_{eq} = L_{50} + (L_{10} - L_{90})^2/56 \quad (2.4)$$

2.4.4 *Traffic Noise Index*

Traffic noise index (TNI) is the A-weighted noise level measured as:

$$\mathrm{TNI} = 4\,(L_{10} - L_{90}) + (L_{90} - 30) \tag{2.5}$$

The term (L_{10} − L_{90}) is called "noise climate (NC)," and the final arbitrary constant is included to yield more convenient numbers. TNI is derived on the assumption that extensive noise level fluctuations over time are the dominant factor in traffic noise annoyance.

2.4.5 *Noise Pollution Level*

According to Robinson (1969) of the British National Physical Laboratory, L_{eq} in itself is an insufficient descriptor of annoyance caused by fluctuating noise, and road traffic noise is a significantly fluctuating noise. The index Noise Pollution Level (NPL) was developed to estimate the dissatisfaction caused by road traffic noise comprising of two terms.

$$\mathrm{NPL} = L_{eq} + (L_{10} - L_{90}) \tag{2.6}$$

2.4.6 *Single Event Noise Exposure Level*

Single event noise exposure level is basically an L_{eq} level normalized to a time period of 1 s. It can be defined as the constant noise level which if maintained for a period of 1 s would produce the same A-weighted sound energy as the actual event itself. It can be converted to equivalent sound pressure levels, L_{eq} as:

$$L_{eq} = SEL + 10\ \log\ (N) - C \tag{2.7}$$

where SEL is average single event sound exposure level of the N events experienced during the specified time interval. The constant term C depends on that time interval; for 24 h, it is 49.4 dB (= 10 × log (number of seconds in 24 h)), and for 1 hour, $C = 35.6$ dB. The sound exposure level (SEL) metric represents all the acoustic energy of an individual noise event as if that event had occurred within a one-second time period. It can be also expressed as [6]:

$$SEL = L_{\mathrm{Aeq,T}} + 10\ \log_{10}(T/T_0), T_0 = 1\ \mathrm{s}. \tag{2.8}$$

2.4.7 Noise Exposure Forecast (NEF)

Total noise exposure at a ground point in the neighborhood of an airport is summation of noise produced by all the different aircraft flying their different flight paths. For a class of aircraft i on a flight path j, the NEF_{ij} at a specific ground point is:

$$\mathrm{NEF}_{ij} = \mathrm{EPNL}_{ij} + 10\log\left(N_{dij} + 16.7N_{nij}\right) - 88 \tag{2.9}$$

where EPNL_{ij} is the effective perceived noise level produced at a point by a particular class of aircraft following a particular flight path, N_{dij} is the number of daytime (7 a.m. to 10 p.m.) operations for each aircraft class i and flight path j, and N_{nij} is the number of night-time (10 p.m. to 7 a.m.) operations for each aircraft class i and flight path j.

2.4.8 Noise and Number Index (NNI)

Noise and number index (*NNI*) is another rating method, defined as:

$$\mathrm{NNI} = (\mathrm{PNL})_{\mathrm{N}} + 15\log_{10}N - 80 \tag{2.10}$$

where $(\mathrm{PNL})_{\mathrm{N}}$ is the average peak noise level of all aircraft during a day. PNL is the peak perceived noise level generated by a single aircraft during the day and N is the number of aircraft events over a 24-h period. The Noise and Number Index scale runs from 0–60 and exposure to aircraft noise reaches an unreasonable level in the range 50–60.

2.5 Machinery Noise

The machinery noise estimation or prediction is generally in terms of sound power level. The estimation of the sound power level and the sound pressure level shall be helpful in planning for the location of machines in factories and industrial units so as to control the combined noise emitted due to various machines. The sound power level emitted by industrial fans, compressors, transformer, electric motors, etc. can be estimated by the following empirical formulations (Eqs. 2.11 to 2.21) recommended by Barron [37]:

2.5.1 Air Distribution System Noise

Noise generated in air distribution systems is one of the issues in design of heating, ventilating, and air conditioning (HVAC) systems. The flow-induced noise generated in elbows may be estimated by the following relation (Bullock [38]):

$$L_W = F_s + 10 \log_{10} f_0 + 10 \log_{10} S + 44.4 \log_{10} u - 54 \qquad (2.11)$$

The quantity f_0 is the octave band center frequency; S is the duct cross sectional area m^2; and u is the velocity of the air upstream of the elbow m/s. The spectrum function F_s is correlated with Strouhal number $N_s = \frac{f_0 D}{u}$, where $D = \sqrt{\frac{4S}{\pi}}$.

2.5.2 Pump Noise

The sound power level for airborne noise from a pump may be estimated from the following correlation within ±3 dB [37]:

$$L_W = K_0 + 10 \log_{10}(hp) \qquad (2.12)$$

The quantity hp is the rated horse power for the pump. For pumps having a rated speed of 1600 rpm or higher, the values of the constant K_0 are as follows:

$K_0 = 98$ dB for centrifugal pump.
$K_0 = 103$ dB for a screw pump.
$K_0 = 108$ dB for a reciprocating pump.

For pumps operating at speeds below 1600 rpm, subtract 5 dB from these values to obtain the K_0 for empirical formulation, $L_W = K_0 + 10\log_{10}(hp)$.

2.5.3 Industrial Fans Noise

The sound power level of noise generated by the fan for any octave band may be estimated from the following correlation (Graham [39]):

$$L_W = L_W(B) + 10 \log_{10}(Q/Q_0) + 20 \log_{10}(P/P_0) + B_T \qquad (2.13)$$

where L_W (B) is the basic sound power level recommended by Graham [39]; Q is the volumetric flow rate through the fan and Q_0 is a reference volumetric flow rate, 0.47195 dm^3/s = 1 ft^3/min; P is the pressure rise through the fan and P_0 is a reference pressure rise, 248.8 Pa = 1 in H_2O; and B_T is the blade tone component, which is zero except for the octave band in which the blade pass frequency lies. For this one octave band, the value of the blade tone component is recommended for different fans. For instance, for centrifugal fan with airfoil blade, $B_T = 3$ dB, while for propeller fan, B_T varies from 5 to 7 dB.

2.5.4 Compressor Noise

The sound power level for centrifugal compressors is calculated as [37]:

$$L_W = 20 \log_{10}(hp/hp_0) + 50 \log_{10}(U_t/U_0) + 81 \tag{2.14}$$

where the term hp is the compressor power input, and $hp_0 = 1$ and $hp = 745.7$ W. The quantity U_t is the blade tip velocity and $U_0 = 243.8$ m/s. For axial compressors, the corresponding correlation is as follows [37]:

$$L_W = 20 \log_{10}(hp/hp_0) + 76 \tag{2.15}$$

The spectrum of noise for centrifugal and axial compressors is broadband, with the peak(maximum) in the sound power level occurring at a frequency of f_m [37]:

$$f_m = 1000\ (U_t/U_0) \text{ for centrifugal compressors, and} \tag{2.16}$$

$$f_m = 2N_b n_r \text{ for axial compressors,} \tag{2.17}$$

where N_b is the number of blades in one stage of the axial compressor and n_r is the rotational speed of compressor.

2.5.5 Transformer Noise

For a self-cooled transformer, the sound power level is related to the kVA rating of the transformer by the following expression [37]:

$$L_W = 45 + 12.5 \log_{10}(\text{kVA}) \tag{2.18}$$

For a forced air-cooled transformer, the following expression applies [37]:

$$L_W = 48 + 12.5 \log_{10}(\text{kVA}) \tag{2.19}$$

Figure 2.6 shows the sound pressure level at a distance of 0.3 m from source on low-voltage side of 100 kVA transformer. Transformer acoustic noise is a hum characterized by spectral spikes at harmonics of the fundamental frequency (100 Hz/

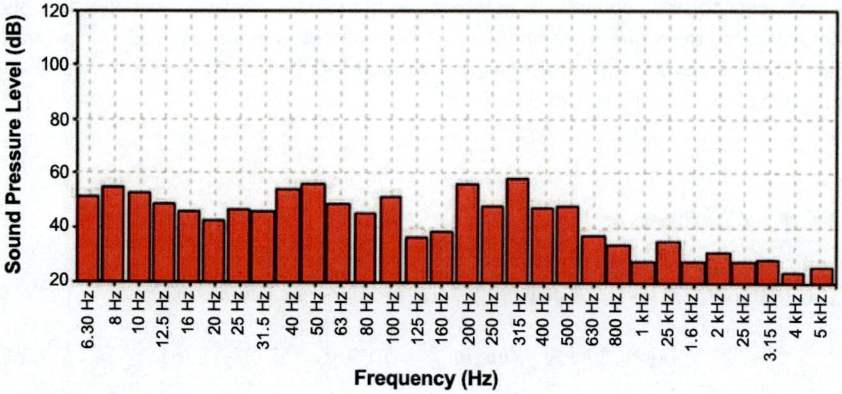

Fig. 2.6 Sound pressure level at a distance of 0.3 m from the source on low-voltage side of 100 kVA transformer

120 Hz) twice the line supply frequency. The low-voltage side produced much higher sound pressure level as compared to the high-voltage side of 53.9 dB (A) at a distance of 0.3 m from the source. The sound power level of the 100 kVA transformer as per IEC 60076-Part 10:2016 was measured to be 59.9 dB(A).

2.5.6 *Electric Motors Noise*

There are two types of electric motors classified on the basis of type of motor cooling. The drip-proof (DRPR) motor cools itself by induced air flow, while the totally enclosed fan-cooled (TEFC) motor uses an internal fan for motor cooling. The sound power level can be calculated from the rated motor horsepower (*hp*) and rotational speed n_r (rpm) as [37]:

$$L_W = 20\ \log_{10}(hp) + 15\ \log_{10}(n_r) - 3 \text{ for } hp \geq 7hp \text{ for drip-proof motors} \tag{2.20}$$

$$L_W = 20\ \log_{10}(hp) + 15\ \log_{10}(n_r) + 13 \text{ for } hp \geq 5\,hp \text{ for TEFC motors} \tag{2.21}$$

2.5.7 *Household Appliances and Equipment Noise*

Home appliance is one of the main sources of indoor noise. Most of the household appliances, such as vacuum cleaners, cooker hoods, refrigerators, shavers, etc., have tonal sounds caused by electric motor or fan. As such, the human auditory system is very sensitive to the tonal components of a signal rather than the broadband

Table 2.2 Sound power levels for some of the equipment and home appliances

Appliance	L_w dB(A)	Equipment	L_w dB(A)
Air conditioner	75	Crane (movable)	115
Clothes dryer/washer	70	Rock drill	125
Refrigerator	50	Transformer	50
Hair dryer	75	Tractor	120

components. There are several psycho-acoustical properties useful to describe the perceptual aspects of the household appliance noises such as loudness, sharpness, tonality, and roughness [40, 41]. However, the sound power level is the widely used parameter to characterize the noise radiated by the household appliances and construction equipment. Table 2.2 shows the sound power levels for some of the equipment and home appliances.

2.6 Impulsive Noise

Impulsive noise is typically an airborne noise of short duration, for example, shooting of a firearm or firework explosion. The Occupational Safety and Health Administration guidance states that exposure to impulsive or impact noise should not exceed a 140 dB peak sound pressure level. Impulsive or impact noises are measured to be much more unsafe to hearing than continuous noises. Impulse noise is single noise bursts with duration of less than 1 s with peak levels 15 dB higher than background noise and differs from steady-state noise by the properties in the time domain. The additional effect of occupational impulse noise on hearing has been shown to be from 5 to 12 dB at 4 kHz audiometric frequency. There had been many risk criteria recommended in different countries. In France, the maximum A-weighted peak level is 135 dB, and in the United Kingdom, the C-weighted peak sound pressure is limited to 200 Pa (140 dB) [42]. This criterion of unweighted 200 Pa (140 dB) is used in the European Union (EU) directive 86/188 and ISO 1999-1990 regardless of the number of impulses. The American Conference of Governmental Industrial Hygienists (ACGIH) recommended a limit of 140 dB C-weighted peak sound pressure level [42]. Noise-induced hearing loss (NIHL) manifests in humans in early stages as an audiometric notch at the frequencies 4–6 kHz. A hearing threshold shift related to intense noise exposure is occasionally temporary in its initial stages and is manifested as temporary threshold shift (TTS). TTS may become permanent threshold shift (PTS) in case of frequent chronic noise exposure. Tinnitus, also called "ringing in ear," often accompaniment of hearing loss may result in a reduced concentration, anxiety, and depression [43].

2.7 Occupational Noise

The Occupational Safety and Health Administration (OSHA), USA, developed noise exposure limits for industrial workers. A noise exposure of 90 dB(A) for 8-h work duration is permitted. Table 2.3 shows the OSHA noise exposure limits for workers. In cases wherein the noise exposure comprises of multiple periods of noise exposure, the combined effect is to be taken.

The Directive 2003/10/EC [44] recommended three levels for occupational settings in terms of daily noise exposure levels and peak sound pressure (maximum value of the "C"-frequency weighted instantaneous noise pressure): first (lower) action level (minimum to provide protection) set at $L_{EX,8h} = 80$ dB(A) and 135 dB (C) p_{peak}; second (upper) action level (mandatory protection) set at $L_{EX,8h} = 85$ dB (A) and 137 dB(C) p_{peak}; and the maximum exposure limits set at $L_{EX,8h} = 87$ dB (A) and 140 dB(C) p_{peak}. The limits are to be transformed using the time-intensity trade-off of 3 dB increase for halving the time.

2.8 Noise Dose

The actual exposure in comparison to the permissible exposure, and for which 100% and above represents exposures that are hazardous. The noise dose is calculated as:

$$\text{Dose (D)} = 100 \times (C_1/T_1 + C_2/T_2 + C_3/T_3 + \ldots + C_n/T_n) \qquad (2.22)$$

where C_n is the total time of exposure at a specified noise level and T_n is the exposure time at which noise for this level becomes hazardous and can be estimated as:

$$T_n = 8/2^{(L-90)/5}(L \text{ is the measured sound level}) \qquad (2.23)$$

Table 2.3 OSHA noise exposure limits for workers

L_{eq} noise level in dB(A)	Time of exposure per day in hours
90	8
92	6
95	4
97	3
100	2
102	1.5
105	1
110	0.5
115	0.25 or less

The noise dosimeter or noise-exposure meter is often used for measurement of the accumulated noise exposure of workers in an industrial environment. The dosimeter is an integrating sound level meter system that can be carried by workers during their normal course of work. The output of a typical dosimeter includes the total noise exposure, the percent of allowable noise dosage, and the peak (maximum) noise exposure level. The Occupational Safety and Health Administration (OSHA) recommends the permissible noise exposure level at 90 dB(A) for an 8-h time-weighted average (*TWA*) using a 5-dB exchange rate. The time-weighted average sound level (*TWA*) is calculated as:

$$TWA = 16.61 \times \text{Log}\,(\text{D}/100) + 90 \tag{2.24}$$

For calculating the *TWA* exposure, all continuous, varying, intermittent, and impulsive sound levels from 80 to 140 dB(A) shall be integrated. In case of an 8-h workshift with the constant noise level over the entire shift, the *TWA* is equal to the measured sound level [45].

2.9 Noise Sources Identification

Noise sources identification (NSI) techniques are used to diagnose noise sources primarily with an objective of identifying the most prominent sub-sources in terms of sound power radiated and sound intensity. Subsequently, the noise sources can be ranked in order of increasing noise emissions and design modifications can be implemented to effectively control the overall noise radiated. There have been several methods of sound mapping based on sound pressure, sound intensity, and selective intensity. The latest techniques involve the array-based techniques using a dozen to up to hundreds of microphones. The main methods are: near-field acoustic holography related techniques, beamforming (phased array technique), refined beamforming (deconvolution methods), moving source beamforming, and spherical beamforming. The technique selected by the user will depend on the required frequency range, the measurement distance, the required resolution, and the area covered by the measurement. The most common method to identify a noise source is to measure and map the sound pressure level (SPL) at various locations. However, sound intensity approach has been the widely used approach for noise sources identification of various noise sources. The acoustic field can be represented with a magnitude and a direction, and thus, it is possible to determine sound power of a source even in the presence of background noise [46]. Near-field acoustic holography permits the analysis of spatial and frequency (or time) dependence of acoustic fields. Beamforming is a recent technique identical to a "camera" in the sense that the array of microphones behaves as a lens. The sound fields can be determined for

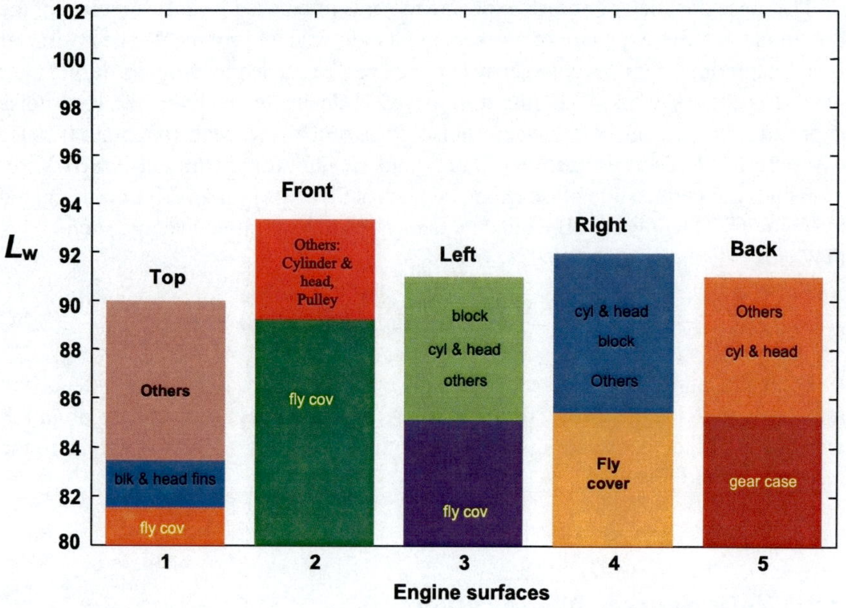

Fig. 2.7 Sound power level, L_w ranking of sources from different engine surfaces using sound intensity technique

objects of many sizes with a wider frequency range and bandwidth by varying the distance between the array and the object [47]. The sound intensity method has the advantage of being able to ignore background noise and is therefore used frequently to identify noise sources, both in the laboratory and in the field. Sound intensity measurements normal to the surface are made from a number of equally spaced points on the surface of the source. The contour plots show the sound field generated by the source. The sound power is the average normal intensity over a surface enclosing the source, multiplied by the surface area. Thus, for the evaluation of the effectiveness of noise reduction methods, the noise radiated by the individual components of machines is to be known, which is obtained by the sound power from the components of a machine. Figure 2.7 shows the typical example of the sound intensity mapping of a diesel engine and sound power level ranking of the various engine surfaces. Thus, it can be helpful in identifying the critical noise sources and also for analyzing the preventive measures for noise control. For instance, on the top control surface, the noise radiating components of the engine are divided into four parts, namely, flywheel cover, block and head cooling fins, rocker cover, and starter motor. Among all these parts, the engine block and head cooling fins radiate the highest acoustic energy. The sound intensity level of rocker cover is the highest in the surface, but due to the small area of the part, the acoustic power is low compared with others. The front surface is mainly divided into three areas for sound power ranking: flywheel cover, cylinder and head cover, and pulley. Among the three areas,

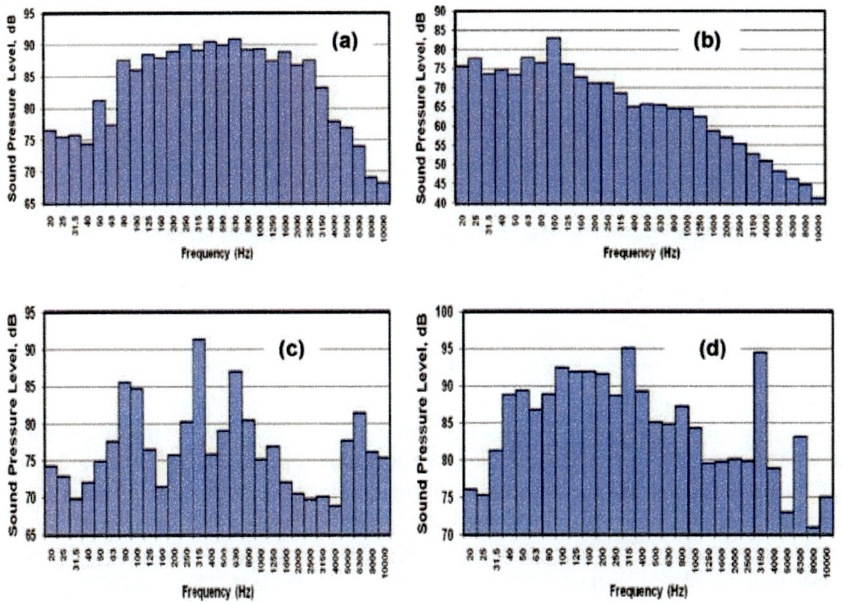

Fig. 2.8 (**a**) Diesel genset noise spectrum with sound level emission of 94.8 dB(A); (**b**) air conditioning noise spectrum with sound level emission of 70.8 dB(A); (**c**) air conditioning noise spectrum with sound level emission of 82.7 dB(A); (**d**) traffic noise spectrum with sound level emission of 74.5 dB(A)

flywheel cover is the major contributor of noise, pulley follows the second place, and the cylinder and head cover had the lowest sound power. The backside sources are the three-gear case (lower), gear case (upper), and cylinder and head, whereby the lower gear case is observed to radiate the highest noise levels. The flywheel cover contributes significant sound power in all surfaces of the test engine, and it is possible to reduce overall sound power of engine effectively by damping and isolation of this part. Thus, the ranking of the noise sources can be made and critical ones can be treated for noise control. The noise exposure levels and the analysis of frequency components of the various critical sources are thus very essential in assessment and control of overall noise radiated by a source.

Once the noise characteristics, that is, amplitude and frequency spectrum, are characterized, preventive measures or corrective actions can be accordingly applied so as to control the noise pollution. A typical example of a laboratory setup wherein the various critical noise sources have been identified based on noise monitoring is presented in Fig. 2.8. Thus, the octave band analysis of these sources shall be helpful in understanding the dominant frequencies and devising suitable control measures for reducing the overall radiated noise by the specific noise sources. Noise sources ranking and simulated noise scenario using noise maps can be very effective in devising the noise control action plans. However, a prerequisite is that for each noise source, a corresponding database is to be built up containing emission data (traffic density, vehicles or aircraft, equipment, industry lay-out, noise levels, octave band spectra, etc.), propagation data (road pavement, rail, ground surface, acoustic properties of building facades, for example), meteorological data, and other data relevant to establish the magnitude of the noise levels at each assessment location [48].

2.10 Sound Absorption and Sound Transmission Loss

2.10.1 Sound Absorption

The sound absorption coefficient is defined as the correlation between the acoustic energy absorbed by a material and the total incident energy incident. There are two methods of measuring sound absorption coefficient: impedance tube method and the reverberation chamber method. The characterization of sound absorption coefficient of materials is needed for noise control applications. Usually, normal incidence impedance and absorption can be obtained from impedance tube measurements. However, practical considerations desire the use of reverberant chambers for the measurement of the random incidence absorption coefficient. The two-microphone transfer function method using an impedance tube as shown in Fig. 2.9 is a commonly used measurement technique as specified in ISO 10534-2 standard and

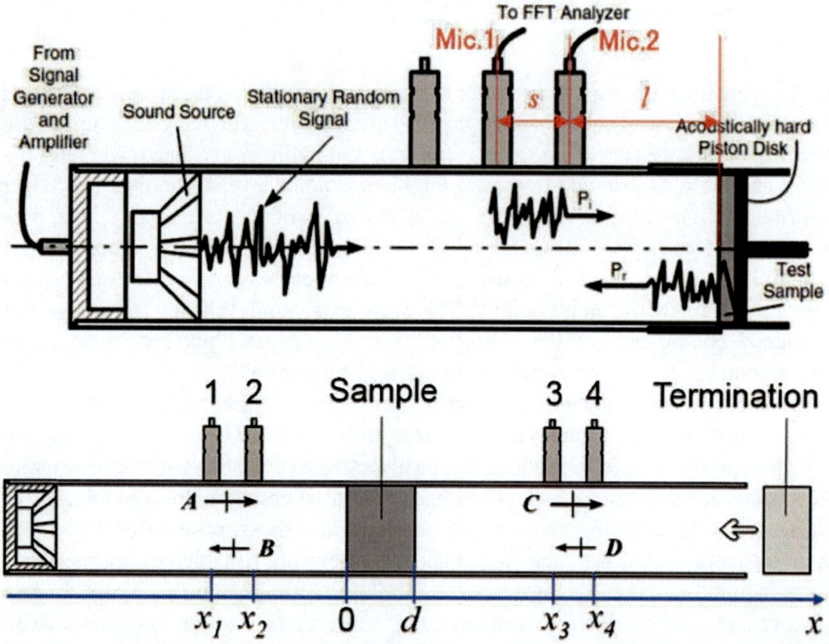

Fig. 2.9 Setup for the impedance tube for sound absorption and sound transmission loss measurement (Kimura et al. [49])

ASTM E1050 standard. The frequency response function between the sound pressure at microphone 2 and the sound pressure at microphone 1 (H_{21}) is measured, and the sound reflection coefficient (R) is calculated using H_{21} by [49]:

$$R = \frac{H_{21} - e^{-jks}}{e^{jks} - H_{21}} e^{2jk(l+s)} \tag{2.25}$$

From the calculation of sound reflection coefficient, the specific impedance ratio ($Z/\rho c$) and sound absorption coefficient is calculated as:

$$\frac{Z}{\rho c} = \frac{1+R}{1-R} \tag{2.26}$$

$$\alpha = 1 - |R|^2 \tag{2.27}$$

where ρ is the density of air, c is the speed of sound in air, k is the wave number of air, l is the distance between microphone 2 and the front of sample, and s is the spacing between the microphones.

The normal incidence sound transmission loss is calculated using a transfer matrix method utilizing a modified four-microphone impedance tube, as shown in Fig. 2.9 [49]. The random incident absorption coefficient α, also known as the Sabine absorption coefficient, is widely used for measurement of sound absorption performance of materials. The random incidence absorption coefficient can be calculated from the following equation [50]:

$$\alpha_s = 55.3\,\frac{V}{S}\left(\frac{1}{c_2 T_2} - \frac{1}{c_1 T_1}\right) - 4V(m_1 - m_2) \tag{2.28}$$

where c_1 and c_2 are the propagation speeds of the sound in air (m/s) for empty and full conditions and T_1 and T_2 are the reverberation times in seconds of the room under empty and full conditions, respectively. The power attenuation coefficients m_1 and m_2 are calculated according to ISO 9613-1 using the atmospheric conditions of the empty and full room correspondingly. The ISO 354 standard recommends a sample area 10–12 m^2 in the reverberant rooms with a volume less than 200 m^3 or, if having a higher volume, the volume is increased by a multiplicative factor of $(V/200)^{2/3}$, which can thus yield larger size samples. Sometimes obtaining large samples of materials (10–12 m^2 or more) to determine the sound absorption coefficient in a reverberation room is not practically feasible. In one such study demonstrated by Carvalho and Sousa, the results obtained for measurements of α_s indicate that there is more variability up to the sample (5 m^2) and, as the area of the sample increases, the sound absorption coefficient values tend to stabilize. Thus, a 5 m^2 sample would be adequate for a reliable measurement of the sound absorption coefficient in a reverberation room if a larger sample is not available. However, with the same area of the sample, but a greater perimeter, there is a gain on the sound absorption coefficient values, demonstrating that the "edge effect" may have a great importance in determining sound absorption coefficient in a reverberation room [51].

2.10.2 Sound Transmission Loss

Sound transmission loss *(STL)* is a measure of the effectiveness of acoustical treatments for engineering applications. The sound insulation properties of partitions are characterized by their transmission loss. The transmission loss is mainly obtained by the two-room method, where the specimen is placed in the common wall between source and receiving room. According to the ASTM E90 standard, the *STL* is defined by:

$$STL = NR + 10\ \log\left(\frac{S}{A}\right) \tag{2.29}$$

where *NR* is the noise reduction between the source room and receiving room, *S* is the sound transmitting area, and *A* is the total absorption of receiving room. The most important requirement is that the sound field in both rooms must be adequately diffuse. Thus, for the measurement of transmission loss, two adjacent reverberation chambers are required [52]. Figure 2.10 shows the sound transmission loss (in dB) of

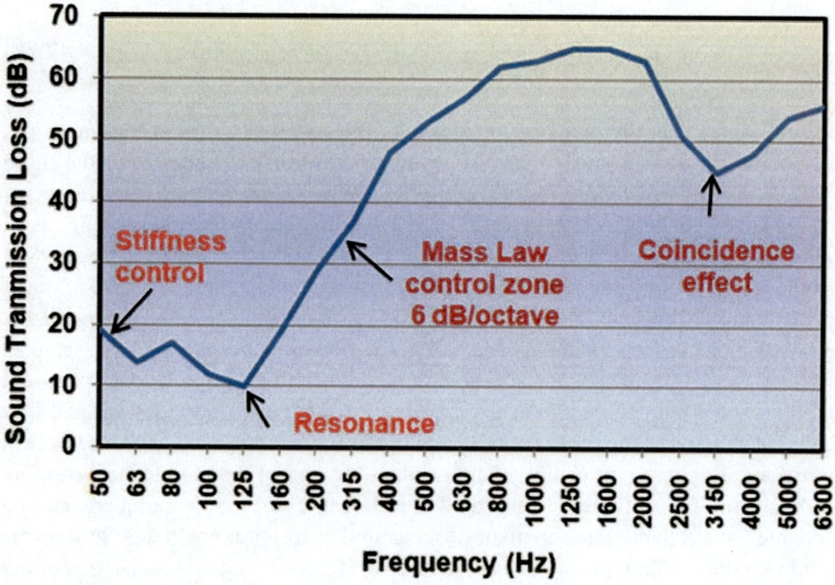

Fig. 2.10 Sound transmission loss (in dB) of a sandwich gypsum panel construction tested in Reverberation chambers showing the different regions

a sandwich gypsum panel construction tested in reverberation chambers of the Institute for Research in Construction, National Research Council Canada [53]. The wall test opening measured 3.05 m × 2.44 m, while the volume of the source room was 65 m^3 and that of receiving room was 250 m^3. The *STL* behavior may be divided into three fundamental regions. At lower frequencies, the static stiffness of partition is vital factor, whereby the resonances may occur and decrease *STL*. In region II (mass-controlled region), the response is governed by the mass of the panel and the curve follows a 6 dB/octave slope. Doubling the mass, or doubling the frequency, outcomes in a 6 dB increase in *STL*. The coincidence between the sound wavelength and the structural wavelength decreases the *STL* at higher frequencies.

The major aspects associated at different regions in the curve are [54]:

- A 6 dB increase in *STL* per doubling of stiffness at low frequencies.
- Damping treatment leads to a reduction in low-frequency resonant responses.
- A 6 dB increase in sound transmission loss per doubling of the mass below the critical frequency.
- A lowering of coincidence frequency with increase in mass.
- An increase in *STL* at critical frequency with the addition of damping.
- An increase in *STL* with added damping and stiffness at high frequencies.

Modern buildings are often required to make use of lightweight structures for optimizing costs and construction time. The use of multilayered structures such as double walls and small amount of absorption in cavity surfaces can considerably improve sound insulation [53]. The sound transmission between rooms is common consideration for noise control engineers, and thus, a precise consideration on the acoustical behavior of rectangular cavities can be very useful in accomplishing the desired noise reduction targets. The theory of sound transmission between two rooms with a common wall presents very interesting facts about how the sound transmission in receiving room can be decreased as [55]:

- Providing more absorption in room with source present and in the other room.
- Increasing the sound transmission loss, that is, enhanced sound insulation of wall.
- Providing less common area as the transmitted power is proportional to intensity striking the wall times wall area.

2.11 Enclosures

An enclosure surrounding a noise source is a special case of two-room problem, where the enclosure is one room and the surrounding space is the second room. While designing the enclosure, mainly two resonances are encountered: one is mechanical resonance of the enclosure panels and the other is the acoustic resonance of air space between the enclosed machine and enclosure walls. The first few panel resonances should not be in the frequency range in which noise attenuation is desired. In case, the sound source radiates low-frequency noise predominantly, then an enclosure with high resonance frequency is desirable, implying a stiff but not massive enclosures. A high-frequency noise source requires a massive enclosure. The use of sound absorbing material in the cavity is helpful in suppressing the resonances [35]. The insertion loss (*IL*) is also correlated with sound transmission loss (*STL*) as [35]:

$$IL = STL + 10 \log_{10}(\overline{\alpha}) \tag{2.30}$$

where $\overline{\alpha}$ is the effective average absorption coefficient within the enclosure.

2.12 Acoustical Materials

Acoustic materials are the materials that absorb, reflect sound, or dampen vibrations with an objective of noise reduction. Acoustical properties of the materials used in different situations should be taken into consideration at the design stage. They usually convert some of the sound energy which strikes them into thermal energy. Some good insulating materials are concrete, thick steel sheets, glass, lead, and thick PVC sheets. Sound transmission loss of some materials at different frequencies is shown in Table 2.4. For two partitions separated by an air gap, minimum structural cross sections between the two using staggered or double studs making them

Table 2.4 Typical values of sound transmission loss of some materials (in dB) (Sharma and Singal [56])

Material	Bulk density (kg/m^3)	Thickness (mm)	Frequency, Hz					
			125	250	500	1000	2000	4000
Corrugated cellulose board	136	10	2	10	12	16	18	–
Wood particle board	800	19	22	18	22	24	23	25
Wood particle board	800	25	23	20	22	23	25	27
Wood particle board composite panel of two layers of material with air gap between	800	59	25	24	28	31	37	32
PVC sheet composite panel of two layers of material with air gap between	1200	20	19	16	19	21	19	23
Cement asbestos	1600	4	11	10	13	19	21	22
Asbestos millboard	1300	5	14	14	19	25	28	27
Calcium silicate slab	275	40	19	20	25	29	26	29
Composite panel plywood sheets with foamed polyurethane between	110	46	12	15	18	21	23	29
Vinyl rubber sheet	650	1.1	4	6	8	10	16	19
Vinyl rubber sheet	2730	0.8	7	10	12	16	21	20

independent enhance the sound insulation. Sound absorbing materials are used to reduce noise levels within spaces, and to control reverberation within spaces. They are usually porous and lightweight foam, fiberglass, cork, and mineral wool. The sound absorption coefficient can be enhanced through various modifications as recommended in one such experimental study in reverberation chambers by Sharma and Singal as [56]:

- With increasing thickness, the absorption coefficient of porous soft materials at any frequency increases.
- On covering the absorbing material with aluminum foil of 0.15 mm thickness, it behaves as a resonant membrane with maximum absorption in low-frequency range (less than 100 Hz),
- With air gap between the porous absorbing material and solid wall backing, the absorption coefficient becomes more or less uniform in mid-frequency range of 500–2000 Hz. The changes in width of air gap is not effective as compared to the increase in thickness of absorptive material,
- On paneling a solid wall with perforated or unperforated sheet with an air gap or absorbing material between the two, mounting behaves like a resonant membrane with maximum absorption in low-frequency range due to an increase in percentage of open area. In case of a 30–35% open area (or perforations), absorption becomes uniformly high over wider frequency range.

The sound absorbing materials can be classified into (i) porous absorbers, (ii) panel or membrane absorbers, (iii) resonant absorbers, and (iv) space or functional absorbers. Porous absorbers have interconnected pores into which the acoustic energy is dissipated. For membrane or panel absorber, the mass is vibrating sheet of membrane and spring is commonly provided by air enclosed in cavity. The resonant absorbers dissipate the acoustic energy with structure vibration and have two common forms: membrane/panel absorbers and Helmholtz absorbers. In case of a Helmholtz absorber, the mass is a plug of air in the opening of a perforated sheet and the spring is usually provided by air enclosed in the cavity [57]. The maximum attenuation occurs at resonance frequency of the mass of air in the neck and spring resistance of air resistance inside the enclosed volume. The resonance frequency of the Helmholtz resonator is given by [57]:

$$f_n = \frac{c}{2\pi}\sqrt{\frac{A}{l_e V}} \tag{2.31}$$

where A is the cross-sectional area of neck, V is the volume of cavity, and l_e is the effective length of the neck. Functional absorbers mean the use of panels or sheets of porous material mounted so that both sides are exposed to sound field; very often, the panels are suspended from the ceilings of the treated enclosure. These are often suspended freely in a room space at some distance from the boundary walls or surfaces and thus provide effective sound absorption in areas where the traditional methods of acoustic treatment are impractical or costly. This technique is often used widely for providing absorption in noisy areas, for example, machine shops. The use of such techniques for studio acoustic treatment offers the possibility of providing a relatively large total amount of sound absorption while minimizing the area of wall surface occupied by the treatment. These absorbers generally show a rising characteristic of absorption coefficient against frequency up to a certain value (typically 500 Hz to 1 kHz), above which the absorption coefficient remains constant [58].

2.13 Noise Management and Control

The specific objectives of noise management and control is to formulate and implement noise limits for the various sources, consideration of noise abatement criteria or a noise abatement goal, and planning of noise impact assessment and its management and control. The noise management policies can be based on [6]:

- *The precautionary principle*: In all cases, where there is rational possibility of health effects due to noise levels, the action plans should consider noise abatement and control irrespective of any scientific or administrative issues.
- *The polluter pays principle*: The economic considerations for managing and controlling the noise levels shall be borne by the polluters responsible for causing noise pollution.
- *The prevention principle*: Action plans should primary be focused on source noise control with prime focus on land-use planning and other alternatives effective for noise control.

An efficient policy management, regulations, and noise abatement goal are very instrumental in accomplishing the desired objectives within a stipulated time period. An integrated approach considering the various aspects of noise control at the source, path and receiver, land-use planning, education, and public awareness toward controlling the noise pollution is very much required especially for areas wherein ambient noise levels are very high. The WHO report on Guidelines for Community Noise recommended that countries adopt the precautionary principle in their national noise policies. WHO recommends the guiding principles of reducing noise exposure as [59, 60]:

- Reduce exposure to noise, while conserving quiet areas.
- Promote interventions to reduce exposure to noise and improve health.
- Coordinate approaches to control noise source and other environmental health risks.
- Inform and involve communities.

The Environmental Noise Directive (END) (2002/49/EC) sets legally binding obligations for environmental noise management and control. The noise mapping exercise is considered as precursor for devising the realization of noise control measures or so called as Noise Action Plans (NAPs). The methods to action planning diverge between various Member States in Europe. The administrative costs had

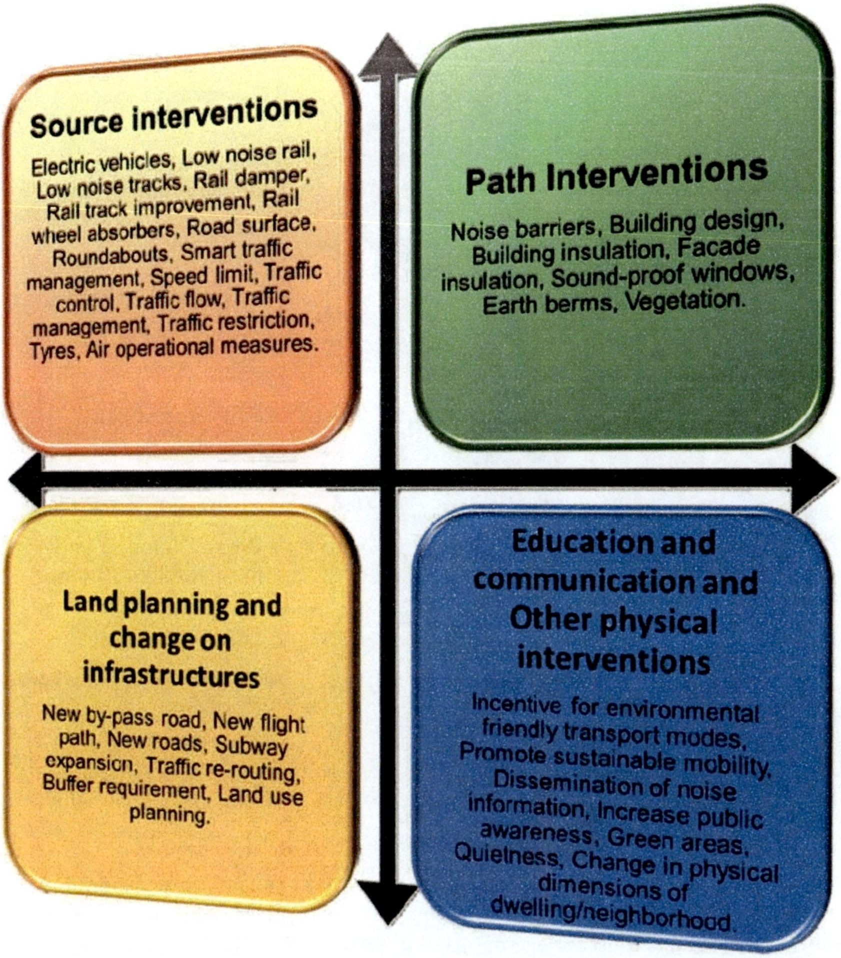

Fig. 2.11 Noise action plans (NAPs) for noise management and control of various noise sources (European Environment Agency report [59])

been estimated as low (€0.15 for noise maps and €0.03 for action plans per citizen, every 5 years). Figure 2.11 summarizes the various NAPs as highlighted in the recent European Environment Agency report [59]. Cost-benefit analysis revealed a favorable cost-benefit ratio of 1:29. The NAPs also depend upon the source-specific conditions as, for instance, in case of aircraft noise, utilization of path measures such as noise barriers is cumbersome, although building insulation is very significant. Therefore, the most predominant measures employed to combat aircraft noise are those at the source (70%) [59].

The measures at source primarily include the traffic management and road surface measures, while the measures at path include noise barriers (73%) and building

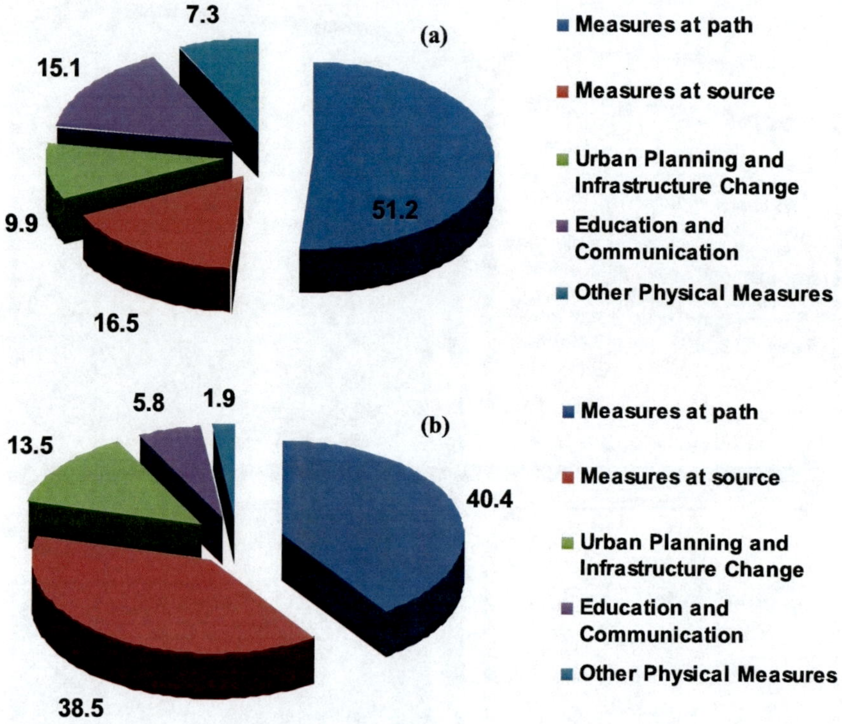

Fig. 2.12 Analysis of noise action plans (NAPs) based on the 2017 reporting round of noise mapping in European Union for: (**a**) agglomeration; (**b**) major roads (EEA reports [59, 60])

sound insulation (22%). The involvement of all stake holders in decision-making process for execution and implementation of NAPs, such as that followed in the United Kingdom, is must for accomplishing the desired goals. Enhancing the green infrastructure and quiet areas (other physical measures) has been also observed in some Member States in recent times. Figure 2.12a and b summarizes the NAPs reported under the END action planning [59, 60]. NAPs pertaining to urban planning and infrastructure change only account for a small percentage (13%).

The cost-effectiveness is one of the most important considerations in selection of NAP. The cost of reducing noise annoyance in a 20-year perspective varies from EUR 16 per person per year to EUR 4200 per year. Measures focused at noise reduction at the source are by far the most cost-effective measure for reducing noise annoyance [61]. Stricter noise limits for vehicles are in particular the most cost-effective measures to reduce noise annoyance. Noise barriers serve to be the most expensive option and are thus a local measure, but not a cost-effective measure on national level. Multiple measures shall be thus indispensable for accomplishing the desired noise reduction goals and bringing ambient levels within the legal limits [61]. Figure 2.13 thus summarizes the paradigms associated with community noise

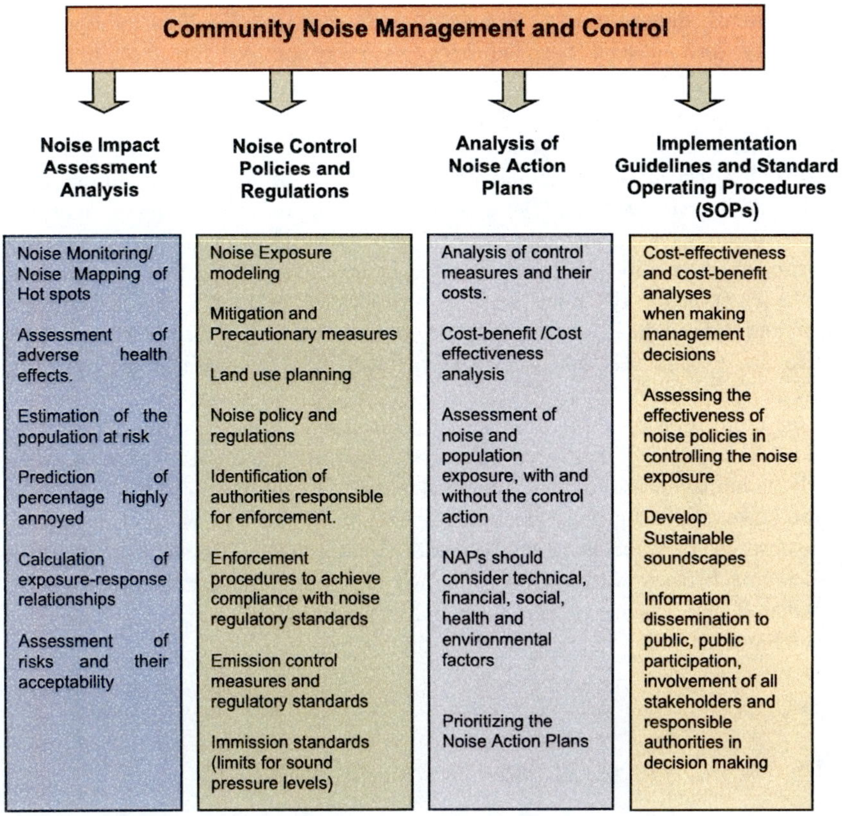

Fig. 2.13 Paradigms associated with community noise management and control

management and control. The community noise management must consider all the aspects pertaining to the noise impact assessment analysis; noise control policies; analysis of noise action plans; and implementation and execution guidelines for effective control of noise pollution in cities. Developing a long-term plan in the field of noise management and control, amendment of the existing legislations if necessary, establishment of noise map directives, installation of noise monitoring systems, and execution of noise control action plans should be given the utmost priority.

2.14 Air and Noise Pollution

Noise pollution is an underrated environmental issue as we cannot see, smell, or taste it unlike other sources: air, water, and land pollution. However, the World Health Organization considers noise pollution not only an environmental nuisance, but also a threat to public health. Traffic noise and air pollution have been associated with cardiovascular disease (CVD). In a recent study carried out by Foraster et al. [62], the correlation between NO_2 and noise ($L_{eq,24\ h}$) was observed to be 0.62. In another study conducted by Danciulescu et al., very good correlation between noise level and concentration of NO_2, SO_2, and CO in the air in two areas of Bucharest city of Europe was observed that was affected mainly by traffic [63]. There have been many such studies that investigated the combined effect of air and noise pollution. Ising et al.'s study concluded that high exposure to traffic noise, especially at nighttime, leads in the long term to an aggravation of bronchitis in children and that this is more affecting the bronchitis symptoms than the exhaust fumes [64]. A study conducted by Davies et al. for 103 roadside sites in the Metro Vancouver region of British Columbia showed the Pearson correlation coefficient for $L_{eq,5min}$ and NO_2 was 0.53, and for $L_{eq,5min}$ and NO_X as 0.64 [65]. Davies et al. [65] recommended that the traffic noise and air pollution measurements and analysis should be conducted for analyzing the correlation of traffic noise and cardiovascular disease, although it is very difficult to separately identify the effects of air pollution and noise pollution exclusively on human health. In one study of Beelen et al. [66], the correlations between traffic noise and background black smoke and traffic intensity on the nearest road were observed to be moderate at 0.24 and 0.30, respectively. The study analyzed the associations between cardiovascular mortality and air pollution and noise together. In Indian scenario also, such combined studies are essentially required for analyzing the effect of air pollution and noise pollution on human health and life.

2.15 Noise, Vibration and Harshness (NVH)

Noise, vibration, and harshness is the study of characteristics and modification of noise and vibration characteristics of vehicles, whereby harshness is a subjective quality and is measured by either via "jury" evaluations or with analytical tools. In jury testing, the sound is replayed to a panel of persons and they are asked to judge the sound quality. The interior NVH deals with noise and vibration experienced by the cabin occupants, whereas the exterior NVH is experienced with noise emitted by vehicles and includes drive-by noise testing. The common sound quality metrics in a vehicle are articulation index (AI), loudness, and sharpness. It is also called sound quality analysis that involves parameters such as loudness, sharpness, sound exposure levels, and others with a primary objective of noise reduction and the inhibition of squeaks and rattle. AI is the measure of intelligibility of voice and is expressed as a percentage of speech units understood by listener. The loudness level refers to perceived sound intensity, while sharpness is the measure that takes into consideration the frequency content of noise. There are many sound quality metrics for example, loudness, sharpness, roughness, fluctuation strength, tonality, etc. However, the choice of which sound quality metrics to apply is specific to individual appliance. Some recent studies revealed that these measures are independently inadequate to define the sound quality [67–70].

References

1. Jamir L, Nongkynrih B, Gupta SK (2014) Community noise pollution in urban India: Need for public health action. Indian J Community Med 39(1):8–12
2. Pandya GH (2002) A comprehensive investigation of noise exposure in and around an integrated iron and steel works. Am Ind Hyg Assoc 63:172–177
3. Sinha S, Sridharan PV (2003) Present and future assessment of noise level in the Neyveli region. J Environ Stud Policy 2(1):1–13
4. Banerjee D, Chakraborty SK, Bhattacharyya S, Gangopadhyay A (2008) Evaluation and analysis of road traffic noise in asansol: an industrial town of Eastern India. Int J Environ Res Public Health 5(3):165–171
5. Gan WQI, Davis HW, Koehoorn M, Brauer M (2012) Association of long-term exposure to community noise and traffic-related air pollution with coronary heart disease mortality. Am J Epidemiol 175(9):898–906
6. Berglund B, Lindvall T, Schwela DH (1999) Guidelines for community noise. World Health Organization, Geneva
7. Environmental Noise Guidelines for the European Region (2018) World Health Organization, Regional Office for Europe, Denmark
8. Sandberg U (2001) Tyre/road noise-myths and realities. In: International congress and exhibition on noise control engineering
9. Yang D, Wang Z, Li B, Luo Y, Lian X (2011) Quantitative measurement of pass-by noise radiated by vehicles running at high speeds. J Sound Vib 330(7):1352–1364
10. Saemann E-U (2012) Holistic approach for rolling sound mitigation. In: 41st International congress and exposition on noise control engineering 10, pp 8275–8286
11. Donavan PR (2012) Investigations of the influence of lower frequency aerodynamic noise on interior cruise and exterior pass-by sound levels. SAE Int J Passenger Cars-Mech Syst 5(3): 1133–1142
12. Bernhard RJ, Wayson RL, Haddock JE, Neithalath N, El Aassar A, Olek J, Pellinen TK, Weiss WJ (2005) An introduction to tire/pavement noise of asphalt pavement. https://www.vaasphalt.org/wp-content/uploads/2012/07/anintroductiontotire-pavementnoiseofasphaltpavement.pdf. Accessed on 10.02.2021
13. European Environmental Agency (2014) Noise in Europe, No. 10/2014
14. European Commission (2011) Report from the Commission to the European Parliament and the Council on the implementation of the Environmental Noise Directive in accordance with Article 11 of Directive 2002/49/EC
15. Zhao C, Wang P, Wang Li, Liu D (2014) Reducing RAILWAY NOISe with Porous sound-absorbing concrete slabs. Adv Mater Sci Eng 206549, 11 p
16. Miedema HME, Vos H (1998) Exposure-response relationships for transportation noise. J Acoust Soc Am 104:3432–3445
17. Moehler U (1998) Community response to railway noise: a review of social surveys. J Sound Vib 120:321–332
18. Knall V, Schuemer R (1983) The differing annoyance levels of rail and road traffic noise. J Sound Vib 87:321–326
19. Fields JM, Walker JG (1982) Comparing the relationships between noise, level and annoyance in different surveys: a railway noise vs. aircraft and road traffic comparison. J Sound Vib 81:51–80
20. Kaku J, Yamada I (1996) The possibility of a bonus for evaluating railway noise in Japan. J Sound Vib 193:445–450
21. Morihara T, Sato T, Yano T (2004) Comparison of dose-response relationships between railway and road traffic noises: the moderating effect of distance. J Sound Vib 277:559–565
22. Ma J (2008) Evaluation and analysis on the effect of noise mitigation measures for railway with ballastless track. Railway Occup Safety Health Environ Prot 35(3):111–115. (Chinese)

23. Diehl RJ, Nowack R, Hölzl G (2000) Solutions for acoustical problems with ballastless track. J Sound Vib 231(3):899–906
24. Basner M, Clark C, Hansell A, Hileman JI, Janssen S, Shepherd K, Sparrow V (2017) Aviation noise impacts: state of the science. Noise Health 19(87):41–50
25. Astley RJ (2014) Can technology deliver acceptable levels of aircraft noise? Internoise 2014, Melbourne, Australia, 16th to 19th November, 2014
26. Darpe AK, Singh SP, Tandon N, Matsagar V (2010) Barrier design for the New Runway at Indira Gandhi International Airport, New Delhi, Delhi Indira Gandhi International Airport (P) Limited, August, 2010, Indian Institute of Technology, Delhi
27. Hooper P, Maughan J, Flindell I, Hume K (2009) Indices to enhance understanding & management of community responses to aircraft noise exposure. OMEGA community noise study
28. International Civil Aviation Organization (ICAO) (2008) Guidance on the balanced approach to aircraft noise management, Doc 9829 AN/451
29. Mahendra PKV, Venugopalachar S (2011) The possible influence of noise frequency components on the health of exposed industrial workers – a review. Noise Health 13(50):16–25
30. Department for Environment, Food and Rural Affairs (DEFRA) (2001) Low frequency noise report. Northern Ireland
31. Emmerich E, Richter F, Reinhold U, Linns V, Linss W (2000) Effects of industrial noise exposure on distortion product otoacoustic emissions (DPOAEs) and hair cell loss of the cochlea – long term experiments in awake guinea pigs. Hear Res 148:9–17
32. Nelson DI, Nelson RY, Barrientos MCPH, Fingerhut M (2005) The global burden of occupational noise-induced hearing loss. Am J Ind Med 48(6):446–458
33. Tekriwal R, Parmar DM, Saxena R (2011) Noise induced hearing loss-a comparison between speech frequency and 4000 Hz frequency. Natl J Physiol Pharm Pharmacol 1(2):79–85
34. Technology for a Quieter America, Washington, DC (2010) The National Academies Press
35. Tandon N, Nakra BC (2017) Vibration and noise engineering. I.K International, New Delhi
36. Sound Level Descriptors, Federal Highway Administration, U.S Department of Transportation, FHWA-HEP-17-053. https://www.fhwa.dot.gov/Environment/noise/resources/fhwahep17053.pdf. Accessed on 10.02.2021
37. Barron RF (2001) Industrial noise control and acoustics. Marcel Dekker
38. Bullock CE (1970) Aerodynamic sound generation by duct elements. Am Soc Heating Refrigerating Air-Conditioning Engineers Trans Part II 76:97–108
39. Graham JB (1972) How to estimate fan noise. Sound Vib 6:24–27
40. Altinsoy ME, Atamer S (2018) Sound label for household appliances. In: Proceedings of the ASME 2018 noise control and acoustics division session presented at Internoise 2018, NCAD 2018, August 26–29, 2018, Chicago, IL, USA
41. Bae SG, Lee WH, Bae MJ (2016) a study on sound characteristics of home appliances noise using psychoacoustics parameters. Int J Eng Technol 8:1606–1610
42. Starck J, Toppila E, Pyykko I (2003) Impulse noise and risk criteria 5:63–73
43. Śliwińska-Kowalska M, Zaborowski K (2017) WHO environmental noise guidelines for the european region: a systematic review on environmental noise and permanent hearing loss and Tinnitus. Int J Environ Res Public Health 14:1139
44. European Commission Directive 2003/10/EC of the European Parliament and of the Council of 6 February, 2003 on the minimum health and safety requirements regarding the exposure of workers to the risks arising from physical agents (noise). European Commission, Brussels, Belgium
45. Occupational Safety and Health Administration, United States Department of Labor. https://www.osha.gov/laws-regs/regulations/standardnumber/1910/1910.95AppA
46. Ginn KB, Haddad K (2012) Noise source identification techniques: simple to advanced applications. In: Proceedings of the acoustics 2012 Nantes conference, 23–27 April 2012, Nantes, France, 1781–86

47. Batel M, Marroquin M (2003) Noise source location techniques-simple to advanced application. Sound Vib 2003:24–36
48. Coelho JLB, Alarcão D (2005) On noise mapping and noise action plans for large urban areas. Forum Acust:1039–1044
49. Kimura M, Kunio J, Schuhmacher A, Ryu Y (2014) A new high-frequency impedance tube for measuring sound absorption coefficient and sound transmission loss, Internoise 2014, Melbourne, Australia, 16th to 19th November, 2014
50. McGrory M, Cirac DC, Gaussen O, Cabrera D (2012) Re-examining the relationship between impedance tube and reverberant room methods. In: Proceedings of acoustics 2012- Fremantle, 21–23 November 2012, Fremantle, Australia
51. Carvalho APO, Sousa MRM (2016) Effect of sample area in reverberant chamber measurements of sound absorption coefficients. In: International Congress on acoustics, ICA 2016 Buenos Aires, 5th to 9th September, 2016
52. Papanikolaou G, Trochides A (1985) Design of a test facility for transmission loss measurement. Appl Acoust 18:315–323
53. Halliwell RE, Nightingale TRT, Warnock ACC, Birta JA (1998) Gypsum board walls: transmission loss data. Internal report IRC-IR-761, March 1998
54. Norton M, Karczub D (2003) Fundamentals of noise and vibration analysis for engineers. Cambridge University Press, Cambridge
55. Lamancusa JS (2001) ME 458 engineering noise control. Transmission of sound through structures. July, 2001
56. Singal SP, Sharma O (1987) Acoustics of buildings, part II: materials and their characterisation. J Res Ind 32:14–24
57. Qiu X (2016) Principles of Sound Absorbers. In: Padhye R, Nayak R (eds) Acoustic textiles, textile science and clothing technology. Springer, Singapore, pp 43–72
58. Taylor EW (1987) Theoretical and practical aspects of the "Functional absorber" method of arranging sound absorbing panels, Research Department report, Engineering Division, The British Broadcasting Corporation. BBC RD 1987/15
59. European Environment Agency Report (2020) Environmental noise in Europe-2020. No. 22/2019. Publications Office of the European Union, Luxembourg
60. ETC/ATNI (2019) Noise Action Plans. Managing exposure to noise in Europe, No ETC/ATNI Report No 8/2019. https://www.eionet.europa.eu/etcs/etc-atni/products/etc-atni-reports/etc-atni-report-8-2019-noise-actionplans-managing-exposure-to-noise-in-europe. Accessed 03.04.2021
61. Conference of European Directors of Roads (CEDR) (2013) Value for money in road traffic noise abatement. CEDR project group road noise. http://www.carreteros.org/explotacion/cedr/3_CEDR.pdf. Accessed 03.04.2021
62. Foraster M, Deltell A, Basagaña X, Medina-Ramón M, Aguilera I, Bouso L, Grau M, Phuleria HC, Rivera M, Slama R, Sunyer J, Targa J, Künzli N (2011) Local determinants of road traffic noise levels versus determinants of air pollution levels in a Mediterranean city. Environ Res 111:177–183
63. Danciulescu V, Bucur E, Pascu LF, Vasile A, Bratu M (2015) Correlations between noise level and pollutants concentration in order to assess the level of air pollution induced by heavy traffic. J Environ Prot Ecol 16:815–823
64. Ising H, Lange-Asschenfeldt H, Moriske HJ, Born J, Eilts M (2004) Low frequency noise and stress: bronchitis and cortisol in children exposed chronically to traffic noise and exhaust fumes. Noise Health 6(23):21–28
65. Davies HW, Vlaanderen JJ, Henderson SB, Brauer M (2009) Correlation between co-exposures to noise and air pollution from traffic sources. Occup Environ Med 66:347–350
66. Beelen R, Hoek G, Houthuijs D, van den Brandt PA, Goldbohm RA, Fischer P, Schouten LJ, Armstrong B, Brunekreef B (2009) The joint association of air pollution and noise from road traffic with cardiovascular mortality in a cohort study. Occup Environ Med 66(4):243–250

67. Salford Innovation Research Centre, Sound quality metrics. http://hub.salford.ac.uk/sirc-acoustics/psychoacoustics/sound-quality-making-products-sound-better/an-introduction-to-sound-quality-testing/sound-quality-metrics/. Accessed on 10.02.2021
68. Fastl H (2001) Neutralizing the meaning of sound for sound quality evaluations. Internoise
69. Abdallah AA, Yang K (2008) Predicting noise, vibration and harshness performance degradation through fatigue analysis. Int J Vehicle Noise Vib 4:269–284
70. Panda K C (2016) Dealing with noise and vibration in automotive industry. Procedia engineering. In: 12th International conference on vibration problems, ICOVP 2015, 144, 1167–1174

Chapter 3
Sound Level Measurements: Standards and Traceability

Abstract This chapter presents an overview and significance of acoustics metrology (sound level, intensity, and power measurements) and the primary and secondary sound standards realized for the establishment of traceability chain in sound level measurements. The methodology of realization of primary and secondary sound standards in national and international perspectives is highlighted. Calibration, as part of an overall measurement management process, reduces the risks associated with measurements, regulatory requirements, and international acceptability. This chapter thus discusses the strategy followed in Indian scenario for the establishment of strong traceability chain in sound standards at CSIR-National Physical Laboratory, the National Metrology Institute (NMI) of India. The dissemination of traceability for sound pressure, sound intensity, and sound power level parameters to the regional laboratories and to the user laboratories, pollution control boards, and industries in India is discussed in detail. Also with the growing demand of installation of Noise Monitoring Terminals for round the clock noise monitoring in the metropolitan cities, the traceability to the national standards of sound pressure is an imperative aspect for reporting the accurate and precise ambient day and night equivalent sound levels.

3.1 Introduction

The calibration of the noise measuring instruments including the sound level analyzers, microphones, sound calibrators, etc. are must for reliable, accurate, and precise measurements especially when the legal considerations (i.e., compliance with legal limits) are involved. Thus, the traceability to the national and international measurement standards is required, which is the responsibility of National Metrology Institute (NMI) of the country. A comprehensive National Measurement System (NMS) includes several facets: units of measurement and national measurement standards, laboratory accreditation, legal metrology, documentary standards, etc. The NMIs serve as national centers of excellence in measurement and as fundamental element of the nationwide technical infrastructure and are able to give a wide range of technical advice, assistance, and support to the government, industry, and the wider community on accurate and precise measurements [1]. CSIR-National

N. Garg, *Environmental Noise Control*,
https://doi.org/10.1007/978-3-030-87828-3_3

Physical Laboratory (CSIR-NPL), India is designated as the National Metrology Institute of the country and bestowed the responsibility of realization, establishment, maintenance, and dissemination of standards at par to the international level through continuous research and development and use of the latest technology. CSIR-National Physical Laboratory (NPL), India being the National Measurement Institute (NMI) of India has the statutory obligation to establish, maintain, and update the national standards of measurement and calibration facilities for different parameters. The primary standard of sound pressure is the utmost accurate and precise standard. The secondary standards are traceable to the primary standards and are used for providing the calibration services to the industry and regional laboratories. The Acoustics and Vibration standard of CSIR-National Physical Laboratory (NPL) of India is responsible for the establishment, validation, maintenance, reproduction, and dissemination of the unit of sound pressure, Pascal (N/m^2). The accuracy and precision of primary standards is periodically verified through participation in key comparsion exercises with leading NMIs of the world. The secondary standards are realized for providing calibration services for the reference and working standard microphones, sound calibrators, pistonphones, multifunction acoustic calibrators, and sound level meters/analyzers. CSIR-NPL, India Calibration and Measurement Capabilities (CMCs) have been appended in Key Comparison Database (KCDB) website since October 2009, which is the highest recognition to any National Metrology Institute of the country for that parameter.

3.2 Primary Sound Standard

The primary standard of sound pressure is maintained through absolute calibration of laboratory standard microphones (LS1P and LS2P) in a coupler cavity by reciprocity technique with an expanded uncertainty of ±0.05 dB to ±0.17 dB in range 31.5 Hz to 25 kHz [2]. The reciprocity calibration appartus of CSIR-NPL, India is Make: Brüel & Kjær (B&K) type 9699 system. The method involves three microphones that are acoustically coupled by air enclosed in cavity with one acting as a source and other as a receiver. The electrical transfer impedance, that is, the ratio of open circuit voltage of receiver microphone to the input current of transmitter microphone for each of three different microphone pairs, is measured. The transmitter current is determined by measuring the voltage across a traceable capacitance connected in series with the transmitter microphone. The sensitivity of each microphone can be evaluated as per IEC 61094-2 [3] solving three equations with three sensitivities as unknown parameters [4]:

$$M_{p,1}^2 = \frac{R_{12}R_{13}}{R_{23}} \frac{V_{o,12}V_{o,13}}{V_{o,23}} \frac{P_{s,23}}{P_{s,12}P_{s,13}} \frac{\kappa_{23}}{\kappa_{12}\kappa_{13}} \frac{Cor_{HW,12}Cor_{HW,13}}{Cor_{HW,23}} C^{-1} \tag{3.1}$$

$$M_{p,2}^2 = \frac{R_{12}R_{23}}{R_{13}} \frac{V_{o,12}V_{o,23}}{V_{o,13}} \frac{P_{s,13}}{P_{s,12}P_{s,23}} \frac{\kappa_{13}}{\kappa_{12}\kappa_{23}} \frac{Cor_{HW,12}Cor_{HW,23}}{Cor_{HW,13}} C^{-1} \tag{3.2}$$

$$M_{p,3}^2 = \frac{R_{13}R_{23}}{R_{12}} \frac{V_{o,13}V_{o,23}}{V_{o,12}} \frac{P_{s,12}}{P_{s,13}P_{s,23}} \frac{\kappa_{12}}{\kappa_{13}\kappa_{23}} \frac{Cor_{HW,13}Cor_{HW,23}}{Cor_{HW,12}} C^{-1} \tag{3.3}$$

where R_{12}, R_{13}, and $R_{23}$are voltage ratio for pair of microphones,$V_{o,\ 12}$is cavity volume plus low-frequency equivalent diaphragm volume of microphone 1 and 2, $P_{s,\ 12}$ is static pressure during measurement with microphones 1 and 2, $\kappa_{12}$is ratio of specific heat capacities of cavity gas with microphones 1 and 2, $Cor_{HW,\ 12}$ is correction factor for heat conduction and wave motion of coupler with microphone 1 and 2, and C is capacitance of series capacitor. A condenser microphone is a reciprocal transducer that can work as a microphone, by converting an acoustic signal to an electrical signal and also as a sound source. The designation LS2P refers to a laboratory standard microphone of mechanical configuration 2 having nearly constant presssure sensitivity as a function of frequency, for example, a Brüel and Kjær (B&K) type 4180 microphone. IEC/TC29 has worked out a standard (IEC61094-4) for the commonly used types of measurement microphone called working standard microphones. The working standard microphones are designated as WS1, WS2, and WS3 depending on their diameters, 23.77 mm, 12.7 mm, and 6.5 mm, respectively.

3.3 Secondary Sound Standard

3.3.1 Pressure-Field Calibration

The secondary standard is utilized for the calibration of working standard microphones, sound calibrators, pistonphones, multifunction acoustic calibrators, and sound level meters. The secondary microphone calibration system of CSIR-National Physical Laboratory (NPL), India is B&K 9721 system that performs the sensitivity calibration by comparison with the reference microphone as per IEC 61094-5 [6]. The two microphones are exposed to same sound pressure level in an active comparison coupler whereby the cavity in the coupler is surrounded by a piezoelectric foil that generates the sound when it is connected to an ac electric signal. Microphone calibration module type 5001 supplies the comparison coupler an adequate signal level for sensitivity calibration [7]. The calibrated data is corrected for ambient pressure, temperature, and air humidity. The comparison measurement is made with the pulse analyzer system as a two-channel measurement with both microphones exposed to the same sound pressure inside the comparison coupler. The system eliminates the influence of preamplifier gain by using the insert voltage facility of the enclosed preamplifiers. The frequency response is determined using an electrostatic actuator (IEC 61094-6 [8]). The actuator measurement is relative which can simulate a constant sound pressure on diaphragm over a wide frequency range. Depending upon the type and application of microphone, free-field or diffuse-field corrections may be added to achieve the required type of frequency response. The reference standard microphone B&K type 4180 is periodically calibrated on primary system. The calibration of sound calibrators is performed on the B&K system with the help of sound calibrator calibration application software type 7794 as per IEC 60942. The system uses a B&K Pistonphone 4228 as a sound source and comparison

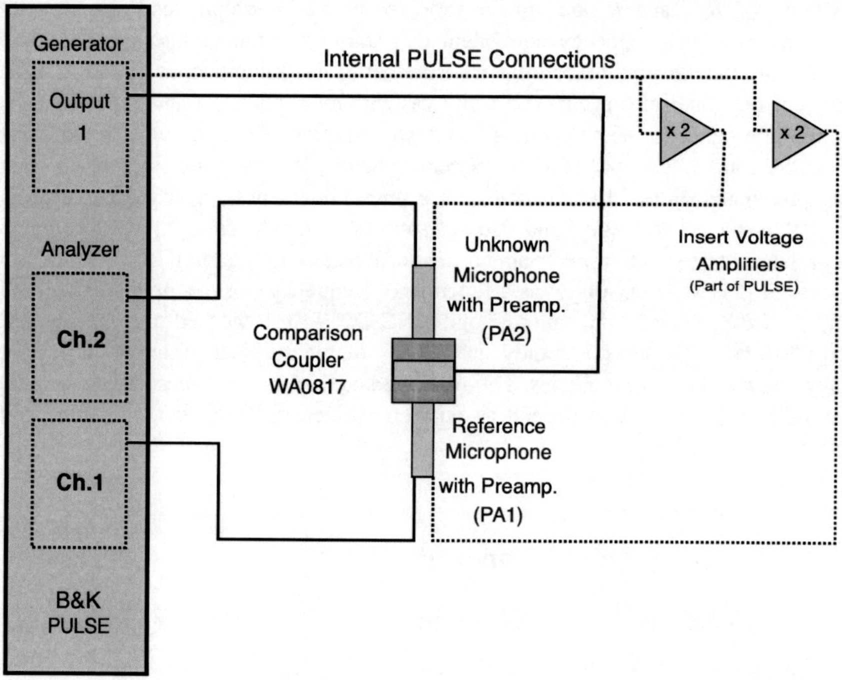

Fig. 3.1 Block diagram for Type 9721 Sensitivity Comparison Measurement. (B&K Manual [7])

method is used for the calibration of test calibrator. The measurement uncertainty ($k = 2$) observed in calibration of measurement microphones ranges from ±0.08 dB to ±0.35 dB in frequency range 31.5 Hz to 20 kHz. The expanded uncertainty in measurement in calibration of sound calibrators and multifunction calibrator ranges from ±0.12 dB to ±0.23 dB in frequency range 31.5 Hz to 16 kHz. The sensitivity of the DUT (device under test) at actual ambient conditions can be calculated as [7]:

$$S_{\text{DUT}} = S_{\text{REF}} + (\text{DUT}_{\text{Acoustic}} - \text{REF}_{\text{Acoustic}}) - (\text{DUT}_{\text{Insert}} - \text{REF}_{\text{Insert}}) + \text{Insert Voltage Gain Deviation} \tag{3.4}$$

where S_{DUT} is the sensitivity of DUT at actual ambient conditions, S_{REF} is corrected sensitivity of reference microphone, $\text{REF}_{\text{Acoustic}}$ is voltage measured on the output of preamplifier in reference channel during acoustic measurement in calibration coupler, $\text{DUT}_{\text{Acoustic}}$ is voltage measured on output of preamplifier in DUT channel, $\text{REF}_{\text{Insert}}$ is voltage measured on output of preamplifier in reference channel during insert voltage calibration measurement, and $\text{DUT}_{\text{Insert}}$ is voltage measured on output of preamplifier in reference channel during insert voltage calibration measurement. Figure 3.1 shows the pictorial view of setup for B&K Type 9721 Sensitivity Comparison Measurement at 250 Hz.

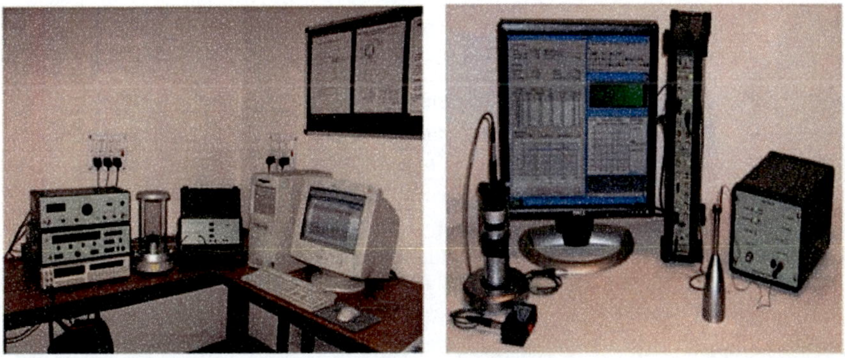

Fig. 3.2 Pictorial view of national Primary Sound Standard (LHS) and Secondary Sound Standard (RHS) realized at CSIR-National Physical Laboratory, New Delhi, India (The National Metrology Institute (NMI) of India)

The sensitivity of a measurement microphone may either be expressed in Volt per Pascal (V/Pa) or in decibel referring to one Volt per Pascal (dB re 1 V/Pa). The term sensitivity generally means the sensitivity at the reference frequency (most often 250 Hz). The frequency response function is the ratio expressed in dB between the sensitivity at a given frequency and that of the reference frequency. Figure 3.2 shows the pictorial view of national Primary Sound Standard and Secondary Sound Standard realized at CSIR-National Physical Laboratory, the NMI of India that disseminates the traceability for sound parameter in the country. It should be noted that pressure-field sensitivity refers to the pressure at the diaphragm only. The free-field sensitivity and diffuse-field sensitivity are defined for pressure applied at both the diaphragm and pressure equalization vent. The higher the sensitivity, the lower the sound pressure levels that may be measured, and conversely, the lower the sensitivity, the higher the sound pressure levels that may be measured. A pressure field is characterized by a sound pressure which has the same magnitude and phase at any position within the field. Pressure fields may be found in enclosures or cavities which are small compared to the wavelength, and such fields occur in couplers applied for testing of earphones or calibration of microphones. A free sound field may be created where sound waves can propagate freely, that is, in a continuous medium without any disturbing objects. A diffuse sound field exists at a given location if the field is created by sound waves arriving more or less simultaneously from all directions with equal probability and level. Free-field microphones have flat free-field frequency response characteristics to sound waves that are perpendicular to the diaphragm, that is, at zero-degree incidence [9].

3.3.2 *Sound Calibrator and Pistonphone*

Sound calibrator uses a low-impedance electrodynamic source to yield a high degree of volume independent operation and tend to be less precise than pistonphone, but are independent of internal cavity volume and ambient pressure corrections. A vibrating piston in a pressure cavity (coupler) produces a sound field of known sound pressure. Sound calibrators (Figs. 3.3) generally produce a sound pressure level of 94 dB and 114 dB at two frequencies: 250 Hz and 1 kHz.

A pistonphone generally produces a sound pressure level of 124 ± 0.2 dB re 20 μPa at the reference conditions. Each pistonphone is individually calibrated and comes complete with adaptors, allowing calibration of 1″, 1/2″, 1/4″, and 1/8″ microphones. Multifunction acoustic calibrators are widely used for verification of microphones, sound level meters, and other instruments. Figure 3.4 shows the pictorial view of B&K 4226 Multifunction Acoustic Calibrator and GenRad Omnical sound calibrator, the technical details of which are readily available from the datasheet provided by manufacturer.

Figs. 3.3 Pictorial view of a Sound Calibrator, B&K 4231 (LHS), Pistonphone, B&K 4228 (in middle), and Norsonic, Nor 1256 Sound Calibrator (RHS)

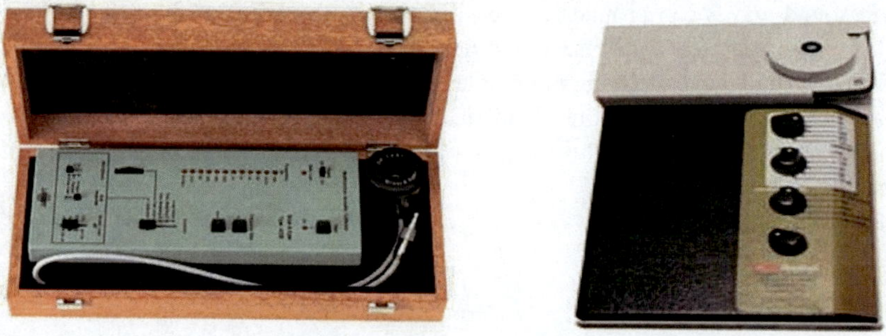

Fig. 3.4 Pictorial view of Multifunction Acoustic Calibrators, (LHS: B&K 4226 and RHS: GenRad Omnical Sound Level Calibrator (125 Hz to 4 kHz, 74 to 104 dB)

3.3.3 Microphone Free-Field Comparison Calibration

The reciprocity technique is the widely used methodology followed by all NMIs that provides the basis for calibration of laboratory standard microphones including a free field. However, the field calibrations are done by using the measurement microphones that may not be suitable for reciprocity calibrations. Besides the free field, reciprocity calibration is time consuming and therefore expensive [10, 11]. The solution to this problem is thus a comparison calibration technique based on direct comparison of a microphone under test with a reference microphone. By definition, the free-field sensitivity implies that a plane wave is impinging on the microphone [12]. The free-field correction is commonly used for evaluating the free-field sensitivity that can be used in field measurements without calibrating the microphone in a free field [10]. However, accurate and precise determination of the free-field correction can only be obtained by applying absolute calibration techniques (reciprocity method) or by using comparison calibration method. A major requirement for this type of calibration is an anechoic chamber.

3.3.3.1 Microphone Comparison Calibration as per IEC 61094-8 Standard

The methodology of comparison calibration is based on the fact that when a calibrated reference microphone and a microphone under calibration are exposed to the same free-field sound pressure sequentially and under the same environmental conditions, then the ratio of their free-field sensitivities for those conditions is given by the ratio of their open-circuit voltages. The free-field sensitivity level (L_{DUC}) of the device under calibration (DUC) re 1 V/Pa can be obtained as [11]:

$$L_{DUC} = L_{REF} + \Delta_M \tag{3.5}$$

where L_{REF} is free-field sensitivity level of the reference microphone and

$$\Delta_M = 20 \log_{10} \frac{|U_{DUC}|}{|U_{REF}|} \tag{3.6}$$

where U_{DUC} and U_{REF} are the open circuit output voltages of the DUC and reference microphone. Fig. 3.5a and b shows the picture of the small anechoic chamber developed by M/s SPEKTRA Schwingungstechnik und Akustik GmbH Dresden, Germany established at M/s Fluid Control Research Institute (FCRI), Palakkad, Kerala, India, a regional NABL (National Accreditation Board for Testing and Calibration Laboratories, India) accredited laboratory in India.

The walls of the small anechoic chamber (1.25 m × 1.25 m × 1.65 m) are made of lightweight sandwich panel (aluminum sheet with polyurethane foam filling). The entire inner surface of the chamber is completely lined with wedge-shaped absorbers

Fig. 3.5 (**a**) and (**b**). Photograph of free-field microphone calibration setup and small anechoic chamber at FCRI, Palakkad, Kerala, India. (Garg et al. 2019 [13]).

of length 35 cm cemented to the walls and ceiling as put on to the bottom surface. The lower cut-off frequency of this chamber derived in theory from its dimensions and the characteristics of the materials used for lining the chamber is 250 Hz. Substitution calibration describes the process where reference microphone is first used to determine the sound pressure at specific point inside a free field and is then replaced by microphone under calibration [13].

3.3.3.2 Microphone Calibration Using Pressure Response

The insert voltage technique can be used to measure the microphone sensitivity and the electrostatic actuator method to get the frequency response. The electrostatic actuator method provides a response according to IEC 61094-6 very similar to the pressure field response of the microphone being tested. The electrostatic actuator response is similar to the pressure field response. In case of free-field and diffuse-field microphones, manufacturers usually specify the correction factors for each specific precision condenser microphone model. The frequency response curve for precision condenser microphones is usually plotted in reference to the sensitivity of the microphone at the reference frequency. Free-field and diffuse-field frequency responses may be obtained by adding corrections to measured responses. Figure 3.6 shows the pressure, free-field (at zero-degree incidence), and diffuse-field response of the microphone [5]. It may be noted here that different types of microphones are available for use in the different sound fields.

In order to measure their frequency responses, a sound source that supplies a constant sound pressure over a wide frequency can serve the purpose best. However, such sources do not exist and thus the electrostatic actuator is a very good substitute. It can generate a constant electrostatic pressure on the flat diaphragm of microphone and thus can be used with all working standard microphones.

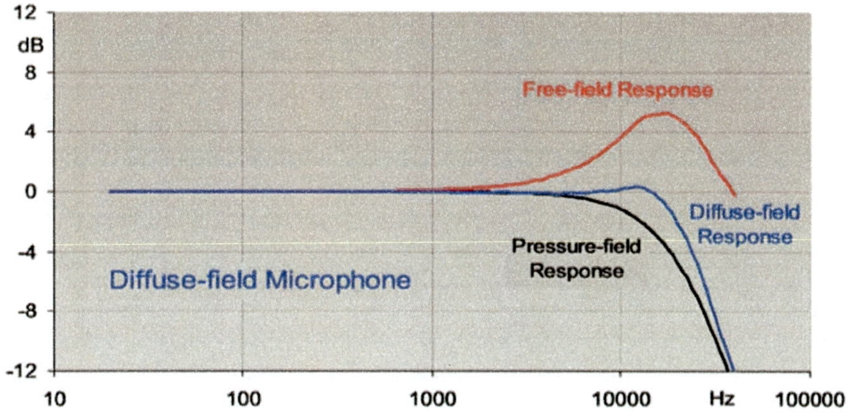

Fig. 3.6 Pressure-field, free-field, and diffuse-field response of microphones.. (Frederiksen 2013, [5])

3.4 Sound Level Meter Calibration

The sound level meter (SLM) testing involves: absolute acoustical sensitivity level, frequency response (measured by using actuator, multifrequency calibrator or sound field), directional response, self-generated noise, time-weighting response, R.M.S. detector response, and overload. The sound level meter calibration is carried out in an anechoic chamber in frequency range 31.5 Hz to 10 kHz. The acoustical setup consists of a signal generator feeding into a power amplifier and then into a loud speaker for generating the required acoustic field. The absolute acoustical sensitivity level indicating the response of the sound level meter to a sinusoidal sound pressure signal at the calibration frequency is measured. The response of the sound level meter to sinusoidal sound pressure signal at the calibration level at 1 kHz and at other frequencies is registered, and thus, the free-field frequency response for the sound level meter is calculated. The expected response in A, C, and linear weightings is measured. The reading of the sound level meter when exposed to single tone-bursts is registered for various burst durations: 200 ms, F; 500 ms, S; 20 ms, I; 5 ms, I; 2 ms, I (F: Fast, I: Impulse). The response of the sound level meter to sequences of repeated tone-bursts when measuring time-weighted sound level is registered for various tone-burst durations. The self-generated noise of sound level meters with microphone and microphone replaced by electrical input signal device is also measured. The details of testing parameters for sound level meter are shown in Appendix II of the book.

3.5 Traceability in Sound Parameter

ISO International Vocabulary of Metrology-Basic and General Concepts and Associated Terms (VIM) defines metrological traceability as: "the property of a measurement result whereby the result can be related to a reference through a documented unbroken chain of calibrations, each contributing to the measurement uncertainty". Calibration to a traceable standard can be used to determine an instrument's bias, precision, and accuracy. Measurement traceability "pyramid" indicates how the measurements ensuring the product quality (on the workshop floor/ industry) are linked to the national standards realized by National Metrology Institutes. Figure 3.7a shows the traceability pyramid at CSIR-National Physical Laboratory of India, showing that the secondary standards are traceable to the primary standard and the working standards in turn traceable to secondary and the measuring instruments traceable to the working standards. Figure 3.7b shows the SI units wall at CSIR-National Physical Laboratory, New Delhi, India showing the realization of seven fundamental units and the SI base units and the corresponding fundamental constants used.

The measurement uncertainty increases as we go down the traceability pyramid. The chain can be extended to any number of links, although each additional link increases the uncertainty of measurement, whereby fewer the links, the less the uncertainty of measurement. Thus, a key concept of measurement technology is that of traceability, which implies that all measurements made with a calibrated instrument or device must have a directly traceable link back to the defined quantity. It ensures that different measurement methods and instruments used produces reliable, repeatable, reproducible, and comparable measurement results. In case of acoustics metrology, the secondary sound standard draws its traceability to the national

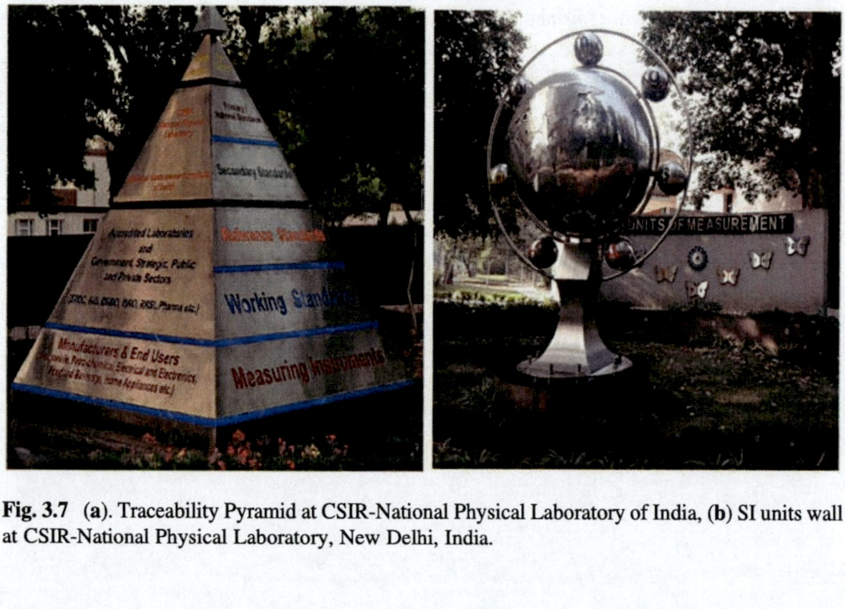

Fig. 3.7 (**a**). Traceability Pyramid at CSIR-National Physical Laboratory of India, (**b**) SI units wall at CSIR-National Physical Laboratory, New Delhi, India.

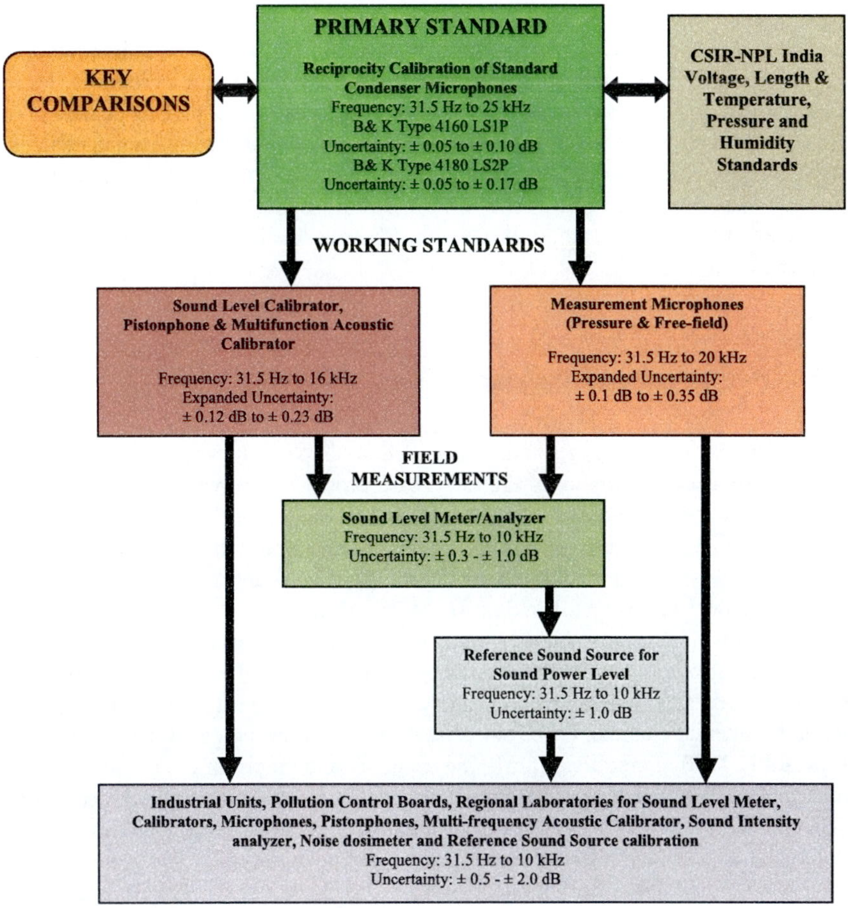

Fig. 3.8 Traceability scenario for sound measurements realized at CSIR-National Physical Laboratory, India (The National Metrology Institute of the Country)

primary standard using reciprocity calibrated reference standard B&K 4180 (LS2P) microphone. Figure 3.8 shows the traceability chain realized for the national standards of sound at CSIR-National Physical Laboratory, India. All the uncertainties stated in the traceability chain are the expanded uncertainties calculated at coverage factor, $k = 2$ and 95% confidence interval.

The primary standard is traceable to the national voltage, temperature, pressure, and humidity standards realized at CSIR-National Physical Laboratory of India. The measurement microphones and sound level calibrators draw their traceability from reference standard microphone calibrated by reciprocity method. The measurement microphones and the calibrator are used to calibrate the sound level meter and reference sound source. Thus, the measurement microhones, sound level meter, and calibrators are used to provide traceability to the industry, pollution control boards, and regional laboratories in India [14]. CSIR-National Physical Laboratory, India participates regularly in international key comparisons with other NMIs to validate the accuracy and precision associated with the national measurement standards of sound pressure. In past, CSIR-NPL succesfully participated in international key comparisons on reference standard microphones and sound calibrators. More recently, CSIR-NPL participated in an inter-comparison of sound level meters in April, 2021 with NMI, Japan as the pilot laboratory and four other participant NMIs. Also, efforts are in progress to extend the measurement frequency range of national primary sound standard from 1 Hz to 25 kHz to provide traceability in the entire measurement frequency range of 1 Hz to 25 kHz to all the stakeholders in the country.

3.6 Measurement Uncertainty

The sensitivity of reference standard microphone is pivotal factor in calibration of working standard microphone. Precise measurement of equivalent volume of the cavity including the equivalent volume of the microphone is pivotal factor in reciprocity method. Table 3.1 shows the typical uncertainty budget for calibration of a reference standard microphone using the primary sound standard: Reciprocity method ($\pm$ 0.05 to $\pm$0.17 dB in 31.5 Hz to 25 kHz).

Variation of environmental conditions from reference value (23 °C, 101.325 kPa, 50% R.H) also influences the sensitivity determination. The maximal deviation between calculated sensitivity level and corresponding sensitivity level measured in environmental chamber is 0.0144 dB over the pressure range 94 kPa to 106 kPa from B&K 4180 microphones [15]. The accuracy and precision in determination of microphone parameters is crucial for the determination of the microphone sensitivity (dB re 1 V/Pa). Figure 3.9 shows the comparison of measurement uncertainty for one-inch and half-inch microphones. The overall uncertainty ($k = 2$) is calculated to be $\pm$0.08 dB to $\pm$0.40 dB in frequency range 31.5 Hz to 20 kHz. Figure 3.10 shows the measurement uncertainty at 250 Hz for various models of B&K microphones used widely in acoustic measurements.

The measurement uncertainty at 250 Hz is varied for various models due to the static pressure and temperature coefficients, gain ratio in comparison measurements, polarization voltage of PULSE analyzer, and across the preamplifier terminals [7]. The measurement uncertainty is influenced by uncertainty in calibration of the reference standard microphone. Figure 3.11 shows the uncertainty calculated for calibration of calibrators, working standard microphones, and LS1P (B&K 4160) and LS2P (B&K 4180) microphones at CSIR-NPL, India [2].

In multichannel microphone systems, wherein the microphones are difficult to access, the verification methods like the insert voltage calibration method or charge injection calibration method provide a quick and easy method of calibration. Insert voltage is a substitution method, wherein the inserted voltage substitutes the open circuit voltage produced by the microphone. The Charge Injection Calibration (CIC) method enables the verification of complete measurement chain. It uses the frequency-independent injection of charge into microphone and preamplifier input circuit and can be used to monitor the preamplifier input resistance at low frequencies and microphone capacitance at mid and high frequencies [9].

Table 3.1 Uncertainty budget of LS2P microphone calibration (Garg and Sharma 2012, [2])

Parameters	Stand Unc.	Prob Dist	31.5 Hz to 63 Hz	≥ 250 Hz to < 2 kHz	≥2 kHz to 4 KHz and ≥ 63 Hz to < 250 Hz	> 4 KHz to 8 kHz	> 8 kHz to 12.5 kHz	16 kHz	20 kHz	25 kHz
Electrical transfer impedance										
Series impedance	0.020	Rect	0.011	0.010	0.011	0.011	0.011	0.011	0.011	0.015
Voltage ratio	0.030	Normal	0.015	0.015	0.015	0.015	0.015	0.02	0.021	0.027
Cross-talk	0.0050	Rect	0.0035	0.0035	0.0035	0.0035	0.0035	0.0035	0.0035	0.0035
Distortion	0.0040	Rect	0.0005	0.0005	0.0005	0.0005	0.0005	0.0005	0.0005	0.0005
Receiver and transmitter ground shield	0.0070	Rect	0.0005	0.0005	0.0005	0.0005	0.0005	0.0005	0.0005	0.0005
Coupler properties										
Coupler volume	0.020	Rect	0.011	0.011	0.011	0.011	0.011	0.011	0.011	0.014
Leakage	0.003	Rect	0.0017	0.0017	0.0017	0.0017	0.0017	0.0017	0.0017	0.0017
Microphone parameters										
Equivalent and front cavity volume	0.020	Normal	0.02	0.02	0.02	0.02	0.03	0.05	0.05	0.05
Microphone cavity depth	0.0007	Rect	0.0004	0.0004	0.0004	0.0004	0.0004	0.0007	0.010	0.0231
Resonance frequency	0.0002	Rect	0.0003	0.0001	0.0001	0.0001	0.0002	0.0002	0.0003	0.0003
Loss factor	0.0010	Rect	0.0008	0.0002	0.0008	0.0008	0.0008	0.011	0.011	0.011
General factors										
Heat conduction	0.011	Normal	0.011	0.006	0.006	0.002	0.002	0.001	0.001	0.001
Radial wave motion	0.010	Normal	0.0000	0.0000	0.0000	0.0020	0.0050	0.011	0.020	0.023
Static pressure	0.0002	Normal	0.0002	0.0002	0.0002	0.002	0.0002	0.0006	0.001	0.008
Temperature	0.0015	Normal	0.0015	0.0015	0.0015	0.0015	0.0015	0.002	0.008	0.008
Relative humidity	0.0002	Normal	0.0002	0.0002	0.0002	0.002	0.0002	0.0006	0.001	0.008

(continued)

Table 3.1 (continued)

Parameters	Stand Unc.	Prob Dist	31.5 Hz to 63 Hz	≥ 250 Hz to < 2 kHz	≥2 kHz to 4 KHz and ≥ 63 Hz to < 250 Hz	> 4 KHz to 8 kHz	> 8 kHz to 12.5 kHz	16 kHz	20 kHz	25 kHz
Rounding error	0.0050	Rect	0.0029	0.0029	0.0029	0.0029	0.0029	0.0029	0.0029	0.0029
Air properties	0.0001	Rect	0.0001	0.0001	0.0002	0.0002	0.0002	0.0003	0.0003	0.0006
Repeatability		Normal	0.0391	0.008	0.023	0.030	0.036	0.042	0.044	0.063
Expanded uncertainty (k = 2) dB			**0.12**	**0.05**	**0.08**	**0.10**	**0.11**	**0.12**	**0.13**	**0.17**

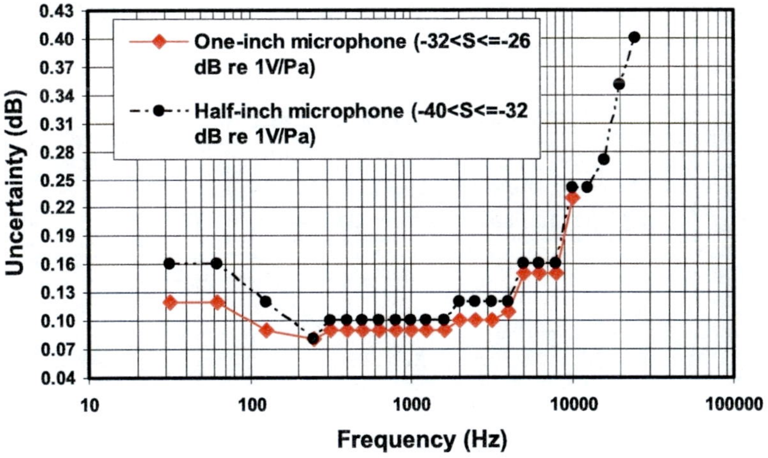

Fig. 3.9 Comparison of measurement uncertainty ($k = 2$, 95% confidence interval) for one-inch and half-inch microphones

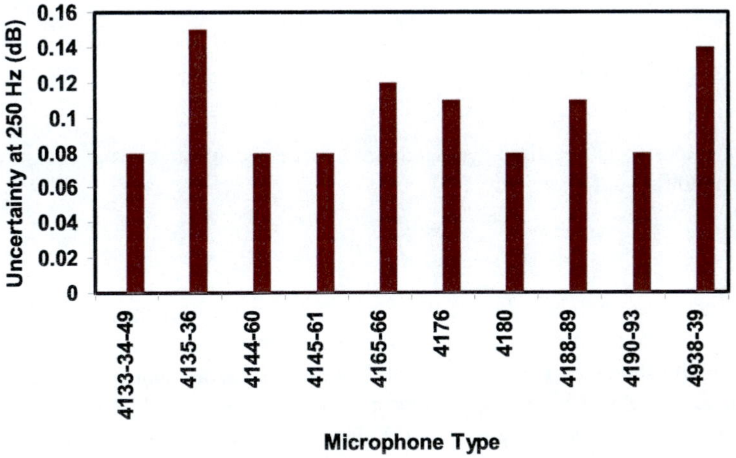

Fig. 3.10 Measurement uncertainty ($k = 2$, 95% confidence interval) at 250 Hz for various B&K microphones

3.7 Sound Intensity

Sound intensity is a vector quantity unlike sound pressure, which is a scalar quantity. There are two methods used for the measurement of sound intensity: one, called the p-p method, that utilizes two phase-matched closely spaced microphones and the second one (p-u method), wherein a microphone and particle velocity sensor is used.

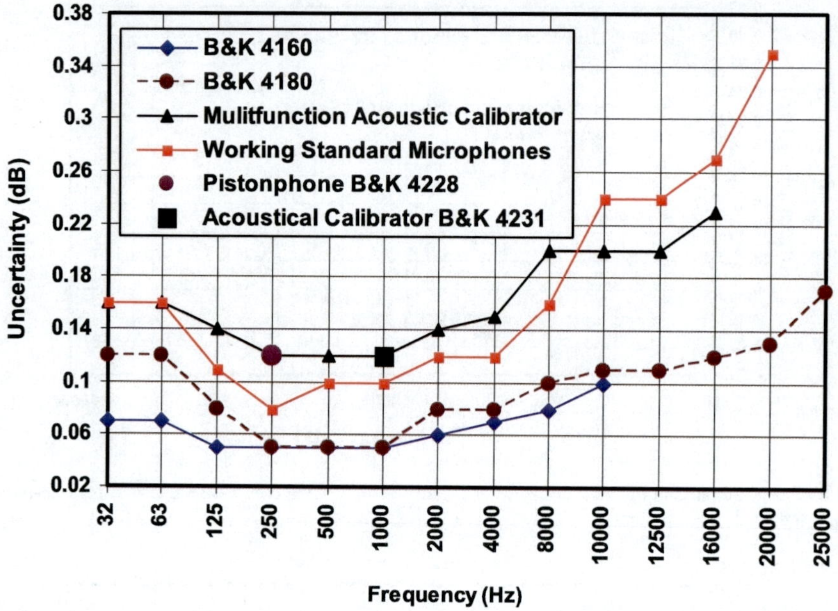

Fig. 3.11 Measurement uncertainty ($k = 2$) for microphones and calibrators

Sound intensity is the time-averaged product of the instantaneous pressure and corresponding particle velocity.

$$I = \overline{p.u} \tag{3.7}$$

where u is the particle velocity. The bar indicates the time-averaging. Thus, to measure the sound intensity, instantaneous sound pressure and corresponding particle velocity should be known. The sound intensity measurement system consists of an intensity probe, an analyzer, and a postprocessing system. The microphones used in the probe should ideally have identical phase response and flat amplitude response as a function of frequency. Microphones of different sizes and spacer lengths can be used in the sound intensity probes depending upon the useful frequency range of the measurement. For instance, a 12 mm spacer can be used for frequency range of 125 Hz to 5 kHz, while a 6 mm from 250 Hz to 10 kHz. The main errors are due to finite difference approximation and phase mismatch, imposing restrictions on the measurement frequency range. Other errors are due to near field, air flow, and statistical errors. The difference in sound pressure level (L_p) and sound intensity level (L_I) at a point called "pressure-intensity index" provides an indication of the accuracy of sound intensity measurements. However, minor phase mismatches between the measurement channels of the system cause a detection of intensity called residual intensity. The difference between the pressure and residual intensity is called residual pressure intensity index. The pressure-intensity index must be 7 dB

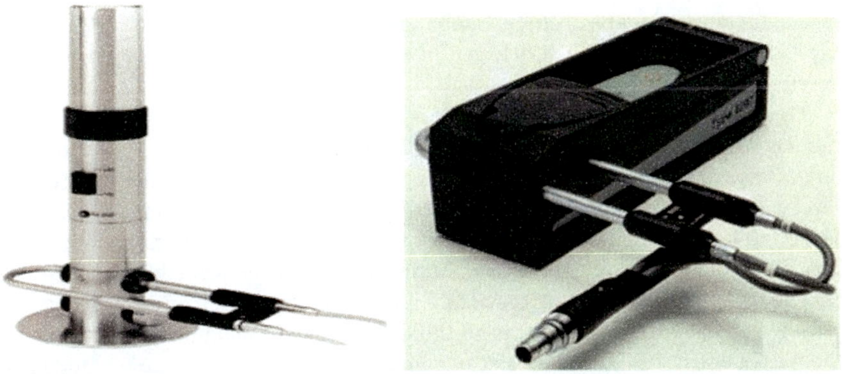

Fig. 3.12 (**a**) Pictorial view of Pistonphone, B&K Type 4228 fitted to the coupler, with both microphones positioned in the upper chamber (LHS); (**b**) arrangement for measuring residual intensity and pressure residual intensity index over a wide range of frequencies using Sound Intensity Calibrator Type 4297 (RHS). (B&K Manual, [17, 18])

smaller than residual pressure-intensity index for an error within 1 dB and 10 dB smaller for an error within 0.5 dB. The dynamic capability of a sound intensity measuring system is residual pressure-intensity index minus 7 dB that gives a limit to pressure intensity index that can be measured accurately [16]. The calibration of a sound intensity measuring device and its microphones involves [17, 18]:

- Sound pressure calibration of individual microphone channels.
- Sound intensity and particle velocity calibration.
- Measurement of pressure-residual intensity index spectrum of the system.

Figure 3.12a shows the Pistonphone, Make: Brüel & Kjær (B&K), Type 4228 fitted to the coupler, with both microphones positioned in the upper chamber of the coupler [17]. The pistonphone produces the same sound pressure level at each microphone, and the microphone channels are calibrated against this known sound pressure level. With this arrangement, the coupler causes a phase change between the sound pressures at the microphones, corresponding to a nominal spacing of 50 mm with no reflections. The phase change between the sound pressures simulates the sound intensity and particle velocity levels so that the pressure-difference signal for the integrator is not zero. Only now is the correct functioning of the instrument confirmed. Figure 3.12b shows an arrangement for measuring pressure-residual intensity index over a wide range of frequencies using Sound Intensity Calibrator Type 4297. The intensity probe is fitted in the calibrator, which delivers broad-band pink noise, so the sound pressure spectrum measured in the coupler is constant (in octave bands) over a wide frequency range. Both microphones are exposed to the same sound pressure, so any intensity detected is residual intensity. It can be shown that, for a given measurement system and frequency, the difference between measured sound pressure level and detected residual intensity level will be a constant. This constant difference is called the pressure-residual intensity index. The pressure-residual intensity index spectrum can be measured with the arrangement shown in Fig. 3.12b by subtracting the detected intensity spectrum from the sound pressure spectrum [18].

The sound power radiated by a source is equal in value to the integral of the scalar product of the sound intensity vector and the associated elemental area vector over any surface totally enclosing the source. In the ISO 9614 methods, the sound power level can be determined using far less restrictive test conditions than imposed by the ISO 3740 family. Limitations of the intensity-based approach is the usable frequency bandwidth of the intensity probes and the time required to perform the measurement. ISO 9614-2: Engineering or survey scanning measurement method allows to calculate the 1/1 and 1/third octave or band-limited weighted sound power level based on the measurement of the component of sound intensity normal to the measurement surface. The intensity probe has to be moved continuously along one or more specified paths surrounding the noise source (the measurement surface enveloping the test object must be divided into segments). The intensity probe then has to be scanned over each segment. ISO 9614-3: Precision scanning measurement method specifies a similar approach to the ISO 9614-2, but allows obtaining precision grade results. In this method, the intensity probe must be moved continuously along one or more specified paths surrounding the noise source: the measurement surface enveloping the object being tested must be divided into segments. The intensity probe has to be scanned over each segment. A set of complementary procedures is described to help indicate the quality of the test and the grade of accuracy. Measurements can be performed either in situ or in special purpose test environments. It is recommended to avoid testing in the presence of variable external noise sources or in variable conditions [19].

3.8 Sound Power Level

Sound power level is the acoustic property of a machine/source and is generally used for noise rating/labeling of products. It is independent of the distance from the source and describes the acoustic strength of the source. Sound power can be determined by performing measurements in free field (anechoic chamber) and diffuse field (reverberation room). ISO 3745 [20] and 3744 [21] are for the measurements in free-field condition. ISO 3745 describes the precision method of measurements in anechoic chamber. The sound power is determined by taking spatial average of sound pressure level measured on equal areas on the enclosing surface. The sound power L_w of a source can be calculated from sound pressure level measurements as:

$$L_w = \overline{L_p} + 10 \log \left(\frac{S}{S_0}\right) + C \tag{3.8}$$

where S_0 is the reference area (1 m^2), S is the area of test sphere, and C is a correction term for the influence of temperature and atmospheric pressure. In case of a free-field over reflecting plane, $S = 2\pi r^2$, r being the radius of test hemisphere. ISO 3744 describes a method with less precision for measurement made in open areas or in large rooms. The measurement surface can be a rectangular box with at least nine sound pressure level measurement points. The ratio of room absorption A to the measurement area S should be greater than 6. The ISO 3744 standard suggests that the surface time-averaged sound pressure level is calculated by correcting the mean time-averaged sound pressure level for background noise and for the influence of the test environment in order to compute the sound power level as:

$$L_w = L_p - K + 10\ \log\left(\frac{S}{S_0}\right) \tag{3.9}$$

where the environment correction K can be calculated as:

$$K = 10\ \log\left(1 + \frac{4}{A/S}\right) \tag{3.10}$$

ISO 3741 describes a precision method of sound power determination in a diffuse field. The average sound pressure level L_p in a room is obtained by sweeping a microphone at steady speed or by an array of three fixed microphones. The sound power of the source is obtained by the direct methods as per ISO 3741:

$$L_w = L_p - 10\ \log_{10}\left(\frac{T}{T_0}\right) + 10\ \log_{10}\left(\frac{V}{V_0}\right) + 10\ \log_{10}\left(1 + \frac{S\lambda}{8V}\right) + 10\ \log_{10}\left(\frac{B}{1000}\right) - 14\ dB \tag{3.11}$$

where L_p is the mean band sound pressure level, T is the reverberation time ($T_0 = 1$ s), V is the volume of room ($V_0 = 1$ m^3), S is the total surface area of room in m^3, $\lambda$is the wavelength at the center frequency, and B is the barometric pressure in mbars. As per the substitution method, the sound source is placed in the room and the spaced average sound pressure level L_p is measured in each frequency band. The sound source is then replaced by a reference sound source, whose sound power output L_{WR} is known. The space-averaged sound pressure level L_{pR} produced by reference source in the room is measured and the sound power level of test source is calculated as [20]:

$$L_w = L_{WR} + \left(L_p - L_{pR}\right) \tag{3.12}$$

B&K 4204 and Nor 278 are such reference sound sources used widely. Figure 3.13 shows the comparison of sound power level of B&K 4204 reference sound source of CSIR-NPL, India and Manufacturer (B&K, Denmark). The reference sound source consists essentially of a specially designed centrifugal fan driven by a powerful asynchronous motor. The motor is an external rotor type, and due to its

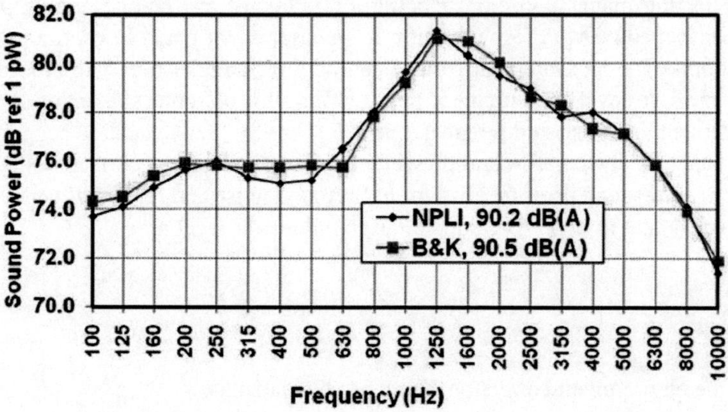

Fig. 3.13 Comparison of sound power level of B&K 4204 reference sound source of CSIR-NPL, India and manufacturer (B&K, Denmark). (Garg and Sharma 2012, [2])

high moment of inertia, it has a very stable speed of rotation. The motor is mounted on a cast aluminum base, shaped to reduce the reflections. The complete assembly of motor and fan is within a cylindrical safety grid fitted with two carrying handles [22].

Table 3.2 shows the various ISO standards for sound power level determination [23]. Thus, the spatially average sound pressure measurements of source under test using a parallelepiped microphone arrangement in reverberant rooms or a spherical microphone arrangement in anechoic or hemi-anechoic rooms can be used to measure the sound power level of a source.

The various ISO standards and methodology used for reporting sound power level measurements as described in Brezas comprehensive study (2019) [24] are summarized in Fig. 3.14 [25]. Sound power can be either directly calculated either by sound pressure or sound intensity measurements or determined by the substitution method utilizing a reference sound source (RSS). The use of different methods with different uncertainties thus inculcates challenges and difficulties for regulatory compliance with legal requirements or standards [25, 26]. In recent times, some National Metrology Institutes have developed their primary sound power standards using a vibration shaker or a loudspeaker [26]. Accurate and precise determination of sound power levels of various noise sources shall facilitate the noise labeling of appliances, type approval noise testing of automobiles, engines and generator sets, legal standards for loudspeakers, public address (PA) systems, horns and sirens, qualification of hemi-anechoic chambers, and determination of room acoustic parameters [25].

Table 3.2 Various methods/standards for the measurement of sound power level

ISO standard	Classification method	Test environment	Noise source characteristics	Measurand
3741	Precision method for reverberation test rooms	Reverberation room-specified requirements	All types of noise sources	A-weighted sound power level; 1/3-octave bands with mid-band frequencies from 100 Hz to 10,000 Hz
3743	Engineering methods for small movable sources in reverberant fields	Special test room (hard walled)	Steady, broad-band, narrow-band, discrete frequency	1/1 or 1/3 octave bands, A-weighted
3744	Engineering methods for an essentially free field over a reflecting plane	Indoors or outdoors, with one or more sound-reflecting planes present on or near which the noise source under test is mounted	All types (steady, non-steady, fluctuating, isolated bursts of sound energy)	1/1 or 1/3 octave bands, A-weighted
3745	Precision methods for anechoic rooms and hemi-anechoic rooms	Anechoic and semi-anechoic room	All types (steady, non-steady, fluctuating, isolated bursts of sound energy)	1/1 or 1/3 octave bands, A-weighted
3746	Survey method using an enveloping measurement surface over a reflecting plane	No special test environment	Steady, broad-band, narrow-band, discrete frequency	A-weighted
9614–1	Precision, engineering or survey discrete measurement points method	In situ or in special purpose test environments	Sound intensity at discrete points	1/1 or 1/3 octave bands, A-weighted
9614–2	Engineering/survey scanning measurement method	Any environment	Sound intensity by uniform scanning	1/1 or 1/3 octave bands, A-weighted
9614–3	Precision scanning measurement method	Specific test environments	All types	1/1 or 1/3 octave bands, A-weighted

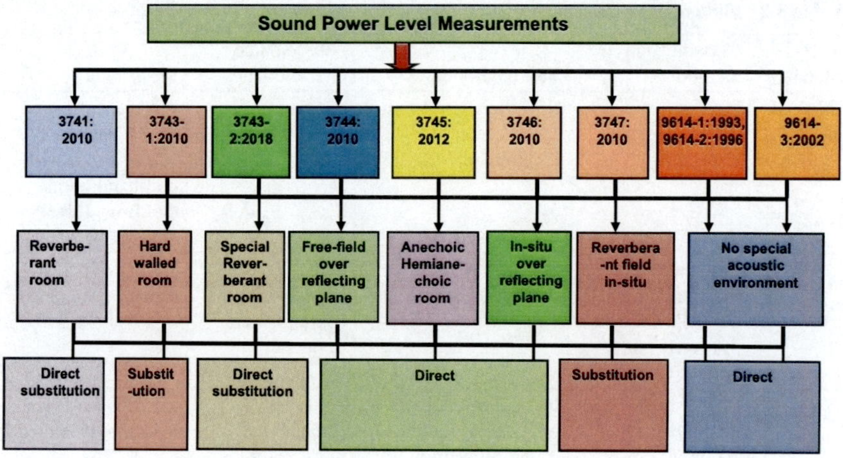

Fig. 3.14 Various ISO standards for sound power level measurements. (Garg et al. 2020, [25])

3.9 Calibration and Measurement Capability

Calibration and Measurement Capability, also called CMC, is available to the customers under normal conditions is defined in BIPM (International Bureau of Weights and Measures, French name: Bureau International des Poids et Mesures) website [27] as:

- As published in the BIPM key comparison database (KCDB) of the CIPM (International Committee for Weights and Measures) MRA (Mutual Recognition Agreement).
- As described in the laboratory's scope of accreditation granted by a signatory to the ILAC (International Laboratory Accreditation Cooperation) Arrangement.

In case of a CMC, the measurement or calibration should be: performed according to a documented procedure and have an established uncertainty budget under the management system of the NMI or the accredited laboratory; performed on a regular basis (including on demand or scheduled for convenience at specific times in the year); and available to all clients. In the Key Comparison Database (KCDB), a CMC is described by the measured quantity and its range, and is characterized by an uncertainty generally given at a 95% level of confidence, together with the method or instrument used, the values of the influence parameters if any, and any other relevant information. ISO/IEC 17025:2017 requires calibration and testing laboratories to have and apply procedures for the estimation of uncertainty of measurement. A National Metrology Institute (NMI) or other Designated Institute (DI) publishing Calibration and Measurement Capabilities (CMCs) in the BIPM Key Comparison Database (KCDB) has two choices for establishing its traceability route to the SI (International System of Units, Systeme Internationale): either via primary realization or via another NMI or DI having relevant CMCs with appropriate uncertainty

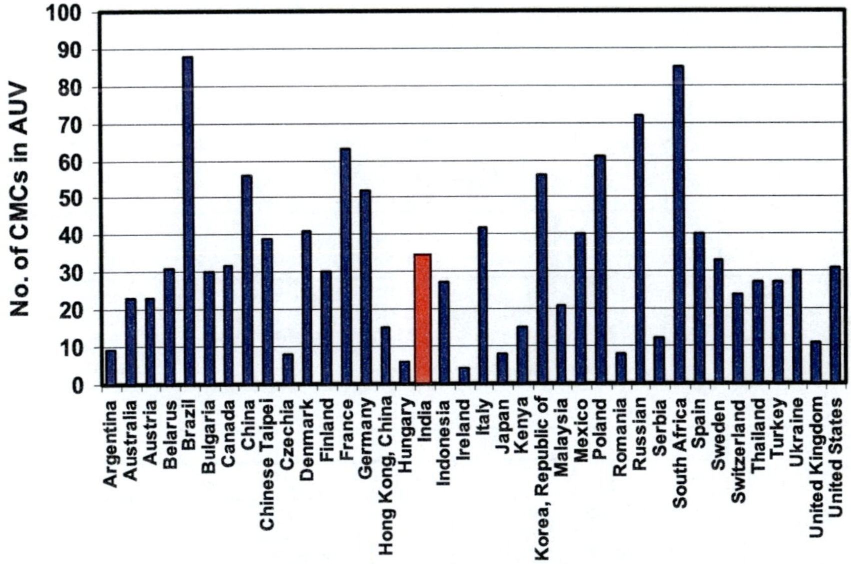

Fig. 3.15 Number of Calibration and Measurement Capabilities (CMCs) of different NMIs enlisted in The BIPM Key Comparison Database (https://www.bipm.org/kcdb/cmc/statistics/public, Accessed on 9th June, 2020)

published in the KCDB or through calibration and measurement services offered by the BIPM, in which case traceability must be declared through the laboratory providing the service [27]. Figure 3.15 shows the comparison of CMCs for some of the NMIs in the field of acoustics and vibration metrology. In case of acoustics metrology, the role of NMI in realization, maintenance, and dissemination of the three acoustic parameters: sound pressure, sound intensity, and sound power level is very important for strengthening the measurement traceability chain throughout the country. The primary standard of sound pressure for pressure field and free-field calibration of reference standard microphones; primary sound power level standard; secondary sound standard for microphone calibration using insert voltage method and frequency response; calibration facility for sound calibrators, pistonphones, multifrequency acoustic calibrators, sound level meter, sound intensity probe, audiometers, dosimeters, reference sound source, etc. are essentially required for establishment of a strong traceability chain as shown in Fig. 3.16 [25]. At present, 23 Calibration and Measurement Capabilities (CMCs) in sound parameter and 11 in vibration parameter exist in Key Comparison Database (KCDB) of CSIR-NPL, India, and thus, these establishments shown in Fig. 3.16 shall not only increase the number of CMCs, but also serve all the stakeholders and achieve *Atmannirbhar Bharat* (or self-reliant India) status in this field.

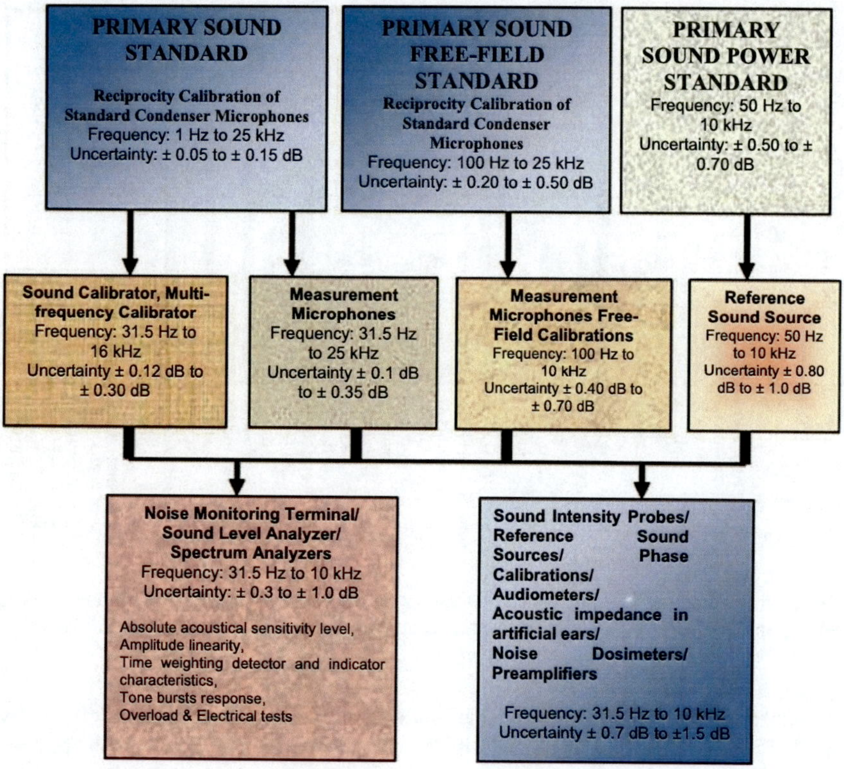

Fig. 3.16 An illustration of traceability chain for sound pressure, sound intensity, and sound power level parameters realized by a National Metrology Institute (NMI). (Garg et al. 2020, [25])

3.10 Measurements with ISO Standards

The accurate acoustical measurements can be only performed when one follows a specific measurement standard recognized internationally. There had been various acoustical measurement standards available for various types of measurements written and moderated by International Organization for Standardization (ISO). Table 3.3 summarizes some of the ISO standards for various noise sources: traffic, railway, aircraft, and industry [28–36]. ISO 9613 Part 1 and 2 standard [28, 29] is generally used for calculating the attenuation of the sound during propagation outdoors. However, there had been various models developed in recent times introducing advanced numerical methods for calculation of sound propagation outdoor. In case of vehicles, there are various classifications such as category L corresponds to motor vehicles with fewer than four wheels, category M corresponds to power-driven vehicles having at least four wheels and used for the carriage of passengers, and category N is for power-driven vehicles having at least four wheels and used for the carriage of goods.

Table 3.3 ISO (International Organization for Standardization) Standards for various noise sources [28–36]

ISO standard	Description/details	Noise source	Comments
ISO 9613-1:1993, Acoustics-Attenuation of sound during propagation outdoors, Part 1: Calculation of the absorption of sound by the atmosphere	An analytical method of calculating the attenuation of sound as a result of atmospheric absorption is recommended	Environmental noise	It applies to an atmosphere with uniform meteorological conditions
ISO 9613-2:1996, Acoustics—Attenuation of sound during propagation outdoors—Part 2: General method of calculation	Describes a method for calculating the attenuation of sound during propagation outdoors in order to predict the levels of environmental noise at a distance	Environmental noise	The method predicts the equivalent continuous A-weighted sound pressure level under meteorological conditions
ISO 1996-2:2017, Acoustics-Description, measurement and assessment of environmental noise, Part 2: Determination of sound pressure levels	Describes how sound levels can be determined by direct measurement, or calculations or by extrapolation of measurement results for assessing environmental noise	All kinds of environmental noise sources	It is used as a reference for noise mapping in European noise directives
ISO 8297:1994, Acoustics - Determination of sound power levels of multisource industrial plants for evaluation of sound pressure levels in the environment—Engineering method	Specifies an engineering method for determining the sound power levels of large multisource industrial plants for evaluation of sound pressure levels in the environment	The method is limited to large industrial plants with multiple noise sources	The method is applicable to industrial areas where most of the equipment operates outdoors, not enclosed by a building
ISO 11819-1:1997, Acoustics—Measurement of the influence of road surfaces on traffic noise—Part 1: Statistical pass-by method	Describes a method of comparing traffic noise on different road surfaces for various compositions of road traffic	Traffic noise on different road surfaces	The method is applicable to traffic traveling at constant speed, that is, free-flowing conditions
ISO 11819-2:2017, Acoustics—Measurement of the influence of road surfaces on traffic noise—Part 2: The close-proximity method	Specifies a method to evaluate different road surfaces w.r.t. their influence on traffic noise, under conditions when rolling noise dominates	Traffic noise	The analysis of the results applies to free-flowing traffic traveling on essentially level roads at constant speeds
ISO 362-1:2015 measurement of noise emitted by accelerating road	Specifies an engineering method for measuring the noise emitted by	Road vehicles of categories M and N	It is intended to reproduce the level of noise generated by the

(continued)

Table 3.3 (continued)

ISO standard	Description/details	Noise source	Comments
vehicles—Engineering method—Part 1: M and N categories	road vehicles of categories M and N under typical urban traffic conditions		principal noise sources during normal driving in urban traffic
ISO 3095:2013 Acoustics-Railway Applications- Measurements of noise emitted by railbound vehicles	Specifies measurement methods for reproducible exterior noise emission levels and spectra for all kinds of vehicles operating on rails or other types of fixed track	Railway exterior noise	It deliberates to describe the noise emission from the unit, minimizing the influence of track
ISO 20906:2009 ACOUSTICS-UNATTENDED MONITORING OF AIRCRAFT SOUND IN THE VICINITY OF AIRPORTS	Describes the application for a permanently installed sound-monitoring system around an airport	Aircraft noise	Standard precisely specifies the requirements for reliable measurements of aircraft sound

3.11 Sound Intensity Mapping and Noise Sources Ranking

Sound intensity can be measured in any sound field and facilitates the measurements to be done directly in situ. It gives a measure of direction as well as magnitude and is thus very useful when locating all the sound sources enabling the radiation patterns of complex vibrating machinery to be studied in situ. The use of sound intensity rather than sound pressure facilitates the in situ measurements, with steady background noise and in the near field of machines. The sound power is the average normal intensity over a surface enclosing the source, multiplied by the surface area. The choice of length of spacer of 6, 8.5, 12, or 50 mm is important, as smaller the spacer, the higher the frequency that can be measured. The phase mismatch in the analyzing system causes a low-frequency limit, and the larger the spacer, the lower the frequency limit it can measure. A 50 mm spacer enables to measure down to 63 Hz and high-frequency limit up to 1.25 kHz. For accuracy to within 1 dB, the wavelength measured must be greater than six times the spacer distance. This corresponds to the following high-frequency limits: 50 mm: up to 1.25 kHz, 12 mm: up to 5 kHz, and 6 mm: up to 10 kHz. However, a single spacer cannot cover the wide frequency range. The effectiveness of noise control method requires the consideration of the noise radiated from the individual components of the machine. The total sound power can be found simply by adding the partial sound powers from all the noise radiating components [37]. Contour and 3-D plots enable a detailed analysis of the sound field generated by a source as shown in Fig. 3.17 for front surface of a diesel engine running at 2000 rpm and 80% load.

The surface of interest is divided into a grid and sound intensity is measured at each point of the grid. The sound power of each region can be measured by

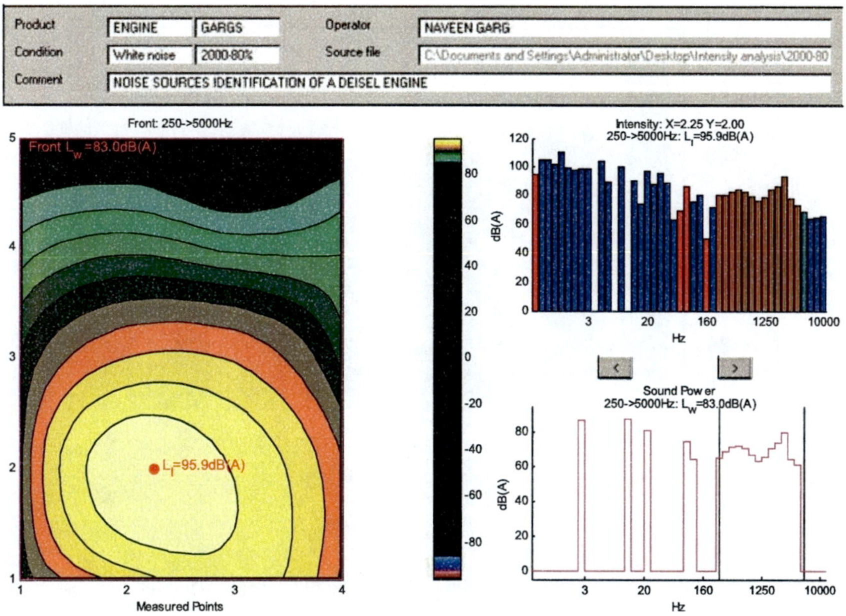

Fig. 3.17 Sound intensity mapping of front surface of diesel engine

multiplying the measured sound intensity with the corresponding surface area of the region. Thus, the sound power contribution of each region to the total sound power of the machine can be ascertained [38]. The sound intensity vectorial distribution can then be displayed for a certain frequency range, and one can superimpose the contour spectral results onto a picture of the object tested allowing for a precise location of noise sources [39]. The noise sources ranking can be thus easily done as shown in Fig. 3.18 for a machine.

3.12 Conclusions

The chapter discusses the functionality, measurement methodology, and uncertainty and the traceability chain established in realization of the primary and secondary sound standards established at CSIR-National Physical Laboratory, India. It is evident that accurate and precise sound measurements are required for evaluation and analysis of sound levels, compliance with respect to the legal limits and for devising noise control and abatement strategies. The strengthening of traceability chain for sound level measurements throughout the country shall be a vital step in mission toward noise pollution control in urban environment. The National Metrology Institute (NMI) role is wholeheartedly dedicated to reduce the measurement uncertainty in primary and secondary standards for achieving highest levels of

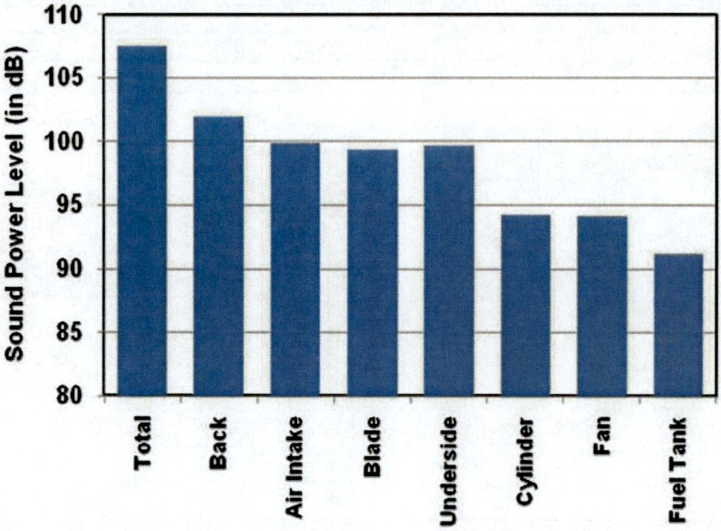

Fig. 3.18 Noise sources ranking of a machine using the sound intensity technique

accuracy and precision and strengthening the traceability chain throughout the country. The Acoustics and Vibration Standards of CSIR-National Physical Laboratory, New Delhi envisage the establishment of Centre of Excellence in Noise and Vibration Monitoring and Control and in Building Acoustics in future with a prime objective of strengthening the measurement traceability chain and contributing towards reducing noise pollution control in the country and achieve Atmannirbhar Bharat (or Self-reliant India) status in this field. It is evident that for reliable sound level measurements using sound level meters and analyzers, the calibration of the instrument along with microphones is very important which can be realized only through the secondary sound standards traceable to the primary standards. Smart phone based app technologies and Noise Monitoring Terminals are also used extensively in recent times for noise monitoring which requires precise calibration and traceability to the national standards for reporting accurate and precise sound level measurements. The strengthening of the measurement traceability chain shall ensure the reliable sound level measurements carried out by industry, user laboratories, state pollution control boards, defense sector in India for global perspectives. In this regard, targeted efforts by NMIs on increasing the number of CMCs in BIPM website and reducing the measurement uncertainty in primary and secondary sound standards by novel measurement methods and techniques [40–43] shall not only strengthen the traceability chain, but also shall strengthen the industrial, academic, research and legal framework of the country. Also, with technological advancements and smart phone based app technologies, the digital transformation in acoustics metrology has to be focused in next decade for fostering the economic and industrial growth.

References

1. National and International needs relating to metrology: International collaborations and the role of the BIPM (1998) Bureau International des Poids et Mesures, A report prepared by the CIPM for the Governments of the Member States of the Convention of the Metre
2. Garg N, Sharma O (2012) Measurement accuracy of secondary standards of sound pressure in comparison to primary standards. MAPAN-J Metrol Soc India 27(4):219–229
3. IEC 61094-2 (1992) Measurement microphones-Part 2-Primary method for pressure calibration of laboratory standard microphones by reciprocity technique. International Electrotechnical Commission, Switzerland
4. Brüel and Kjær (B&K) User Manual for Reciprocity Calibration Apparatus Type 9699, (1997)
5. Frederiksen E (2013) Acoustic metrology-an overview of calibration methods and their uncertainties. Int J Metrol Qual Eng 4:97–107
6. IEC 61094-5 (2001) Measurement microphones-Part 5: Method for pressure calibration of working standard microphones by comparison. International Electrotechnical Commission, Switzerland
7. Brüel and Kjær (B&K) (2011) User Manual for Microphone calibration system Type 9721
8. IEC 61094-6 (2001) Measurement microphones-Part 6: Electrostatic actuators for determination of frequency response. International Electrotechnical Commission, Switzerland
9. Brüel and Kjær (B&K) Microphone Handbook, Technical Documentation, Vol. 1, March, 2019, BE 1447 –1
10. Barrera-Figueroa S, Rasmussen K, Jacobsen F (2007) On experimental determination of the free-field correction of laboratory standard microphones at normal incidence. Metrologia 44: 57–63
11. Barrera Figueroa S, Torras Rosell A, Rasmussen K, Jacobsen F, Henríquez V C (2012) A practical implementation of microphone free-field comparison calibration according to the standard IEC 61094–8. Proceedings Internoise 2012, Proceedings Acoustical Society of America
12. IEC International standard 61094-1 (2000) Measurement Microphones, Part 1: Specifications for Laboratory Standard Microphones
13. Garg N, Surendran P, Dhanya MP, Chandran AT, Asif M, Singh M (2019) Measurement uncertainty in microphone free-field comparison calibrations. MAPAN-J Metrol Soc India 34: 357–369
14. Garg N (2015) Establishing a traceability chain for sound pressure and vibration amplitude measurements. NCSLI Measur J Measur Sci 10:68–78
15. Li W, Wong GSK, Hanes P, Ohm WS (2005) Measurement of sensitivity level pressure correction for LS2P laboratory standard microphones. Metrologia 42:45–48
16. Tandon N, Nakra BC (2017) Vibration and noise engineering. I. K. International Pvt. Ltd., New Delhi
17. Brüel and Kjær (B&K) Manual, Sound Intensity Calibrator Type 3541
18. Brüel and Kjær (B&K) Manual, Sound Intensity Calibrator Type 4297
19. White paper on a guide to measuring sound power. An overview of international standards, Siemens PLM Software, www.siemens.com/plm. Accessed on 10.02.2021
20. ISO 3745 (2012) Acoustics-Determination of sound power levels and sound energy levels of noise sources using sound pressure-Precision methods for Anechoic rooms and hemi-anechoic rooms
21. ISO 3744 (2010) Acoustics-Determination of sound power levels and sound energy levels of noise sources using sound pressure-Engineering methods for an essentially free field over a reflecting plane
22. Brüel and Kjær (B&K) Manual, Reference Sound Source, Type 4204
23. Barnard AR (2014) Quantifying Acoustic Sources Through Sound Power Measurements. Sound and Vibration. Materials Reference Issue, 8–11. http://www.sandv.com/downloads/1407barn.pdf. Accessed on 17.12.2020

24. Brezas S (2019) Investigation on the dissemination of unit watt in airborne sound and applications, Aachener Beiträge zur Akustik, Bd. 32 Logos Verlag Berlin, ISBN 978-3-8325-4971-8, Berlin 2019, http://publications.rwth-aachen.de/record/772281/files/772281.pdf. Accessed on 10.02.2021
25. Garg N, Chauhan BS, Singh M (2020) Realization and dissemination of unit watt in airborne sound: measurement methodology, sound emission regulations and implications 35, 601-612. MAPAN-J Metrol Soc India. https://doi.org/10.1007/s12647-020-00417-x
26. Wittstock V, Kurtz P, Arina R (2016) Realization, Dissemination and Application of the unit watt in airborne sound; EURAMET, v1.0; JRP REPORT: SIB56; INRIM Italy, LNE France, PTB Germany, SP Sweden, TUBITAK Turkey; September 2016, 1–22
27. Bureau International des Poids et Mesures (BIPM), http://www.bipm.org. Accessed on 16 Dec 2020
28. ISO 9613-1 (1993) Acoustics-Attenuation of sound during propagation outdoors, Part 1: Calculation of the absorption of sound by the atmosphere
29. ISO 9613-2 (1996) Acoustics-Attenuation of sound during propagation outdoors, Part 2: General method of calculation
30. ISO 1996-2 (2017) Acoustics-Description, measurement and assessment of environmental noise, Part 2: Determination of sound pressure levels
31. ISO 8297 (1994) Acoustics-Determination of sound power levels of multisource industrial plants for evaluation of sound pressure levels in the environment-Engineering method
32. ISO 11819-1 (1997) Acoustics-Measurement of the influence of road surfaces on traffic noise - Part 1: Statistical Pass-By method
33. ISO 11819-2 (2017) Acoustics – Measurement of the influence of road surfaces on traffic noise – Part 2: The close-proximity method
34. ISO 362-1 (2015) Measurement of noise emitted by accelerating road vehicles-Engineering method-Part 1: M and N categories
35. ISO 3095 (2013) Acoustics-Railway Applications- Measurements of noise emitted by railbound vehicles
36. ISO 20906 (2009) Acoustics-Unattended monitoring of aircraft sound in the vicinity of airports
37. Brüel and Kj�r (B&K) Manual (1993) Sound Intensity, Sound Intensity (br0476) (bksv.com) (Accessed on 13.02.2021)
38. Tandon N (1999) Sound intensity technique and its applications in noise control. PINSA 65(1): 51–60
39. Batel M, Marroquin M, Hald J, Christensen JJ, Schuhmacher AP, Nielsen TG (2003) Noise source location techniques-simple to advanced applications. Sound and Vibration, March 2003, 24–36
40. Garg N, Kumar A, Sanjid MA, Chaudhary KP, Maji S (2014) Reaffirmation of measurement uncertainty in pressure sensitivity determination of LS2P microphones by reciprocity method. Measurement 51:281–288
41. Garg N, Maji S (2019) Analysis of the effect of microphone parameters in reciprocity calibrations using Taguchi method. MAPAN-J Metrol Soc India 30:179–190
42. Garg N, Yadav S, Aswal DK (2019) Monte Carlo simulation in uncertainty evaluation: strategy, implications and future prospects. MAPAN-J Metrol Soc India 34:299–304
43. Garg N, Chauhan BS (2020) Measurement uncertainty in vibration calibration in frequency range of 5 Hz to 10 kHz. MAPAN-J Metrol Soc India 35(3):397–405

Chapter 4
Environmental Noise Control Regulations

Abstract This chapter presents a retrospective and prospective view of noise policies and regulations followed across the globe. The Noise Pollution (Regulation and Control) Rules of India and those followed in other countries are discussed. The chapter suggests the inclusion of noise limits for construction activities, household appliances, and presents the implications of noise policies in national and international scenario. The noise limits for domestic appliances, motor vehicles, and construction equipment at the manufacturing stage are discussed. A National Planning Policy Framework for Environmental Noise Management and Control in Indian perspective to effectively control noise pollution in urban cities and in the proposed Smart Cities is suggested.

4.1 Introduction

Noise pollution has become a serious problem for the society. In India, with expanding vehicular population traffic noise levels have increased, which can cause serious health effects. The World Health Organization (WHO) recognized noise as one of the major pollutants affecting the human health. The major sources of noise pollution are: road traffic, rail, aircraft noise, construction noise, industrial noise emitted, honking noise from vehicles, noise emitted from household appliances, loudspeakers, community processions, etc. Road traffic noise has been observed to be the major source of community noise pollution. A conservative estimate of the social cost of traffic noise is €40 billion per year, of which 90% is from cars and trucks that represent a loss of 0.4% of total European Union (EU) GDP each year—equivalent to about one-third of the societal cost of road accidents [1]. An EU study looking into the benefits of stricter new standards for vehicle noise has concluded that strict standards to produce an effect equivalent to that of halving the traffic, would outweigh the costs of developing and introducing quieter vehicles by over twenty times [2]. Thus, the noise pollution legislations such as limits, policies, and ordinances for various noise sources are essential for controlling the noise pollution in urban agglomerations.

N. Garg, *Environmental Noise Control*,
https://doi.org/10.1007/978-3-030-87828-3_4

4.2 Noise Pollution (Regulation and Control) Rules of India

The Noise Pollution (Regulation and Control) Rules, 2000 have come into force on the date of their publication in the Official Gazette [3, 4]. The noise levels in any area/zone shall not exceed the ambient air quality standards in respect of noise as specified in Table 4.1. The authority shall be responsible for the enforcement of noise pollution control measures and the due compliance of the ambient air quality standards in respect of noise.

Daytime shall mean from 6.00 a.m. to 10.00 p.m., while the nighttime shall mean from 10.00 p.m. to 6.00 a.m. Silence zone is an area comprising not less than 100 meters around hospitals, educational institutions, courts, religious places, or any other area which is declared as such by the competent authority. Mixed categories of areas may be declared as one of the four above mentioned categories by the competent authority. The Ministry of Environment, Forests and Climate Change has recently notified the Noise Pollution (Regulation and Control) Amendment Rules, 2017 (effective from tenth August 2017) to amend the Noise Pollution (Regulation and Control) Rules, 2000. The amended rules have given the power to state governments to declare silence area/zone which is less than 100 meters around hospitals, educational institutions, and courts. The term "festive occasion" has been specifically defined in the amended rules. The ambient noise standards for coal mines are discussed in Appendix III of the book.

4.2.1 Restrictions on the Use of Loud Speakers/Public Address System and Sound Producing Instruments

A loud speaker or a public address system shall not be used except after obtaining written permission from the authority. The use of loud speakers or public address system and the like during night hours (between 10.00 p.m. to 12.00 midnight) on or during any cultural or religious festive occasion of a limited duration not exceeding 15 days in all during a calendar year is permitted. The noise level at the boundary of the public place, where loudspeaker or public address system or any other noise source is being used, shall not exceed 10 dB (A) above the ambient noise standards for the area or 75 dB (A) whichever is lower. The peripheral noise level of a privately

Table 4.1 Ambient air quality standards in respect of noise in India (CPCB India [3])

Area code	Category of area/zone	Limits in dB(A) L_{eq} [a]	
		Daytime	Nighttime
A	Industrial area	75	70
B	Commercial area	65	55
C	Residential area	55	45
D	Silence zone	50	40

[a] L_{eq} denotes the time weighted average of the sound level in decibels in *A*-weighting

owned sound system or a sound producing instrument shall not, at the boundary of the private place, exceed by more than 5 dB (A) the ambient noise standards specified for the area in which it is used.

4.2.2 Restrictions on the Use of Horns, Sound-Emitting Construction Equipment, and Bursting of Firecrackers

No horn shall be used in silence zones or during nighttime in residential areas except during a public emergency. Sound-emitting firecrackers shall not be burst in silence zone or during nighttime. Sound-emitting construction equipment shall not be used or operated during nighttime in residential areas and silence zones.

4.2.3 Consequences of Any Violation and Complaints to Be Made

Whoever in any place covered under the silence zone/area commits any such offense shall be liable for penalty under the provisions of the Act. The authority shall act on the complaint and take action against the violator in accordance with the provisions of these rules and any other law in force.

4.2.4 Noise Limits for Vehicles at Manufacturing Stage

The Noise Limits for vehicles applicable at manufacturing stage applicable from first April 2005 [4] has been recommended in Table 4.2. The test method to be followed shall be IS: 3028–1998.

4.2.5 Noise Limits for Generator Sets Run with Diesel

4.2.5.1 Noise Limit for Diesel Generator Sets (up to 1000 KVA) Manufactured on or After the First January 2005

The maximum permissible sound pressure level for new diesel generator (DG) sets with rated capacity up to 1000 KVA, manufactured on or after the first January 2005 shall be 75 dB(A) at 1 meter from the enclosure surface. The diesel generator sets should be provided with integral acoustic enclosure at the manufacturing stage itself. Noise from DG set shall be controlled by providing an acoustic enclosure or by treating the room acoustically, at the users end. The acoustic enclosure or acoustic

Table 4.2 Noise limits for vehicles applicable at manufacturing stage applicable from first April 2005 (CPCB India [4])

S. no.	Type of vehicle	Noise limits in dB(A)
1	**Two wheeler**	
1.1	Displacement up to 80 cc	75
1.2	Displacement more than 80 cc but up to 175 cc	77
1.3	Displacement more than 175 cc	80
2.0	**Three wheeler**	
2.1	Displacement up to 175 cc	77
2.2	Displacement more than 175 cc	80
3.0	**Vehicles used for carriage of passengers and capable of having not more than nine seats including the driver seat**	74
4.0	**Vehicles used for carriage of passengers having more than nine seats including the driver seat, and a maximum gross vehicle weight (GVW) of more than 3.5 tonnes**	
4.1	With an engine power less than 150 KW	78
4.2	With an engine power of 150 KW or above	80
5.0	**Vehicles used for carriage of passengers having more than nine seats including the driver seat: vehicles used for carriage goods**	
5.1	With maximum GVW not exceeding 2 tonnes	76
5.2	With maximum GVW greater than 3 tonnes but not exceeding 3.5 tonnes	77
6.0	**Vehicles used for transport of goods with a maximum GVW exceeding 3.5 tonnes**	
6.1	With an engine power less than 75 KW	77
6.2	With an engine power of 75 KW or above but less than 150 KW	78
6.3	With an engine power of 150 KW or above	80

treatment of the room shall be designed for minimum 25 dB (A) insertion loss or for meeting the ambient noise standards, whichever is on the higher side (if the actual ambient noise is on the higher side, it may not be possible to check the performance of the acoustic enclosure/acoustic treatment). The DG set shall be provided with proper exhaust muffler with insertion loss of minimum 25 dB (A).

4.2.6 *Firecrackers' Noise*

The manufacture, sale, or use of firecrackers generating noise level exceeding 125 dB(AI) or 145 $dB(C)_{pk}$ at 4 meters distance from the point of bursting shall be prohibited. For individual firecracker constituting the series (joined firecrackers), the abovementioned limit be reduced by 5 log 10 (N) dB, where $N =$ number of crackers joined together [4]. The cracker's noise is every year tested at CSIR-National Physical Laboratory, India, to ascertain whether their noise levels at 4 meters distance from the point of bursting.

Table 4.3 Ambient air quality standards with respect to noise in airport noise zone

Type of airports	Limits in dB(A) L_{eq}	
	Daytime	Nighttime
Busy airports	70	65
All other airports excluding proposed airports	65	60

4.2.7 Airport Noise Limits

The noise standards (issued by Ministry of Environment and Forests, India, dated 18th June 2018) within the overall boundary of airports shall be applicable as industrial areas, that is, daytime 75 dB (A) L_{eq} and nighttime 70 dB (A) L_{eq} as per the Noise (Regulation and Control) Rules 2000 and shall be measured at different points of airport boundary and then averaged (Table 4.3). These standards will not be applicable to a civil airport which has less than 15,000 aircraft movement annually. For the purpose of noise management at airports, a busy airport shall be defined as "a civil airport which has more than 50,000 aircraft movements per year (a movement being a take-off or a landing)" excluding those purely for training purposes on light aircraft. Other airports include an airport having more than 15,000 but less than 50,000 aircraft movements annually, while proposed airports are those airports that are not functional yet and are under development.

4.2.8 Noise Regulations for Horns and Sirens

The noise regulations for horns and sirens are shown in Table 4.4.

4.3 Environmental Noise Regulations Followed in Various Countries

Traffic noise seriously affects human health, well-being, and quality of life in many ways. EU research shows the benefits of stricter new standards on vehicle noise would outweigh the costs of developing and introducing quieter vehicles by over 20 times [2]. Vehicles alone can easily be 6 dB quieter than today, if only effective standards are set [5]. Figure 4.1 summarizes the various regulations that are to be considered in recent times for framing a national noise policy for noise pollution management and control. The consideration of all the noise sources in totality and setting up of legal limits in conjunction with framing legal guidelines for action against the violation of noise regulations, devising guidelines for noise monitoring, mapping, abatement, and control and building sound insulation regulations is a vital requirement in recent times to arrest the noise pollution in any country.

Table 4.4 Noise regulations with respect to horns and sirens

Vehicular noise sources	Vehicular noise regulations: Norms notified by	Vehicular noise regulations: To be implemented and complied by	Standard code	Noise limits/regulations
Horn as a component	Specification for horns, etc. (manufacturing and installation) by MoRTH	Ministry of Road Transport and Highways (MoRTH) (traffic police)	IS 1884–1992	85 dB for type 1 horns 90 dB for type 2A horns 100 dB for type 2B horns and 105 dB for type 3 horns
Horn installation			IS 15796–2008 (AIS 014)	Maximum sound pressure level when determined in accordance with method of test shall meet the following: (a) between 83 dB (A) for horns fitted on two wheelers or three wheelers of a power less than or equal to 7 kW and (b) between 93 dB (A) to 112 dB(A) for the horn fitted on all other motor vehicles including tractor, two wheelers and three wheelers of a power greater than 7 kW
Sirens			AIS 125 (National Ambulance Code)	Noise limits for sirens fitted on ambulance are in the range of 110 dB (A) to 120 dB (A).

Worldwide, noise-induced hearing impairment is the most widespread irreversible occupational hazard. The WHO guidelines (1999) had recommended critical health effects and sound level limits for specific environments [6]. The guidelines values consider all the identified adverse health effects for the specific environment. For instance, for hospital, ward rooms, indoors, the legal limit of 30 dB(A) L_{Aeq} and 40 dB(A) L_{Amax} is recommended for 8 hours duration in nighttime. In case of ceremonies, festivals, and entertainment events, the 100 dB(A) L_{Aeq} for 4 hours duration and 110 dB(A) L_{Amax} is recommended [6]. For public addresses, indoors and outdoors, the 85 dB(A) L_{Aeq} for 1 hour duration and 110 dB(A) L_{Amax} is recommended. Adverse health effects are observed at the level above 40 dB $L_{night, outside}$, such as self-reported sleep disturbance and environmental insomnia, and thus these values are lowest observed adverse effect level (*LOAEL*) 40 dB $L_{night,outside}$ for night noise [7]. Above 55 dB, the cardiovascular effects become the major public health concern, which are likely to be less dependent on the nature of the noise. An

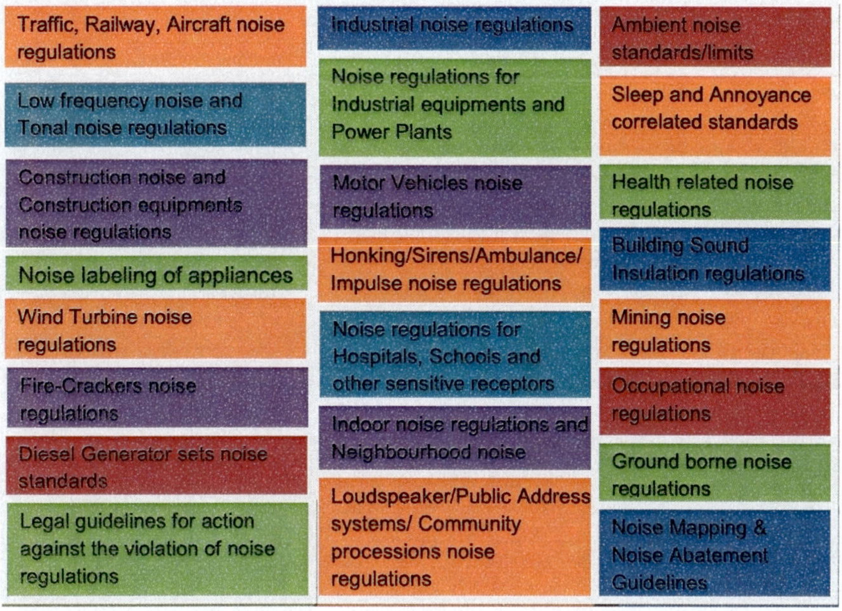

Fig. 4.1 Regulations/noise limits required for various noise sources for framing a national noise policy for noise pollution management and control

interim target (IT) of 55 dB $L_{\text{night,outside}}$ is recommended in the situations where the achievement of Night Noise Guidelines (*NNG*) is not feasible in the short run for various reasons. *The Night Noise Guidelines for Europe (2009) report recommends that due to differences in legal systems it is hard to predict what the actual effect of a certain limit value will be. It could be a relatively high value but rigidly enforced, or a very low value with no legal binding whatsoever*. The report recommends a threshold of 42 dB $L_{\text{night, outside}}$ for self-reported sleep disturbance. Table 4.5 shows the residential traffic noise criteria for various countries [8]. Many countries place a more stringent limit on the night-time L_{Aeq} level from road traffic noise than on the day-time level. L_{Aeq} is measured or calculated over a variety of periods, from 24 hours to an L_{Aeq} (1 hour), with some countries, including Denmark, Hong Kong, and the USA, assuming that a single criterion will assure that both daytime and nighttime noise levels will be satisfactory. The European Union Directive introduced average day-evening-night noise levels, L_{den} as the indicator of annoyance due to road traffic noise. In some countries like Australia, there is a similarity in the residential guideline/criteria/target noise levels at 1 m from the facades of buildings for new roads within any one metric, that is, for those that use $L_{\text{A10,18h}}$ the level is 63 dB (A), while for those that use $L_{\text{Aeq,T}}$ the daytime level is 55 dB(A) and the nighttime level is 50 dB(A) [9]. The I-INCE report from TSG 6 [10] takes a worldwide approach in considering the strategies for managing exposure to community noise from the immission perspectives and advocates a clear understanding

Table 4.5 Residential traffic noise criteria of various countries (NSW Road Noise Policy [8])

	Daytime criteria	Nighttime criteria
Austria	60 dB(A) L_{eq}; planned roads in quite areas, 55 dB (A) L_{eq}	50 dB(A) L_{eq}; planned roads in quite areas, 45 dB (A) L_{eq}
Canada	Outdoor 55 dB(A) $L_{Aeq,16h}$ indoor 40–45 dB(A) $L_{Aeq,16h}$	Outdoor 50 dB(A) $L_{Aeq,8h}$ indoor 40–45 dB(A) $L_{Aeq,8h}$
Denmark	50–65 dB(A) $L_{Aeq,24h}$ depending upon land use	
France	60–65 dB(A) $L_{Aeq,16h}$ (6 a.m. to 10 p.m.)	55–60 dB(A) $L_{Aeq,8h}$ (10 p.m. to 6 a.m.)
Germany	64 dB(A) $L_{Aeq,16h\ (6\ a.m.\ to\ 10\ p.m.)}$	54 dB(A) $L_{Aeq,8h\ (10\ p.m.\ to\ 6\ a.m.)}$
Greece	Residences 70 dB(A) $L_{A10,18h}$; 67 dB(A) L_{Aeq} (8 a.m. to 8 p.m.)	
Hong Kong	70 dB(A) $L_{A10,\ 1h}$ dwellings	
Italy	65 dB(A) $L_{Aeq,16h}$	55 dB(A) $L_{Aeq,8h}$
Japan	50–60 dB(A) $L_{Aeq,16h}$	40–50 dB(A) $L_{Aeq,8h}$
Member states of European Union	58–62 dB(A) L_{Aeq}	48–55 dB(A) L_{Aeq}
UK	55–72 dB(A) L_{Aeq} (7 a.m.–11 p.m.)	45–66 dB(A) L_{Aeq} (11 p.m.–7 a.m.)
USA	52 dB(A) (interior), 67 dB(A) (exterior) $L_{Aeq,\ (1\ hr)}$ or 55 dB(A) (interior), 70 dB(A) (exterior) L_{A10}	
Switzerland	60 dB(A) $L_{Aeq,(15\ hr)}$ reference level	50 dB(A) $L_{Aeq,(9\ hr)}$ reference level
The Netherlands	48–58 dB(A) $L_{den\ (7\ a.m.–7\ p.m.,\ 7\ p.m.–11\ p.m.,\ 11\ p.m.–7\ a.m.)}$ (new motorways); 53–68 dB(A) $L_{den\ (7\ a.m.–7\ p.m.,\ 7\ p.m.–11\ p.m.,\ 11\ p.m.–7\ a.m.)}$ (modification of existing motorways)	
Spain	Proposed criteria of 55–75 dB (A) L_{eq}	Proposed criteria of 45–65 dB (A) L_{eq}

of the relationship between noise emission and immission. Noise emission is the term used to describe the output sound from a source. It is the noise immission being received by the community that is the factor for comparison with the applicable criteria [11].

4.3.1 Motor Vehicle Noise Regulations

ISO 362-1:2015 [12] describes vehicles into various categories: L, M, M1, M2, M3, N, N1, N2, and N3. The European legislation No. 540/2014 defines the limit values for three phases: Phase 1 applicable for new vehicle types from first July 2016, Phase 2 applicable for new vehicle type from first July 2020 and for first registration from first July 2022, and Phase 3 applicable for new vehicle type from first July 2024 and for first registration from first July 2026. For the first phase, the limits are defined from the range 72 to 80 dB for M category of vehicle and from 72 to 82 dB for N category of vehicle. In second phase, the noise limits vary from 70 to 78 dB for M category, while for the N category, it varies from 71 to 81 dB. In the third phase, the noise limits vary from 68 to 77 dB for M category and 69 to 79 dB for N category. Europe has the most stringent regulations on tire noise. In 2009, Regulation (EC) 661/2009 and Regulation (EC) 1222/2009 required revised noise limits (more stringent) for tire approval and mandatory tire labeling beginning from November 2012 as shown in Fig. 4.1 [13]. The European commission had proposed to reduce the noise produced by passenger cars, light commercial vehicles, buses, light trucks, coaches, and trucks in hope to reduce road vehicle noise by around 25%. The new regulations have introduced: new test method that better reflects current driving behavior; lower noise limit values; additional sound emission provisions (ASEP), which will be included in type approval procedure as preventive requirements intended to cover driving conditions in real traffic [14]. The tire approval process is in UN/ECE R117/2011 (2011) and UN/ECE R51/2011 (2011) requiring the measurement of rolling resistance, wet grip, and pass-by noise. The National Highway Traffic Safety Administration (NHTSA) in the USA requires tire labeling according to the Uniform Tire Quality Grading (UTQG) standard, including wet braking traction performance, predicted tread wear life, and high temperature endurance without tire noise requirements [15]. This labeling includes a sticker that must indicate: the fuel efficiency class (letter A to G, A—the best, G—the worst); the wet grip class (letter A to G); at this stage the regulation only provides classification for wet grip performance for C1 tires, but intends to introduce a classification for C2 tires and eventually C3 tires once tests are available and the external rolling noise measured value (in decibels). The external rolling noise (N) refers to the noise of the tires experienced outside the car and hence not by the driver or passengers. The limit values which determine the external noise class of tires are determined in Part C of Annex II of Regulation (EC) No 661/2009, and the absolute values vary between C1, C2, and C3 tires. For instance, for C1 tire, the maximum limit is 74 dB(A), while for C2 tire is 73 dB(A) and that for C3 tire is 75 dB [16] (Fig. 4.2).

4.3.2 *Railway Noise Regulations*

In some calculations or measuring methods (e.g., German Schall 03), a general bonus for rail noise is included [17]. Railway noise is often seen as less annoying than other noise sources. The general discount is between 3 and 10 dB in different countries [18]. Some studies recommend that a rail noise bonus is not justifiable both at evening and nighttime, but only eligible during the day and not in the night [17–19]. One such study utilizing a telephonic survey ($N = 1643$) investigated the relationship between railway noise and sleep medication intake and the impact of railway noise events on motility parameters. The study proposed that the legislation using a railway bonus will underestimate the noise impact by about 10 dB (A) L_{den} under the conditions comparable with those in the survey study [19]. The European Noise Policy with respect to rail transport has set the pass-by noise limits as shown in Table 4.6. The standards are described in terms of $L_{\mathrm{Aeq,TP}}$ which is averaged sound level measured during pass-by at 7.5 m from track center and 1.2 m height [20]. The

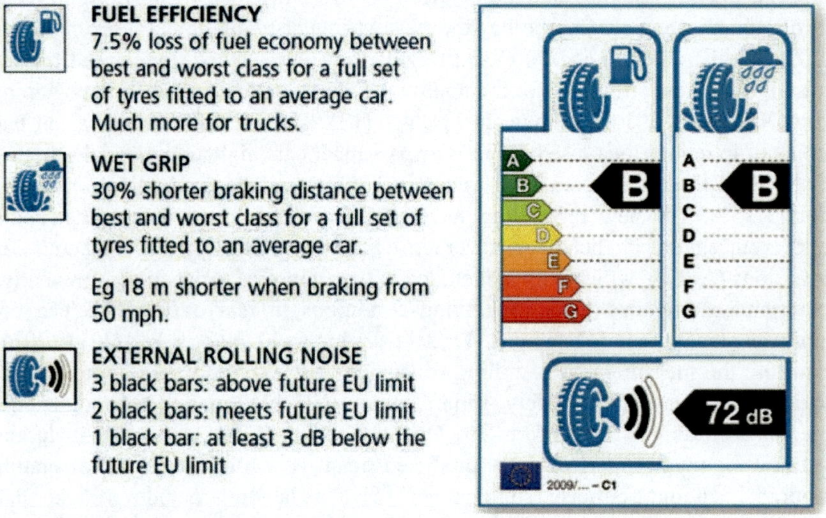

Fig. 4.2 Tire noise labeling as recommended in EU regulation 1222/2009 [13]

Table 4.6 Limits values for pass-by noise (Commission Regulation (EU) No 1304/2014 [20])

Category of the rolling stock subsystem	$L_{Aeq,TP}$ at 80 km/h in dB	$L_{Aeq,\ TP}$ at 250 km/h in dB
Electric locomotives	84	99
Diesel locomotives	85	n.a
EMUs	80	95
DMUs	81	96
Coaches	79	n.a
Wagons	83	n.a

American Public Transit Association (APTA) developed noise and vibration design goals for community exposures as part of its "Guidelines for Design of Rail Transit Facilities." The APTA procedures are based upon the maximum A-weighted sound level (L_{Amax}) of a single vehicle pass-by. The Federal Transit Administration (FTA) methods are based upon the cumulative effect of the pass-by maximum sound level, pass-by duration, number of pass-bys, and times of day of the pass-bys in terms of the nighttime and day-night average sound level [21].

The APTA procedures are based upon the maximum A-weighted sound level (L_{Amax}) of a single vehicle pass-by. TCRP Report 23 of APTA (American Public Transport Authority) provides following values as guidelines for maximum airborne noise from train operations as shown in Table 4.7 [21]. The APTA guidelines for low-density urban residential areas recommend a 40–50 day and 35–45 night L_{50} dB (A) (sound level exceed 50% of the time); high-density urban residential areas as 50–60 day and 45–55 night L_{50} dB(A) and commercial areas as 60–70 L_{50} dB(A). The design-goal sound levels are generally applicable at the near side of the nearest

Table 4.7 Maximum airborne noise from train operations as per APTA Manual [21]

Community area category	Description	Single event maximum noise level design goal in dB (A)		
		Single family dwelling	Multi-family dwelling	Commercial buildings
I	Low density residential	70	75	80
II	Average density	75	75	80
III	High density residential	75	80	85
IV	Commercial	80	80	85
V	Industrial/ highway	80	85	85

dwelling or occupied building or in residential areas at 15 m from the track centerline, whichever is farther [21].

4.3.3 *Aircraft Noise Regulations*

Aircraft noise can be calculated in a number of ways: (1) single event metrics: measurements taken to describe the noise occurring during one noise event, such as an aircraft over-flight, (2) exposure metrics: used to provide a description of the type of noise exposure experienced over a given period of time, (3) supplementary metrics: measurements often used in conjunction with the above, to provide a more meaningful depiction of the potential impact of noise exposure. Many such studies have led to the definition of different noise indicators identified by various noise characteristics: spectrum, impulsiveness, and high levels. The *WECPNL* has been used earlier in some countries like Japan, China, and Korea. The "Störindex *Q*" was also used for aircraft traffic [23]. The "Time Above Threshold, TAX" represents the time over which a threshold level X is exceeded by aircraft noise, during a defined reference time period, and it is expressed in seconds or minutes. The indicator was developed by the Federal Aviation Administration (FAA 2007) like a secondary metric for assessing airports impact; the typical threshold value was set up at 60 or 65 dB [24]. L_{eq} type indicators are widely used and take into account the number of noise events, the noise energy and duration of those events. More recently introduced indicators, such as N70 and PEI/AEI are useful in communicating the impact of aircraft noise but are subject to limitations and do not replace L_{eq} type indicators that remain the basis of aircraft noise impact assessment in the UK and internationally [22]. Day-evening-night sound level, L_{den}, is recommended by the EU as an indicator for the assessment of environmental noise. Standard noise metrics are essentially based on long-term averages and use noise contours to depict areas exposed to different measures of noise, for example L_{den} in EU, L_{Aeq} in the UK or L_{dn} in the USA. In Australia, supplementary noise descriptors have been developed that includes flight movement charts to which was added information on respite, number of noisy events (N70), Person Event Index, and Average Individual Exposure (*AIE*). Often the long-term average noise contours are presented with supplementary metrics such as N70, Person Event Index (*PEI*), Average Individual Exposure (*AIE*), and Time Above (*TA*) to give a more transparent and more complete picture of aircraft noise exposure [25, 26]. The descriptors are primarily used in environmental assessments as communication tools. In some cases, such as the evaluation of annoyance from low noise emissions by aircraft, a good correlation is observed between the SEL and the people's annoyance. Presently, the European Commission (EC 2000) supports the L_{night} indicator arguing that $L_{\max}$ or *SEL* do not consider the number of events, and a common method to calculate an average long-term $L_{\max}$ or *SEL* is still missing [24]. In North America, the N70 is now in relatively common use as a supplementary metric. In Austria, the mediation agreement for the expansion of Vienna Airport uses N65 metrics as one of the agreed controls. Also, in Sweden, the N70 and N80 as formal legislative tools for aircraft noise management is proposed. Thus, the traditional contours and metrics based on long-term averages can be supplemented with shorter-term targeted descriptors to better reflect community experience and associated noise impact. Table 4.8 summarizes the different aircraft noise metrics used as summarized in Environmental Research and Consultancy Department, Civil Aviation Authority, London report [22] and PARTNER-COE-2011-004 report [25].

4.3.3.1 Respite from Aircraft Noise

Respite from aircraft noise is related to the predictable breaks from aircraft overflights. The measurement of respite is by clock hours in Australia: if an hour has no aircraft movements, it is counted as respite but even a single aircraft movement disqualifies that hour. Respite is then presented as a percentage of the total hours of operation at the airport (typically curfew and nighttime periods with no activity are not counted as hours of operation to avoid inflating the figure). A Respite Working Group has been set up at Heathrow airport for providing management and assessment of respite from aircraft noise [27]. The concept is useful and effective strategy for providing break from aviation noise, although an effective communication strategy such as an appropriate noise indicator is to yet to be devised for its successful implementation [28].

4.3.4 Wind Turbine Noise Regulations

The analysis of wind turbine noise is considered an important aspect during the planning of wind farms. L_{Aeq} metrics is the most commonly used metrics for regulating wind farm noise. Some countries use a metric derived from L_{Aeq} metric like L_{r}, L_{dn}, L_{den}, and L_{night}, but include a penalty for times of the day with increased

Table 4.8 Different aircraft noise metrics used

Noise metrics	Empirical formulation	Used in countries
Noise and number index (*NNI*)	$NNI = L_{PN\ max} + 15\log10N - 80$, where *N* is the number of aircraft, $L_{PN\ max}$ is the logarithmic average aircraft noise level (PNdB) heard. An *NNI* of 55 indicates a high annoyance area and $NNI = 35$ indicates the threshold of annoyance.	UK
Noise exposure forecast (*NEF*)	$NEF = EPNL + 10\ \mathrm{Log}_{10}\ (N_D + 16.7\ N_N) - 88$ dB For effective perceived noise level (EPNL), the reference duration is 10 seconds, whereas *SEL*s are normalized to one second.	Canada, Hong Kong, Spain, and Greece
Noise exposure forecast	$ANEF = EPNL + 10\ \mathrm{Log}10\ (N_D + 4N_N) - 88$ dB	Australia
Supplementary metrics	Person-event index (*PEI*) is a sum of the number of people exposed to events exceeding 70 dB(A) sound level multiplied by the number of such events. Average individual exposure is calculated as *PEI*/Total exposed population. Others are N70 contours, time above (TA) metrics for the time duration for which aircraft noise exceeds a certain dB limit	
Kosten index	Until 2003, the Netherlands employed the Kosten Index, Ke; a noise metric founded on L_{Amax}	Netherlands
Weighted equivalent continuous perceived noise level (*WECPNL*)	$WECPNL = L_A + 10\log(N_2 + 3\ N_3 + 10\ (N_1 + N_4)) - 27$ where L_A denote the maximum A-weighted sound pressure levels, N_1 is the number of events registered during nighttime (00:00–07:00) and N_2 is the number of events registered during daytime (07:00–19:00). N_3 and N_4 are the number of events during evening time (19:00–22:00) and nighttime (22:00–24:00).	Defined by ICAO for aircraft noise exposure, Japan, China, and Korea

sensitivity like part of day (L_r in Germany), the evening (L_{den}), and the night (L_{den} and L_{dn}). Besides the metrics L_r, Denmark also uses the metrics L_{pALF} (in the frequency range 10 Hz to 160 Hz) for the low-frequency noise indoors. Possible tone issues associated with wind turbine were addressed in number of governing jurisdictions. The tonal analysis incorporated a 5 dB penalty added to established threshold limits in some cases. Germany is an exception as it uses a penalty of 3 or 6 dB depending upon its distinctiveness. The limit values of these jurisdictions show a range of 35 to 55 dB. Countries like France, Sweden, South Australia, and New Zealand have more stringent noise limits for rural areas with relatively low background noise levels than for residential areas. The Danish limits apply for wind speed of 6 m/s and 8 m/s at a height of 10 m with a maximum limit of 39 dB for residential areas. The German standard imposes a limit of 35 dB(A) for residential areas and recommends that wind turbines should be located at a distance of 750 m to 1000 m away from residential areas [29].

4.4 Construction Noise Regulations

No legislation or regulations on a national level exists regarding construction noise in different countries and is mostly regulated on a local level. Most of the regulations primarily focus transportation noise and ambient noise levels [30–38]. Some countries have general accepted guidelines that provide recommendations for the local authorities [39]. Construction noise and vibration limits are essential, particularly those related to digging and piling operations for underground metro train stations, etc. Roberts (2009) study showed the standard noise limits for general construction at residential receptors in terms of $L_{\mathrm{Aeq,15\ min}}$ [40]. The piling and digging operations for metro stations not only cause vibrations which may induce cracks in building but also produce an airborne sound through building elements, which may cause annoyance to the residents. The noise labeling of construction equipment should be made mandatory as there are sometimes practical complications involved in enforcement of standard noise limits for construction equipment especially in residential areas [41]. However, communication in time about noisy construction activities (kind of activity, duration, part of the day) can be helpful to prevent severe complaints from the neighborhood due to relatively noisy activities [39]. The intrusive criteria for construction noise can be adapted as recommended in the New South Wales (NSW) industrial noise policy (INP) as [42]: The noise management level (in $L_{\mathrm{Aeq,15min}}$) for works during the recommended standard hours (Monday to Friday: 7 a.m. to 6 p.m., Saturday: 8 a.m. to 1 p.m., no work on Sundays or public holidays) is the rating background level (L_{A90}) + 10 dB(A). In cases, wherein the noise levels are higher, all feasible and reasonable work practices to minimize noise impacts need to be implemented. For works outside the recommended standard hours, the noise management level is rating background level + 5 dB(A). The highly noise-affected level of 75 dB(A) L_{Aeq} is the level above which there may be strong community reaction to noise and all measures should be implemented to reduce noise. For protection of the amenity and sleep of people when they are at home (Internal): 40 dB(A) L_{Aeq} (15 min) for evening (6 p.m. to 10 p.m.) and 35 dB(A) L_{Aeq} (15 min) for nighttime (10 p.m. to 7 a.m.) has been recommended [42].

4.5 Noise Labeling of Domestic Appliances

Sound labeling of appliances shall be a helpful tool for creating awareness among manufacturers and customers about increasing the product quality by reducing noise emitted from the domestic appliances [43]. EU passed legislation on noise emission of household appliances includes Noise Directive 86/594/EEC (a community legislation in force regarding measurement and declaration of noise emission from household appliances of any kind) and 92/75/EEC, which introduced the obligation of labeling of appliances. Labeling is considered as a powerful tool as it helps the manufacturers to develop better products and to promote them and also help the consumer to identify the product and provide consumers with accurate information about different aspects. The sound power level is the generally used criteria required for sound labeling. Sound power level measurements, L_{WA}, are measured according to ISO 3744 [44] and ISO 3746 [45]. The European Union Ecolabel scheme re-launched in September 2000 was a voluntary scheme and as such there are no regulations to oblige manufacturers to apply for the label. The noise criteria are below 56 dB(A) for all the domestic products except for washing machine when spinning and vacuum cleaners (recommended level of 76 dB(A)) [42]. The noise limits should preferably be considered in terms of sound power level, which is constant for a given source and is independent of the acoustic environment [46]. Table 4.9 shows the proposed noise limits for domestic appliances and construction equipment at manufacturing stage that may be considered in Indian perspectives [47]. Noise control in home appliances has to be economic, simple, and

Table 4.9 Proposed noise limits for domestic appliances and machines (Garg et al. [47])

Noise source	Maximum permitted sound power level, dB (A)
Refrigerator and freezer	55
Washing machine, spin dry	72
Microwave oven	70
Power transformer	65
Electrical fan (desk fan, stand fan, wall fan, ceiling fan)	70
Vacuum cleaner and dryer	70
Indoor heater	55
Air conditioners (split, indoor)	72
Air coolers	72
Hair dryer	75
Portable diesel generator sets	86
Food processor	90
Food blender	100
Lawn movers, welding generators, compressors	110
Handheld concrete breakers	120
Heavy earth moving machinery	120

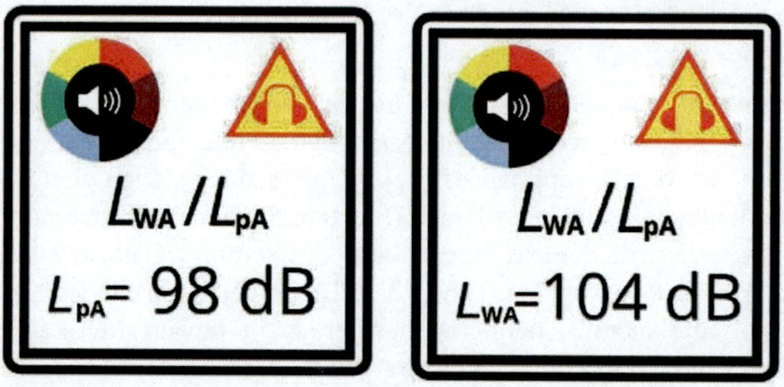

Fig. 4.3 Proposed noise labels in terms of sound pressure level (L_{pA}) or sound power level (L_{WA}) parameter for various machines, domestic appliances, and other noise sources (Garg et al. [47])

easily implemented as there exist a global competition among manufacturers. An overall reduction of radiated noise levels of 6 dB and 10 dB was obtained in the domestic dryer and the vacuum cleaner, respectively, by using the developed jute felt acoustical blanket. Green materials like jute fiber and its composite possesses more significant acoustical attenuation properties as well as flammability properties than glass fiber and as such using jute lined closed enclosure along with jute dissipative silencer can accomplish the desired noise abatement goals [48]. Figure 4.3 shows a typical illustration of noise labeling of noise sources in terms of sound pressure level or sound power level parameter [47].

The proposed noise labels can be used by manufacturers on various appliances and machines that will be very helpful for product standardization and to ascertain compliance with the legal recommended limits. A color code can also be used symbolically for depicting the range of sound levels emitted. These labels can be either in terms of sound power level or sound pressure level parameters as sound power level measurements require a very specialized equipment and environment such as anechoic chamber or a sound intensity probe, which all the laboratories and manufacturers cannot afford. Ultimately, there is a critical requirement for devising specific legal noise limits and procedure for periodic compliance monitoring of vehicular horns, sirens, loudspeakers, and public address systems and other community noise sources in India for noise pollution prevention and control. The legal obligations for the inclusion of noise labels on the appliances and machines similar to that followed for energy consumption shall be a very crucial step in developing better and silent solutions and reducing the health hazards associated with noise pollution.

4.6 Low-Frequency Noise (LFN) Criteria

Low-frequency noise from gas turbines, compressors, heat pumps, ventilation systems, etc. had been always an issue among the researchers for consideration. The assessment and prediction of annoyance due to LFN is not simple. The A-weighting is the commonly used metrics for assessment of noise. However, the A-weighted sound pressure level alone is not successful in assessing the response to LFN (and to infrasound) as A-weighting network significantly decreases the contribution of low-frequency energy in a sound due to the reduced loudness sensitivity of our hearing at low frequencies. Some studies recommend that a minimum (C-A) difference of 20 dB is necessary to ascertain the presence of an LFN problem. However, a greater difference may be permissible at low A-weighted levels, as the (C-A) difference for low levels of background noise may exceed 20 dB without causing complaint. The difference between (C-A) exceeding 15 dB indicates potential infrasound LFN problem situation. (C-A) level difference is an appropriate starting metric for identifying a potential infrasound LFN problem [49]. The "C" weighting is more suitably used for assessment of higher noise levels. As a comparison, at 250 Hz, the C-weighting is zero and at 63 Hz, the weighting is only −0.8 dB [49]. Some studies proposed C-weighted SPLs supplementary to the A-weighted site criteria which are listed in Table 4.10 [50].

In some countries like in Sweden, Denmark, The Netherlands, the United Kingdom, and Poland, exposure criteria are based on frequency analysis in one-third octave bands in various frequency ranges from 8 Hz to 250 Hz. In majority of cases, measured sound pressure levels are compared with the criterion curves. According to Danish method, the $L_{pA,LF}$ (averaged over 10 minutes) in dwellings should not exceed 20 dB in the evening and night period or 25 dB during the day. On the other hand, in Germany, if noise is not tonal, then the maximum acceptable level of $L_{A(10–80\ Hz)}$ is 35 dB and 25 dB for day/night period, respectively [51]. The Swedish recommendation for assessing LFN at workplaces is based on one-third octave band measurements in the frequency range of 25 to 200 Hz and three criterion curves (S40, S60, and S80) representing A-weighted sound pressure level groups 40 dB, 60 dB, and 80 dB. The three measuring methods were proposed by Pawlaczyk-Łuszczyńska M et al. (2006) as follows [51]:

Table 4.10 Maximum allowable dB(C) levels at residential areas to minimize infrasound noise and vibration problem (Hessler [50])

	For normal suburban/urban residential areas, daytime residual level, L_{90} > 40 dBA	For very quiet suburban or rural urban residential areas, daytime residual level, L_{90} < 40 dBA
For intermittent day-time only or seasonal source operation	70	65
Extensive or 24/7 source operation	65	60

(a) Method I: a frequency analysis in one-third octave bands from 10 to 250 Hz.
(b) Method II: the determination of equivalent continuous A-weighted sound pressure level (SPL) in the frequency range of 10–250 Hz and introduction of the penalty for tonal character of noise using the following formula:

$$L_{\mathrm{A10-250\ Hz}} = 10\ \log \sum_{f} 10^{0.1\left(L_f + k_{fA}\right)} + K_1 \mathrm{dB} \tag{4.1}$$

where L_f is the sound pressure level in the one-third octave bands from 10 to 250 Hz, k_{fA} is the relative response of the A-weighting frequency characteristics for the f-th one-third octave band in dB; K_1 is the penalty for tonal character of noise.

(c) Method III: based on an equivalent continuous A-weighted sound pressure level corrected due to presence of low frequencies (K_2) and tonal character of noise (K_1), expressed as:

$$L_{\mathrm{A,LFN}} = L_{\mathrm{Aeq,Te}} + K_1 + K_2 \mathrm{dB} \tag{4.2}$$

where $L_{\mathrm{Aeq,Te}}$ is the equivalent continuous-weighted sound pressure level in dB, K_1 is the penalty for tonal character of noise, and K_2 is the penalty for presence of low-frequency component in the spectrum [51, 52]. A recent study by Alves et al. [53] in Portugal focused on assessing the impacts of low-frequency noise emitted by high-voltage lines and power poles recommended that the measurement and assessment of discomfort due to low-frequency noise should be made with an appropriate filter or without any filter applied, that is, in linear mode. Table 4.11 shows the criterion for the assessment of discomfort due to low-frequency noise based on exhaustive literature review by Alves J A et al. (2018) [53].

Broner (2011) recommended criteria for assessment of LFN based on review of many case histories and the literature. In case, the measuring LFN sound pressure level is fluctuating at least ±5 dB (C), then a penalty of 5 dB (C) to the proposed criterion in Table 4.12 is recommended. The noise levels to be recorded are the maximum and minimum C-weighted sound pressure levels using the fast time weighting, the L_{C10} and L_{C90} levels for the purpose of providing an indication of the level of fluctuation of the LFN. A minimum duration of 3–5 minutes should be used so as to ensure that the low-frequency sound level is sampled accurately [49].

In the BS 4727–3 (1995) and IEC 60050–801:1994 standards, 16 Hz (some 20 Hz) is considered to be the lower limit of the low-frequency region. According to these standards, sound becomes inaudible (infrasonic) below 20 Hz. Leventhall (2007) proposed that the low-frequency range should be considered from 10 to 100 Hz and possibly extended even further to 5 to 200 Hz [54]. Nakamura and Tokita (1981) measured the detection, annoyance, displeasure, oppressiveness, and vibration thresholds in the frequency range from 5 to 700 Hz. The threshold levels beyond 700 Hz and up to 1000 Hz were obtained by using a cubic spline extrapolation method [55]. The PATNER report (2007) validated the Tokita and Nakamura

Table 4.11 Criterion for the assessment of discomfort due to low-frequency noise (Alves et al. [53])

	Sound pressure level (dB)							
One-third octave bands	Swedish	Dutch	Polish	British	HTL	S40	S60	S80
10			80.4	92				
12.5			73.4	87				
16			66.7	83				
20		74	60.5	74	78.5			
25		64	54.7	64	68.7	70	80	90
31.5	56	55	49.3	56	59.5	61	71	81
40	49	46	44.6	49	51.1	54	64	74
50	43	39	40.2	43	44	48	58	68
63	41.5	33	36.2	41.5	37.5	46.5	56.5	66.5
80	40	27	32.5	40	31.5	45	55	65
100	38	22	29.1	38	26.5	43	53	63
125	36		26.1	36	22.1	41	51	61
160	34		23.4	34	17.9	39	49	59
200	32		20.9		14.4	37	47	57
250			18.6		11.4			

Table 4.12 Criteria for assessment of LFN (Broner [49])

Sensitive receiver		Range	Criteria L_{eq} dB (C)
Residential	Nighttime or plant operation 24/7	Desirable	60
		Maximum	65
	Daytime or intermittent (1–2 hours)	Desirable	65
		Maximum	70
Commercial/office/ industrial	Nighttime or plant operation 24/7	Desirable	70
		Maximum	75
	Daytime or intermittent (1–2 hours)	Desirable	75
		Maximum	80

annoyance thresholds as predictors of annoyance due to low-frequency aircraft noise. The study recommends that when high levels of LFN are present, Tokita and Nakamura thresholds with C-weighted sound exposure level (L_{CE}) metric should be used as an indicator of potential for LFN annoyance. C-weighted sound exposure level, L_{CE}, metrics was observed to better correlate with the subjective data than metrics specifically designed to quantify low-frequency noise impact. Figure 4.4 shows the Nakamura's and Tokita threshold for the perception of low-frequency noise. It shows the nominal levels of outdoor sound that would elicit a particular response from occupants inside a house [56].

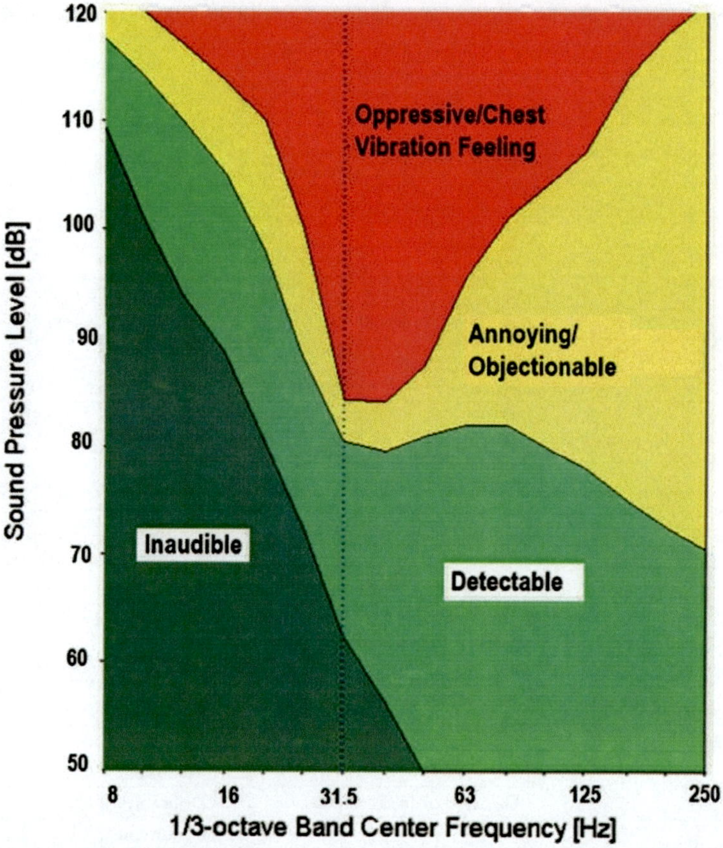

Fig. 4.4 Nakamura's and Tokita threshold for the perception of low-frequency noise (PARTNER project report [56])

4.7 Noise Pollution Index (*NPI*)

As multi-exposure, an example of noise index combining both road and railway sources are proposed in a recent study. This index, named *NPI* (noise pollution index) is based on the use of dose responses curves and an aggregation method proposed by Miedema (discussed elaborately in Chap. 9). It is observed that % *A* (annoyance) reaches a maximum of 63.9% for highest noise level of 75 dB(A) for combined effect of road traffic and railway noise using Miedema (2004) formulation, thus a six-step closed scale is determined by normalizing the percentage of annoyed people (%*A*) exposed to road plus railway sources by a factor 10. Table 4.13 gives an example of a calculation of the *NPI* index obtained from values listed in Table 4.14. *NPI* can be computed for intermediate values of noise levels using equation:

Table 4.13 *NPI* due to simultaneous exposure to road and railway sources, (Heimann et al. [57])

$L_{\text{den-road}}$ (dB)	$L_{\text{den-railway}}$ (dB)						
	45	50	55	60	65	70	75
45	1	1	1	2	3	4	5
50	1	1	2	2	3	4	5
55	2	2	2	2	3	4	5
60	3	3	3	3	3	4	5
65	4	4	4	4	4	4	5
70	5	5	5	5	5	5	5
75	6	6	6	6	6	6	6

Table 4.14 Interpretation of the *NPI* values (Heimann et al. [57])

NPI value	Exposure to noise
1	Very low
2	Low
3	Moderate
4	Pronounced
5	High
6	Very High

$$\%A_{\text{road+rail}} = f\left(L_{\text{den,eq,road+rail}}\right) \tag{4.3}$$

NPI can be used to represent multi-sources exposure on a large-scale map. It can be computed using isophones curves (L_{den} or L_{n}), with a step of 1 dB obtained for each type of sources [57].

There have been some studies such as one reported by Gorai and Pal [58] for an iron ore mining complex that intends to calculate noise impact index using the subjective response of the community. This method is based on a social survey for the assessment of a quantity defined as the sound level weighted population (*SLWP*), which measured the total number of people expected to be highly annoyed by the noise among the entire population and is expressed as [58]:

$$SLWP = \sum [P_i W(L_{dn})_i] \tag{4.4}$$

where P_i is the population distribution function and $W(L_{dn})_i$ is the fractional impact or weighting function that characterizes the severity of the noise impact as a function of day-night sound level L_{dn}. Noise impact index (*NII*) had been calculated as the sound level weighted population divided by the total population that exists in the study zone.

$$NII = \frac{SWLP}{P_{total}} \tag{4.5}$$

The benefit of using such descriptor is that it accounts the subjective response to noise. Also, it can be used for noise impact assessment studies by comparing the existing index and indicative index calculated on the basis of the existing noise levels and the permissible noise levels respectively [58]. The details of *NII* and its possible use in Indian perspectives similar to the Air Quality index (*AQI*) are discussed in detail in Appendix IV of the book.

4.8 Indoor Noise Regulations

The indoor noise standards have been generally recommended in terms of balanced noise criteria curves (NCB). However, different legislations pertaining to the indoor noise exists in different countries. Traditionally, the Nordic countries have specified the indoor noise limits as A-weighted continuous equivalent sound pressure level, $L_{pAeq,24h}$, and to the nighttime limits as $L_{pA,max}$. In Indian scenario, the National Building Codes (NBC, 2016) recommend the indoor noise limits as shown in Table 4.15.

The sound pressure level in rooms varies considerably at low frequencies as due to standing waves, and the variation may be 20–30 dB at low frequencies. For low frequencies, it is thus not adequate to describe sound in a room from measurements in a single position. There seems to be agreement that, for assessment of annoyance, the measurement result should mainly reflect high-sound level areas of the room. The level that is exceeded in 10% of the space of a room, L_{10}, is proposed as a rational and objective target for a measurement method by Pedersen et al. (2007). The national methods used in Sweden and Denmark for low-frequency noise measurements use measurements near corners of the floor plan [60]. In Brazil, noise levels according to NBR 10152:198 recommend a 35–45 dB(A) L_{eq} for bedrooms and 40–50 dB(A) L_{eq} for the living rooms [61]. The National Building

Table 4.15 Acceptable indoor noise levels for various buildings (NBC [59])

S. No	Location	Noise level in dB(A)
1	Auditoria and concert halls	20–25
2	Radio and TV studios	20–25
3	Cinemas	25–30
4	Music rooms	25–30
5	Hospitals	35–40
6	Apartments, hotels, and homes	35–40
7	Conference rooms, small offices, and libraries	35–40
8	Court rooms and class rooms	40–45
9	Large public offices, banks, and stores	45–50
10	Restaurants	50–55

Codes of India recommend that areas that are exposed to 80–90 dB(A), the building should be located at a distance of about 30 m from the road, but a distance of 45 m or more where possible, should be aimed at for greater relief from the noise.

4.9 Implications of Environmental Noise Regulations and Policies

4.9.1 International Scenario

The European Environment Agency (EEA) published a report titled "Noise in Europe 2014" to assess the scale of the noise challenge, the actions that are being taken and may need to be considered in the future [62, 63]. An estimated 125 million people are affected by environmental noise levels above 55 dB L_{den}, almost 20 million adults are annoyed and further eight million suffer sleep disturbance due to environmental noise exposure. Road traffic is the most dominant source of environmental noise. The thresholds for noise mapping and action planning as required by the END are 55 dB L_{den} (day, evening, and night periods) and 50 dB L_{night} [64]. The END describes quiet area as: an area delimited by the competent authority, that is, not exposed to a value of L_{den} greater than a certain value set by the Member State, from any noise source [64].

The recent report of International Institute of Noise Control Engineering (I-INCE) [65] examined options for supplementing or replacing the day-evening-night sound level (L_{den}) by other metrics as a predictor of noise impacts, especially annoyance from transportation. *Community Tolerance Level* and *Soundscape* are two such non-acoustic measures that appear to be related, may be used to characterize community response to transportation noise. Community Tolerance Level includes the non-A-weighted acoustic variables and the non-acoustic variables by combining the totality of the variables into the one-number descriptor, CTL. A recently adopted International Standard ISO 12913-1:2014 (pertaining to definition and conceptual framework) has as its purpose the enabling of a broad international consensus on the definition of soundscape. This new ISO standard distinguishes the perceptual construct (soundscape) from the physical phenomenon (acoustic environment), and clarifies that soundscape exists through human perception of the acoustic environment. The fact that CTL represents community contexts with a single, decibel-denominated number shows that a soundscape can be described with just one parameter that can be adapted to prospectively evaluate CTL in a community setting [64]. The standard defines the term "soundscape" as acoustic environment as perceived or experienced and/or understood by a person or people, in context, soundscape is different from acoustic environment, as the former represents the perceptual construct, while the latter refers to a physical phenomenon. Most important, soundscape exists through human perception of the acoustic environment. The soundscape approach focused with the objective to create (urban) environments of high acoustic quality is creative [66]. Community noise ordinances and national noise policies are currently being revised in many parts across the globe. They need to be coordinated and consistent with the regulations at all other levels, namely regional and national for achieving an effective enhancement in the quality of the environment [67]. Community noise criteria and goals must be used as tools in land-use and transport planning [68]. Greater harmonization of road traffic noise policy is a desirable goal for future prospects [69]. In Switzerland, 1.6 million people, that is, every fifth inhabitant, are exposed to harmful or disturbing road traffic noise during the day; 70,000 people are affected by excessive railway noise and 65,000 by excessive aircraft noise. Traffic noise incurs the annual costs of CHF 1.8 million approximately. The federal authorities have planned for setting the noise-reduction programs with an estimated cost of over CHF 4 billion [70]. Not only for the developed nations, the developing countries have become increasingly aware of noise and have implemented new policies for noise abatement and control in recent times [63].

4.9.2 National Perspective

Noise policy has both global and local aspects. The noise pollution control in a community can only be effectively accomplished by a shared responsibility approach. Different levels of responsibilities can be developed: local, regional, national, and federal. Community noise is a local problem. Noise policy has both global and local aspects. A noise policy that only covers local issues, or one that only considers the global aspect, cannot be considered inclusive and is unlikely to achieve the desired results [71]. Noise regulations must then consider all levels of action, sharing the necessary responsibilities for each type of noise source and thus the noise regulations or policies at different levels must be coordinated and consistent. In Indian scenario, the thrust areas on development of master plan are: mixed land-use, urbanizing land-use, and transportation [72]. Measures to reduce noise at source are generally more cost effective than those designed to hamper its propagation [73]. One such Danish case study clearly illustrates that porous asphalt is far more

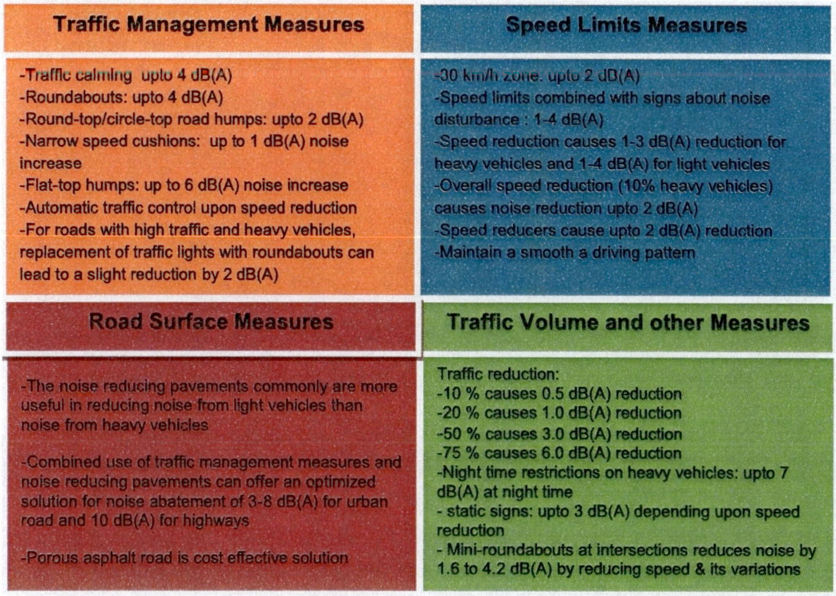

Fig. 4.5 Summary of measures for reducing traffic noise based on review of studies [81–87]

cost effective than home insulation or noise barrier construction, which are 3–10 times more expensive [74]. The traffic management measures have a positive impact not only on the noise reduction but also on air quality and road safety. A 50% reduction of the traffic volume results in a 3 dB(A) reduction in noise levels, regardless of the absolute number of vehicles, providing that speed, traffic composition, and driving pattern is unchanged [75]. In one such study reported by Ausejo et al. (2011) [76], the 50% reduction in heavy vehicles, which consist of a maximum of 10% of all traffic volume, decreases the noise levels by up to 2 dB(A),whereas a 50% reduction of all the vehicles decreases the noise by up to 3 dB(A). The best long-term solution can be realized by combining the global noise abatement measures (i.e., traffic management) and local measures (i.e., noise barriers). The average cost of a noise barrier is around €300 per m^2 depending on its construction and materials used. This is around €2.4 million for a barrier 4 m high and 1 km along both sides of a road [77]. Of these measures, the cost effectiveness of quieter tires is greater as several studies reported that tire/road noise reduction comes at zero cost [78, 79]. Financial incentives may need to be introduced to speed up the development of low-noise tires. One means of doing so may be to levy a tax on tires or introduce some other type of financial incentive proportional to assigned noise level [80]. The various such measures are assimilated in Fig. 4.5 based on the summary from various studies focused on the objective of reducing traffic noise (Ellebjerg, 2005 [81]; Bendtsen et al., 1998 [82]; Anderson, 2003 [83] and FEHRL report 2006 [84]; DRI report 2004 [85]; Campolieti and Bertoni, 2009 [86] and Hyden and Várhelyi, 2000, [87]).

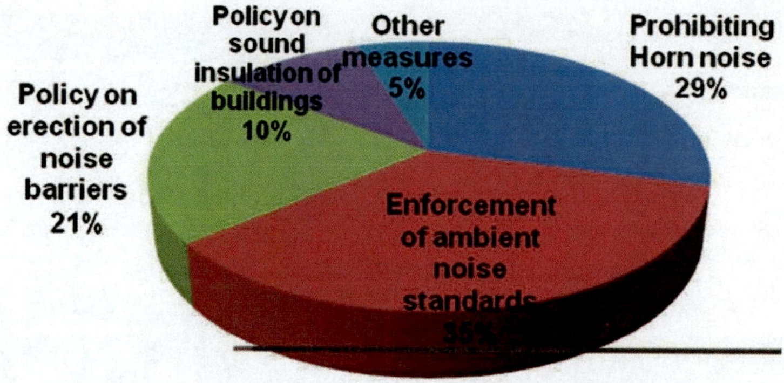

Fig. 4.6 Survey results on prominent legal measures for environmental noise control

Thus, various efforts in coordination with many legal and state departments and relevant ministries to develop a national plan shall be a great step to combat noise pollution [88]. The various policy issues and legal measures such as noise standards/ limits for various noise sources; effective enforcement of ambient noise standards; policy for noise screening by installing noise barriers across the sensitive receptors; implementation of an effective traffic management policy in conjunction with implementation of soundscape approach (discussed in details in Chap. 7) for urban development shall be indispensable for reducing noise pollution. Figure 4.6 shows the results of a survey conducted by Garg et al. (2019) among researchers in noise control about their views for the prominent legal measures for noise pollution control in Indian cities [89]. Policy on prohibition of vehicular honking, enforcement of ambient noise standards, and installation of noise barriers for sensitive receptors was endorsed by the majority.

Noise labeling of domestic appliances and construction equipment; type approval noise testing of vehicles periodically like emission testing; imposing noise limits for construction activities; and framing a comprehensive land-use policy considering noise issues can be some proactive steps in reducing noise pollution. Aircraft noise is one of the most significant local objection to airport expansion and construction. A balanced approach is necessary for controlling the aircraft noise. Establishment of Noise Pollution Grievances Redressal and Monitoring system for registering noise complaints and a public helpline number for noise complaint cell such as that established by Government of Delhi, India (Helpline no. 155271, website: *www.ngms.delhi.gov.in*) shall be a great step. A noise pollution control committee and a Nodal officer of noise pollution for Delhi police for mitigating noise pollution in Delhi city had been also formulated in compliance with the directions of the Hon'ble National Green Tribunal (NGT). The development of guidelines w.r.t noise for construction sites had been already initiated by Delhi Pollution Control Committee (DPCC), Delhi in January, 2022. Also, the Central Pollution Control Board (CPCB) of India is focused towards developing the guidelines for noise emissions associated with loudspeakers and public address systems for mitigating the noise pollution. The National Committee on Noise Pollution Control (NCNPC) is the apex committee in India that looks into the various aspects of noise pollution control. Thus, the future policy reviews need to ensure a greater harmonization of noise policies across different states, and should ensure alignment with the current state of knowledge. Above all, the awareness of the public and stakeholders is a key component in the prevention and control of community noise pollution. CSIR-National Physical Laboratory, New Delhi recently organized a one day International workshop in December, 2021 on Noise Pollution Monitoring, Mapping and Control to sensitize the issue of noise pollution and associated health hazards with a theme on Noise Mapping and Control. A national policy framework on noise pollution control in conjunction with noise awareness campaigns shall be a proactive step for mitigating the noise pollution and associated health hazards. Figure 4.7 shows the proposed illustrative plan for a National Planning Policy Framework (NPPF) for Environmental Noise Management and Control to effectively control noise pollution

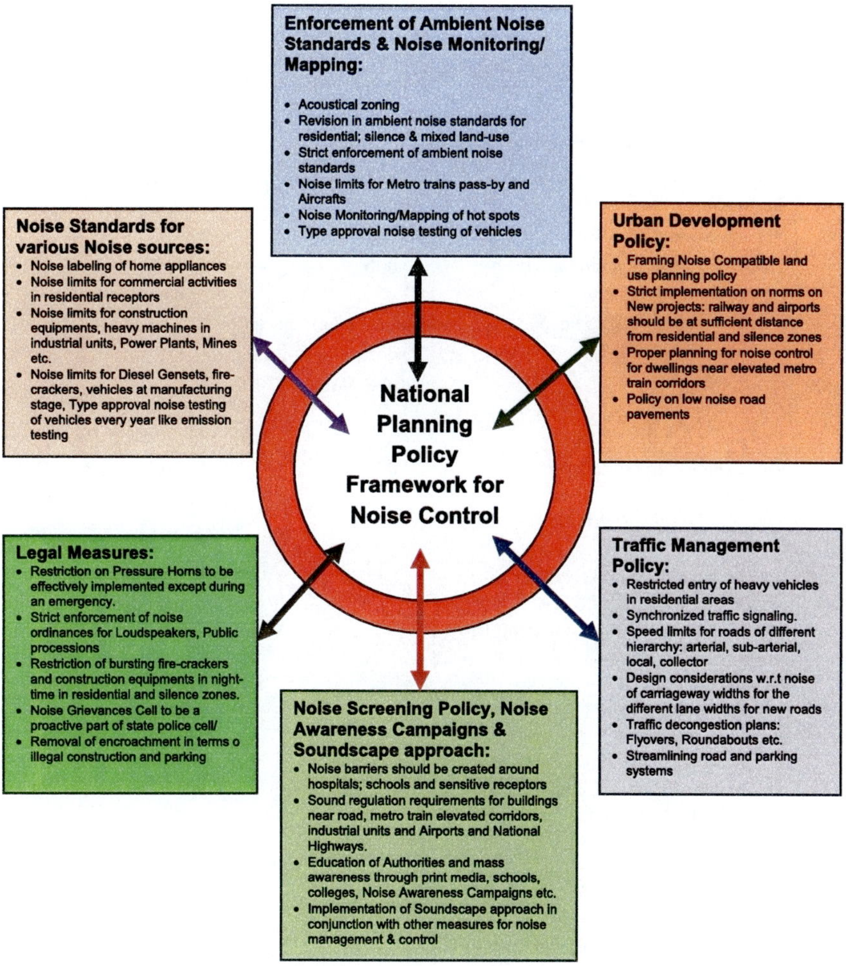

Fig. 4.7 Proposed plan for a national planning policy framework for environmental noise management and control in Indian perspective to effectively control noise pollution in urban cities and proposed Smart Cities in next two decades

4.9.2.1 Compensation for Violation of Noise Pollution Rules

A recent scale of compensation was formulated and enforced by CPCB in March 2020 in directions of the Hon'ble National Green Tribunal (NGT) of India for the various noise sources. Table 4.16 enlists the scale of compensation (in Indian Rupees) and actions to be taken by the designated authority against the defaulters for violation of Noise Pollution (Regulation and Control) Act, 2000. The time limits have been prescribed for the bursting of firecrackers on various occasions as: Diwali:

Table 4.16 Scale of compensation (in Indian Rupees) for violation of noise rules [90] (1 US $ = 75.29 Indian Rupees as on dated 15.02.2022)

Sl. no.	Violation of noise rules	Compensation (in Indian rupees) to be paid by defaulter		Action to be taken designated authority
1.	Use of loud speakers/public address system	10,000		Seizure
2.	Violation w.r.t generator set norms			
(a)	DG sets more than 1000 KVA	1,00,000		Seizure of DG sets
(b)	DG sets 62.5 to 1000 KVA	25,000		
(c)	DG sets up to 62.5 KVA	10,000		
3.	Violation of use of sound-emitting construction equipment	50,000		Seizure/sealing of equipment
4.	Bursting of firecrackers beyond the prescribed time limit prescribed			
(a)	Violation by individual/household	Residential/ commercial/ mixed zone	Silence zone	
		1000	3000	
(b)	Violation during possession a. Public rallies b. Barat during marriage religious event	10,000	20,000	Fine to be paid by organizer
(c)	First violation within a fixed premises: Function organized by RWA, marriage function, public function, institution function, banquet hall, open ground functions	20,000		Fine to be paid by organizer and owner of the place
(d)	Second violation within a fixed premises: Function organized by RWA, marriage function, public function, institution function, banquet hall, open ground functions	40,000		
(e)	More than two violation within a fixed premises: Function organized by RWA, marriage function, public function, institutional function, banquet hall, open ground functions	1,00,000 and action under EP act		Sealing of premises

8.00 p.m. to 10.00 p.m.; Gurpurb: 4.00 a.m. to 5.00 a.m. and 9.00 p.m. to 10.00 p.m. and New Year and Christmas: 11.55 p.m. to 12.30 a.m [90].

The degree of compliance for compliance testing of firecrackers for noise emissions as recommended is minimum 90%, that is, not more than 1 out of every 10 samples tested shall fail. If the degree of compliance is less than 90%, the lot shall be considered to have failed. In India, the West Bengal Pollution Control Board (WBPCB) has a stringent standard, that is, the approved sound limit is 90 dB (AI) at 5 m distance from point of bursting [91]. The European Directive 2013/29/EU recommended a maximum impulsive sound level of 120 dB(A) for a safe distance of 1 m for fireworks which present a very low hazard and are intended for use in confined areas and 120 dB(A) for a safe distance of 15 m for fireworks, which present a medium hazard, which are intended for outdoor use in large open areas [92]. Eventually for the color and light-emitting fireworks developed in recent times, the stringent limit is proposed to be 90 dB(AI) at 4 m from the point of bursting for avoiding any health impact especially to the children and older generation. In a recent verdict in October 2018, the apex court of India advocated the use of green firecrackers only with reduced air emissions particularly the suspended particulate matter. Consequently, the Council of Scientific and Industrial-National Environmental Engineering Research Institute (CSIR-NEERI) has taken the initiative of developing green crackers with reduced noise emissions in collaboration with Ministry of Environment, Forest and Climate Change (MoEF and CC) and other sister CSIR laboratories in India. Green crackers have explicitly shown reduction in residue, smoke, and quantity of hazardous metals without compromising on the performance [93].

The violation of noise rules guidelines recommended that the sound-emitting firecrackers shall not be burst in silence zone or during the nighttime in each zone. Also, the sound-emitting construction equipment shall not be operated or used during the nighttime in residential areas or in silence zones.

4.10 Environmental Noise Guidelines and Health Endpoints

The effect of environmental noise on human health has been a serious concern globally. Consequently, the link between the exposure and the outcome has to be well defined for prevention and well-being. The WHO Regional Office for Europe has recently developed some guidelines for protecting human health from exposure to environmental noise originating from various sources [94]. Table 4.17 summarizes the threshold for various noise sources and health effects as recommended for the European region [7, 94, 95]. The Guidelines Development Group for WHO Regional Office for Europe has recommended using the term "environmental noise" instead of community noise that account for all noise sources except occupational noise source in workplaces. The health effects considered are: annoyance; cardiovascular and metabolic effects; cognitive impairment; effects on sleep; hearing impairment and tinnitus; adverse birth outcomes; and quality of life, mental health,

Table 4.17 Summary of threshold or limits for various noise sources and health effects as recommended for the European region [7, 94, 95]

Noise sources and health effects	Acoustic indicator	Threshold or limit, dB	Reference
Noise sources			
Road traffic noise	L_{den} and L_{night}	53 dB L_{den} and 45 dB L_{night}	WHO Environmental Noise Guidelines for the European Region (2018) [94]
Railway noise	L_{den} and L_{night}	54 dB L_{den} and 44 dB L_{night}	
Aircraft noise	L_{den} and L_{night}	45 dB L_{den} and 40 dB L_{night}	
Wind turbine noise	L_{den}	45 dB L_{den}	
Leisure noise	Yearly average $L_{Aeq,24h}$	70 dB $L_{Aeq,24h}$	
Health effects			
Annoyance disturbance	L_{den}	42 dB	European environmental Agency technical report no. 11/2010 [95]
Hypertension	L_{den}	50 dB	
Ischemic heart diseases	L_{den}	60 dB	
Reported health	L_{den}	50 dB	
Reported awakening	$SEL_{indoors}$	53 dB	
Self-reported sleep disturbance, environmental insomnia	$L_{night,\ outside}$	42 dB	Night Noise Guidelines for Europe (2009) [7]
Biological effects (EEG awakening), sleep quality	$L_{Amax,\ inside}$	35 dB	

and well-being. The grading of recommendations, assessment, development, and evaluations (GRADE) approach was followed for assessing the quality of evidence [95, 96].

The nighttime noise may have more of an impact on cardiovascular health than daytime noise, and noise exposure at night is a particular problem because sleep is disturbed [97, 98]. It is evident that future longitudinal studies are essentially required to establish comprehensive evidences between the various health end points and the noise exposure for prevention and well-being. For instance, in a study reported by van Kempen et al. [99], a comprehensive evidence was established for a relative risk (RR) of 1.08 (95% CI: 1.01–1.15) per 10 dB (L_{den}) for the association between road traffic noise and the incidence of ischemic heart diseases (IHD).

4.11 Noise-Induced Hearing Loss (NIHL) and Damage Risk Criteria (DRC)

The exposure to sound levels above 85 dB initially manifests a temporary hearing loss known as temporary threshold shift (TTS) that may have a faster resolution within 10–15 days of exposure. However, a repeated or sustained exposure of noise to hair cells and associated nerve fibers leads to degenerative changes and TTS becomes permanent threshold shift (PTS) or sometimes called as noise-induced PTS

(NIPTS) [100]. The exposure to sudden impulsive noise is more detrimental than exposure to the steady-state noise [101]. The fact that frequencies around 4 kHz are most affected by noise is most likely due to resonance frequency of the outer ear/ear canal as well as the mechanical properties of middle ear [102]. With further noise exposure, the notch gets deeper and wider eventually involving frequencies such as 2 kHz, 1 kHz, and 0.5 kHz [103, 104]. The left ear is more adversely affected by noise, even in the presence of symmetric noise exposure [105]. The damage risk criteria for impulse sounds show that 10 impulses the peak levels should not exceed 149 dB (linear, peak) at the ear. The A-weighted impulse level should not exceed 125 dB (A, imp) at the ear for 10 impulses [106]. The Research and Technology Organization of North Atlantic Treaty Organization (NATO) criteria recommended that for both impulse noise from rifles, and blast from explosions and large-caliber weapons, the critical noise exposure level that should not be exceeded is [107]:

- For impulses from rifles this critical level is 116 dB(A), SEL *per impulse*, measured in free field at the location of the ear.
- For impulses from blasts, this critical level is 135 dB(A), SEL *per impulse.*
- Applying equal energy principle to impulse sounds from blasts, $N > 100$, N impulses together, the exposure limit becomes 143 dB(A), SEL or $L_{eq} = 98.4$ dB(A).

Price and Kalb developed an Auditory Hazard Assessment Algorithm for Humans (AHAAH) model. Exposure to 500 ARUs (auditory risk units) is considered unsafe and 200 ARUs is considered safe [108]. Doses greater than 500 ARUs produce permanent hearing loss. The limit should be reduced to 200 ARU for daily occupational exposures [109]. The unweighted 200 Pa (140 dB) is used in European Union (EU) directive 86/188 and ISO 1999-1990 regardless of the number of impulses [110]. The lower frequency noise exposures are typically less hazardous than that make a sharp crack [111]. However, the relationship between the damage potential and source is a complex function of frequency content, pressure, and time and thus the variety of approaches may be considered until the most conservative criterion are established [111, 112].

4.12 Occupational Noise Regulations

Occupational noise exposure has been analyzed to be a serious health hazard as it may be associated with an increased incidence of hyperglycemia, with the highest risk observed at 31.5 Hz [113]. It is plausible that long-term noise exposure may produce adverse changes in blood glucose [114] and other serious health hazards like hypertension and other cardiovascular effects [115]. Thus, the occupational noise regulations and limits are very important for avoiding the associated health hazards. The commonly accepted 8-h average A-weighted sound pressure level (permissible exposure limit, PEL) is based on the exposure during a typical daily work shift. The relationship between noise level and exposure duration is commonly known as the exchange rate. The “dose-trading relation” or “trading ratio” is the number of decibels by which the sound pressure level may be decreased or increased for a doubling or halving of the duration of exposure. There is no international consensus regarding the exchange rate and it varies between countries from 3 to 5 dB. A PEL of 85 dB(A) and the 3-dB exchange rate are currently used by many countries [116]. For instance, in case of a 3-dB exchange rate, noise exposure time is halved for each 3-dB increase in noise level; while for a 5-dB exchange rate, the exposure time is halved for each 5-dB increase. The National Institute for Occupational Safety and Health (NIOSH) recommends that a hearing conservation program be established for workers exposed to an 8-h time-weighted average of 85 dB (A) [117]. Some studies had reported that 3 dB overestimates the risk of NIHL and that 5 dB is a better fit, but also overestimates the risk, particularly for exposures above 100 dB (A) [118]. In the European Union, the occupational noise based on the directive 2003/10/EC, the noise hazardousness is described in three levels: lower action level, upper action level, and the noise limit level. Lower action level is described as the occupational environment if the $L_{Aeq,8h}$ reaches 80 dB(A) or a peak pressure value of 135 dB(C). If noise reaches a $L_{Aeq,\ 8h}$ of 85 dB(A) or has a peak sound pressure value of 137 dB(C), it is described as an upper action limit level. Finally, the ceiling threshold for acceptable noise exposure is an $L_{Aeq,\ 8h}$ of 87 dB (A) or a peak sound pressure limit of 140 dB(C) [119]. In India, the occupational permissible exposure limit for 8 h time weighted average is 90 dB(A). NIHL has been a compensable disease since 1948 in India. It was only in year 1996 that the first case got compensation [120]. It may be noted that occupational noise regulations do not protect completely from NIHL, but rather strike a balance between hearing conservation and economic development [121]. Thus, noise exposures from the occupational and other noisy sources are potential risk factors and should be controlled for public health interventions especially in urban environment [122].

4.13 NIHL Due to Emergency Vehicles Sirens and Horns

Occupational noise had been primarily observed to be the most common cause for hearing loss among the individuals exposed to higher noise exposure levels across the globe. Sixteen percent of the disabling hearing loss in adults (over four million DALYs) worldwide is attributed to occupational noise, ranging from 7% to 21% in the various subregions [123]. The noise-induced hearing loss is evident from many studies for those who have been driving the emergency vehicles with sirens and horns, etc. More recently in May 2021, the Manipur state government in India had called for muting sirens of ambulances in the state in a bid to reduce anxiety and panic amid the COVID-19 pandemic second wave in April and May 2021. Sirens should be activated only if traffic is high and roads are blocked. Hearing handicap is usually denoted as an average hearing threshold level (HTL) of greater than 25 dB (A) for both ears at selected frequencies [124]. A person is considered functionally deaf with 85–90 dB(A) hearing loss [125]. There had been many studies in past few years in India that had evidently reported the noise-induced hearing loss to those experiencing high noise exposure levels. A study conducted among auto drivers in central Bangalore using a structured questionnaire and audiometric evaluation reported the prevalence of hearing loss among the auto rickshaw drivers was found to be as high as 40% [126]. Another cross-sectional study that included 30 loco pilots and 50 control subjects shows mild hearing impairment in loco pilots exposed to higher noise levels. A positive correlation between hearing loss and occupation was suggested [127]. A similar audiometric survey of 1000 Indian Air Force (IAF) personnel of various trades, age, and exposure groups carried out to analyze the prevalence and severity of noise-induced hearing loss (NIHL) revealed an overall incidence of NIHL of 22.9% in the personnel with a higher prevalence among the technical trades (26.18%) as compared to non-technical trades (12.5%) [128]. One such study focused on ascertaining the hearing impairment of professional drivers (90 males) in Kolkata city of India indicated that the audiometric threshold levels were the highest for drivers with more than 10 years of noise exposure, lowest for office workers and intermediate for drivers with less than 10 years of noise exposure. Evidently, the periodic checking of the hearing loss of professional drivers and associated staff those who are exposed to noisy environments is imperative [129]. The bus drivers had been observed to be more vulnerable to the auditory and non-auditory ill effects of prolonged exposure to high intensity sound levels that resulted in deterioration of hearing capacity and increased in blood pressure [130]. Even for those involved in agriculture sector particularly the farm workers driving the tractors producing high sound levels more than 90 dB(A) had been observed that the tractor driving significantly increased the hearing threshold levels and the risk is even higher for the tractor drivers with $\geq$15 years of driving experience [131]. Also, the disc jockeys (DJs) players are at higher risk for developing occupational NIHL [132]. Thus, it is vital that periodic checking of hearing loss has to be conducted for such exposed individuals and preventive measures such as use of ear muffs should be widely exercised for preventing the hearing impairment. Also, the strict legislations and their enforcement is the need of hour for safeguarding the auditory health of exposed workers in India.

4.14 Conclusions

The chapter presents various standards, limits and regulations for all the noise sources that contribute to the environmental noise pollution. The various Indian and international standards are discussed in detail. The source-based sound standards are very crucial for preventing the noise pollution and developing new designs and innovations for reducing noise emissions and exposure. Occupational noise exposure also is a very serious concern and has been a prime factor in contributing to the noise-induced hearing loss. In Indian scenario, the source-based sound standards are yet to be developed for various sources such as loudspeakers, public address systems, domestic appliances, mining machinery, power transformers, vehicular horns/sirens, and agricultural and industrial machinery. In case of loudspeakers and public address systems, sound limiters have been employed to control the noise emissions in some parts of India. However, the concept of sound limiters find very limited usage as it can easily be tampered and it is difficult to set limits particularly in the absence of source standards for loudspeakers and public address systems. Evidently, there is huge need of standards for all the noise sources and development and formulation of the noise standards in conjunction with strict enforcement strategy that shall be a vital step for controlling the noise emissions. Also, the type approval testing facilities and laboratories needs to be developed and enhanced further in various parts of the country for catering to the growing needs of industry and pollution control bodies for the successful enforcement of the source-based noise standards.

References

1. den Boer LC, Schroten A (2007) Traffic noise reduction in Europe, CE Delft, March 2007
2. TNO (2011) VENOLIVA-Vehicle Noise Limit Values, Final report for EC DG Enterprise and Industry
3. Noise Pollution (Regulation and Control) Amendment rules (2010). http://moef.gov.in/rules-and-regulations/environment-protection/noise-pollution/
4. Central Pollution Control Board, India. website: http://cpcb.nic.in
5. Quiet Please (2010) Better health through strong EU regulation of road and rail traffic noise. T&E – European Federation for Transport and Environment, Belgium, May, 2010
6. Berglund D, Lindvall T, Schwela D (1999) Guidelines for community noise. World Health Organization, Geneva
7. World Health Organization (2009) Night noise guidelines for Europe. WHO Regional Office for Europe, Copenhagen
8. NSW Road Noise Policy (2011) Department of Environment, Climate Change and Water. NSW, Sydney
9. Burgess M, Macpherson J (2016) Overview of Australian road traffic noise policy 2016. Acoustics Australia 44:227–234
10. I-INCE Technical Study Group on Community Noise (2011) Environmental Noise Impact Assessment and Mitigation. I-INCE Publication Number: 11–1, TSG 6, March, 2011
11. Lang WW, ten Wolde T (2006) A global approach to noise control policy. Noise Control Eng J 54(5):290–346
12. International Standard Organization, ISO 362 (2015) Measurement of noise emitted by accelerating road vehicles-Engineering method-Part 1: M and N categories
13. Introduction to the Tyre Labelling regulation (2013) Practical guidance for UK suppliers. Tyre Industry Federation, April
14. Miloradović D, Glišović J, Lukić J (2017) Regulations on road vehicle noise-trends and future activities. Mobility Vehicle Mech 43:59–72
15. Tan L (2018) Literature review of tire-pavement interaction noise and reduction approaches. J Vibroeng 20(6):2424–2452
16. Review study on the Regulation (EC) No 1222/2009 on the labelling of tyres (2016) European Commission, DG ENER C.3, Belgium, March, 2016
17. Reducing Railway Noise Pollution (2012) Directorate general for internal policies, IP/B/TRAN/FWC/2010–006/LOT4/C1/SC2, Policy Department Structural and Cohesion Policies, European Parliament, Brussels

18. Möhler Z (2010) Noise bonus for railways? ZEUS GmbH. Study on behalf of Federal Environment Agency
19. Lercher P, Brink M, Rudisser J, Van Renterghem T, Botteldooren D, Baulac M, Defrance J (2010) The effects of railway noise on sleep medication intake: results from the ALPNAP-study. Noise Health 12(47):110–119
20. Commission Regulation (EU) No 1304/2014 of 26 November 2014 on the technical specification for interoperability relating to the subsystem 'rolling stock-noise' amending Decision 2008/232/EC and repealing Decision 2011/229/EU. L 356/421 Official Journal of the European Union
21. Staiano MA (2003) Noise-impact estimates per FTA and APTA criteria. J Sound Vib 267: 407–418
22. Jones K, Cadoux R (2009) Metrics for aircraft noise. Environmental research and consultancy department (ERCD) report, 0904, January
23. Survey of Legislation (2009) Regulations and Guidelines for control of community noise, Final Report of the I-INCE Technical Study Group on Noise Policies and Regulations. TSG 3, I-INCE Publication: 09–1, July
24. Pronello C, Camusso C (2012) A review of transport noise indicators. Transport Rev Trans Transdics J 32(5):599–628
25. More S R (2011) Aircraft noise characteristics and metrics. A PARTNER project 24 report, PARTNER-COE-2011-004, July 2011
26. Porter N, Knowles A (2016) Key issues in aviation noise management. Acoustics Bull 2016: 52–57
27. Porter N, Knowles A, Fisher N, Southgate D (2014) The next generation of supplementary aviation noise metrics and their use in managing aviation noise. Internoise 2014, Melbourne, Australia, 16th to 19th November, 2014
28. Porter N, Norman R (2017) Respite from aviation noise. 12th ICBEN congress on noise as a public health problem, 18th to 22nd June, 2017, Zurich
29. Koppen E, Fowler K (2015) International Legislation for wind turbine noise. Euronoise, 2015, 31st May to 3rd June, 2015, Maastricht, 321–326
30. Garg N, Sinha AK, Gandhi V, Bhardwaj RM, Akolkar AB (2016) A pilot study on establishment of ambient noise monitoring network across the major cities of India. Appl Acoust 103: 20–29
31. Garg N, Kumar A, Saini PK, Maji S (2014) A retrospective view of ambient noise standards in India: status and proposed revision. Noise Control Engg J 63(3):266–278
32. U.S Department of Housing and Urban Development (1984) Environmental Criteria and Standards, 24 CFR Part 51, V 12 July, 1979, amended by 49FR 880, 6 January, 1984
33. Federal Interagency Committee on Noise (FICON) (1992) Federal agency review of selected airport noise analysis issues. Washington DC, August, 1992
34. Mestre V, Schomer P, Fidell S, Berry B (2011) Technical support for Day/Night average sound level (DNL) replacement metrics research. DOT/FAA/AEE/2011–02, June, 2011
35. Schomer PD (2005) Criteria for assessment of noise annoyance. Noise Contr Eng J 53(4): 132–144
36. ISO 1996-1 (2003) Acoustics-description, measurement and assessment of environmental noise – part 1: basic quantities and assessment procedures
37. Schomer PD (2002) On normalizing DNL to provide better correlation with response. Sound Vibrat:14–23
38. Central Pollution Control Board (2018) Status of Ambient noise level in Indian in 2017. CPCB report, NANMN/05/2017, August, 2018
39. Granneman JH (2013) Construction noise: overview of regulations of different countries. Internoise 2013, 15th to 18th September, 2013, Innsbruck, Austria
40. Roberts C (2009) Construction noise and vibration impact on sensitive premises. In Proceedings of Acoustics 2009, 23rd to 5th November, 2009, Australia

41. Garg N, Maji S (2016) A retrospective view of noise pollution control policy in India: status, proposed revisions and control measures. Curr Sci 111(1):29–38
42. Interim Construction Noise Guideline (2009) Department of Environment and Climate Change NSW, Sydney, DECC 2009/265. https://www.epa.nsw.gov.au/-/media/epa/corporate-site/resources/noise/09265cng.pdf. Accessed on 18 Feb 2021
43. Maluski S, Churchill C, Cox T J (2004) Sound quality testing and labeling of domestic appliances in the U.K. Internoise 2004, August 22nd -25th, 2004, Czech Republic
44. ISO 3744 (2010) Acoustics-Determination of sound power levels and sound energy levels of noise sources using sound pressure-Engineering methods for an essentially free field over a reflecting plane
45. ISO 3746 (2010) Acoustics-Determination of sound power levels and sound energy levels of noise sources using sound pressure-Survey method using an enveloping measurement surface over a reflecting plane
46. Tian J, Jiao F, Lu Yadong (2011) The latest development of low noise household appliances in China, proceedings International Institute of Noise Control Engineering, 5th to 6th July, 2011, Paris
47. Garg N, Chauhan BS, Singh M (2020) Realization and dissemination of unit watt in airborne sound: measurement methodology, sound emission regulations and implications. MAPAN J Metrol Soc India 35, 601-612. https://doi.org/10.1007/s12647-020-00417-x
48. Fatima S, Mohanty AR (2012) Noise control of home appliances-the green way. Noise Vibrat Worldw:26–34
49. Broner N (2011) A simple outdoor criteria for assessment of low frequency noise emission. Acoustics Australia 39:1–11
50. Hessler GF Jr (2005) Proposed criteria for low frequency industrial noise in residential communities. J Low Freq Noise Vibrat Active Contr 24(2):97–105
51. Pawlaczyk-Łuszczyńska M, Szymczak W, Dudarewicz A, Śliwińska-Kowalska M (2006) Proposed criteria for assessing low frequency noise annoyance in occupational settings. Int J Occup Med Environ Health 19(3):185–197
52. Shehap AM, Shawky HA, El-Basheer TM (2016) Study and assessment of low frequency noise in occupational settings. Archives of Acoustics 41(1):151–160
53. Alves JA, Silva LT, Remoaldo PC (2018) Impacts of low frequency noise exposure on well-being: a case-study from Portugal. Noise Health 20(95):131–145
54. Leventhall G (2007) Effects of ultrasound and infrasound relevant to human health. Prog Biophys Mol Biol 93(1–3):130–137
55. Nakamura S, Tokita Y (1981) Frequency characteristics of subjective responses to low frequency sound. International Conference on Noise Control Engineering, Nederlands Akoetisch Genootschap, Delft, The Netherlands, pp 735–738
56. Hodgdon K K, Atchley A A, Bernhard R J (2007) PARTNER low-frequency noise study. Report no. PARTNER-COE-2007-001, April, 2007
57. Heimann D, de Franceschi M, Emeis S, Lercher P, Seibert P (eds) (2007) Air Pollution, Traffic Noise and Related Health Effects in the Alpine Space-A Guide for Authorities and Consulters. ALPNAP comprehensive report. Università degli Studi di Trento, Dipartimento di Ingegneria Civile e Ambientale, Trento, Italy, 335 pp
58. Gorai AK, Pal AK (2006) Noise impact assessment of an iron ore mining complex. Noise Control Eng J 54(6):352–359
59. National Building Code (NBC) of India (2016) Volume 2, Bureau of Indian Standards
60. Pedersen S, Møller H, Waye KP (2007) Indoor measurements of noise at low frequencies – problems and solutions. J Low Freq Noise Vibr Act Contr 26(4):249–270
61. Zannin PHT, Ferraz F (2016) Assessment of indoor and outdoor noise pollution at a university hospital based on acoustic measurements and noise mapping. Open J Acoustics 6:71–85
62. European Environment Agency (2014) Noise in Europe 2014. EEA report No. 10/2014. December, 2014

63. Schwela D (2017) Review of environmental noise policies and economics in 2014–2016. 12th ICBEN Congress on Noise as a Public Health Problem, ICBEN 2017, 18th–22nd June, 2017, Zurich, 1–13
64. European Environment Agency report (2016) Quiet areas in Europe The environment unaffected by noise pollution EEA Report No 14/2016
65. I-INCE Technical Study Group on Metrics for Environmental Noise Assessment and Control (TSG 9) (2015) Supplemental Metrics for Day/Night average sound level and Day/Evening/ Night average sound level. I-INCE Publication Number: 2015–1, April, 2015
66. Aletta F, Kang J, Axelsson O (2016) Soundscape descriptors and a conceptual framework for developing predictive soundscape models. Landsc Urban Plan 149:65–74
67. Coelho JLB (2007) Community noise ordinances. Handbook of noise and vibration control (MJ Crocker, ed). Wiley, pp 1525–1532
68. Kihlman T, Kropp W (2001) City traffic noise-a local or global problem? Noise Control Eng. Journal 49(4):165–169
69. Burgess M, Macpherson J (2016) Overview of Australian road traffic noise policy 2016. Acoustics Australia 44:227–234
70. Swiss Federal Council report (2015) Environment Switzerland
71. Lang WW (2003) Global versus local issues in noise control policy. Noise Vibrat Worldw 34(2):17–22
72. Central Pollution Control Board (2017) Delhi's ambient noise levels influenced by traffic flow – case studies, Control of Urban Pollution Series, CUPS/87/2017–18, September, 2017
73. Ohm (2006) Acoustical classification and conformity checking of road surfaces: costs, benefits and cost-effectiveness of low-noise surfaces, COWI, Stakeholder Workshop
74. Danish Road Institute (DRI) (2005) Cost-benefit analysis on noise-reducing pavements Copenhagen
75. Bendtsen H (2005) Traffic management and noise reducing pavements: recommendations on additional noise reducing measures Silvia project deliverable
76. Ausejo M, Tabacchi M, Recuero M, Asensio C, Pagán R, Pavón I (2011) Design of a noise action plan based on a road traffic noise map. Acta Acustica United With Acustica 97:492–502
77. Witteveen+Bos (2004) Barrier positioning – phase 1: possible solutions, phase 2: acoustic effects and costs (in Dutch)
78. Sandberg U (2006) Financial incentives to encourage the use of low-noise Tyres. VTI (Swedish National Road and Transport Research Institute)
79. Nijland HA, van Kempen EEMM, Van Wee GP, Jabben J (2003) Costs and benefits of noise abatement measures. Transp Policy 10(2):131–140
80. den Boer LC, Schroten A (2007) Traffic noise reduction in Europe. CE Delft (March 2007)
81. Ellebjerg L (2005) Effectiveness and benefits of traffic flow measures on noise control. European Commission DG research report, work package H1: methods for noise control by traffic management
82. Bendtsen H, Michelsen L, Reif K, Reiff L (1998) Vejtrafik og støj-en grundbog. Copenhagen: Danish Road Directorate, Report 146
83. Andersen B (2003) Støjudsendelse fra biler på vejnettet. Kgs. Lyngby: Danish Transport Research Institute. Report 2. In Danish with English summary
84. Forum of European National Highway Research Laboratories (FEHRL) report (2006) Guidance manual for the implementation of low-noise road surfaces. Editor: Phil Morgan (TRL), Sustainable road surfaces for traffic noise control, 2006/02
85. Danish Road Institute (2004) Traffic management and noise reducing pavements: recommendations on additional noise reducing measures. Ministry of transport and energy, Copenhagen
86. Campolieti D, Bertoni D (2009) The action plan for noise reduction in Modena: methods, effects and perspectives. Radiat Prot Dosim 137(3–4):252–255
87. Hyden C, Várhelyi A (2000) The effects on safety, time consumption and environment of large scale use of roundabouts in an urban area: a case study. Accid Anal Prev 32:11–23

88. Egyptian Environmental Affairs Agency (EEEA) report (2016) The National Plan to combat noise and reduce its sources. Ministry of Environment, Egypt
89. Tiwari SK, Kumaraswamidhas LA, Garg N (2019) Modeling of ambient noise levels in urban environment. Paper 191, International Conference on Advanced Production and Industrial Engineering, IACPIE 2019, DITE, Delhi, India, 20th to 21st December, 2019
90. CPCB scale of compensation for violation of noise rules, https://greentribunal.gov.in/sites/default/files/news_updates/CPCB%20Report%20in%20O.A.%20No.%20519%20of%202016%20With%20O.A.%20No.%20496%20of%202018.pdf. Accessed on 18 Feb 2021
91. Central Pollution Control Board (CPCB) Report (2017) Status of Pollution abatement measures. Firecrackers (Patakas), CUPS/88/2017–18
92. Directive 2013/29/EU of the European Parliament and of the Council of 12 June 2013 on the harmonisation of the laws of the Member States relating to the making available on the market of pyrotechnic articles
93. Sinha S, Goyal K (2020) Fireworks: requisite for forensics evolution from conventional to green crackers in India-a perspective. J Forensic Res 11(3) https://doi.org/10.37421/2157-7145.11.455
94. Environmental Noise Guidelines for the European Region (2018) World Health Organization. Regional Office for Europe, Denmark
95. European Environment Agency (2010) Good practice guide on noise exposure and potential health effects. European Environment Agency, Copenhagen
96. Jarosińska D, Héroux ME, Wilkhu P, Creswick J, Verbeek J, Wothge J, Paunovic E (2018) Development of the WHO environmental noise guidelines for the European region: an introduction. Int J Environ Res Public Health 15:813. https://doi.org/10.3390/ijerph15040813
97. Science for Environment Policy 2015 Thematic Issue: Noise impacts on health. European Commission's Directorate-General Environment, Issue 47
98. Münzel T, Gori T, Babisch W, Basner M (2014) Cardiovascular effects of environmental noise exposure. Eur Heart J 35(13):829–836
99. van Kempen E, Casas M, Pershagen G, Foraster M (2018) WHO environmental noise guidelines for the European region: a systematic review on environmental noise and cardiovascular and metabolic effects: a summary. Int J Environ Res Public Health 15:379. https://doi.org/10.3390/ijerph15020379
100. Taneja MK (2014) Noise-induced hearing loss. Indian J Otol 4(20):151–154
101. Suvorov G, Denisov E, Antipin V, Kharitonov J, Pyykkö SI, Toppila E (2001) Effects of peak levels and number of impulses to hearing among forge hammering workers. Appl Occup Environ Hyg 16(8):816–822
102. Pierson LL, Gerhardt KJ, Rodriguez GP, Yanke RB (1994) Relationship between outer ear resonance and permanent noise-induced hearing loss. Am J Otolaryngol 15:37–40
103. Coles RR, Lutman ME, Buffin JT (2000) Guidelines on the diagnosis of noise induced hearing loss for medicolegal purposes. Clin Otolaryngol Allied Sci 25(4):264–273
104. Hong O, Kerr MJ, Poling GL, Dhar S (2013) Understanding and preventing noise-induced hearing loss. Dis Mon 59(4):110–118
105. Le TN, Straatman LV, Lea J, Westerberg B (2017) Current insights in noise-induced hearing loss: a literature review of the underlying mechanism, pathophysiology, asymmetry, and management options. J Otolaryngol Head Neck Surg 46:41
106. Smoorenburg GF (1993) Risk of noise-induced hearing loss following exposure to Chinese firecrackers. Audiology 32:333–343
107. North Atlantic Treaty Organization Report (2003). Reconsideration of the effects of impulse noise. Research and Technology Organization, Technical report, TR-017
108. Price GR, Kalb JT (1991) Insights into hazard from intense impulse from a mathematical model of the ear. J Acoustical Soc Am 90:219–227
109. Fedele PD, Binseel MS, Kalb JT, Price GR (2013) Using the Auditory hazard Assessment Algorithm for Humans (AHAAH) with hearing protection software, release MIL-STD-1474E. U.S. Army Research Laboratory, Report no. ARL-TR-6748

110. Starck J, Toppila E, Pyykko I (2003) Impulse noise and risk criteria. Noise Health 5:63–73
111. Price GR (2007) Predicting mechanical damage to the organ of Corti. Hear Res 226:5–13
112. Flamme GA, Liebe K, Wong A (2009) Estimates of the auditory risk from outdoor impulse noise I: firecrackers. Noise Health 11(45):223–230
113. Chang TY, Yu TY, Liu CS, Young LH, Bao BY (2020) Occupational noise exposure and its association with incident hyperglycaemia: a retrospective cohort study. Sci Rep 10:8584. https://doi.org/10.1038/s41598-020-65646-1
114. Recio A, Linares C, Banegas JR, Diaz J (2016) Road traffic noise effects on cardiovascular, respiratory, and metabolic health: an integrative model of biological mechanisms. Environ Res 146:359–370
115. Skogstad M, Johannessen HA, Tynes T, Mehlum IS, Nordby K-C, Lie A (2016) Systematic review of the cardiovascular effects of occupational noise. Occup Med 66(1):10–16
116. Arenas JP, Suter AH (2014) Comparison of occupational noise legislation in the Americas: an overview and analysis. Noise Health 16:306–319
117. NIOSH (1998) Criteria for a recommended standard in services. Education and Information Division National Institute for Occupational Safety and Health, Cincinnati
118. Dobie RA, Clark WW (2014) Exchange rates for intermittent and fluctuating occupational noise: a systematic review of studies of human permanent threshold shift. Ear Hear 35:86–96
119. Official Journal of the European Union Directive 2003/10/EC of the European Parliament and of the Council of 6 February 2003: on the Minimum Health and Safety Requirements Regarding the Exposure of Workers to the Risks Arising from Physical Agents (Noise) Official Journal of the European Union
120. Nandi SS, Dhatrak SV (2008) Occupational noise-induced hearing loss in India. Indian J Occup Environ Med 12(2):53–56
121. Sheppard A, Ralli M, Gilardi A (2020) Salvi R (2020) occupational noise: auditory and non-auditory consequences. Int J Environ Res Public Health 17:8963. https://doi.org/10.3390/ijerph17238963
122. Lewis RC, Gershon RRM, Neitzel RL (2013) Estimation of permanent noise-induced hearing loss in an urban setting. Environ Sci Technol 47(12):6393–6399
123. Nelson DI, Nelson RY, Concha-Barrientos M, Fingerhut M (2005) The global burden of occupational noise-induced hearing loss. Am J Ind Med 48(6):446–458
124. Prince MM, Stayner LT, Smith RJ, Gilbert SJ (1997) A re-examination of risk estimates from the NIOSH occupational noise and hearing survey (OHNS). J Acoust Soc Am 101(2):950–963
125. ASHA (1981) On the definition of hearing handicap. ASHA 23(4):293–297
126. Anil H T, Arunima S (2020) Noise induced hearing loss in autorickshaw drivers in Bangalore: a cross sectional study. Ind J Otolaryngol Head Neck Surg. https://doi.org/https://doi.org/10.1007/s12070-020-01829-6
127. Saxena A (2018) Noise induced hearing loss in Indian railway loco pilots: are we aware? Otolaryngol Open J 4(2):18–21
128. Nair S, Kashyap RC (2009) Prevalence of noise induced hearing loss in Indian air force personnel. Med J Armed Forces India 65(3):247–251
129. Majumder J, Mehta CR, Sen D (2009) Excess risk estimates of hearing impairment of Indian professional drivers. Int J Ind Ergon 39(2009):234–238
130. Balaji R, Rajasegaran R, John NA, Venkatappa US (2016) Hearing impairment and high blood pressure among bus drivers in Puducherry. J Clin Diagn Res 10(2):CC08–CC10
131. Khadatkar A, Mehta CR (2018) Effect of age and duration of driving on hearing status of Indian agricultural tractor drivers. J Low Freq Noise, Vibrat Active Cont 37(4):1037–1044
132. Pandya ND, Shah CJ, Mehta HB, Sharma S, Kota K (2021) A study of occupational noise-induced hearing loss using pure tone audiometry in disc jockeys of Bhavnagar. Natl J Physiol Pharm Pharmacol 11(09):1–4

Chapter 5
Environmental Noise Pollution: Evaluation and Analysis

Abstract The chapter presents an overview of measurement strategies carried out across the globe for evaluation and analysis of ambient noise levels in urban cities. The Central Pollution Control Board (CPCB), New Delhi, India, had initiated a National Ambient Noise Monitoring Network (NANMN) with an objective of collecting real-time noise monitoring data from major cities of India. The chapter discusses the ambient noise levels measured in past 10 years from 2011 to 2020 and presents the implications and conclusions derived from the analysis of 24 × 365 h noise monitoring data available from the 70 sites in Indian perspectives. The noise monitoring data reported by CPCB, India, during the odd-even rule implementation, Diwali festival, and Sundays is also discussed and compared with business-as-usual days. The chapter also presents the comparison of long-term versus short-term and temporal noise monitoring strategies to understand the best practicable and economical option (BPEO) in Indian perspectives. The conclusions presented in this chapter shall be helpful in understanding the noise scenario, seasonal variations, effect of odd-even, and lockdown on environmental noise levels, selecting BPEO for noise monitoring of larger parts of the country, revision in ambient noise standards, devising alternative models for ambient noise levels predictions and forecasting, evaluating uncertainty in noise measurements, and taking suitable control measures for noise control at noisy hotspots among the seven major cities of India.

5.1 Introduction

Noise pollution has been a serious issue globally nowadays. Every country in the world is concerned toward the serious health effects of noise emitted from the expanding number of vehicles, industries and other sources. Thus, the long-term noise monitoring is fundamentally needed to monitor the high ambient noise levels and planning of suitable noise abatement measures for control. The European Environmental Noise Directive 2002/49/EC requires the values of acoustic parameters, L_{den} and L_{night} representative of a year [1]. The noise maps should present noise levels expressed in the harmonized indicators L_{den} and L_{night}. The WHO estimates that, in Western Europe alone, one million healthy life-years are lost

N. Garg, *Environmental Noise Control*,
https://doi.org/10.1007/978-3-030-87828-3_5

annually to environmental noise [2]. Other estimates put the external cost of noise-related health issues in the EU between 0.3 and 0.4% of GDP [3], and 0.2% in Japan [4]. Studies in the USA and Europe also demonstrated the relationship between environmental noise and real estate markets, with housing prices falling with the increase in noise levels [5–7]. Long-term noise monitoring has been rarely reported in Indian scenario until the establishment of a large monitoring network in year 2011 [8, 9]. The Central Pollution Control Board (CPCB), India, initiated the process of developing National Ambient Noise Monitoring Network (*NANMN*), a follow-up of Section 5.2.8 (IV) of National Environmental Policy (NEP) 2006, through which it was decided to include ambient noise as a regular parameter for monitoring in specified urban areas [10, 11].

5.2 Environmental Noise Monitoring Studies Across the Globe

Noise pollution has been a serious concern globally in recent years. A recent study found that Guangzhou in China had the worst noise pollution, while Zurich in Switzerland had the least. The Worldwide Hearing Index was created by digital hearing app founders Mimi Hearing Technologies GmbH. They analyzed the hearing test results of 200,000 of their users. Delhi was the second worst city for noise pollution, followed by Cairo, Mumbai, Istanbul, and Beijing. The top five quietest cities were all in Europe: Zurich, Vienna, Oslo, Munich, and Stockholm. Typical sources of noise pollution are transport, such as road, rail and air traffic, construction and industry, and radios and televisions blaring in shops, restaurants, and bars. The data collected from over 200,000 hearing tests worldwide ranked average levels of hearing loss for people living in 50 global cities. The average hearing loss index was observed to have 64% positive correlation with noise pollution levels in the cities. The results of hearing loss index in each city and the noise pollution were mapped in the range between 0 and 1. The Mimi study found that the average city dweller has a hearing loss equivalent to 10–20 years older than their actual age. Residents of Vienna, Austria, have the least amount of hearing loss proportionate to their age, with an average hearing loss of 10.59 years. Citizens of Delhi, India, have the most, with a hearing age of 20 years older [12, 13]. Noise monitoring and mapping studies have been carried out in different parts of the world to analyze the potential effects of ambient noise levels on human health and also for devising the control measures. Table 5.1 summarizes the methodology and conclusions derived from these studies conducted across the globe [14–34]. It can be observed from these studies that every country is suffering from the serious problem of accentuated noise levels in the urban cities, which have to be controlled for minimizing the adverse health effects. Large-scale noise exposure modeling is used in epidemiological research projects as well as for noise mapping and strategic action planning. Within the SiRENE (Short and Long Term Effects of Transportation Noise Exposure) project, a large-scale

Table 5.1 Summary of some noise monitoring studies carried out in different parts of the world [14–34]

Researcher	Country	Methodology/strategy followed	Conclusions/remarks
Arana et al. [14]	Spain	Noise mapping of urban agglomerations in Spain.	Comparing across all European agglomerations, Germany had the lowest exposure to traffic noise and Spain the highest.
Zannin et al. [15]	Brazil	Evaluation of noise pollution in six urban parks located in the city of Curitiba.	The most noise-polluted parks in Curitiba were identified.
Murphy et al. [16]	Ireland	Estimates of the extent of environmental noise exposure from road transport on residents and workers in central Dublin.	The results show that the extent of noise exposure in Dublin is considerable, and in relative terms, it is worse for the night-time period.
Seong et al. [17]	USA	US DoT Noise Model implemented in Sound PLAN. After analyzing noise levels with raster, vector and facade maps, the human exposure to high noise levels was estimated.	Results show that 48% of the total county population ($N = 870{,}166$ residents) is potentially exposed to 55 dB (A) or higher noise levels during daytime. About 9% of the population is potentially exposed to 67 dB(A) or higher.
Skineer et al. [18]	UK	The surveys involved a total of 1160 24-h noise measurements at a sample of dwellings in the UK, and over 5500 in-depth interviews with a sample of the UK adult population.	A more strategic approach to the control of environmental noise both nationally and in London is developed.
Piccolo et al. [19]	Italy	Extensive measurements of the main indexes for noise pollution (L_{eq}, L_1, L_{10}, L_{50}, L_{90}, L_{99}) and of the traffic flow and composition were carried out for city of Messina (Italy).	Results indicate that main roads of Messina are overloaded by traffic flow during day-time period; more than 25% of residents should be highly disturbed by road traffic noise.
Jamrah et al. [20]	Jordan	This study investigated traffic noise pollution in Amman at 28 locations that cover most of the city of Amman.	The measured noise level exceeded the 62 dB(A) acceptable limit at most of the locations. The CRTN prediction model was successful in predicting noise levels at most of the locations.
Yilmaz et al. [21]	Turkey	Noise measurements were taken in the morning, at noon and in the evening to determine noise pollution all over the city.	From the measured noise values, a map of noise pollution was created for Erzurum.
Li et al. [22]	Beijing	Traffic noise surveying and analysis was performed along three main roads in the Beijing urban area.	Main roads are overloaded by traffic flow during daytime and noise levels due to road traffic along these roads are above

(continued)

Table 5.1 (continued)

Researcher	Country	Methodology/strategy followed	Conclusions/remarks
			relevant environmental standards by 5 dBA.
Morillas et al. [23]	Spain	An urban noise survey was conducted in the city of Cáceres (Extremadura, Spain).	90% of noise measurements surpassed an L_{eq} of 65 dB (A) during working hours.
Zannin et al. [24]	Brazil	The equivalent sound level values, $L_{Aeq,2hr}$ were measured and tabulated for 1000 locations spread over the urban zones of the city of Curitiba.	It has been found out that 93.3% of the locations displayed the day equivalent sound levels over 65 dB(A).
Alberola et al., [25]	UK	The database consists of 2 week's noise recordings at each of 50 separate locations in residential areas affected mainly by road traffic noise.	The results showed increased variability (standard deviations) at the lower values of either logarithmic or arithmetic mean L_{Aeq} over the time periods investigated.
Olayinka [26]	Nigeria	Forty two different locations throughout Ilorin were selected to monitor the background noise level, peak noise level and dominant noise sources at these locations.	Only 6 locations out of 42 were under normally acceptable situation.
Björk [27]	Kuopio, Finland	A total of 7581 h of continuous noise data was collected, $L_{eq,1h}$, and L_{max} and L_{AE} for the noise events were measured.	The time patterns of the noise levels during different seasons and the importance of different noise sources was investigated.
Tsai et al. [28]	Tainan, Taiwan	This study analyzed the spatial characteristics of urban environmental noise by using noise maps produced at 345 noise monitoring stations in Tainan, Taiwan.	The results of noise exposure analysis showed that over 90% of the Tainan City population were exposed to unacceptable noise.
Gaja [29]	Spain	The aim of this study was to determine the appropriate measuring time in order to obtain a 24-h noise level suitable to represent the annual equivalent level.	If the sampling strategy involves measurements on randomly chosen days, then at least 6 days should be used.
Scatolini et al. [30]	Brazil	Measuring sites were chosen from 62 and 72 DNL (day-night-level) noise contours, in urban sites compatible with residential use. Fifteen sites were monitored for at least 168 hours without interruption or seven consecutive days.	Only four of the 15 sites assessed presented aircraft operations as a clear cause for the noise annoyance.
Gunes et al. [31]	Turkey	Noise levels were measured on 28 different points between mornings (8:00–9:00), noon (12:	The noise levels of the points increased in the faculty

(continued)

Table 5.1 (continued)

Researcher	Country	Methodology/strategy followed	Conclusions/remarks
		00–13:00), and evening (17:00–18:00) hour in Namik Kemal University Campus (Tekirdag, Turkey).	buildings, intersections, classrooms, library, dining hall, car parks.
Mohammadi [32]	Iran	This paper described the reaction to environmental noise in the city of Kerman. A total of 250 questionnaires were processed.	The finding of this study revealed that the main isolated noise sources were traffic (50%), street noise (34%), and construction (16%).
Leao et al. [33]	Australia	The participatory process proposed, and currently under implementation and test, to characterize the levels of exposure to traffic noise of residents living in the vicinity of highways in the City of Boroondara (Victoria, Australia) using the 2Loud? Application (mobile).	It seems that although technology is moving fast and our current smartphones have the capacity of performing complex and good quality data collection, there is still need for research on how to channel this data to those responsible for managing environmental issues.
Gozalo et al. [34]	Chile	Multiple regression models were developed for two cities for urban variables: street location, land-use, street geometry, road traffic control, public and private transportation.	A total of 73.3% and 81.3% of streets in two cities exceeded the day time sound levels of 55 dB. Urban variables considered were observed to have higher correlation with noise levels.

nationwide assessment of Switzerland's road, railway, and aircraft noise exposure was conducted for the year 2011. Noise was thereby modeled at facade points of all buildings in Switzerland, resulting in a total number of 54,300,00 receiver locations at 1,813,000 buildings. The noise exposure was modeled in an hourly resolution and typical noise exposure metrics such as L_{day}, L_{night}, L_{evening}, and L_{den} were derived [35]. Noise pollution is so pervasive in cities that citizens and policymakers tend to believe that it is unavoidable. The recent HARMONICA (HARMOnised Noise Information for Citizens and Authorities) project has developed and published a new index, called the Harmonica index, as well as innovative and educational tools to better inform the public and to assist decision-making. Hourly Harmonica index = background noise subindex (*BGN*) + peak noise subindex (*EVT*). The Harmonica index, therefore, is divided into two parts: A component related to the background noise: *BGN* subindex: $BGN = 0.2\ (L_{\text{A95eq}} - 30)$, where L_{A95eq} is the equivalent background noise level during the 1-h period, the background noise being evaluated every second by the noise level exceeded 95% of the time during the previous 10-min period. An event-related component: *EVT* subindex, which represents the acoustic energy provided by noise peaks that emerge above the background noise: $EVT = 0.25\ (L_{\text{Aeq}} - L_{\text{A95eq}})$, where L_{Aeq} is the equivalent noise level during a one-hour period. The Harmonica index gives a score of 0 to 10, rounded to one decimal place. The higher the score, the poorer the acoustic environment [36].

In recent years, the traditional approach of noise monitoring is also replaced by the use of latest technology such smartphones and wireless sensing units. Smartphones have become omnipresent and one of the main communication resources for human beings. Their characteristics have also experienced a substantial improvement as they now integrate multiple sensors able to convert the smartphone into a flexible and multipurpose sensing unit. The combined use of multiple smartphones endowed with several types of sensors gives the possibility to monitor a certain area with fine spatial and temporal granularity, a procedure typically known as crowdsensing. The sampling rate and the selected buffer size can have a significant impact on the accuracy of noise level estimations, being that estimation errors can vary from 1 to 12% in the best cases [37]. Wireless sensor network is another such approach followed in recent years. The WSU is comprised of a microcontroller unit that enables on-board computations, a wireless transceiver that uses Zigbee protocol for data transmission, and an external peripheral board that houses the microphone transducer. It contains 34 digital I/O pins and 21 analog input pins, allowing it to interface with numerous external peripherals [38]. Figure 5.1a and b shows the prototype wireless sensing unit employed for noise measurements.

Tables 5.1 and 5.2 discuss some of the noise monitoring studies carried out using conventional approach, smartphone applications, and wireless sensing units [38–46]. It can be thus observed that the new technological advancements pertaining to the use of smartphone application and wireless sensing units can be used to map the larger parts of the urban agglomerations in urban cities. Eventually, the low-cost smart sensor wireless units connected through Internet of Things (IoT) architecture are being developed that are very useful in acquisition and transmission of noise monitored data and shall be very helpful for noise monitoring of the larger parts of the urban cities [46].

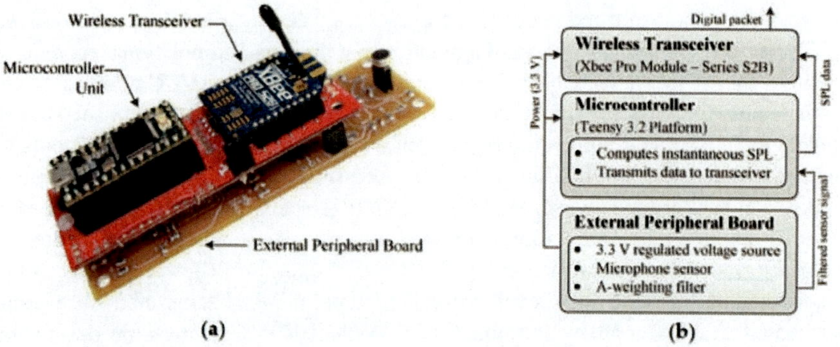

Fig. 5.1 (**a**) Wireless sensing unit prototype with microcontroller, wireless transceiver, and external peripheral board; (**b**) block diagram of the system (Peckens et al. [38])

Table 5.2 Summary of some noise monitoring studies carried out using smartphones and wireless sensing units [38–46]

Researcher	Country	Methodology/strategy followed	Conclusions/remarks
Peckens et al. [38]	USA	A network of Wireless Sensing Units was deployed for long-term noise monitoring in Holland, Michigan, USA, on the campus of Hope College.	The study showed that through the hardware and software validation the WSU can be used as an appropriate substitution for the traditional methods of noise measurement.
Kumar et al. [39]	India	The system involves a client application on smartphones that records noise, processes the information, and communicates to a server and shares the information on Google Maps.	It is evident from the study that the area under study is exposed to higher noise levels as compared to the standards laid down by the Central Pollution Control Board (CPCB), New Delhi, India.
Pődör and Révész [40]	Hungary	Participatory noise mapping utilizing smartphone-based measurement data.	Study found out that, with some limitations, mobile phones are suitable for making noise measurements. While they are not capable of taking accurate noise measurements, they can be used for differential measurements.
Kardous and Shaw [41]	NOISH, Cincinnati	The initial study examined 192 *apps* on the iOS and Android platforms and found four iOS *apps*. This study evaluated the same four apps using external microphones.	The study showed that the use of external calibrated microphones greatly enhances the accuracy and precision of smartphone-based noise measurements.
Grubeša et al. [42]	Croatia	The study analyzes the Mobile Crowd Sensing (MCS) method and compares the results obtained with this method with the results from the standard sound level meter.	The research showed that the MCS method can make noise mapping easier, cheaper, and less time-consuming and the accuracy and precision of measurements obtained by using calibrated smartphones are acceptable.
Maisonneuve et al. [43]	Paris	The goal of this Noise Tube project was to turn GPS equipped mobile phones into noise sensors that enable citizens to measure their personal exposure to noise in their everyday environment.	The approach and the supporting rationale as well as a prototype implementation for Noise Tube project was discussed.
Murphy and King [44]	USA	Maps generated using the traditional approach with those generated using smartphone-based measurement data were compared.	The study tests the accuracy of smartphone-based measurements against simultaneous measurements taken using traditional sound level meters in the field.

(continued)

Table 5.2 (continued)

Researcher	Country	Methodology/strategy followed	Conclusions/remarks
Kivelä and Hakala [45]	Finland	Wireless sensor network (Ci Net Noise) is used to perform environmental monitoring for various shaped and sized areas in different surrounding.	The study showed that the area-based measurement method offers a flexible way to conduct noise measuring.
Anachkova et al. [46]	Skopje	A wireless noise sensor unit design for noise monitoring toward the realization of the Internet of Things (IoT) and Smart city concept.	Low cost smart sensor unit showed good performance for noise data acquisition and transmission.

5.3 Environmental Noise Indices

Noise pollution is pervasive in cities and thus the easy-to-understand information pertaining to the environmental noise shall be very instrumental for community. Thus, there is a continuous research on the development of new environmental noise indices that are more representative of the community perception toward the environmental noise. The holistic evaluation of urban places and accounting the perceptive effects is the major objective for designing the environmental noise indicator. The criterion for selecting the environmental noise indicators had been "validity," "practical applicability," "transparency," "enforceability," and "consistency" [47, 48]. The European Commission Position paper (2000) on EU noise indicators suggested the major classifications of the noise indicators as [47]:

- *Basic indicators*: basic indicators contain only a few assumptions about the link with effects and can considered to be purely physical quantities, useful when dealing with complaints and unusual situations.
- *Composite indicators*: composite indicators are derived from the basic by combining them in one or more ways: day/evening /night weightings, averages over longer periods.
- Useful for overall policy to reduce the number of affected persons.
- Applicable in land-use planning, mapping zoning and related activities related to noise control.
- Simplify estimation of dose/response effects.
- Simplify noise mapping.
- Can more easily be understood by the public, politicians, planners, etc.
- *Complex indicators*: complex indicators cover things such as combined exposure with other noise sources or even other sources of annoyance, and/or take into account the impact on the population. Number over xx dB, area over xx dB, conflict maps, population exposure indicator, etc. are some of the complex indicators used.

- Comparison between countries, regions, cities or parts of a city.
- Comparison of different kinds of sources, thus making it easier to estimate levels in situations where two or more different sources occur simultaneously.

The three types of indicators (basic, composite, and complex) can be used for different purposes, depending on the aggregation level in space and time. The recent HARMONICA (HARMOnised Noise Information for Citizens and Authorities) project had developed and published a new index called the Harmonica index in order to better inform the public and to assist decision-making. This new noise pollution index is different to existing indicators as it takes into account both the continuous and the sporadic nature of noise: including background noise and noise peaks caused by aircraft, rail traffic, etc. The indicator is based on a public survey ranging from 0 to 10. The new index is demonstrated on the website *www.noiseineu.eu*. It is calculated as: Hourly Harmonica index = Background noise subindex (*BGN*) + peak noise subindex (*EVT*). The Harmonica index, therefore, is divided into two parts: a component related to the background noise: *BGN* subindex: $BGN = 0.2\ (L_{A95eq} - 30)$, where L_{A95eq} is the equivalent background noise level during the one-hour period, the background noise being evaluated every second by the noise level exceeded 95% of the time during the previous 10-min period. An event-related component: *EVT* subindex, which represents the acoustic energy provided by noise peaks that emerge above the background noise: $EVT = 0.25\ (L_{Aeq} - L_{A95eq})$, where L_{Aeq} is the equivalent noise level during a one hour period. The Harmonica index gives a score of 0 to 10, rounded to one decimal place. The higher the score, the poorer the acoustic environment. The graphical representation of the index is in two parts, a triangle (representing the event) on the top of a rectangle (representing the background noise), in order to clearly translate its two components: The color of the index (green/orange/red) as shown in Table 5.3 indicates how the noise environment compares to the WHO's environmental quality objectives and the values recognized as critical for noise. These colors also take the time of day into account, as people are more sensitive to noise at night [36].

The correlation analysis in order to ascertain the correlation coefficient for different combination of the Harmonica indices and the noise indicators showed the high values of the Pearson correlation coefficient and adequacy of Harmonica indices for environmental noise assessment [49]. It has been evidently reported that

Table 5.3 Color code representation of the harmonica index [36]

Colour	Day (from 6 am to 10 pm)	Night (from 10 pm to 6 am)
(green)	between 0 and 4	between 0 and 3
(orange)	between 4 and 8	between 3 and 7
(red)	over 8	over 7

the sound type, user characteristics, and other factors play a vital role in acoustic comfort level. Furthermore, it had been shown that the environmental noise annoyance only depends on approximately 30% of the physical facets of parameters such as acoustic energy [50]. Thus, the concept of "soundscape" is different from "acoustic environment" as it relates to perceptual constructs rather than just physical phenomena. Soundscape research represents a paradigm shift in the field of environmental noise in that it combines physical, social, and psychological approaches and considers environmental sounds as a "resource" rather than "waste." The sound quality has primarily three aspects: (1) stimulus-response compatibility, which is the functional aspect of a sound; (2) pleasantness of sounds, which is based on an instantaneous overall impression emerging from various sound attributes as well as individual preferences and experience; and (3) identifiability of sounds or sound sources [51]. Psychoacoustic factors, including loudness, fluctuation strength or roughness, sharpness, and pitch strength [52], have been analyzed to be successful for the assessment of sound quality. According to Aletta et al. [53], there are eight sorts of potential soundscape descriptors to choose from: noise annoyance, pleasantness, quietness or tranquility, music-likeness, perceived affective quality, restorativeness, soundscape quality, and appropriateness. A specific smartphone-based perceptive and acoustic data in Paris for about 3400 questionnaires correlated the sound quality indicator (SQ) as [54]:

$$SQ = 19.08 - 0.19L_{50} - 0.06\left(L_{A10} - L_{A90}\right) \tag{5.1}$$

The sound unpleasantness was calculated as [54]:

$$\text{Sound unpleasantness} = -8.08 + 0.19L_{A50} + 0.06L_{A10} - 0.06L_{A90} \tag{5.2}$$

A study by Axelsson et al. proposed a measurement system for soundscape quality that consists of two-dimensional space defined by the attributes: pleasant, exciting, eventful, chaotic, unpleasant, monotonous, uneventful, and clam as eight vectors separated by 45° in a circumplex model of soundscape perception [55]. Energetic indicators cannot solely explain all the sound pleasantness variance and in evaluating the fluctuating sounds and induced noise annoyance [56, 57]. These indicators inform about the total sound level, but they give the same value regardless the temporal structure of the sound environments. The spectral dimension plays a significant role as low-frequency noise increases the annoyance [58]. Also, the A-weighting does not fulfill the perceptive requirements and L_{eq} is more relevant than L_{Aeq}. L_{eq} is the objective variable that explains the higher level of variation when a sound environment is perceived as very unpleasant. If the objective variables are ordered based on their degree of explanation of the variability of the subjective variable of higher unpleasantness, from highest to lowest, the order would be L_{eq}, L_{Aeq}, sharpness, and loudness [59]. Conventionally L_1, L_{10}, L_{50}, and L_{90} are used to give approximate indications of the maximum, intrusive, median, and background sound levels [60]. The L_{50} indicator has been observed to have a good correlation with the perceived sound intensity and sound pleasantness [54]. The L_{10} indicator describes the contribution of the loudest events and outperforms L_{Aeq} to describe the perception of high noise levels. Some descriptors like the difference between A and C-weighted equivalent continuous sound levels L_{Ceq}-L_{Aeq} describe the amount of low frequencies, while some like L_{Amax}, maximum A-weighted noise level; NA, number of events above threshold, time above threshold (e.g., NA80, number of events above 80 dB(A), TA80, time above 80 dB(A)) correlate with the awakening probability as the number of high noise level events may affect the sleep motility. The normalized time and frequency second derivative, $TFSD_{mean,4kHz}$ and $TFSD_{mean,500Hz}$ are good for discriminating the biophonic sounds and anthropogenic sounds in urban environments [61]. The emergences and noise variation indicators such as the intermittency ratio (IR) defined as the ratio between the sound energy contribution of the events and the overall contributions during the measurement period are sometimes helpful for correlating with the health effects. IR values were considered in the range of 0–100%, whereby 0% IR means that no single event occurs 3 dB above background and 100% describes that all energy is produced by individual noisy events. For instance, highly intermittent nocturnal noise is correlated with increased risk of cardiovascular diseases. In case of road traffic noise, IR has an additional effect on % HA and can explain shifts of the exposure-response curve of up to about 6 dB between low IR and high IR exposure situations [62]. Nocturnal intermittent noise exposure had been observed to be more relevant than daytime exposure for acute cardiovascular diseases, whereas heart failure was observed to be most strongly associated with continuous daytime noise. This suggested that for acute diseases sleep is an important mediator for health consequences of transportation noise [62]. Can and Gauvreau study [63] for the district of Marseille, France, based on clustering analysis revealed the three indicators: L_{50A}, the standard deviation, $\sigma_{Leq,A}$ and the sound gravity spectrum, $SGC_{[50\ Hz\ -\ 10\ kHz]}$ cover the dimensions of sound environment: sound level, temporal variability, and its frequency spectrum, respectively. This study enunciated that using the three indicators augment the perceptive studies by delimiting the spatial extent of each typical sound environment [63]. However, aggregating the complex noise indicators into a single dimensionless indicator could be an optimal option in this regard. Figure 5.2 shows the classification of the various environmental noise indicators in terms of descriptive power based on the recent study presented by Asensio et al. [61].

5.4 Environmental Noise Monitoring in India

CPCB, India, initiated the process of developing National Ambient Noise Monitoring Network (*NANMN*), a follow-up of Section 5.2.8 (IV) of National Environmental Policy (NEP)-2006, through which it was decided to include ambient noise as a regular parameter for monitoring in specified urban areas. The real-time noise monitoring network, *NANMN* program was established with an objective of getting the real-time continuous noise monitoring data. The phase I of the Real Time National Ambient Noise Monitoring Network was established in year 2011 that

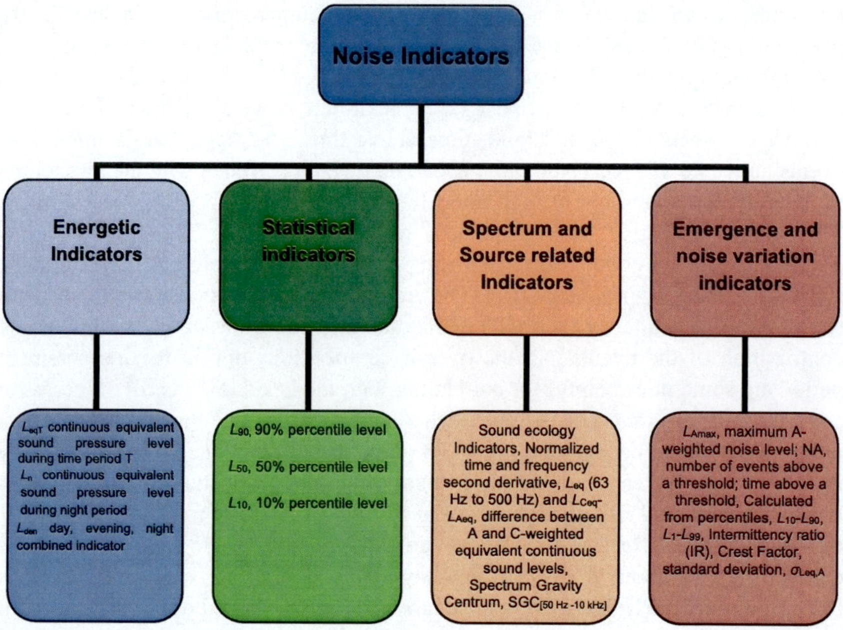

Fig. 5.2 Classification of various environmental noise indicators in terms of descriptive power. (Based on Asensio et al. [61])

covers 35 locations in seven metro cities (Delhi, Hyderabad, Kolkata, Mumbai, Lucknow, Bengaluru, and Chennai). Subsequently, 35 more noise monitoring stations were established in 2014 with each city having 10 monitoring stations presently. The 70 locations cover 25 commercial sites, 16 residential sites, 17 sites in silence zones, and 12 sites in industrial zone. The Noise Monitoring Network so established is unique and one of the largest of its kind across the globe [64–67].

5.4.1 Establishment of NANMN *Project*

The noise monitoring data analyzed are reported from CPCB noise monitoring stations established under *NANMN* program and situated at 35 different locations all over India since the year 2011. Figure 5.3 shows the schematic diagram of the ambient noise monitoring network. The real-time ambient noise monitoring network consists of [65]:

- *Automatic Data Collecting Remote Stations* comprising of noise monitoring sensor, data logger, internal Global Positioning System (GPS), microprocessor, protection circuits, communication modem, power source including battery and charge regulator for solar panel on mains AC power supply.

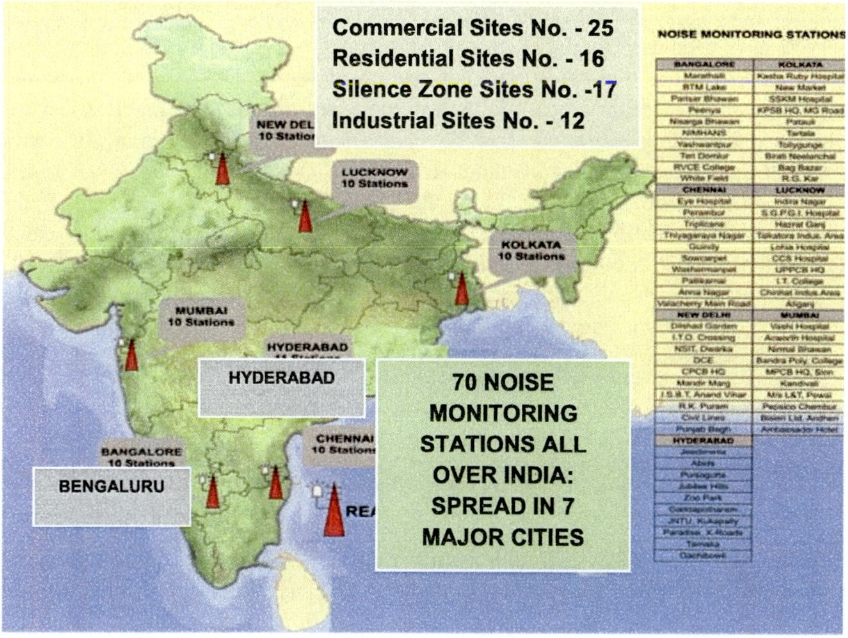

Fig. 5.3 Establishment of diversified *NANMN* project with 70 noise monitoring stations installed in the seven major cities of India. (http://www.sgsweather.com [68])

- *Automatic Communication options for data communication* comprising of Transmission Control Protocol/Internet Protocol, TCP/IP (Telnet; Simple Mail Transfer Protocol, SMTP; File Transfer Protocol, FTP, etc.), short SMS messages to mobile telephones for notices or alerts, Global System for Mobile Communication (GSM) modem, General Packet Radio Services (GPRS) modem, fiber optics, Radio-modem, Ethernet, Serial digital interface at 1200 baud (SDI-12), local area wireless technology, such as Wi-Fi, Bluetooth, satellite (internal or external transmitter).
- *Central Processing Station* comprising of hot redundant server, user workstation, software for data visualization and processing.
- *Web-Enabled Software* for uploading on internet or sharing among CPCB and State Pollution Control Boards only to the authorized personnel.

The Noise Remote Terminal (NMT) is a standalone operating remote terminal consisting of a sound level meter traceable to the national standards for continuously measuring the ambient noise. Each terminal consists of a high-quality outdoor microphones (Model: 41CN) connected to an advanced acoustic signal processing unit and a high resolution data logger. The Noise Processor 2000 NP receives and digitizes the output single generated by the microphone and computes the acoustic parameters such as L_{eq}, time, frequency weighting A and C, and time- weighting Slow (S) and Fast (F). For dissemination of the real-time information to public, bright LED display boards are installed at various locations [65]. The high-end dual servers have been used for data reception, processing, visualization, communication, and archiving. The servers are in hot redundant mode ensuring that under no condition data reception process fails as in the event of failure of main server, the alternate hot redundant server takes over the entire process automatically with minimal time delay. Thus, the Central Receiving Station gets the data from all the remote stations through GPRS mode, processes the noise data, and generates different reports. The modem allows transmitting the data to central station called the National Noise Monitoring Centre, *NNMC* located at the CPCB Headquarters, Parivesh Bhawan, New Delhi, where the dynamic noise maps are prepared and presented with reliable and precise data updated in real-time, fed by continuous noise measurements from all sites where a remote acoustic terminal is installed [65].

5.4.2 *Evaluation and Analysis of* NANMN *Data*

The day equivalent level, L_{day}, and the night equivalent level, L_{night}, are calculated from the 24 h noise data for each day of the year. The day-time means from 6.00 a.m. to 10.00 p.m., while the night time means from 10.00 p.m. to 6.00 a.m. It may be noted that the ambient air quality standards in respect of noise are prescribed by CPCB, India in terms of L_{day} and L_{night} as enlisted in Table 4.1 of Chap. 4. The value of L_{day} and L_{night} is calculated as:

$$L_{\text{day},n} = 10 \log \left[\frac{1}{n} \sum_{i=1}^{n} 10^{0.1\left(L_{\text{day},i}\right)} \right] \tag{5.3}$$

$$L_{\text{night},n} = 10 \log \left[\frac{1}{n} \sum_{i=1}^{n} 10^{0.1\left(L_{\text{night},i}\right)} \right] \tag{5.4}$$

where n is the number of days or nights included in the long-term and $L_{\text{day},\,i}$ and $L_{\text{night},\,i}$ are the ith corresponding A-weighted equivalent level for the considered period. Tables 5.4, 5.5, and 5.6 show the annual average ambient levels, L_{day} and L_{night} for 70 noise monitoring stations installed across seven major cities in India for past 10 years (2011–2020) [69–72]. The monthly standard deviation is also reported along with the annual average value for all the sites. The analysis of the annual average noise levels in the past 7 years can be thus very helpful in understanding the noise scenario, ascertaining the compliance with the ambient standards and comparison of the environment noise levels in the different zones.

Table 5.4 Annual average ambient levels, L_{day} and L_{night} for 35 noise monitoring stations installed across seven major cities in India for past 4 years (Garg et al. [69])

Name of location	City	Area characteristics	2011		2012		2013		2014	
			L_{day}	L_{night}	L_{day}	L_{night}	L_{day}	L_{night}	L_{day}	L_{night}
Dilshad Garden	Delhi	Silence	52.0 ± 0.9	50.0 ± 1.4	52.0 ± 1.1	49.0 ± 2.1	51.0 ± 1.1	49.0 ± 2.4	52.0 ± 0.9	48.0 ± 1.6
CPCB HQ.		Commercial	64.0 ± 2.0	53.0 ± 1.4	62.0 ± 1.0	53.0 ± 1.3	63.0 ± 0.8	53.0 ± 1.0	65.0 ± 1.7	54.0 ± 1.6
DTU, Bawana		Silence	52.0 ± 1.3	48.0 ± 2.1	51.0 ± 0.9	49.0 ± 3.2	52.0 ± 1.7	49.0 ± 3.0	52.0 ± 1.1	48.0 ± 2.5
ITO		Commercial	73.0 ± 0.6	71.0 ± 1.0	72.0 ± 4.0	69.0 ± 5.3	74.0 ± 0.7	73.0 ± 0.4	74.0 ± 1.0	73.0 ± 1.4
NSIT Dwarka		Silence	56.0 ± 1.3	54.0 ± 0.8	56.0 ± 0.7	54.0 ± 1.1	56.0 ± 0.5	53.0 ± 0.9	56.0 ± 1.4	53.0 ± 1.8
Gomti Nagar	Lucknow	Silence	71.0 ± 0.8	61.0 ± 1.5	63.0 ± 0.9	55.0 ± 1.1	66.0 ± 2.2	57.0 ± 1.6	69.0 ± 1.4	61.0 ± 2.0
Hazrat Ganj		Commercial	67.0 ± 0.9	58.0 ± 1.0	72.0 ± 0.5	61.0 ± 1.0	72.0 ± 0.5	62.0 ± 1.3	72.0 ± 0.5	61.0 ± 1.5
Indira Nagar		Residential	54.0 ± 1.2	47.0 ± 2.9	53.0 ± 1.1	47.0 ± 3.0	54.0 ± 1.4	48.0 ± 3.6	57.0 ± 0.9	49.0 ± 4.7
PGI Hospital		Silence	55.0 ± 2.5	49.0 ± 2.8	58.0 ± 1.2	52.0 ± 3.6	60.0 ± 1.4	53.0 ± 3.0	62.0 ± 1.3	55.0 ± 3.5
Talkatora Industrial Area		Industrial	63.0 ± 0.4	55.0 ± 1.6	64.0 ± 0.7	56.0 ± 1.6	63.0 ± 0.5	56.0 ± 1.9	64.0 ± 1.2	57.0 ± 2.0
Kasba Gole Park	Kolkata	Industrial	64.0 ± 1.2	59.0 ± 1.3	65.0 ± 1.6	61.0 ± 2.6	68.0 ± 3.5	64.0 ± 4.7	70.0 ± 2.5	67.0 ± 2.9
New Market		Commercial	67.0 ± 0.5	60.0 ± 1.4	67.0 ± 0.7	59.0 ± 1.4	68.0 ± 0.5	60.0 ± 1.6	70.0 ± 2.3	65.0 ± 5.2
Patauli		Residential	55.0 ± 1.0	49.0 ± 2.0	55.0 ± 1.0	49.0 ± 3.2	55.0 ± 1.6	48.0 ± 6.2	55.0 ± 1.3	49.0 ± 4.5
SSKM Hospital		Silence	61.0 ± 0.4	54.0 ± 0.9	62.0 ± 0.8	56.0 ± 1.8	62.0 ± 1.2	57.0 ± 1.9	62.0 ± 1.1	56.0 ± 1.7
WBPCB HQ		Commercial	62.0 ± 0.6	56.0 ± 1.3	61.0 ± 0.7	54.0 ± 1.1	62.0 ± 1.4	55.0 ± 1.4	64.0 ± 0.6	58.0 ± 0.9
AS HP	Mumbai	Silence	66.0 ± 1.2	59.0 ± 1.5	65.0 ± 1.0	59.0 ± 0.3	65.0 ± 0.8	60.0 ± 1.5	66.0 ± 2.0	60.0 ± 0.9
Bandra		Commercial	70.0 ± 0.5	68.0 ± 0.8	69.0 ± 0.7	67.0 ± 1.9	69.0 ± 0.4	67.0 ± 0.5	70.0 ± 0.5	67.0 ± 0.7
MPCB HQ.		Commercial	67.0 ± 0.6	63.0 ± 0.5	66.0 ± 0.5	63.0 ± 0.7	68.0 ± 1.6	65.0 ± 2.0	71.0 ± 0.6	68.0 ± 1.0
Thane MCQ		Commercial	62.0 ± 1.8	53.0 ± 2.3	62.0 ± 0.7	55.0 ± 1.9	62.0 ± 1.2	55.0 ± 1.4	64.0 ± 1.0	56.0 ± 1.3
Vashi Hospital		Silence	68.0 ± 1.7	58.0 ± 1.4	69.0 ± 0.9	57.0 ± 2.7	69.0 ± 0.8	57.0 ± 0.8	69.0 ± 1.5	59.0 ± 3.6
Abids	Hyderabad	Commercial	72.0 ± 0.5	63.0 ± 0.9	72.0 ± 0.9	63.0 ± 1.9	72.0 ± 0.8	64.0 ± 2.1	74.0 ± 1.9	65.0 ± 2.5
Jeedimetla		Industrial	62.0 ± 0.5	56.0 ± 1.4	63.0 ± 1.2	56.0 ± 2.1	63.0 ± 1.3	56.0 ± 1.6	65.0 ± 0.6	58.0 ± 0.8

(continued)

Table 5.4 (continued)

Name of location	City	Area characteristics	2011		2012		2013		2014	
			L_{day}	L_{night}	L_{day}	L_{night}	L_{day}	L_{night}	L_{day}	L_{night}
Jubilee Hills		Residential	58.0 ± 1.0	50.0 ± 1.7	56.0 ± 0.7	49.0 ± 0.5	56.0 ± 0.6	49.0 ± 1.2	57.0 ± 1.6	49.0 ± 1.2
Punjagutta		Commercial	76.0 ± 0.6	71.0 ± 1.0	75.0 ± 0.5	70.0 ± 0.5	76.0 ± 1.7	71.0 ± 1.3	79.0 ± 0.7	73.0 ± 0.5
Zoo Park		Silence	53.0 ± 1.5	48.0 ± 2.8	54.0 ± 1.8	48.0 ± 2.0	54.0 ± 1.4	49.0 ± 1.1	56.0 ± 1.2	50.0 ± 2.2
BTM	Bengaluru	Residential	66.0 ± 0.4	56.0 ± 0.4	66.0 ± 0.5	56.0 ± 1.0	66.0 ± 0.8	56.0 ± 0.8	66.0 ± 0.7	57.0 ± 1.1
Marathahalli		Commercial	56.0 ± 1.9	53.0 ± 1.8	54.0 ± 0.7	52.0 ± 0.6	57.0 ± 2.1	54.0 ± 2.8	59.0 ± 0.7	56.0 ± 0.8
Nisarga Bhawan		Residential	56.0 ± 3.0	47.0 ± 1.8	56.0 ± 2.0	47.0 ± 1.9	56.0 ± 1.9	48.0 ± 1.6	56.0 ± 1.5	49.0 ± 1.4
Parisar Bhawan		Commercial	66.0 ± 1.1	58.0 ± 0.7	65.0 ± 0.3	57.0 ± 1.6	65.0 ± 0.7	57.0 ± 0.8	65.0 ± 0.8	57.0 ± 0.5
Peeniya		Industrial	56.0 ± 1.6	53.0 ± 2.6	56.0 ± 1.2	49.0 ± 1.2	58.0 ± 1.1	53.0 ± 2.3	58.0 ± 0.8	55.0 ± 2.0
Eye Hospital	Chennai	Silence	64.0 ± 0.6	51.0 ± 1.2	62.0 ± 1.5	52.0 ± 3.1	64.0 ± 1.5	53.0 ± 2.2	61.0 ± 3.9	53.0 ± 1.2
Guindy		Industrial	76.0 ± 0.6	71.0 ± 1.1	75.0 ± 1.1	71.0 ± 1.3	75.0 ± 1.0	71.0 ± 1.5	76.0 ± 1.9	72.0 ± 1.2
Perambur		Commercial	68.0 ± 0.9	58.0 ± 0.8	69.0 ± 1.2	58.0 ± 1.2	68.0 ± 0.5	57.0 ± 0.7	69.0 ± 1.3	58.0 ± 0.8
T. Nagar		Commercial	73.0 ± 0.5	62.0 ± 1.1	73.0 ± 0.3	62.0 ± 1.0	74.0 ± 1.0	64.0 ± 2.0	75.0 ± 0.9	67.0 ± 1.6
Triplicane		Residential	68.0 ± 0.4	55.0 ± 1.0	68.0 ± 0.5	56.0 ± 0.8	68.0 ± 0.5	56.0 ± 0.7	68.0 ± 1.4	57.0 ± 2.0

Table 5.5 Annual average ambient levels, L_{day} and L_{night} for 35 noise monitoring stations installed across seven major cities in India for the years 2015–2020

Name of location	City	Area characteristics	2015		2016		2017		2018		2019		2020	
			L_{day}	L_{night}	L_{day}	L_{night}	L_{day}	L_{night}	L_{day}	L_{night}	L_{day}	L_{night}	L_{day}	L_{night}
Dilshad Garden	Delhi	Silence	57.3 ± 3.3	54.0 ± 5.6	53.0 ± 1.4	51.0 ± 2.7	54.0 ± 1.5	53.0 ± 0.4	64.1 ± 1.1	57.0 ± 0.6	68.7 ± 2.6	68.4 ± 2.8	70.4 ± 3.5	70.6 ± 3.5
CPCB HQ.		Commercial	69.2 ± 1.8	59.1 ± 1.5	66.0 ± 0.9	57.0 ± 0.8	66.0 ± 0.5	57.0 ± 1.5	66.1 ± 0.4	58.0 ± 0.7	66.5 ± 0.5	56.7 ± 0.7	64.5 ± 2.3	67.4 ± 2.3
DTU, Bawana		Silence	63.0 ± 6.9	62.0 ± 11	57.0 ± 3.5	57.0 ± 7.6	56.0 ± 2.0	52.0 ± 3.1	56.2 ± 1.0	52.3 ± 2.0	62.2 ± 2.9	57.9 ± 2.3	57.2 ± 3.7	54.8 ± 3.3
ITO		Commercial	74.0 ± 0.9	70.0 ± 1.2	75.0 ± 4.2	71.0 ± 4.3	72.0 ± 0.9	67.0 ± 0.7	73.1 ± 0.4	69.9 ± 0.6	75.1 ± 3.1	73.5 ± 3.7	79.8 ± 4.4	75.6 ± 2.7
NSIT Dwarka		Silence	60.0 ± 2.0	56.4 ± 2.1	58.0 ± 1.0	55.0 ± 1.9	57.0 ± 0.6	54.0 ± 1.0	57.4 ± 0.5	55.6 ± 1.0	57.9 ± 0.9	54.2 ± 4.0	57.4 ± 1.9	54.3 ± 1.8
Gomti Nagar	Lucknow	Silence	–	–	72.0 ± 6.2	64.0 ± 6.4	67.0 ± 1.4	59.0 ± 1.3	66.1 ± 0.6	61.8 ± 2.8	66.9 ± 1.0	60.2 ± 1.1	64.5 ± 2.1	70.9 ± 4.1
Hazrat Gunj		Commercial	73.3 ± 0.8	64.1 ± 1.5	74.0 ± 0.9	66.0 ± 2.3	73.0 ± 5.2	69.0 ± 4.8	71.5 ± 2.6	68.0 ± 2.2	74.0 ± 4.1	73.2 ± 4.5	68.0 ± 3.3	70.5 ± 4.3
Indira Nagar		Residential	61.9 ± 2.8	56.0 ± 4.7	79.0 ± 6.2	77.0 ± 8.4	57.0 ± 1.1	51.0 ± 2.2	55.3 ± 4.0	49.6 ± 4.0	50.6 ± 1.9	44.7 ± 3.0	76.5 ± 4.5	75.8 ± 4.8
PGI Hospital		Silence	63.9 ± 1.0	58.0 ± 2.3	65.0 ± 1.7	59.0 ± 2.9	63.0 ± 3.7	59.0 ± 4.4	60.4 ± 3.1	53.9 ± 3.0	66.7 ± 2.1	67 ± 3.7.0	81.7 ± 3.7	80.1 ± 5.2
Talkatora Industrial Area		Industrial	67.3 ± 0.5	61.0 ± 3.1	68.0 ± 0.9	61.0 ± 2.5	67.0 ± 2.0	61.0 ± 2.7	67.4 ± 2.4	62.1 ± 3.3	63.7 ± 3.4	62.3 ± 4.4	67.1 ± 4.6	68.9 ± 8.7
Kasba Gole Park	Kolkata	Industrial	81.0 ± 6.7	81.0 ± 7.1	81.0 ± 9.7	83.0 ± 4.2	73.0 ± 4.4	73.0 ± 4.2	68.0 ± 2.6	68.3 ± 2.6	64.4 ± 5.2	61.1 ± 3.8	69.5 ± 3.2	66.5 ± 4.0
New Market		Commercial	80.0 ± 1.1	79.0 ± 1.0	87.0 ± 0.5	87.0 ± 0.8	78.0 ± 0.7	78.0 ± 0.8	72.7 ± 2.1	72.7 ± 1.9	72.9 ± 2.8	76.3 ± 4.0	68.8 ± 5.6	67.6 ± 5.1
Patauli		Residential	70.0 ± 1.3	69.0 ± 2.2	76.0 ± 1.4	76.0 ± 4.3	67.0 ± 2.4	67.0 ± 4.3	65.6 ± 2.9	65.2 ± 3.0	73.4 ± 5.2	72.4 ± 5.7	70.6 ± 4.9	74.4 ± 6.0
SSKM Hospital		Silence	63.9 ± 7.3	58.0 ± 8.2	64.0 ± 4.9	59.0 ± 4.2	67.0 ± 4.3	60.0 ± 4.2	72.4 ± 3.4	73.8 ± 2.7	70.4 ± 3.2	69.6 ± 3.9	64.7 ± 6.1	63.0 ± 5.3
WBPCB HQ		Commercial	64.0 ± 9.9	58.1 ± 1.2	62.0 ± 4.4	56.0 ± 4.6	63.0 ± 4.2	57.0 ± 4.6	66.1 ± 3.2	63.9 ± 3.5	63.4 ± 0.9	57.0 ± 1.1	66.4 ± 4.9	65.8 ± 6.3

(continued)

Table 5.5 (continued)

Name of location	City	Area characteristics	2015		2016		2017		2018		2019		2020	
			L_{day}	L_{night}	L_{day}	L_{night}	L_{day}	L_{night}	L_{day}	L_{night}	L_{day}	L_{night}	L_{day}	L_{night}
AS HP	Mumbai	Silence	77.0 ± 1.6	75.0 ± 1.3	74.0 ± 1.1	74.0 ± 1.0	74.0 ± 1.2	74.0 ± 3.8	70.4 ± 4.4	70.5 ± 4.8	–	–	71.7 ± 11.5	69.8 ± 10.8
Bandra		Commercial	68.0 ± 1.3	64.0 ± 1.7	70.0 ± 7.4	69.0 ± 9.0	70.0 ± 4.6	70.0 ± 4.7	64.8 ± 3.2	65.0 ± 3.4	67.5 ± 4.6	66.9 ± 4.7	74.7 ± 5.3	74.4 ± 5.0
MPCB HQ.		Commercial	71.0 ± 0.9	70.0 ± 0.7	69.0 ± 2.6	67.0 ± 5.9	66.0 ± 6.3	61.0 ± 12.9	69.0 ± 2.3	74.7 ± 7.5	70.2 ± 0.5	67.3 ± 3.3	72.9 ± 4.6	64.1 ± 5.1
Thane MCQ		Commercial	64.0 ± 7.6	57.0 ± 9.4	73.0 ± 1.9	75.0 ± 2.1	61.0 ± 1.9	58.0 ± 2.3	61.0 ± 4.0	57.1 ± 2.7	69.1 ± 0.5	69.1 ± 1.1	67.9 ± 0.9	68.6 ± 1.3
Vashi Hospital		Silence	69.0 ± 1.2	59.0 ± 1.5	70.0 ± 2.2	65.0 ± 4.2	78.0 ± 1.9	83.0 ± 4.7	72.1 ± 2.3	68.5 ± 3.9	75.8 ± 7.2	74.2 ± 7.7	69.4 ± 4.7	69.1 ± 5.1
Abids	Hyderabad	Commercial	77.6 ± 0.8	69.7 ± 1.8	53.0 ± 2.0	51.0 ± 2.6	74.0 ± 0.8	66.0 ± 1.5	74.4 ± 0.8	68.2 ± 1.3	73.1 ± 0.7	65.7 ± 1.4	72.7 ± 4.7	67.4 ± 4.7
Jeedimetla		Industrial	68.0 ± 0.6	63.0 ± 0.9	57.0 ± 1.3	57.0 ± 3.6	74.0 ± 1.5	85.0 ± 1.6	80.0 ± 1.0	80.1 ± 4.6	72.6 ± 5.3	73.2 ± 6.9	71.7 ± 4.0	70.0 ± 4.6
Jubilee Hills		Residential	60.0 ± 0.5	53.2 ± 0.8	58.0 ± 3.3	55.0 ± 5.2	61.0 ± 0.7	55.0 ± 1.3	61.4 ± 1.2	55.9 ± 1.2	59.7 ± 1.2	53.8 ± 1.1	74.4 ± 4.0	69.1 ± 4.6
Punjagutta		Commercial	80.0 ± 1.8	75.0 ± 3.0	66.0 ± 2.6	57.0 ± 3.5	72.0 ± 0.6	65.0 ± 1.0	70.7 ± 1.4	64.8 ± 1.4	69.8 ± 0.9	62.9 ± 0.8	68.2 ± 2.3	61.8 ± 2.5
Zoo Park		Silence	56.0 ± 0.7	51.0 ± 1.5	75.0 ± 2.0	71.0 ± 1.7	56.0 ± 1.9	50.0 ± 1.5	55.2 ± 0.8	50.4 ± 1.0	55.8 ± 3.4	61.4 ± 3.3	69.0 ± 8.9	68.1 ± 8.8
BTM	Bengaluru	Residential	65.8 ± 3.7	58.0 ± 5.0	66.0 ± 2.9	62.0 ± 2.3	66.0 ± 3.4	66.0 ± 4.0	66.6 ± 3.6	65.9 ± 3.6	64.7 ± 3.0	64.7 ± 3.3	62.5 ± 5.3	62.5 ± 4.6
Marathahalli		Commercial	59.0 ± 0.5	57.2 ± 0.9	56.0 ± 1.1	56.0 ± 1.2	64.0 ± 0.5	63.0 ± 1.1	67.4 ± 4.5	66.2 ± 5.2	73.1 ± 3.7	72.7 ± 7.2	71.8 ± 6.1	71.4 ± 6.4
Nisarga Bhawan		Residential	58.0 ± 2.6	52.0 ± 4.3	58.0 ± 1.5	53.0 ± 4.6	57.0 ± 1.0	50.0 ± 1.0	75.2 ± 3.7	70.0 ± 4.1	76.4 ± 4.0	79.1 ± 8.0	80.2 ± 8.8	77.4 ± 8.6
Parisar Bhawan		Commercial	66.7 ± 0.8	59.0 ± 1.1	67.0 ± 1.9	62.0 ± 3.1	67.0 ± 2.3	61.0 ± 2.4	67.4 ± 1.0	61.6 ± 1.0	67.2 ± 0.5	62.1 ± 0.6	64.3 ± 2.6	58.7 ± 3.4
Peeniya		Industrial	61.0 ± 1.2	59.0 ± 1.2	64.0 ± 0.5	57.0 ± 2.4	66.0 ± 9.3	60.0 ± 7.9	64.1 ± 1.1	57.0 ± 0.6	63.4 ± 0.8	58.4 ± 1.4	62.6 ± 2.5	60.9 ± 2.7
Eye Hospital	Chennai	Silence	67.0 ± 3.1	60.0 ± 4.3	67.0 ± 1.8	59.0 ± 2.3	67.0 ± 0.0	62.0 ± 3.3	66.4 ± 6.3	58.7 ± 3.3	69.2 ± 3.8	61.0 ± 4.3	72.8 ± 6.9	69.7 ± 5.3
Guindy		Industrial	79.8 ± 1.0	75.7 ± 2.5	79.0 ± 0.6	76.0 ± 1.9	76.0 ± 0.4	72.0 ± 2.0	76.8 ± 0.4	74.9 ± 1.4	80.1 ± 3.5	78.2 ± 4.8	82.8 ± 6.8	83.6 ± 7.1
Perambur		Commercial	72.0 ± 1.5	65.0 ± 4.7	73.0 ± 1.2	71.0 ± 4.5	75.0 ± 3.7	76.0 ± 4.5	75.8 ± 1.3	76.8 ± 5.7	74.4 ± 2.3	71.6 ± 2.7	63.1 ± 3.9	59.4 ± 5.4
T. Nagar		Commercial	76.6 ± 0.4	69.3 ± 0.7	76.0 ± 1.6	68.0 ± 5.2	75.0 ± 3.0	67.0 ± 1.0	75.8 ± 1.3	70.2 ± 2.0	77.3 ± 0.9	73.0 ± 1.9	80.3 ± 6.5	75.9 ± 4.0
Triplicane		Residential	73.0 ± 3.1	68.0 ± 5.6	73.0 ± 5.8	72.0 ± 7.3	69.0 ± 10.4	58.0 ± 4.6	67.1 ± 3.5	63.8 ± 3.5	60.9 ± 3.1	59.0 ± 4.1	70.0 ± 4.6	68.9 ± 5.0

Table 5.6 Annual average ambient levels, L_{day} and L_{night} for additional 35 noise monitoring stations installed in November 2014 across seven major cities for the years 2015–2020

Name of location	City	Area characteristics	2015		2016		2017		2018		2019		2020	
			L_{day}	L_{night}	L_{day}	L_{night}	L_{day}	L_{day}	L_{night}	L_{night}	L_{day}	L_{night}	L_{day}	L_{night}
Civil Lines	Delhi	Commercial	73.0 ± 5.3	73.0 ± 5.3	69.0 ± 1.0	70.0 ± 1.9	66.0 ± 4.7	65.0 ± 5.7	61.2 ± 0.5	58.8 ± 0.6	61.7 ± 0.6	58.3 ± 0.5	60.6 ± 1.0	57.3 ± 1.1
R. K. Puram		Residential	66.0 ± 6.6	59.0 ± 7.6	65.0 ± 5.0	62.0 ± 7.8	62.0 ± 2.9	56.0 ± 4.9	62.8 ± 2.5	60.3 ± 4.4	66.8 ± 2.9	61.6 ± 1.0	63.3 ± 3.1	59.2 ± 3.3
Anand Vihar		Commercial	70.0 ± 3.3	68.0 ± 5.6	68.0 ± 0.8	67.0 ± 4.1	68.0 ± 1.4	65.0 ± 1.4	66.6 ± 1.3	65.9 ± 2.9	65.6 ± 0.4	61.9 ± 0.6	65.5 ± 5.0	60.6 ± 4.2
Mandir Marg		Silence	76.0 ± 5.8	73.0 ± 4.8	76.0 ± 12.8	61.0 ± 9.4	66.0 ± 7.6	49.0 ± 1.8	61 ± 3.9	62.6 ± 2.3	65.4 ± 2.6	54.8 ± 3.0	58.3 ± 3.5	56.6 ± 2.5
Punjabi Bagh		Residential	86.0 ± 1.1	81.0 ± 2.3	59.0 ± 1.0	53.0 ± 1.7	64.0 ± 5.7	62.0 ± 8.5	59.8 ± 1	53.4 ± 1.7	61.3 ± 1.1	52.4 ± 1.3	57.3 ± 1.2	51.7 ± 1.2
Chinhat	Lucknow	Industrial	68.6 ± 5.5	63.0 ± 9.1	68.0 ± 3.4	62.0 ± 5.2	65.0 ± 2.1	56.0 ± 1.8	66.7 ± 1.7	57.9 ± 2.2	67.3 ± 2.4	60.0 ± 4.0	67.1 ± 3.5	56.5 ± 4.8
IT College		Silence	65.0 ± 4.1	57.4 ± 1.3	65.0 ± 2.0	59.0 ± 1.9	68.0 ± 3.5	62.0 ± 4.6	65.3 ± 1.0	70.0 ± 3.8	64.7 ± 1.0	59.5 ± 1.3	64.7 ± 2.3	71.5 ± 2.8
CSS Airport		Commercial	74.7 ± 8.6	66.1 ± 6.0	76.0 ± 0.9	67.0 ± 0.8	76.0 ± 0.8	67.0 ± 4.7	59.8 ± 1.2	70.0 ± 3.8	76.3 ± 0.9	67.0 ± 1.1	62.7 ± 2.1	58.3 ± 2.7
RSC Aliganj		Commercial	63.0 ± 0.8	58.1 ± 0.7	66.0 ± 0.7	61.0 ± 0.7	66.0 ± 0.5	60.0 ± 0.6	76.7 ± 0.5	70.9 ± 2.6	65.1 ± 0.6	57.6 ± 0.6	73.6 ± 4.2	65.0 ± 4.6
Vibhuti Khand		Residential	63.4 ± 1.3	57.0 ± 2.1	64.0 ± 3.3	60.0 ± 4.8	64.0 ± 0.9	60.0 ± 2.2	66.4 ± 2.8	59.9 ± 1.7	63.9 ± 0.7	62.2 ± 1.3	62.7 ± 1.7	57.5 ± 1.4
Birati N.	Kolkata	Residential	63.2 ± 2.6	58.0 ± 4.2	67.0 ± 5.9	70.0 ± 3.6	64.0 ± 2.8	58.0 ± 3.6	80.4 ± 1.8	80.1 ± 3.5	79.1 ± 8.9	77.9 ± 10.6	66.9 ± 3.5	63.3 ± 3.8
R G Kar		Silence	70.0 ± 8.6	63.0 ± 6.0	64.0 ± 0.9	60.0. ± 1.7	64.0 ± 0.8	60.0. ± 1.7	66.5 ± 1.9	60.4 ± 1.3	63.7 ± 3.3	62.2 ± 3.0	68.7 ± 3.1	71.8 ± 3.0
Tollygunge		Commercial	66.8 ± 0.5	62.9 ± 0.7	67.0 ± 0.4	63.0 ± 0.6	67.0 ± 2.4	63.0 ± 0.6	67.2 ± 0.4	64.1 ± 0.4	66.8 ± 0.4	62.9 ± 0.6	65.5 ± 1.5	61.6 ± 1.4
Bag Bazar		Residential	78.0 ± 4.4	70.0 ± 2.5	78.0 ± 3.5	69.0 ± 1.0	75.0 ± 0.9	68.0 ± 1.0	75.3 ± 1.1	70.7 ± 1.8	87.2 ± 4.6	90.1 ± 3.9	77.1 ± 9.5	74.6 ± 11.2
Tartala		Industrial	64.7 ± 1.1	61.1 ± 1.7	65.0 ± 1.2	62.0 ± 1.3	65.0 ± 1.6	62.0 ± 1.3	69.3 ± 1.2	67.2 ± 1.2	75.3 ± 1.9	73.1 ± 1.6	68.1 ± 2.0	64.4 ± 2.0
M&M Kandivali	Mumbai	Industrial	62.1 ± 1.1	55.8 ± 1.6	66.0 ± 4.5	69.0 ± 11.5	62.0 ± 0.7	57.0 ± 2.7	61.1 ± 0.7	54.8 ± 1.2	60.4 ± 1.0	52.6 ± 1.5	60.8 ± 2.9	53.8 ± 2.4
Ambassador Hotel		Commercial	73.1 ± 0.9	68.5 ± 1.3	73.0 ± 0.8	68.0 ± 0.8	72.0 ± 0.9	67.0 ± 1.1	72.5 ± 0.6	68.1 ± 0.8	74.5 ± 2.9	70.3 ± 3.2	74.2 ± 3.8	69.6 ± 5.1
L&T Powai		Industrial	60.3 ± 1.2	57.4 ± 1.5	60.0 ± 1.2	57.0 ± 1.9	59.0 ± 1.7	55.0 ± 2.7	58.4 ± 1.6	53.6 ± 2.3	59.4 ± 2.7	55.1 ± 3.2	62.2 ± 2.8	60.0 ± 4.8
Pepsico Chembur		Residential	67.0 ± 4.3	63.0 ± 7.8	64.0 ± 1.0	58.0 ± 2.1	64.0 ± 1.6	60.0 ± 4.8	66.8 ± 2.6	60.6 ± 2.9	68.1 ± 1.9	61.3 ± 1.4	76.5 ± 4.5	75.4 ± 4.6
Andheri		Industrial	71.3 ± 1.4	67.6 ± 0.8	72.0 ± 2.7	67.0 ± 0.8	71.0 ± 1.6	68.0 ± 1.6	71.7 ± 2.1	67.7 ± 1.5	78.9 ± 3.1	78.0 ± 4.0	80.1 ± 5.3	79.6 ± 5.7
Tarnaka	Hyderabad	Residential	60.2 ± 1.1	54.2 ± 3.2	69.0 ± 2.7	70.0 ± 5.6	67.0 ± 5.6	62.0 ± 7.6	73.5 ± 2.8	72.6 ± 3.4	81.4 ± 10.7	81.7 ± 11.2	76.6 ± 6.4	70.0 ± 5.6

(continued)

Table 5.6 (continued)

Name of location	City	Area characteristics	2015		2016		2017		2018		2019		2020	
			L_{day}	L_{night}	L_{day}	L_{night}	L_{day}	L_{day}	L_{night}	L_{night}	L_{day}	L_{night}	L_{day}	L_{night}
Gaddapothram		Industrial	65.0 ± 2.0	59.0 ± 3.5	65.0 ± 3.8	62.0 ± 5.1	78.0 ± 4.4	75.0 ± 3.9	75.1 ± 4.5	67.8 ± 5.3	83.3 ± 5.0	74.0 ± 6.9	77.3 ± 4.8	75.2 ± 5.2
Gachibowli		Silence	60.9 ± 2.2	55.0 ± 3.4	68.0 ± 4.9	67.0 ± 4.8	63.0 ± 5.6	60.0 ± 3.4	61.0 ± 2.3	56.1 ± 3.1	60.0 ± 2.5	58.2 ± 2.7	60.9 ± 2.4	71.7 ± 5.5
Paradise		Commercial	79.1 ± 0.9	74.9 ± 1.0	76.0 ± 3.1	61.0 ± 1.8	76.0 ± 4.6	75.0 ± 7.7	84.6 ± 3.5	85.6 ± 3.5	81.4 ± 2.5	80.3 ± 2.9	78.0 ± 2.7	73.9 ± 3.3
Kukatpalli		Commercial	70.3 ± 1.1	66.4 ± 1.0	59.0 ± 2.2	53.0 ± 2.7	71.0 ± 1.1	66.0 ± 1.7	70.6 ± 1.1	67.5 ± 0.8	69.9 ± 0.7	66.7 ± 0.8	68.2 ± 2.4	65.4 ± 3.1
Yeshwantpur	Bengaluru	Commercial	72.1 ± 0.5	63.4 ± 0.7	72.0 ±0.5	64.0 ± 0.8	72.0 ± 0.5	64.0 ± 0.9	72.0 ± 0.6	65.5 ± 0.8	72.1 ± 0.4	64.0 ± 0.6	70.2 ± 2.1	64.0 ± 2.2
R.V.C.E		Silence	60.1 ± 0.7	54.1 ± 0.7	60.0 ± 0.7	55.0 ± 0.8	60.0 ± 1.2	55.0 ± 0.7	59.6 ± 0.9	53.9 ± 0.6	67.4 ± 0.9	65.9 ± 2.6	80.9 ± 3.0	80.9 ± 4.2
Whitefield		Industrial	67.1 ± 0.7	61.2 ± 0.6	68.0 ± 1.7	62.0 ± 0.2	72.0 ± 5.6	66.0 ± 4.8	66.9 ± 1.1	72.5 ± 4.2	66.2 ± 0.8	61.1 ± 1.1	63.9 ± 1.9	58.6 ± 2.1
Dolmur		Residential	66.0 ± 3.3	60.0 ± 4.5	64.0 ± 1.8	59.0 ± 1.9	68.0 ± 5.7	61.0 ± 5.8	64.5 ± 1.8	59.2 ± 2.5	64.3 ± 1.8	59.3 ± 2.6	62.9 ± 1.5	58.1 ± 2.7
Nihmans		Silence	63.0 ± 3.4	60.0 ± 5.1	63.0 ± 2.9	65.0 ± 8.1	67.0 ± 6.8	66.0 ± 9.2	66.6 ± 5.6	64.5 ± 3.8	72.0 ± 3.9	74.1 ± 3.0	70.1 ± 5.4	74.9 ± 4.0
Pallikarnai	Chennai	Commercial	73.8 ± 0.6	67.2 ± 0.6	77.0 ± 4.3	71.0 ± 0.5	78.0 ± 4.2	77.0 ± 6.7	72.6 ± 1.9	72.5 ± 2.2	84.4 ± 4.9	78.6 ± 5.0	72 ± 5.4.0	72.6 ± 9.9
Velachery		Residential	67.0 ± 3.2	67.0 ± 9.4	68.0 ± 3.8	70.0 ± 10.4	74.0 ± 6.0	69.0 ± 7.7	77.3 ± 4.9	76.2 ± 8.2	65.5 ± 1.8	59.7 ± 4.1	64.4 ± 2.6	67.6 ± 2.0
Washermanpet		Commercial	70.0 ± 1.4	67.0 ± 5.7	72.0 ± 1.9	65.0 ± 2.6	72.0 ± 2.0	66.0 ± 5.7	74.4 ± 2.8	72.9 ± 4.2	75.6 ± 4.4	71.5 ± 4.3	74.0 ± 4.4	74.1 ± 5.3
Anna Nagar		Silence	67.0 ± 1.7	63.0 ± 4.6	68.0 ± 2.5	66.0 ± 7.9	66.0 ± 1.4	60.0 ± 3.0	67.1 ± 1.7	61.5 ± 3.4	75.0 ± 4.3	77.8 ± 4.0	73.4 ± 5.7	72.8 ± 7.8
Sowcarpet		Residential	66.6 ± 1.2	60.0 ± 2.7	70.0 ± 3.2	69.0 ± 6.8	68.0 ± 3.4	65.0 ± 5.7	66.9 ± 1.4	62.9 ± 3.1	62.3 ± 3.5	60.2 ± 3.0	65.1 ± 4.7	60.7 ± 3.6

5.5 Evaluation and Analysis of Environmental Noise Levels

5.5.1 Environmental Noise Scenario: Cluster Analysis

The environmental day and night equivalent noise levels can be analyzed either on city-wise or zone-wise basis. A cluster-based zone-wise analysis would be quite suitable for understanding the noise scenario in each zone, comparison with the legal limits and devising appropriate noise control action plans. Table 5.7 shows the frequency distribution of annual average L_{day} and L_{night} values observed in dB for the 70 sites spread across the seven major cities of India for past 3 years. The cluster analysis of the environmental noise levels based on zone-wise revealed the following important observations:

- The analysis of year 2020 noise monitored data shows that 88% of the commercial sites, 62.6% of the residential sites, 70.5% of the silence zone sites, and 74.9% of the industrial zone sites registered day equivalent noise levels between 60 and 75 dB(A). Seventy-six percent of the commercial sites registered night equivalent levels between 60 and 75 dB(A), while 56.4% of the residential sites, 70.6% of the silence zone sites, and 66.7% of the industrial sites registered night equivalent noise levels between 55 and 70 dB(A). Thus, it is evident that majority of the sites in all four zones showed day equivalent noise levels between 60 and 75 dB(A), while the majority of sites in all zones except the commercial zone showed night equivalent levels between 55 and 70 dB(A).
- The 60 dB(A) L_{day} limit can be seen as *NOAEL* (no observed adverse effect level) for the correlation between road traffic noise and myocardial infarction (MI); the risk of MI increases incessantly for noise levels higher than 60 dB(A) [73, 74]. It can be observed that 60 dB L_{day} limit was met by only four sites for the year 2020, while seven sites met the limit in year 2019 and nine sites in the year 2018. The Interim Target (IT) of 55 dB L_{night} as recommended in EU Night Noise Guidelines report [74] was met by only four sites in year 2020, seven sites in year 2019 and eight sites in year 2018.
- The majority of sites (seven out of ten) in Chennai city showed the day sound levels above 70 dB(A) and six sites showed night equivalent sound levels above 70 dB (A). However, for the Delhi city, only two sites showed day and night equivalent levels $\geq$ 70 dB(A). Also, for the Mumbai and Hyderabad cities, six sites showed day equivalent levels $\geq$ 70 dB(A) and five sites showed night equivalent levels $\geq$ 70 dB(A). Overall, 32 sites (45.7%) comprising of 12 commercial, 5 industrial, 7 in silence zone, and 8 residential were the Most Exposed Urban Sites with day equivalent sound levels $\geq$70 dB(A). Thirty sites (42.9%) comprising of 9 commercial, 4 industrial, 11 in silence zone and 6 residential were the Most Exposed Urban Sites with night equivalent sound levels $\geq$70 dB(A).
- The majority of annual 24-h equivalent sound levels, $L_{\mathrm{Aeq,24h}}$ values as shown in Table 5.8 range from 60 to 75 dB(A) for commercial and industrial zone sites, while for the residential and silence zone sites, the majority of $L_{\mathrm{eq,24h}}$ values range from 55 to 70 dB(A). For instance, 84% of the observations of $L_{\mathrm{eq,24h}}$ values range from 60 to 75 dB(A) for commercial zone sites, and 83.2% values lie in

Table 5.7 Frequency distribution of annual average L_{day} and L_{night} values observed in dB for the 70 sites spread across the seven major cities of India for past 3 years

Variation of L_{day} and L_{night} values in dB (A)	Percentage of noise monitoring locations in various zones																							
	Commercial						Residential						Silence						Industrial					
	2018		2019		2020		2018		2019		2020		2018		2019		2020		2018		2019		2020	
	L_{day}	L_{night}	L_{day}	L_{night}	L_{day}	L_{night}	L_{day}	L_{night}	L_{day}	L_{night}	L_{day}	L_{night}	L_{day}	L_{night}	L_{day}	L_{night}	L_{day}	L_{night}	L_{day}	L_{night}	L_{day}	L_{night}	L_{day}	L_{night}
$40 < L_{eq} \leq 45$	0	0	0	0	0	0	0	0	0	6.3	0	0	0	0	0	0	0	0	0	0	0	0	0	0
$45 < L_{eq} \leq 50$	0	0	0	0	0	0	0	6.3	0	0	0	0	0	0	0	0	0	0	0	0	0	0	0	0
$50 < L_{eq} \leq 55$	0	0	0	0	0	0	0	6.3	6.3	12.5	0	6.3	0	23.5	0	12.5	0	11.8	0	16.7	0	8.3	0	8.3
$55 < L_{eq} \leq 60$	4.0	12.0	0	16.0	0	16.0	12.5	18.8	6.3	18.8	6.3	18.8	23.5	23.5	12.5	18.8	17.6	5.9	8.3	16.7	8.3	25.0	0	25.0
$60 < L_{eq} \leq 65$	12.0	20.0	8.0	20.0	20.0	20.0	18.8	25.0	37.5	31.3	31.3	18.8	23.5	29.4	25.0	25.0	23.5	41.2	16.7	8.3	33.3	25.0	33.3	16.7
$65 < L_{eq} \leq 70$	28.0	36.0	36.0	24.0	32.0	32.0	37.5	18.8	18.8	0	18.8	18.8	35.3	11.8	37.5	25.0	17.6	23.5	41.7	33.3	16.7	0	33.3	25.0
$70 < L_{eq} \leq 75$	40.0	24.0	32.0	28.0	36.0	24.0	6.3	12.5	6.3	6.3	12.5	18.8	17.6	11.8	18.8	12.5	29.4	5.9	8.3	16.7	8.3	25.0	8.3	0
$75 < L_{eq} \leq 80$	12.0	4.0	16.0	8.0	8.0	8.0	18.8	6.3	12.5	12.5	25.0	18.8	0	0	6.3	6.3	0	0	25.0	0	16.7	16.7	8.3	16.7
$80 < L_{eq} \leq 85$	4.0	0	8.0	4.0	4.0	0	6.3	6.3	6.3	6.3	6.3	0	0	0	0	0	11.8	11.8	0	8.3	16.7	0	16.7	8.3
$85 < L_{eq} \leq 90$	0	4.0	0	0	0	0	0	0	6.3	6.3	0	0	0	0	0	0	0	0	0	0	0	0	0	0

Table 5.8 Frequency distribution of annual average 24-h equivalent sound levels, $L_{Aeq,24h}$ in dB (A) of seven major cities in past 2 years

Variation of single value noise descriptors	Commercial		Residential		Silence		Industrial		Overall	
	2019	2020	2019	2020	2019	2020	2019	2020	2019	2020
$45 < L_{eq} \leq 50$ dB(A)	0	0	6.3	0	0	0	0	0	1.4	0
$50 < L_{eq} \leq 55$ dB(A)	0	0	0	0	0	0	0	0	0	0
$55 < L_{eq} \leq 60$ dB(A)	0	4.0	18.8	6.3	12.5	21.4	16.7	8.3	10.1	7.5
$60 < L_{eq} \leq 65$ dB(A)	24.0	32.0	31.3	43.8	31.3	35.7	41.7	33.3	30.4	34.3
$65 < L_{eq} \leq 70$ dB(A)	28.0	44.0	12.5	25.0	31.3	21.4	0.0	33.3	20.3	28.4
$70 < L_{eq} \leq 75$ dB(A)	36.0	8.0	6.3	0.0	12.5	14.3	16.7	8.3	20.3	14.9
$75 < L_{eq} \leq 80$ dB(A)	4.0	12.0	12.5	6.3	6.3	0	16.7	8.3	8.7	7.5
$80 < L_{eq} \leq 85$ dB(A)	8.0	0	6.3	12.5	0	7.1	8.3	8.3	5.8	6.0
$85 < L_{eq} \leq 90$ dB(A)	0	0	6.3	6.3	0	0	0	0	1.4	1.5

Table 5.9 Frequency distribution of annual average sound levels, L_{dn} in dB(A) of seven major cities in past 2 years

Variation of single value noise descriptors	Commercial		Residential		Silence		Industrial		Overall	
	2019	2020	2019	2020	2019	2020	2019	2020	2019	2020
$45 < L_{dn} \leq 50$ dB(A)	0	0	0	0	0	0	0	0	0	0
$50 < L_{dn} \leq 55$ dB(A)	0	0	6.3	0	0	0	0	0	1.4	0
$55 < L_{dn} \leq 60$ dB(A)	0	0	6.3	6.3	0	0	0	0	0	1.4
$60 < L_{dn} \leq 65$ dB(A)	4.0	4.0	12.5	6.3	12.5	17.6	16.7	8.3	10.1	8.6
$65 < L_{dn} \leq 70$ dB(A)	24.0	24.0	43.5	25.0	43.8	5.9	41.7	33.3	36.2	21.4
$70 < L_{dn} \leq 75$ dB(A)	24.0	36.0	6.3	12.5	18.8	43.8	0.0	25.0	14.5	21.4
$75 < L_{dn} \leq 80$ dB(A)	32.0	16.0	12.5	25.0	12.5	15.0	16.7	16.6	18.8	25.7
$80 < L_{dn} \leq 85$ dB(A)	8.0	20.0	6.3	25.0	12.5	5.9	25.0	8.3	13.0	15.7
$85 < L_{dn} \leq 90$ dB(A)	8.0	0	6.3	0	0	11.8	0	8.3	4.3	5.7

range of 60–75 dB(A) for industrial zone sites. The percentage of observations ranging from 55 to 70 dB(A) for residential sites is observed to be 75.1% and that for silence zone sites is 78.5%. Overall, for the 70 noise monitoring sites, 77.6% of the observations for $L_{Aeq,24h}$ values lie between 60 to 75 dB(A).

- The majority of annual day-night average sound levels, L_{dn} values as shown in Table 5.9 range from 65 to 80 dB(A) for commercial, residential, and industrial zone sites, while for the silence zone sites, the majority of L_{dn} values range from 60 to 75 dB(A). For instance, 76% of the observations of L_{dn} values range from 65 to 80 dB(A) for commercial zone sites, 62.5% values lie in range of 65–80 dB for the residential zone sites, 74.9% values lie in range of 65–80 dB(A) for industrial zone sites, and 67.3% values lie in range of 60–75 dB(A) for the silence zone sites. Overall, for 70 noise monitoring sites, 68.5% of observations for L_{dn} values lie between 65 and 80 dB(A).

5.5.2 *Monthly Day and Night Equivalent Noise Levels*

The analysis of monthly day and night equivalent noise levels for the 70 sites under consideration for past 3 years revealed the following inferences:

- Majority of monthly day equivalent sound levels as shown in Table 5.10 varied from 60 to 75 dB(A) for commercial zone and industrial sites, while for the residential and silence zone sites, the majority of day equivalent levels varied from 55 to 70 dB(A). The 60 dB(A) L_{day} limit was experienced by very few sites as only 20.4% of the sites experienced the day equivalent levels up to 60 dB(A). It can be observed that commercial zone sites had the least probability, while silence zone sites had the highest probability of day equivalent levels up to 60 dB(A). Overall, 67.9% of the observations showed that day equivalent noise levels varied from 60 to 75 dB(A) in 2020.
- Majority of monthly night equivalent levels as shown in Table 5.11 varied from 55 to 70 dB(A) for commercial zone and industrial sites, while for the residential and the zone sites, the majority of night equivalent levels varied from 50 to 65 dB (A). For the year 2020, 69.2% of the observations showed that the night equivalent levels varied from 55 to 70 dB(A) for commercial zone sites; 72.4% of the observations showed that the night equivalent levels varied from 50 to 65 dB (A) for residential zone sites; 55.5% of the observations showed that the night equivalent levels varied from 50 to 65 dB(A) for the silence zone sites and 57.6% of the observations showed that the night equivalent sound levels varied from 55 to 70 dB(A) in industrial zone sites. Overall, 59.2% of the observations showed that night equivalent noise levels varied from 55 to 70 dB(A) in the year 2020.
- It can be observed from Table 5.11 that all the monthly L_{night} values exceeded the ambient noise standards for residential and silence zone sites. Only 27.1% of the observations in residential zone and 27.5% of the observations in silence zone met the Interim Target (IT) of 55 dB recommended in EU Night Noise Guidelines report [74]. In case of commercial zone sites, only 10.0% of the observations met the IT of 55 dB. Only 17.4% of L_{night} values in the industrial zone met the 55 dB limit. Overall, 80.3% of the observations exceeded the IT of 55 dB, which indicates that 55 dB L_{night} limit is an ambitious target for all the sites under consideration.

Table 5.10 Frequency distribution of monthly L_{day} values observed in dB for the 70 sites spread across the seven major cities of India for the past 3 years

Variation of L_{day} values in dB	Frequency distribution (in %)														
	Commercial Zone			Residential Zone			Silence Zone			Industrial Zone			Overall 70 sites		
	2018	2019	2020	2018	2019	2020	2018	2019	2020	2018	2019	2020	2018	2019	2020
45 dB<L_{day}≤50 dB	0	0.7	0.7	1.0	4.2	0	0.5	2.9	2.9	0	0.7	0	0.4	2.0	1.0
50 dB<L_{day}≤55 dB	1.0	0	0.7	1.6	5.7	4.2	5.4	5.4	12.3	0	1.4	2.8	2.0	2.9	4.7
55 dB<L_{day}≤60 dB	3.7	1.7	7.4	13.0	18.8	21.4	30.9	34.3	22.1	8.3	14.6	10.4	13.2	15.7	13.2
60 dB<L_{day}≤65 dB	16.7	17.3	26.5	40.1	34.4	38.5	19.1	17.6	24.5	24.3	23.6	27.8	23.9	22.4	29.0
65 dB<L_{day}≤70 dB	25.0	34.0	32.2	26.6	18.2	16.1	32.4	25.5	14.7	39.6	31.3	31.9	29.6	27.9	24.2
70 dB<L_{day}≤75 dB	37.7	28.0	17.4	5.7	3.1	5.7	7.8	4.4	12.7	13.9	9.7	15.3	19.0	13.5	14.7
75 dB<L_{day}≤80 dB	12.3	13.3	10.7	8.3	5.2	7.8	2.0	4.9	6.4	11.1	11.1	5.6	8.7	9.0	8.1
80 dB<L_{day}≤85 dB	2.3	4.3	3.7	2.6	1.0	5.7	0.5	1.5	2.5	0.7	2.1	1.4	1.7	2.5	3.5
85 dB<L_{day}≤90 dB	1.3	0.7	0.7	1.0	4.7	0	0	0.5	1.5	2.1	4.9	4.9	1.1	2.3	1.4

Table 5.11 Frequency distribution of monthly L_{night} values observed in dB for the 70 sites spread across the seven major cities of India for the past 3 years

Variation of L_{night} values in dB	Frequency distribution (in %)														
	Commercial Zone			Residential Zone			Silence Zone			Industrial Zone			Overall 70 sites		
	2018	2019	2020	2018	2019	2020	2018	2019	2020	2018	2019	2020	2018	2019	2020
40 dB< L_{night} ≤ 45 dB	0	0	0.3	2.1	2.1	1.0	1.0	0.5	0.5	0	0	0	0.7	0.6	0.5
45 dB< L_{night} ≤50 dB	0	0.3	1.0	1.0	4.2	2.1	9.3	6.9	10.8	0	0.7	2.8	2.5	2.9	3.9
50 dB< L_{night} ≤ 55 dB	2.7	0.3	8.7	18.2	21.4	24.0	24.5	17.6	16.2	13.2	12.5	14.6	13.3	11.9	15.3
55 dB< L_{night} ≤ 60 dB	16.7	16.3	22.8	29.7	27.6	30.2	29.4	36.8	22.1	24.3	23.6	19.4	24.0	25.6	23.7
60 dB< L_{night} ≤ 65 dB	23.0	26.0	23.2	20.8	15.6	18.2	20.1	17.6	17.2	27.1	20.1	23.6	22.5	20.8	18.4
65 dB< L_{night} ≤ 70 dB	34.7	32.7	23.2	17.2	11.5	9.4	8.3	5.9	7.8	16.7	20.8	16.0	21.2	18.2	17.1
70 dB< L_{night} ≤ 75 dB	14.7	13.3	10.7	6.3	1.0	5.7	5.4	4.4	8.8	13.2	14.6	11.8	10.2	8.5	9.3
75 dB< L_{night} ≤ 80 dB	4.0	6.7	8.4	1.0	5.2	7.3	1.5	6.4	13.2	2.8	2.1	6.3	2.5	5.5	8.9
80 dB< L_{night} ≤ 85 dB	3.3	4.0	1.7	2.1	2.6	1.6	0.5	0.5	2.0	2.1	1.4	1.4	2.1	2.4	1.7
85 dB< L_{night} ≤ 90 dB	0.7	0.3	0	1.0	4.7	0.5	0	0.5	1.5	0.7	4.2	4.2	0.6	2.0	1.2

5.5.3 *Difference Between Day and Night Equivalent Noise Levels*

The analysis of monthly and annual average ($L_{\text{day}} - L_{\text{night}}$) values observed in dB for the 70 sites for past decade is analyzed for the following reasons:

- Analyze the difference of day and night equivalent levels, L_{day} and L_{night} levels to ascertain the severity of night noise levels in comparison to the day ambient noise levels.
- Ascertain whether 10 dB night time adjustment is appropriate or not.
- Recommend a suitable single value noise descriptor as followed in European continent also in Indian scenario for correlating the noise annoyance and for developing exposure-effect relationship and correlating with health aspects such as annoyance, hypertension, coronary heart disease, etc.

Figure 5.4a and b shows the frequency distribution of monthly ($L_{\text{day}} - L_{\text{night}}$) values observed in dB for the 70 sites spread across the seven major cities of India for the past 3 years. Table 5.12 shows the statistical analysis for ($L_{\text{day}} - L_{\text{night}}$) values observed in dB for the 70 sites. The analysis of ($L_{\text{day}} - L_{\text{night}}$) for the overall 70 sites shows that 75.7% of observations show a difference between 0 and 10 dB(A) in year 2020. Only 1.4% of the observations show a difference between 10 and 20 dB for all the 70 sites. Overall, it can be observed that majority of observations pertaining to ($L_{\text{day}} - L_{\text{night}}$) levels in past 10 years lie between 0 and 10 dB(A).

Table 5.13 shows the statistical analysis for monthly ($L_{\text{day}} - L_{\text{night}}$) values observed in dB for the commercial, residential, industrial, and silence zones sites exclusively for the past 2 years. It can be observed that majority of the monthly difference between the day and night equivalent levels lie between 0 and 10 dB(A), which indicates that the night noise levels are as severe as the day equivalent noise levels. Only 1.3–2.5% of the observations show a difference $\geq$ 10 dB(A), which indicates the non-suitability of 10 dB night time adjustment to compensate for the sleep interference and other disruptions used in calculation of day-night average sound level descriptor in Indian perspectives particularly for these metropolitan cities considered in the study. These observations also suggest that 5 dB evening time correction in day-evening-night average sound level descriptor is not justified. Thus, the 24-h equivalent continuous sound level, $L_{\text{Aeq,24h}}$, would rather be more suitable as it is a common way of expressing day-night average sound level without 10 dB night adjustment. In Indian scenario, the ambient noise standards are recommended for day equivalent sound levels and night equivalent sound levels. However, for correlating the noise annoyance, one single noise descriptor would be rather more rational as followed in European continent [75].

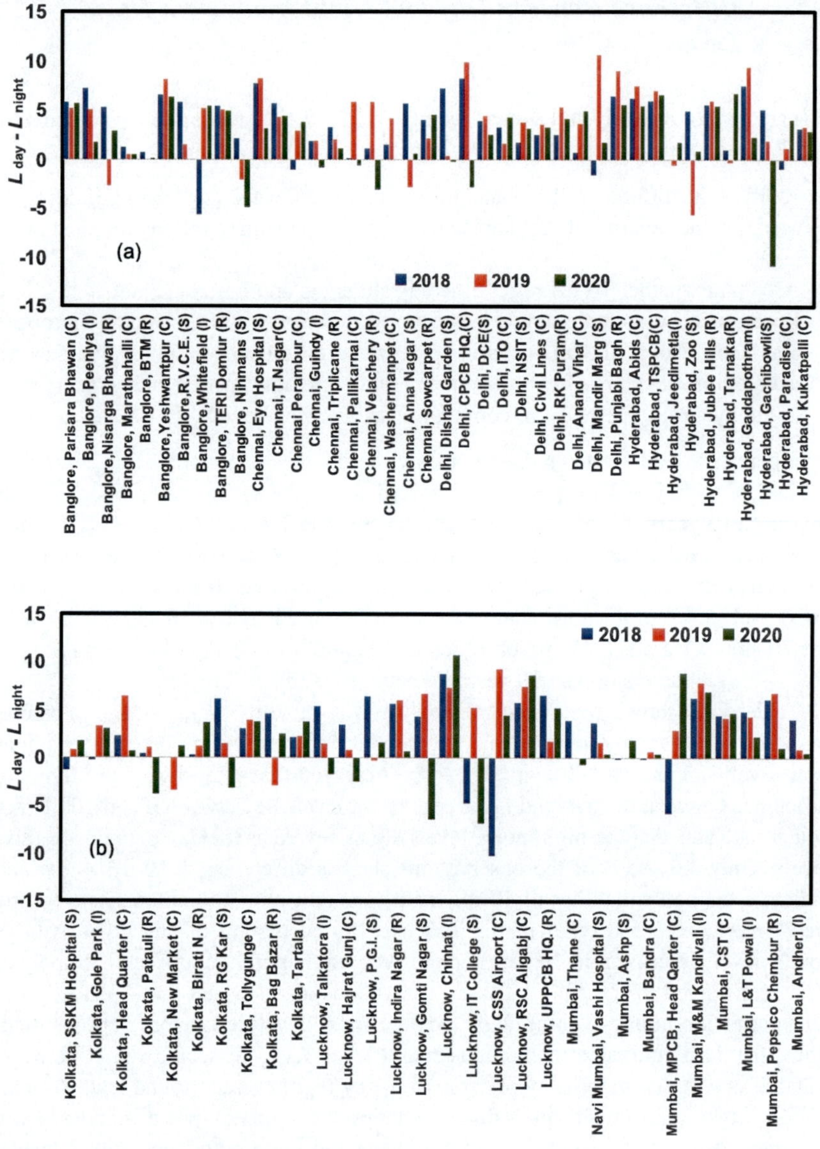

Fig. 5.4 **(a, b)** Annual average ($L_{day} - L_{night}$) in dB at 70 locations spread across three major cities in the past 3 years

5.5.4 Correlations Between Noise Descriptors

The correlation analysis of various single noise descriptors is done so as to develop empirical formulations for converting one indices to the other exclusively for the four different zones based on the past 3 years noise monitored data. Table 5.14

Table 5.12 Frequency distribution (in %) of difference of annual average ($L_{day} - L_{night}$) values observed in dB for the 70 sites spread across the seven major cities of India

Variation of difference ($L_{day} - L_{night}$) values in dB	2011	2012	2013	2014	2015	2016	2017	2018	2019	2020
$-15 < (L_{day} - L_{night}) \leq -10$	0	0	0	0	0	0	1.4	1.4	0	1.4
$-10 < (L_{day} - L_{night}) \leq -5$	0	0	0	0	0	0	1.4	2.9	1.4	2.9
$-5 < (L_{day} - L_{night}) \leq 0$	0	0	0	0	0	20.0	10.0	14.3	13.0	18.6
$0 < (L_{day} - L_{night}) \leq 5$	34.3	28.6	40.0	31.4	50.7	47.1	47.1	45.7	52.2	58.6
$5 < (L_{day} - L_{night}) \leq 10$	54.3	57.1	48.6	57.1	35.0	30.0	37.1	35.7	31.9	17.1
$10 < (L_{day} - L_{night}) \leq 15$	11.4	14.3	11.4	11.4	14.3	2.9	1.4	0	1.4	1.4
$15 < (L_{day} - L_{night}) \leq 20$	0	0	0	0	0	0	1.4	0	0	0

Table 5.13 Frequency distribution of difference of monthly ($L_{day} - L_{night}$) values observed in dB for commercial, residential, silence, and industrial zones sites exclusively

Variation of difference (L_{day} - L_{night}) values in dB	Commercial Zone		Residential Zone		Silence Zone		Industrial Zone	
	2019	2020	2019	2020	2019	2020	2019	2020
$-15 < (L_{day} - L_{night}) \leq -10$ dB	0.3 %	0.0 %	0.0 %	0.5 %	0.0 %	1.0 %	0.0 %	1.4 %
$-10 < (L_{day} - L_{night}) \leq -5$ dB	2.1 %	1.0%	1.2 %	1.0 %	0.5 %	1.5 %	2.1 %	0.7 %
$-5 < (L_{day} - L_{night}) \leq 0$ dB	17.0 %	10.3 %	7.1 %	6.3 %	14.4 %	8.0 %	15.3 %	9.7 %
$0 < (L_{day} - L_{night}) \leq 5$ dB	48.5 %	46.0 %	28.6 %	37.2 %	42.6 %	55.3 %	41.7 %	51.4 %
$5 < (L_{day} - L_{night}) \leq 10$ dB	32.1 %	41.3%	56.0 %	53.9 %	39.8 %	31.7 %	38.9 %	34.7 %
$10 < (L_{day} - L_{night}) \leq 15$ dB	0.0 %	1.3 %	7.1 %	1.0 %	2.8 %	2.5 %	2.1 %	2.1 %

Table 5.14 Correlation of noise descriptors for various zones

Long-term noise descriptors	
Commercial Zone	L_{night}=1.06 L_{day} - 7.77; R^2=0.74; *S.E*=2.79 dB
	L_{dn}=1.05 L_{day} - 0.20; R^2=0.85; *S.E*=1.53 dB
	$L_{eq,24h}$=1.02 L_{day} - 2.39; R^2=0.97; *S.E*=1.12 dB
Residential Zone	L_{night}=0.87 L_{day} + 2.34; R^2=0.68; *S.E*=2.09 dB
	L_{dn}=0.96 L_{day} + 4.72; R^2=0.89; *S.E*=1.03 dB
	$L_{eq,24h}$=0.99 L_{day} - 0.54; R^2=0.99; *S.E*=0.89 dB
Silence Zone	L_{night}=0.94 L_{day} - 1.0; R^2=0.67; *S.E*=2.32 dB
	L_{dn}=1.01 L_{day} + 1.97; R^2=0.90; *S.E*=1.31 dB
	$L_{eq,24h}$=1.01 L_{day} - 1.47; R^2=0.99; *S.E*=0.78 dB
Industrial Zone	L_{night}=0.95 L_{day} - 0.80; R^2=0.78; *S.E*=1.87 dB
	L_{dn}=0.97 L_{day} + 4.83; R^2=0.89; *S.E*=1.45 dB
	$L_{eq,24h}$=0.99 L_{day} - 0.26; R^2=0.98; *S.E*=0.69 dB
Overall	L_{night}=1.01 L_{day} – 5.02; R^2=0.75; *S.E*=2.55 dB
	L_{dn}=1.03 L_{day} + 1.15; R^2=0.89; *S.E*=1.59 dB
	$L_{eq,24h}$=1.01 L_{day} – 1.72; R^2=0.98; *S.E*=0.83 dB
Short-term statistical noise descriptors	
Overall	$L_{A10} = 0.975 L_{Aeq} + 5.30$; R^2=0.83; *S.E*=1.79 dB
	$L_{A50} = 0.921 L_{Aeq} + 3.60$; R^2=0.71; *S.E*=2.12 dB
	$L_{A90} = 0.919 L_{Aeq} + 2.30$; R^2=0.78; *S.E*=1.57 dB

shows the correlations of noise descriptors for various zones. It can be observed that $L_{\mathrm{Aeq,24h}}$ as good correlation with day equivalent noise levels for each of the four different zones. However, there may be some uncertainties associated with inter-conversion of one noise descriptor to other especially in high honking zone areas, etc. The empirically described conversion rules between the environmental noise exposure metrics shall be helpful for epidemiological meta-analyses or reviews, when results from different studies are pooled and need to be related to one common exposure metric.

5.5.5 *Overall Noise Scenario: Conclusions and Outlook*

Figures 5.5, 5.6, 5.7, and 5.8 show the day equivalent annual average levels, night equivalent annual average levels, 24-h equivalent annual average levels, and day-night annual average sound levels for the past 3 years for seven major cities of India.

The L_{day} and L_{night} levels observed for 70 sites for the year 2020 revealed that only nine industrial sites met the ambient noise standards. Thus, no site lying in residential zone, or in commercial zone, or in silence zone qualifies the ambient noise standards. Table 5.15 shows the status of compliance of various sites with ambient noise standards in past 10 years (2011–2020). It can be observed that neither residential site nor any silence zone site ever met the ambient noise standards. These observations thus suggest a retrospective and prospective view of ambient standards limits particularly for the residential and areas under silence zone in Indian scenario. The recent study by Garg et al. proposed amendments in ambient noise standards of India based on single-noise descriptor proposed $L_{\mathrm{Aeq,24h}}$ of 70 dB(A) for industrial zone; 65 dB (A) $L_{\mathrm{Aeq,24h}}$ for commercial area and mixed residential and commercial zones; 60 dB(A) for residential zone, and 55 dB(A) for silence zone [76]. In accordance with these limits, 17 sites (4 residential sites, 7 industrial sites and 6 commercial sites) for year 2019 and 19 sites (1 residential zone site, 9 industrial sites, and 9 commercial sites) for the year 2020 met these criteria.

5.5.6 *Noise Limit Exceedance Factor* (NEF)

The noise limit exceedance factor (*NEF*) is another method for the assessment of increment in noise levels with respect to the permissible noise limits. The average exceedance factor (*AEF*) indicates the average noise pollution level. Both these parameters are calculated as [77]:

$$NEF = C_{\mathrm{O}}/C_{\mathrm{I}} \tag{5.5}$$

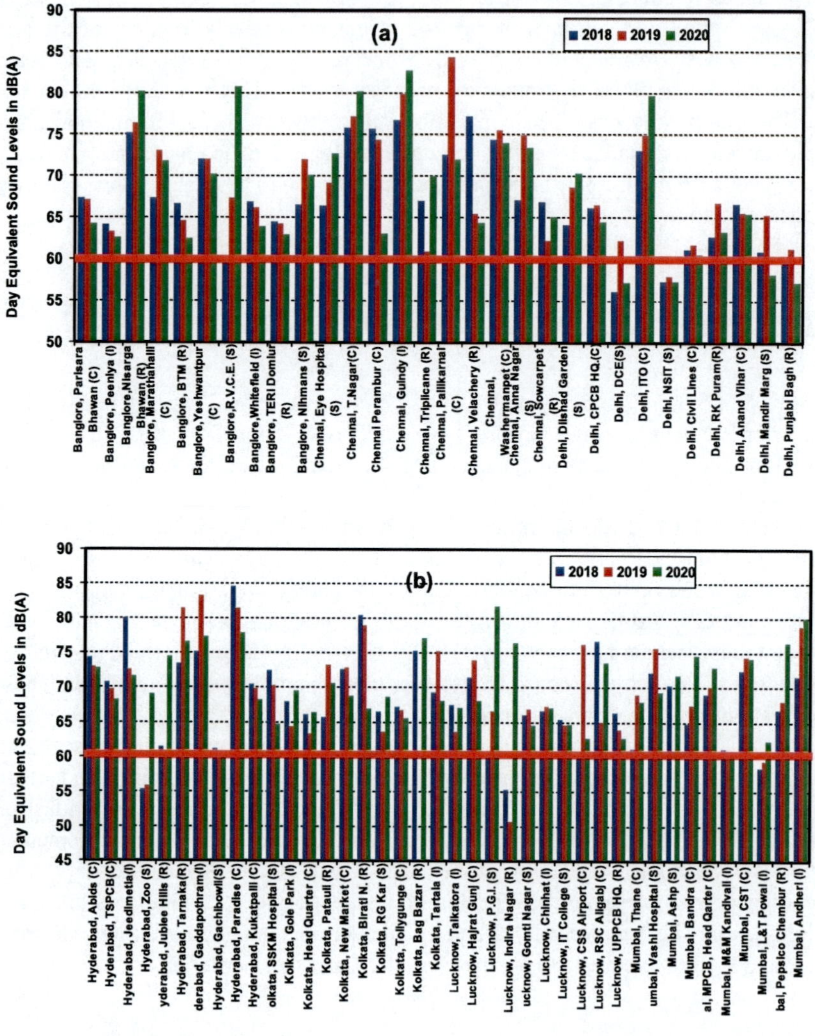

Fig. 5.5 (**a**, **b**) Annual average day equivalent sound levels, L_{day} in dB(A) observed in the past 3 years for seven major cities of India

where C_0 is the observed noise level and C_1 is the permissible limit of noise level. The average exceedance factor (*AEF*) is the average of *NEF* over the number of locations. Table 5.16 shows the annual average exceedance factor for different areas/ zone in year 2020. It can be observed that silence zone has the *AEF* of 1.4 for day equivalent levels and 1.7 for night equivalent levels, followed by residential zones (*AEF* 1.3–1.5) and commercial zone (*AEF* 1.1–1.2). The industrial zone was observed to have the least value of *AEF* as 0.9.

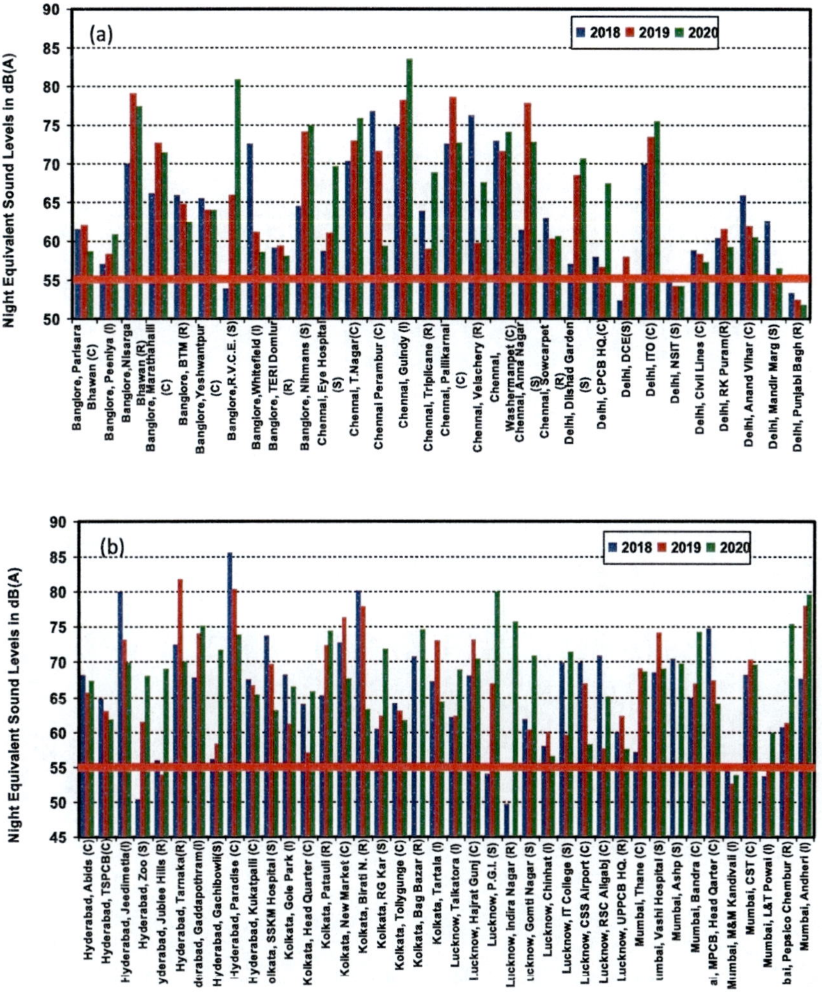

Fig. 5.6 (**a**, **b**) Annual average night equivalent sound levels, L_{night} in dB(A) observed in the past 3 years for seven major cities of India

A recent study reported by Kalawapudi et al. [78] for the nine major cities of Mumbai Metropolitan Region (MMR) covering a total of 153 noise monitoring locations showed that the silence zones with *AEF* 1.6–1.8 are the noisiest zones followed closely by residential (*AEF* 1.5–1.6) and commercial (*AEF* 1.3–1.4) zones. Silence zones that are not isolated from commercial or residential zones have been the worst affected areas as high *AEF* of 1.6 for day equivalent levels and 1.8 for night equivalent levels signify extreme violation of permissible limits.

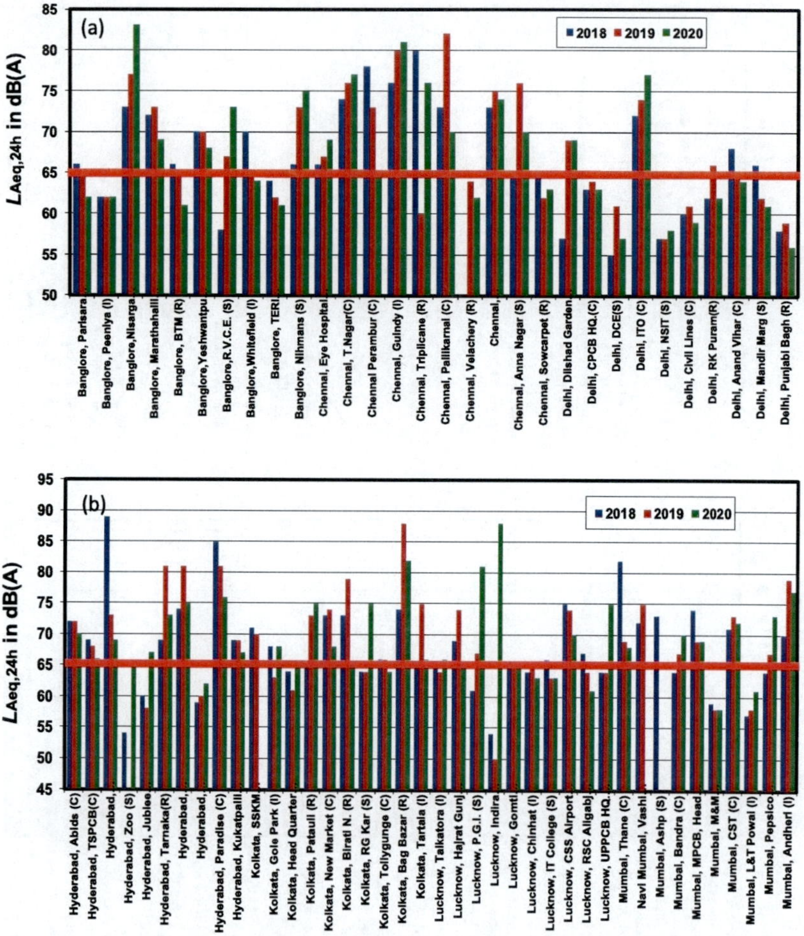

Fig. 5.7 **(a, b)** Annual average 24 h equivalent sound levels, $L_{Aeq,24h}$ in dB(A) observed in the past 3 years for seven major cities of India

5.6 Analysis of Ambient Noise Levels of Delhi City

The Delhi Pollution Control Committee (DPCC) has installed 26 Noise Monitoring Terminals in Delhi city for continuous noise monitoring since year 2020. The 26 stations include: 9 commercial sites, 7 residential sites, 5 sites lying in the silence zone, and 5 industrial sites. Figures 5.9 and 5.10 show the monthly average day and night equivalent sound levels, $L_{Aeq,24h}$ and L_{dn} sound levels for the month of September 2020 [79]. The maximum standard deviation of the monthly day equivalent levels was observed to be 4.9 dB(A), while the maximum standard deviation

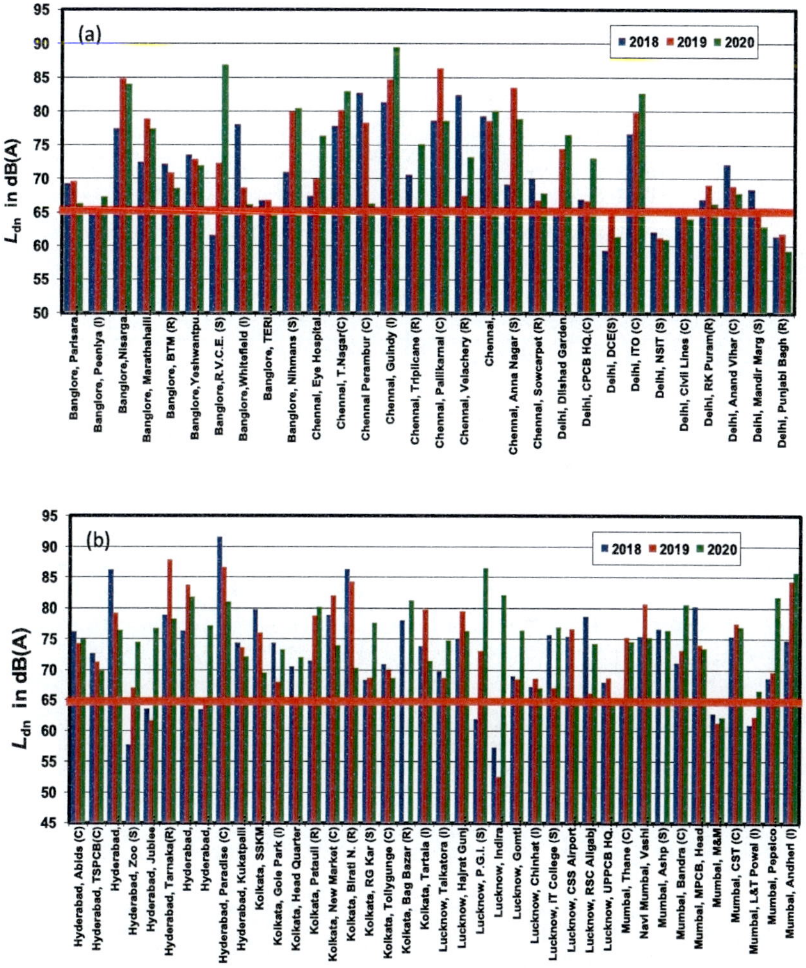

Fig. 5.8 (**a**, **b**) Annual average day-night average sound levels, L_{dn} in dB(A) observed in the past 3 years for seven major cities of India

for night equivalent noise levels was observed to be 3.9 dB(A). It can be observed that 14 sites (7 commercial, 5 industrial, and 2 residential) complied with day-time ambient noise limits and 9 sites (5 industrial and 4 commercial) complied with the night-time ambient noise limits. No residential or silence zone site met the ambient noise limits. Table 5.17 shows the frequency distribution of monthly average sound levels. It can be observed that the majority of sites (21 sites) registered day equivalent sound levels in range 55–65 dB(A), 23 sites (88.5%) had night equivalent noise levels in range 50–60 dB(A), 19 sites (73.1%) showed $L_{Aeq,24h}$ sound levels in range 55 to 65 dB(A), and 20 sites (77%) showed L_{dn} levels in range 55–65 dB(A).

Table 5.15 Status of compliance of various sites with ambient noise standards of India

Year	No of sites meeting the ambient noise criteria	Name of the sites
2011	7 sites (3 commercial and 4 industrial)	CPCB Head Quarters, Thane, Marathali, Peeniya, Jeedimetla, Talkatora and Kasba Gole Park
2012	8 sites (4 commercial and 4 industrial)	CPCB Head Quarters, West Bengal Head Quarters, Thane, Marathali, Peeniya, Jeedimetla, Talkatora and Kasba Gole Park
2013	8 sites (4 commercial and 4 industrial)	CPCB Head Quarters, West Bengal Head Quarters, Thane, Marathali, Peeniya, Jeedimetla, Talkatora and Kasba Gole Park
2014	5 sites (1 commercial and 4 industrial	CPCB Head Quarters, Peeniya, Jeedimetla, Talkatora and Kasba Gole Park
2015	10 industrial sites	Talkatora, Jeedimetla, Peeniya, Chinhat, Tartala, M&M Kandivali, L&T Powai, Andheri, Whitefield and Gaddapothram
2016	12 sites (10 industrial and 2 commercial)	Talkatora, Jeedimetla, Peeniya, Chinhat, Tartala, M&M Kandivali, L&T Powai, Andheri, Whitefield, Gaddapothram, Kukatpalli and Abids.
2017	8 industrial sites	Talkatora, Peeniya, Chinhat, Tartala, M&M Kandivali, L&T Powai, Andheri and Whitefield.
2018	9 industrial sites	Peeniya, Gole Park, Tartala, Talkatora, Chinhat, M&M Kandivali, L&T Powai, Andheri and Gaddapothram.
2019	8 sites (7 industrial and 1 residential)	Whitefield, Gole Park, Talkatora, Peeniya, Indira Nagar, Chinhat, Kandivali and L&T Powai
2020	9 industrial sites	Peeniya, Whitefield, Jeedimetla, Gole Park, Tartala, Talkatora, Chinhat, Kandivali and L&T Powai

Table 5.16 Annual average exceedance factor (*AEF*) for different zones in year 2020

Category of area/zone	L_{day}			L_{night}		
	Maximum value	Minimum value	*AEF*	Maximum value	Minimum value	*AEF*
Industrial Area (n=12)	1.1	0.8	0.9	1.2	0.8	0.9
Commercial Area (n=25)	1.2	0.9	1.1	1.4	1.0	1.2
Residential Area (n=16)	1.5	1.0	1.3	1.7	1.1	1.5
Silence zone (n=17)	1.6	1.1	1.4	2.0	1.4	1.7

Overall, nine sites comprising of five industrial and four commercial met the ambient noise standards for day and night time limits. It can be observed that 17 sites (65.4%) met the 55 dB(A) L_{night} limit, while only 3 sites met the 50 dB(A) L_{night} limit. Seventeen sites only met the criteria of 60 dB(A) L_{day} limit. One site, namely,

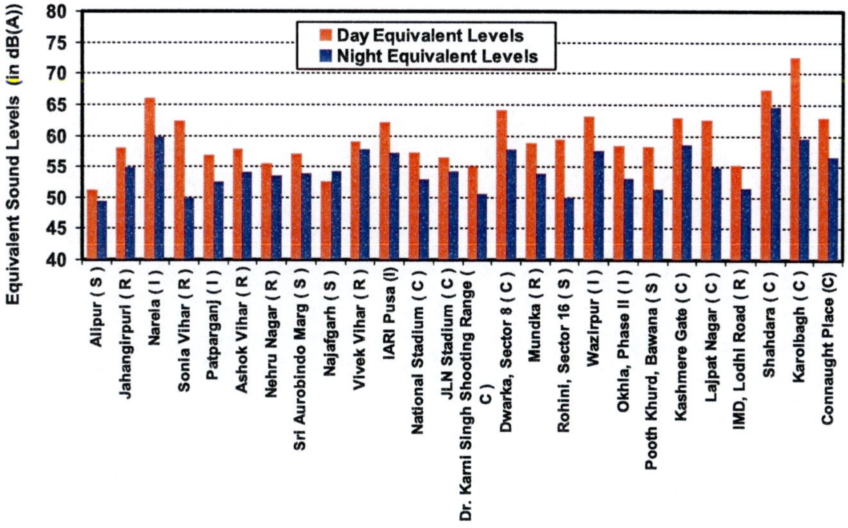

Fig. 5.9 Monthly average day and night equivalent noise levels for 26 noise monitoring stations in Delhi city for September 2020

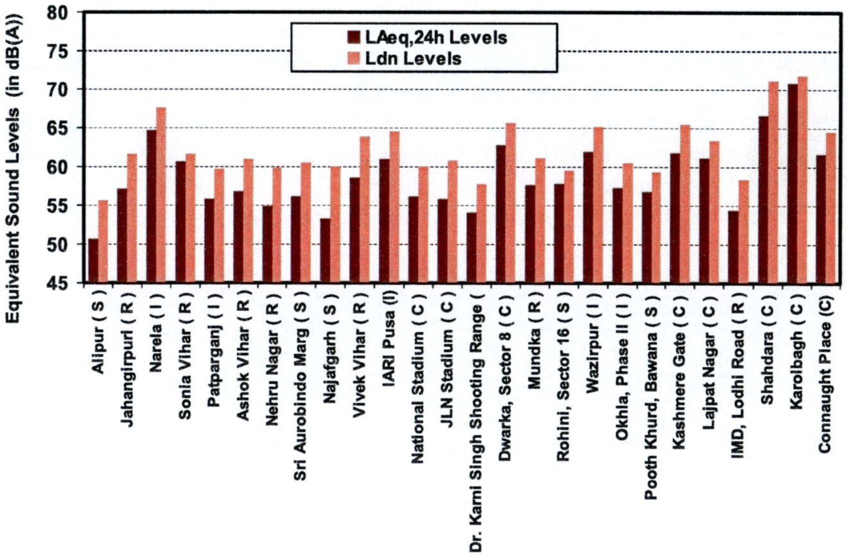

Fig. 5.10 Monthly average $L_{\text{Aeq,24h}}$ and L_{dn} levels for 26 noise monitoring stations in Delhi city for September 2020

Table 5.17 Frequency distribution of single noise descriptors for 26 stations in Delhi city

Variation of noise descriptors	Percentage of Noise Monitoring Stations			
	L_{day}	L_{night}	$L_{Aeq,24h}$	L_{dn}
$45 < L_{eq} \leq 50$ dB(A)	0	7.7	0	0
$50 < L_{eq} \leq 55$ dB(A)	7.7	57.7	19.2	0
$55 < L_{eq} \leq 60$ dB(A)	53.8	30.8	42.3	30.8
$60 < L_{eq} \leq 65$ dB(A)	26.9	3.8	30.8	46.2
$65 < L_{eq} \leq 70$ dB(A)	7.7	0	3.8	15.4
$70 < L_{eq} \leq 75$ dB(A)	3.8	0	3.8	7.7
$75 < L_{eq} \leq 80$ dB(A)	0	0	0	0

the Karol Bagh site, was the Most Exposed Urban Site (MEUS) with L_{day} levels greater than 70 dB(A).

Twenty-two sites (84.6%) met the criteria of $L_{Aeq} \leq 62$ dB(A), while 21 sites (80.8%) registered L_{dn} sound levels ≤ 65 dB(A). The analysis of difference in day and night equivalent sound levels showed that 15 sites registered difference between 0 and 5 dB(A) and 8 sites registered difference between 5 and 10 dB(A). Only two sites (one residential and one commercial) showed difference greater than 10 dB(A), while one silence zone site showed a difference less than 0 to 5 dB(A). The average difference for commercial zone sites is observed to be 5.7 dB(A), 4.4 dB(A) for residential sites, 5.3 dB(A) for industrial zone sites, and 3.9 dB(A) for silence zone sites.

5.7 Ambient Noise Levels on Diwali Festival

Diwali (or Deepawali) is a festival of light that is celebrated with great enthusiasm all over India and comes usually in the month of October or November. On this occasion, burning of firecrackers and illumination as a symbol of joy and prosperity is usually followed in every part of the country. The firecrackers are usually burst during the night hours, but during the day time also many people explode crackers. However, air and noise pollution caused due to burning of firecrackers is very harmful for the environment. There had been only a few studies such as by Garg et al. [80] that focused on the impact of Diwali festival on the noise pollution at many sites simultaneously on the day of Diwali festival in India. It may be noted that noise standards for firecrackers have been formulated as prohibition of manufacture, sale, or use of firecrackers generating noise levels exceeding 125 dB(AI) or 145 dB$(C)_{pk}$ at 4 m distance from the point of bursting. For individual firecracker constituting the series, the limit may be reduced by $5\log_{10}(N)$ dB, where N is number of crackers joined together. The cracker's noise of different brands of firecrackers is tested every year at CSIR-National Physical Laboratory, India, so as to ascertain whether the noise criteria are met or not [81]. A recent study reported on noise emissions of indigenously made green crackers developed to reduce air emissions analyzes the impact of fireworks noise on ambient noise scenario and presents a prospective view of the legal noise regulations for the green crackers in India [82]. In India, the Central Pollution Control Board (CPCB) protocol for ambient noise monitoring on Diwali festival day recommended the duration of six hours (18.00 to 24.00 hours), which can't be compared with the night equivalent level limits. Thus, an appropriate noise limit shall be helpful for legal bodies to take immediate actions for noisy spots [82]. Figure 5.11a and b shows the sound pressure levels emitted by various crackers developed by Indian manufacturers measured at a distance of 4 m and 1.3 m above the ground [80, 83].

The measurements were done using Brüel & Kjær (B&K) Precision Impulse Sound level meters with B&K ¼ inch microphones traceable to the national standards of sound pressure. A simple relation between A-weighted impulse and C-weighted peak sound pressure level for ground shots with pulse duration more than 25 ms (at distance of 4 m) can be analyzed as [81]:

$$\mathrm{dB(C)_{pk}} = \mathrm{dB(AI)} + 31 \tag{5.6}$$

The above relation shows the average difference for various type of the shots and a maximum difference between two indicators of 34 dB to a minimum of 9 dB had been observed for some firecrackers. For the garland (commonly called as ladi or chain) type of firecrackers, which are widely used in Diwali festivals in Indian scenario, the relationship between the two indicators is derived by Sharma et al. [81]:

$$\mathrm{dB(AI)} = SEL + 0.03\ (T/N) + 22 \tag{5.7}$$

$$\mathrm{dB(C)_{pk}} = \mathrm{dB(AI)} + 25 \tag{5.8}$$

where SEL is the sound exposure level, T is time duration of garland burst in ms, and N is the number of crackers in the garland.

The evaluation and analysis of noise pollution due to the burning of firecrackers is also essentially required for understanding the noise scenario on festival days and devising prior suitable measures for prevention and control. The analysis of ambient noise levels at the 70 sites in October 2015 and the day before, after, and on the day of Diwali festival reveals that nine sites out of 70 sites (12.9%) under consideration met the ambient noise standards for each case. These sites were: Whitefield site in Bangalore city; Jeedimetla and Gaddapotharam sites in Hyderabad city; Tartala site in Kolkata city; Talkatora and Chinhat sites in Lucknow city; and Kandivali, Powai, and Andheri sites in Mumbai city. It was observed that 15 sites on Diwali day

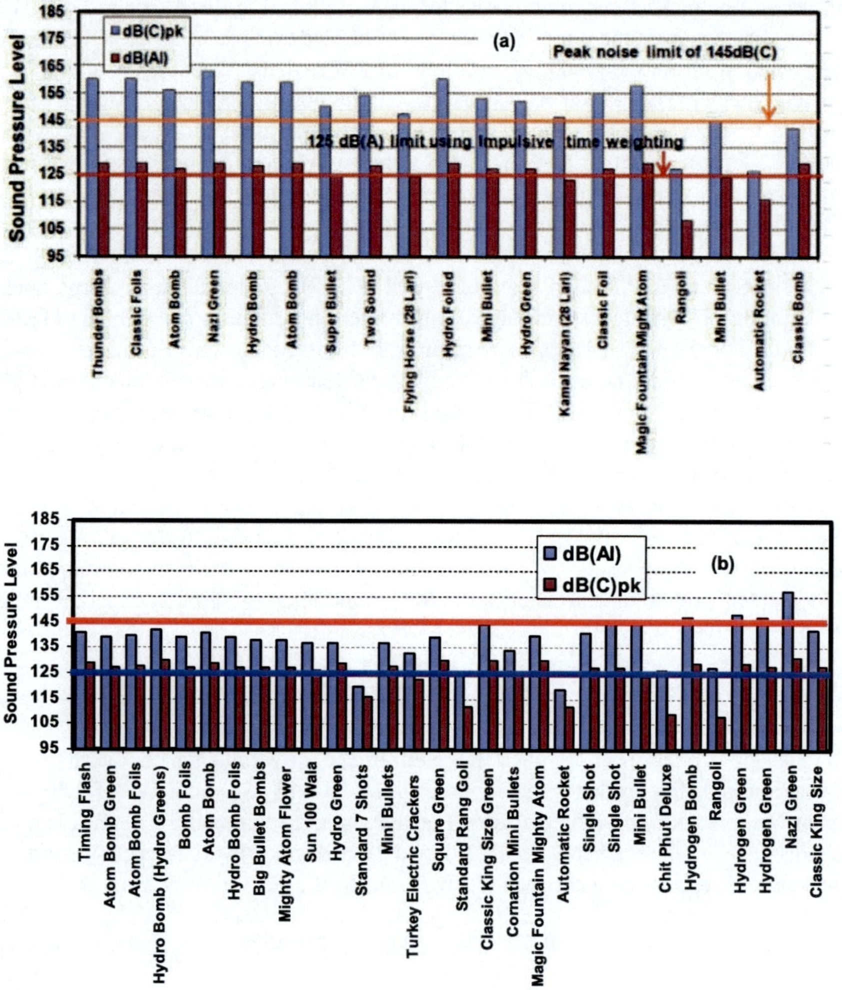

Fig. 5.11 (**a, b**) Sound pressure levels emitted by various crackers developed by Indian manufacturers measured at a distance of 4 m and 1.3 m above the ground [80, 83].

registered night noise levels higher than the day levels, while for October 2015 monthly average noise levels; only two sites had higher night noise levels than the day levels. Thus, these observations indicate that the festival had a significant impact on ambient noise levels especially during the night hours. Figure 5.12a and b shows the increment in L_{day}, L_{night} and $L_{\text{Aeq,24h}}$ levels in dB(A) for the seven cities (70 sites) on the day of Diwali festival with respect to the annual average day and night levels of year 2015. The comparison of noise levels observed on the day of Diwali festival w.r.t previous month average noise levels showed that 55 sites (78.6%) registered an increment in L_{day} by 0–10 dB(A), while 47 sites (67.1%) registered an increment in L_{night} by 0–10 dB(A).

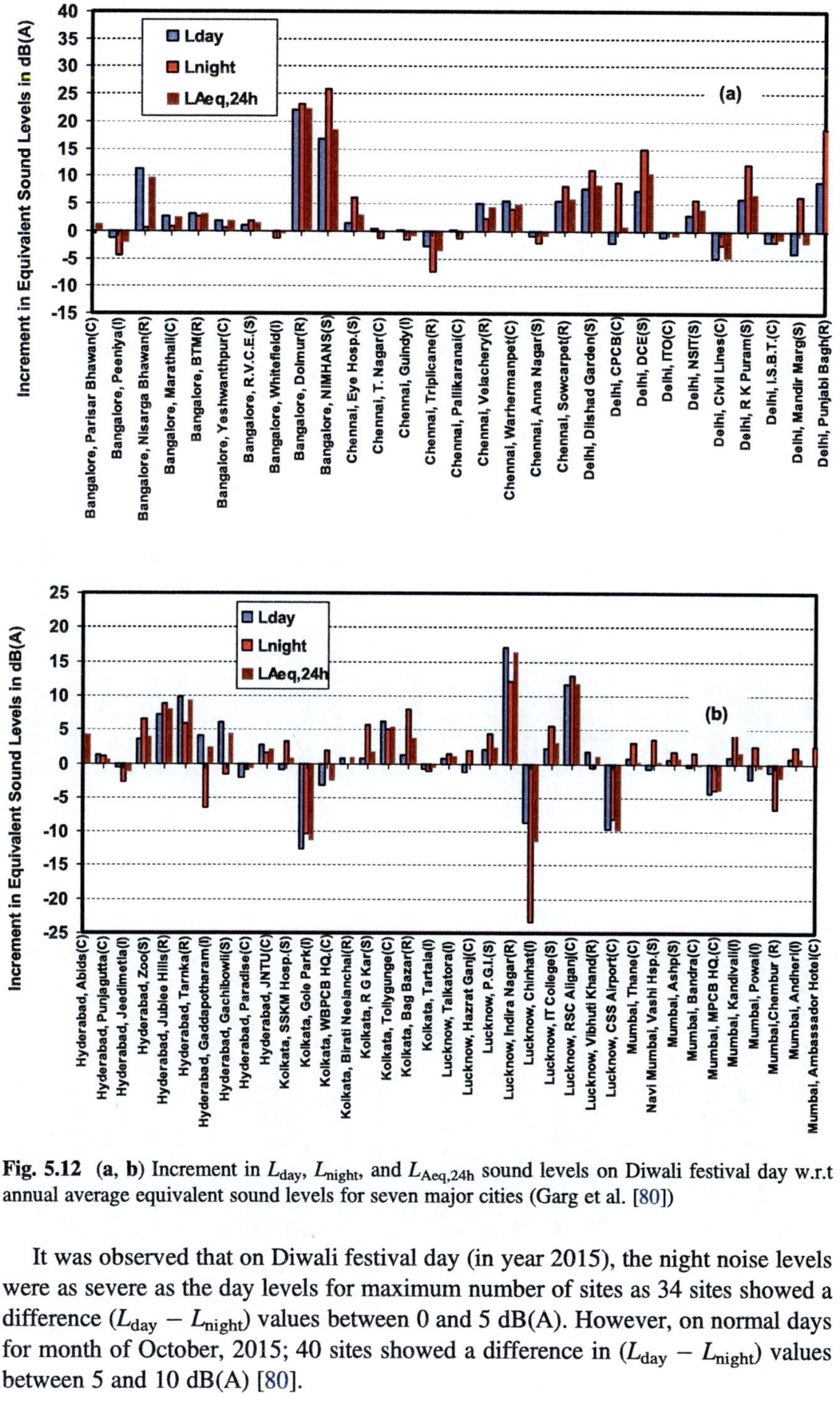

Fig. 5.12 (**a**, **b**) Increment in L_{day}, L_{night}, and $L_{Aeq,24h}$ sound levels on Diwali festival day w.r.t annual average equivalent sound levels for seven major cities (Garg et al. [80])

It was observed that on Diwali festival day (in year 2015), the night noise levels were as severe as the day levels for maximum number of sites as 34 sites showed a difference ($L_{day} - L_{night}$) values between 0 and 5 dB(A). However, on normal days for month of October, 2015; 40 sites showed a difference in ($L_{day} - L_{night}$) values between 5 and 10 dB(A) [80].

The Supreme Court in October 2018 imposed stringent rules in National Capital Region (NCR) pertaining to the sale and use of firecrackers. Firecrackers with reduced emission and decibel levels and green crackers shall be allowed to be sold during all festivals that too with strict time limits. On Diwali, firecrackers could be burst between 8 and 10 p.m., while people can burst firecrackers on the Christmas and the New Year days between 11:55 p.m. and 12:30 a.m. and that on Gurpurab (4.00 a.m. to 5.00 a.m. and 9.00 p.m. to 10 p.m.). Joined firecrackers (series or laris) were banned and the fireworks that comply with permissible limits of 125 dB(A) or 140 $dB(C)_{pk}$ at 4 m from the point of bursting were permitted to be sold. Bursting of firecrackers is not allowed in silence zones. The court has also directed all states to explore feasibility of community cracker bursting at pre-identified areas/fields. In year 2019, the police registered 371 cases, seized 3764 kg of firecrackers, and 210 people were arrested for bursting prohibited firecrackers in National Capital Region (NCR) on Diwali festival day (27.10.2019). Figure 5.13a and b shows the

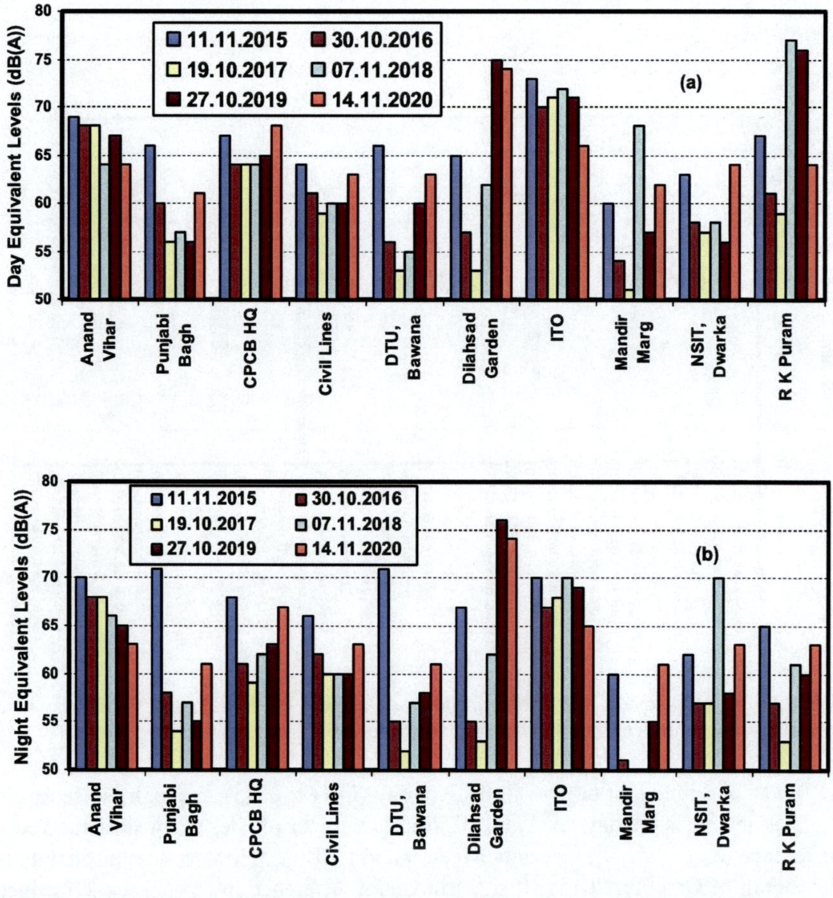

Fig. 5.13 **(a, b)** Ambient day and night equivalent noise levels on Diwali days for Delhi city since past 5 years [67]

day equivalent noise levels and night equivalent noise levels for Diwali days for ten sites in Delhi city since past 5 years based on CPCB reports [84, 85]. The following observations are derived from the analysis of 5 years noise monitored data by CPCB (CPCB reports, 2019 and 2020 [84, 85]) on Diwali and pre-Diwali days as:

- No site meets the ambient noise standards since past five years on the day of Diwali festival and pre-Diwali day. The difference in day and night ambient noise levels was up to 2 dB(A) for year 2020 and varied from −2 to 2 dB(A) for nine sites in year 2019. Majority of the sites (seven in the past 2 years) had day equivalent sound levels higher than the night equivalent sound levels on festival day.
- The comparison of day and night equivalent noise levels on the Diwali day in the past 5 years showed that seven sites registered increased day equivalent sound levels on Diwali day in year 2020 as compared to year 2015. Nine sites registered increased night equivalent noise levels in the past 5 years on the Diwali day.

Figure 5.14a and b shows the comparison of the day and night equivalent noise levels on the Diwali day in comparison with the pre-Diwali day. The pre-Diwali days

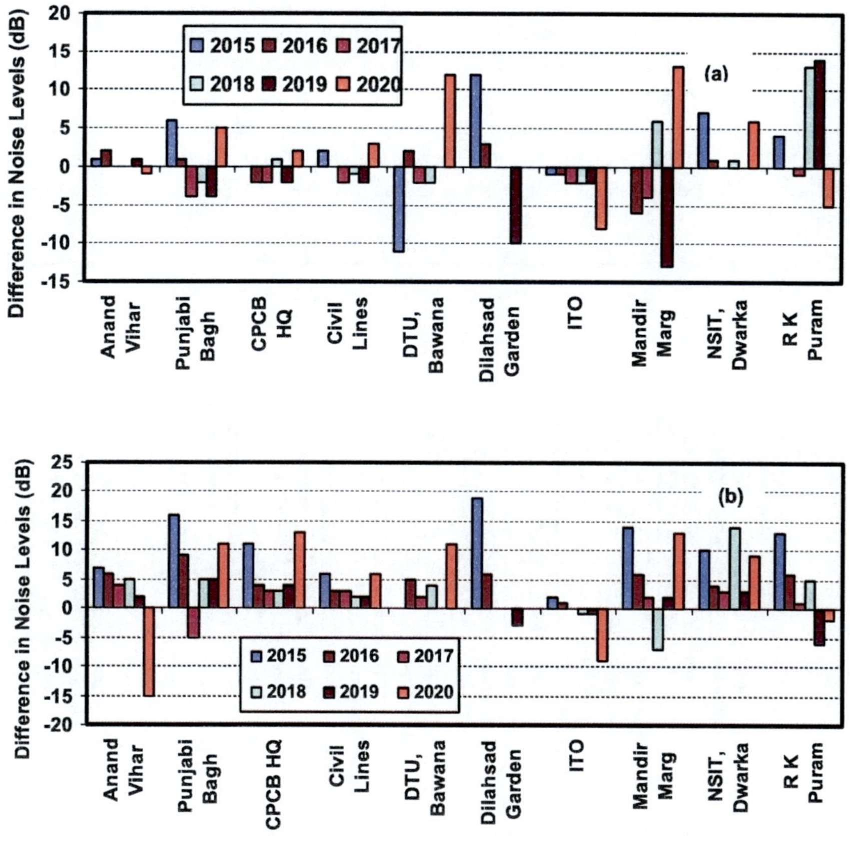

Fig. 5.14 (**a**, **b**) Increment in day and night equivalent noise levels on Diwali days in comparison to pre-Diwali days for Delhi city since past 5 years

considered were 05.11.2015, 24.10.2016, 12.10.2017, 01.11.2018, 21.10.2019, and 09.11.2020. The day and night equivalent noise levels had increased in year 2020 for six sites as compared to pre-Diwali day. The day and night equivalent sound levels had increased up to 13 dB(A) on the Diwali day.

5.8 Ambient Noise Levels on Ganesh Festival

Ganesh Chaturthi is a 10-day Hindu festival celebrated to honor the elephant-headed God Ganesha's birthday. Maharashtra Pollution Control Board had been carrying out study of the ambient noise levels in the state of Maharashtra during the period of Ganesh Festival for more than 9 years as a continuous process. Noise monitoring was carried out at 132 locations in Maharashtra for 5 days period during Ganesh Festival in September 2019 from 6 p.m. to 12 a.m. for each location which comprises of residential, commercial, and silence zone. Figure 5.15 shows the equivalent noise levels for night (6.00 p.m. to 12.00 a.m.) on some specific days during Ganesh festival for Mumbai city in year 2016 [86]. There was a significant decrease in the noise levels observed in almost all areas when compared to previous years' noise levels. The decrease in noise pollution was also a result of the awareness and active role of police and administration in enforcement of the noise regulations [86]. The contribution of festival on hourly noise levels is evident from the hourly noise levels variation. The precise contribution of festival is however not clear as the noise levels

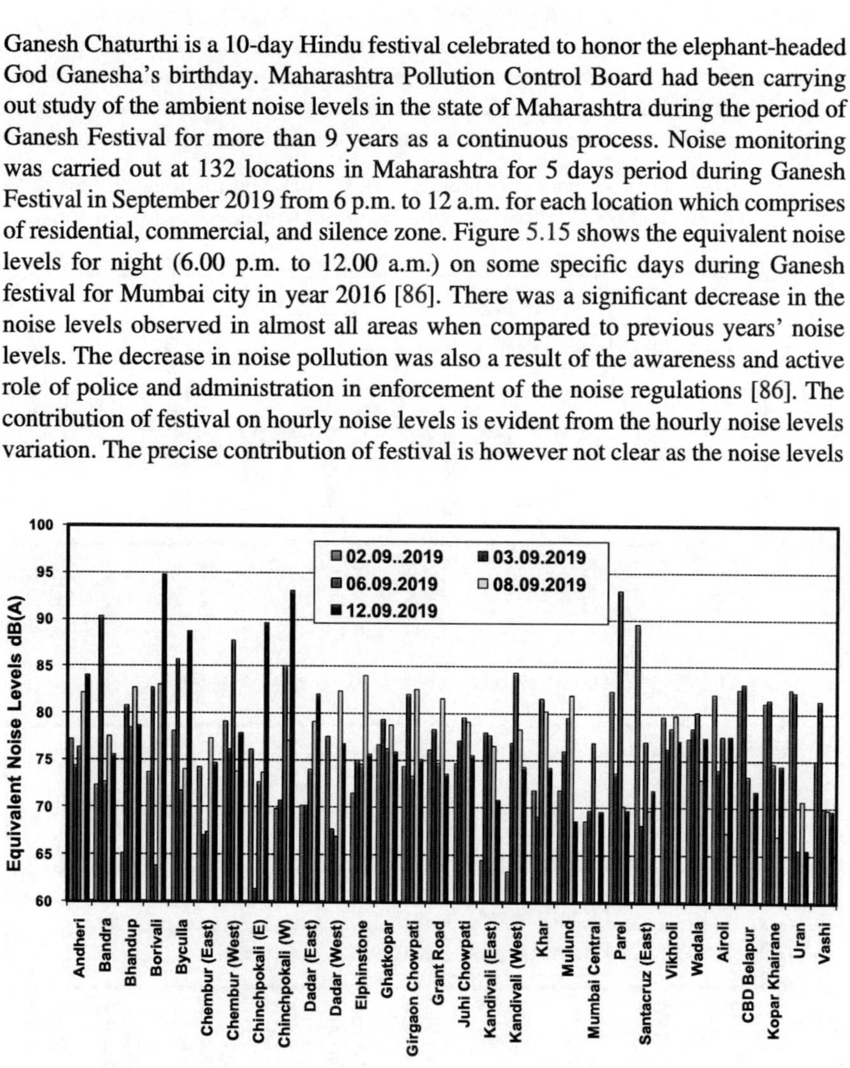

Fig. 5.15 Equivalent noise levels observed at various locations in Mumbai and Navi Mumbai city during the Ganesh festival in year 2019 (MPCB report [86])

were due to other activities also like traffic, construction, household, etc. A recent study by Kalawapudi et al. [87] reported the reduction in noise levels by 28.5 dB (A) during the festival when compared to previous 2 years. The highest noise levels up to 73.7 dB(A) were recorded, while in past 2 years, noise levels (L_{eq}) were recorded up to 102 dB and 80.5 dB, respectively. This was attributed to the restriction imposed during the pandemic, on processions to mark arrival and immersion of idols, guidelines to avoid crowd during daily worship, very less pandals, and limited number of people celebrating the event in pandals which led to significant reduction in noise levels.

5.9 Ambient Noise Levels on Sundays

The recent CPCB report (2017) [88] on "How quiet are Sundays" raised an interesting issue in Indian perspectives on the status of ambient noise levels on Sunday's as there is high probability of less vehicular density on Sunday's comparatively to the business-as-usual (BAU) days. All Central and the State government offices, private organizations, schools, colleges, and institutions remain closed on Sundays. An analysis of the ambient noise levels on Sundays in comparison to the BAU was conducted for ascertaining the difference in noise levels in two situations. The noise monitored data acquired under the *NANMN* project for year 2017 for three cities—Bengaluru, Chennai, and Delhi—were utilized for such analysis. The difference in average day and night equivalent levels on Sunday's in a month with respect to the average monthly day and night equivalent levels (excluding Sunday's ambient levels) was analyzed. Table 5.18 shows the difference in ambient noise levels on Sundays with respect to average ambient noise levels on the other days for Bengaluru city. For instance, there were five Sundays in January 2017 and thus, the average ambient noise levels on these 5 days was calculated and compared with the average ambient noise levels for the rest of 26 days in January 2017. The difference shows a very interesting observation that the difference is marginal and in some cases only a maximum difference of 7 dB(A) is observed. For instance, the Peeniya industrial site showed a difference ranging from 4 to 7 dB(A), which implied that in some months of year 2017, the ambient noise levels were 4 to 7 dB(A) higher than the Sundays. Also, in some cases like in March 2017, the average ambient noise levels on Sunday was 2 dB(A) higher than the BAU's. Tables 5.19 and 5.20 show the difference in ambient noise levels on Sundays with respect to average ambient noise levels on the other days for Chennai and Delhi city. The difference is again observed to be marginal in case of Chennai and Delhi cities. These observations thus suggest that the ambient noise levels on Sunday's are not much different as compared to BAUs. This may be due to flexible working system of some multinational companies opened on even Sundays. Also, as obvious, the difference in night levels are expected to be negligible for BAU's and Sundays, which is confirmed from the analysis presented for the three cities showing that the night noise levels are not much different to a larger extent for the three cities on both the situations. However,

Table 5.18 Difference in ambient noise levels on Sunday's with respect to BAU's for Bengaluru city in year 2017

	BTM (R)		Marathali (C)		Nihmans (S)		Nisarga Bhawan (R)		Parisar Bhawan (C)		Peeniya (I)		RVCE (S)		Dolmur (R)		Whitefield (I)		Yeshwantpur (C)	
Month	Day	Night	Day	Night	Day	Night	Day	Night	Day	Night	Day	Night	Day	Night	Day	Night	Day	Night	Day	Night
Jan	−1	0	−1	1	−2	−1	0	1	−2	3	−4	−1	−3	−1	0	−1	−2	−5	−1	0
Feb	−1	0	−2	2	−2	−1	1	0	−2	1	−5	−5	−4	−3	−3	−1	−3	−3	−1	0
March	0	0	−2	0	−1	0	1	2	0	3	2	2	−1	0	−1	0	−2	−3	0	0
April	0	0	−1	0	−2	0	1	0	−1	1	0	1	0	−1	−1	2	−2	−2	−1	0
May	2	1	−2	2	−1	0	2	0	−1	1	1	0	0	−1	−1	0	−2	−2	−1	0
June	0	0	1	0	0	1	−1	1	−2	1	−4	0	0	1	0	1	−2	−1	−1	0
July	0	−1	−2	0	−2	1	0	0	−3	1	−6	0	−2	0	−2	0	−3	−2	−2	1
Aug	0	−1	−1	0	−2	1	0	0	−2	0	−4	0	−2	0	−1	0	−2	−1	−2	−1
Sep	1	1	1	0	1	0	1	2	−2	1	−5	+1	−3	1	−3	−1	−4	−4	−1	0
Oct	1	0	−3	0	−5	−2	−4	−1	−2	1	−7	−1	−3	0	−1	−1	−2	−2	1	−1
Nov	0	0	−1	−1	−7	−3	0	2	0	1	−4	0	−3	0	−1	−1	−1	−3	−1	−1
Dec	0	0	3	0	−5	1	0	−1	−1	3	−5	−1	−2	−2	0	1	−3	−3	−2	0

Table 5.19 Difference in ambient day and night noise levels on Sunday's with respect to BAU's for Chennai city in year 2017

	Anna Nagar (S)		Eye Hospital (S)		Guindy (I)		Pallikarnai (C)		Perambur (C)		Sowcarpet (R)		T. Nagar (C)		Triplicane (R)		Velachery (R)		Washermanpet (C)	
Month	Day	Night	Day	Night	Day	Night	Day	Night	Day	Night	Day	Night	Day	Night	Day	Night	Day	Night	Day	Night
Jan	0	−3	−1	0	0	0	−1	0	1	0	0	0	−1	0	0	0	0	−1	−1	0
June	0	1	0	−1	−1	−1	−2	−1	1	−1	0	1	−1	0	−1	−4	0	0	−1	0
Nov	1	2	−1	0	−1	−1	−4	−1	0	4	0	−1	−3	0	2	−2	3	−7	−2	0

Table 5.20 Difference in ambient day and night noise levels on Sunday's with respect to BAU's for Delhi city in year 2017

	Anand Vihar		Civil Lines		CPCB		DCE		Dilshad Garden		ITO		Mandir Marg		NSIT		Punjabi Bagh		R K Puram	
Month	Day	Night	Day	Night	Day	Night	Day	Night	Day	Night	Day	Night	Day	Night	Day	Night	Day	Night	Day	Night
Jan	0	0	−1	1	−2	0	−2	−1	0	0	−1	1	0	−1	−1	0	−2	0	0	1
Feb	−1	0	−1	0	−3	0	2	2	0	0	−1	1	−4	−1	−1	0	−1	0	−3	−1
March	0	1	0	0	−2	1	−1	0	1	1	−1	1	3	1	1	1	1	1	0	1
April	0	0	−1	0	−3	0	−1	0	0	0	−1	1	0	2	−1	−1	−2	−1	−3	−1
May	0	2	−1	0	−3	−1	−1	3	0	0	−2	1	0	0	−2	1	0	–	−1	1
June	0	0	−1	0	−3	−1	1	0	0	0	−2	0	−1	−1	−1	0	−1	−1	−2	−1
July	0	1	0	0	−3	0	−1	0	0	0	−2	1	−5	−1	0	−1	−1	−1	−2	0
Aug	0	0	−3	0	−3	0	−3	−1	0	0	−2	1	−5	0	−1	−1	−3	0	−3	−1
Sep	0	0	−3	0	−3	−1	−2	0	0	0	−1	0	−2	1	0	−1	−2	3	−3	−1
Oct	0	0	−3	0	−3	0	−4	−1	−1	0	−1	0	−2	−3	−1	0	−2	−2	−2	−1
Nov	−1	0	−3	2	−3	0	−2	0	0	0	−1	1	−5	−6	−1	0	−2	0	−6	−1
Dec	0	0	−1	1	−3	0	−2	−1	0	0	−1	0	−3	0	−1	0	−2	0	−2	0

a comprehensive analysis including the comparison of vehicular density in both the situations and correlation with the difference in ambient day levels shall present a broader picture.

5.10 Odd-Even Rule Implications on Ambient Noise Levels

The vehicular population on the roads is one of the major environmental issues faced by every country in the world as it creates air and noise pollution, which is harmful to the community. In Indian scenario, every metropolitan city has been facing this problem since past two decades. Thus, it is very important to focus on the suitable effective measures to control the vehicular density and arrest the increasing air and noise pollution levels in these cities. The Government of Delhi decided to implement odd-even vehicular restriction for the month of January starting from January first to January 15th, 2016, so as to control the increasing air pollution in the city. According to the rule, cars with even numbered registration plates will ply on even dates, while those with odd numbers will ply on the odd dates between 08:00 and 20:00 h. This rule does not apply on Saturdays and Sundays and all taxis, passenger cars operating on CNG, and electric power cars with only women passengers and all motorized two wheelers. Overall, the rule was a great step is reducing the traffic congestion on the roads, but there has been no definitive answer whether the pollution levels have come down during the period or not. Additional transport arrangements were made by Delhi government such as running of 9000 buses (4500 DTC; 1500 Cluster and 3000 private); 82,000 autos; 79,600 taxis; and also the carrying capacity of Delhi Metro trains was enhanced from 26 Lakhs to 32 Lakhs across all the lines [89]. The second phase of odd-even restrictions was started in April 2016 implemented from 15th April to 30th April 2016 again with an objective of reducing the air pollution levels in the Delhi city. The present analysis is focused on the ambient noise levels during the implementation of odd-even restrictions in the city for January and April 2016 at ten sites in the city. Although the major thrust behind the implementation of odd-even scheme was to control the air pollution, yet the present analysis only focuses on noise pollution as noise pollution is also one of the important concerns apart from air pollution for the metropolitan cities like Delhi. Table 5.21 shows the difference in average ambient levels between odd-even day in first phase and normal day for L_{day}, L_{night}, and $L_{\text{Aeq,24h}}$ parameters for ten noise monitoring stations installed in Delhi city under *NANMN* pilot project. It can be thus observed that for only three sites, namely, R K Puram; Mandir Marg; and DTU, Bawana, there was a reduction of ambient noise levels by 1.8 dB, 1.9 dB, and 2.6 dB, respectively. The other sites, namely, Anand Vihar, Dilshad Garden, ITO, and NSIT Dwarka, registered an increase in day ambient noise levels during the odd-even restrictions. However, this increase is not at all associated with the odd-even restrictions, which is confirmed by the fact that even the night levels for these sites were higher for the days wherein odd-even restrictions are imposed in comparison to the normal days [90].

Table 5.21 Difference in average ambient levels between odd-even day in first phase and normal day for L_{day}, L_{night}, and $L_{Aeq,24h}$ parameters in Delhi city (Garg et al. [90])

		Difference in noise levels in dB(A)		
		January, 2016		
Name of location	Area characteristics	L_{day}	L_{night}	$L_{Aeq,24h}$
Civil Lines	Commercial	−0.2	−0.3	−0.2
RK Puram	Residential	−1.8	−2.8	−2.0
Anand Vihar	Commercial	0.2	0.2	0.2
Mandir Marg	Silence	−1.9	0.4	−1.0
Punjabi Bagh	Residential	−0.4	1.0	0.0
Dilshad Garden	Silence	0.3	0.9	0.6
CPCB HQ.	Commercial	−0.4	0.5	−0.2
DTU, Bawana	Silence	−2.6	−1.1	−2.1
ITO	Commercial	1.3	1.1	1.2
NSIT, Dwarka	Silence	0.4	1.3	0.7

Table 5.22 Difference in average ambient levels between odd-even day in second phase and normal day for L_{day}, L_{night}, and $L_{Aeq,24h}$ parameters in Delhi city (Garg et al. [90])

		Difference in noise levels in dB(A)		
		April 2016		
Name of location	Area characteristics	L_{day}	L_{night}	$L_{Aeq,24h}$
Civil Lines	Commercial	−0.6	0.4	−0.2
RK Puram	Residential	−0.4	0.8	0.1
Anand Vihar	Commercial	0.1	0.2	0.1
Mandir Marg	Silence	−1.4	−1.3	−1.3
Punjabi Bagh	Residential	−1.0	−0.8	−0.9
Dilshad Garden	Silence	−0.3	3.8	0.7
DTU, Bawana	Silence	0.1	2.4	0.7
ITO	Commercial	−1.4	−1.7	−1.5
NSIT, Dwarka	Silence	−0.1	5.5	1.3

Table 5.22 shows the difference in average ambient levels between odd-even day in second phase and normal day for L_{day}, L_{night}, and $L_{Aeq,24h}$ parameters for ten noise monitoring stations installed in Delhi city. The data for CPCB site was however not available for few days only due to some maintenance work. It can be thus observed that for three sites, namely, Mandir Marg, Punjabi Bagh, and ITO, there was a reduction in day equivalent sound levels.

The analysis of vehicular population at some sites during the days wherein odd-even restrictions were imposed by Goel et al. [91] showed that during the odd-even experiment, car flow rates per hour on different roads decreased by 9–17% in parallel with approximately similar increases in bus flow and auto rickshaw rates and significant increase in motorized two wheeler flow rates. The study also revealed that increase in flow of vehicles other than cars will offset the decrease in the emissions from cars. Also, another reason may be attributed to the

Table 5.23 Average ambient levels between odd-even days and normal days for L_{day} and L_{night} levels in Delhi city

Name of Location	Area characteristics	Ambient noise levels in dB(A) for normal days		Ambient noise levels in dB(A) for odd-even days	
		November, 2019		November, 2019	
		L_{day}	L_{night}	L_{day}	L_{night}
Civil Lines	**Commercial**	58.0	58.2	60.4	58.1
R. K. Puram	**Residential**	65.5	57.7	59.3	53.4
Anand Vihar	**Commercial**	65.9	63.0	65.7	62.3
Mandir Marg	**Silence**	60.3	53.4	54.6	46.8
Punjabi Bagh	**Residential**	57.5	51.0	56.8	50.5

increase in number of cars converted to CNG as from September to December 2015; 5463 vehicles have been converted to CNG; while from January to April 14, 2016; 12,126 cars have been converted to CNG [92]. These studies in conjunction with the ambient noise levels analysis show that for the ten sites under consideration; there is a marginal decrease in the ambient noise levels.

Another odd-even exercise was started in November 2019 from 4th November 2019 to 15th November 2019 in Delhi city with the same rules and strategy as was followed earlier. The odd-even rule was followed for 9 days except for 10th, 11th, and 12th November 2019 as tenth November was Sunday and 11th and 12th November were relaxed due to a festival (Guru Nanak Jayanti). Table 5.23 shows the average ambient noise levels for five locations in Delhi city for odd-even days and for normal days (averaged for 15th October 2019 to second November 2019).

The difference in average day and night ambient noise levels between odd-even days and normal days was analyzed and found to be inconclusive due to lack of noise data from large number of locations in Delhi city and also due to unavailability of statistical data on comparison of vehicular counts in both the situations. Table 5.24 shows the difference in average ambient levels between odd-even days and normal days for the L_{day} and L_{night} levels in Delhi city for five locations. It can be observed that RK Puram site and Mandir Marg site showed an appreciable reduction in day equivalent noise levels during the odd-even rule. However, this reduction may not be due to odd-even implementation which is confirmed by the fact that the night equivalent noise levels at both these sites were also reduced during odd-even implementation. One site namely Civil Lines commercial site showed an increment in day equivalent noise levels by 2.3 dB(A).

Table 5.24 Difference in average ambient levels between odd-even days and normal days for L_{day} and L_{night} levels in Delhi city

Name of location	Area characteristics	Difference in noise levels in dB(A) November 2019	
		L_{day}	L_{night}
Civil Lines	Commercial	2.3	−0.2
RK Puram	Residential	−6.2	−4.3
Anand Vihar	Commercial	−0.2	−0.7
Mandir Marg	Silence	−5.7	−6.5
Punjabi Bagh	Residential	−0.7	−0.5

Table 5.25 Difference in average ambient levels between odd-even days and Sundays and normal day and Sundays for L_{day} and L_{night} levels in Delhi city

Name of location	Area characteristics	Difference in noise levels in dB(A) for normal days and Sundays		Difference in noise levels in dB(A) for odd-even days and Sundays	
		L_{day}	L_{night}	L_{day}	L_{night}
Civil Lines	Commercial	−0.9	−0.2	1.4	−0.4
RK Puram	Residential	8.6	4.9	2.3	0.7
Anand Vihar	Commercial	0.7	0.8	0.4	0.2
Mandir Marg	Silence	8.3	7.2	2.5	0.7
Punjabi Bagh	Residential	2.4	−0.6	1.7	−1.1

Table 5.25 shows the difference in average ambient levels between odd-even days and Sundays and normal days and Sundays (average of 3rd and 10th November 2019). The two sites namely, RK Puram residential site and Mandir Marg silence zone site showed an appreciable reduction in day and night equivalent noise levels on Sundays as compared to the normal days. However, an exact counting of the vehicles during the odd-even days and normal days shall present an exact correlation of the decrease in ambient noise levels with decrease in the number of vehicles.

5.11 Environmental Noise Levels During the COVID-19 Lockdown in India

The entire world had been facing the wide spreading of Coronavirus (COVID 19) disease since the starting of the year 2020. The World Health Organization declared it as a "global pandemic" on 11th March 2020. Subsequently, many countries imposed complete lockdown so as to prevent the spreading of this disease at an alarming rate. During the lockdown, the reduced air and noise pollution levels had been observed in different parts of the world. Reductions from 4 dB to 6 dB L_{den} were observed for 21 noise monitoring stations with highly dominant road noise in south of France [93]. A significant reduction in hourly average equivalent sound and hourly minimum sound levels was observed at all the 12 stations in Dublin, Ireland, during the lockdown period and this can be attributed to reductions in both road and air traffic movements [94]. Aletta and Osborn have reported an average reduction in the A-weighted day-evening-night sound level (L_{den}) of 7.6 dB observed on the Paris road network. The aircraft noise emissions for the Paris Charles De Gaulle airport dropped by 21.5 dB(A) L_{den} [95]. Asensio et al. [96] reported a reduction which ranged 4–6 dB(A) due to COVID-19 lockdown in Madrid, Spain. In India, a 14 hours voluntary public curfew on 22nd March 2020 called as national Janta curfew was observed followed by a much targeted nationwide lockdown initially for 3 weeks starting from 24th March to 14th April 2020, extended up to third May 2020 followed by further extension till 31st May 2020. The government of India announced a "Janta Curfew" from 7 a.m. to 9 p.m. on Sunday, 22nd March to stop the spread of coronavirus disease, wherein people were urged not to step out of their homes except for those working in essential services like healthcare, security, or media. The lockdown rules were imposed by the Ministry of Home Affairs (MHA), India so as to prevent the rapid outbreak of pandemic. Also, all places of social gatherings, schools, colleges, offices, and transportation systems including rail, road, and air industries were restricted with only few exceptions for necessary services only. Figures 5.16 and 5.17a and b report the average ambient levels, L_{day} and L_{night} for the 70 sites in the seven major cities witnessed during the normal days and lockdown period [64]. The ambient noise levels in pre-lockdown period (Business-As-Usual or normal) days were analyzed for the period of seventh March 2020 to 21th March 2020, while the ambient noise levels in the lockdown period were analyzed for the period of 23rd March 2020 to 7th April, 2020. Figure 5.18a and b shows the difference in average day equivalent noise levels, L_{day} and night equivalent noise levels, L_{night} in dB(A) during the lockdown period and annual average ambient day and night equivalent noise levels observed in year 2019 [64]. Some important conclusions derived from the witnessed ambient noise levels during the lockdown period and Janta curfew days were [64]:

- The analysis of ambient day and night equivalent noise levels in lockdown period with annual average 2019 levels show that 57 sites (81.4%) showed a decrement in both day and night equivalent noise levels. Sixty-one sites (87.1%) showed a decrement in day equivalent noise levels, while 55 sites (78.6%) showed decrease in night equivalent noise levels. Forty-three sites (61.4%) showed a decrement in day levels $\geq$5 dB(A) and 38 sites (54.3%) showed decrement in night equivalent noise levels $\geq$5 dB(A). Twenty-two sites showed a decrement in day equivalent noise levels $\geq$10 dB(A), while 18 sites showed a decrement in night equivalent noise levels $\geq$10 dB(A). It was observed that the majority of the commercial and industrial zone sites registered decrement in the ambient day and night equivalent noise levels.
- It was also revealed that only 19 sites (27.1%) complied with the ambient noise limits for the lockdown period. No residential zone or silence zone site met the ambient noise standards in lockdown period. Interestingly, some of the sites partially complied with either day or night noise limits. Thirty-three sites

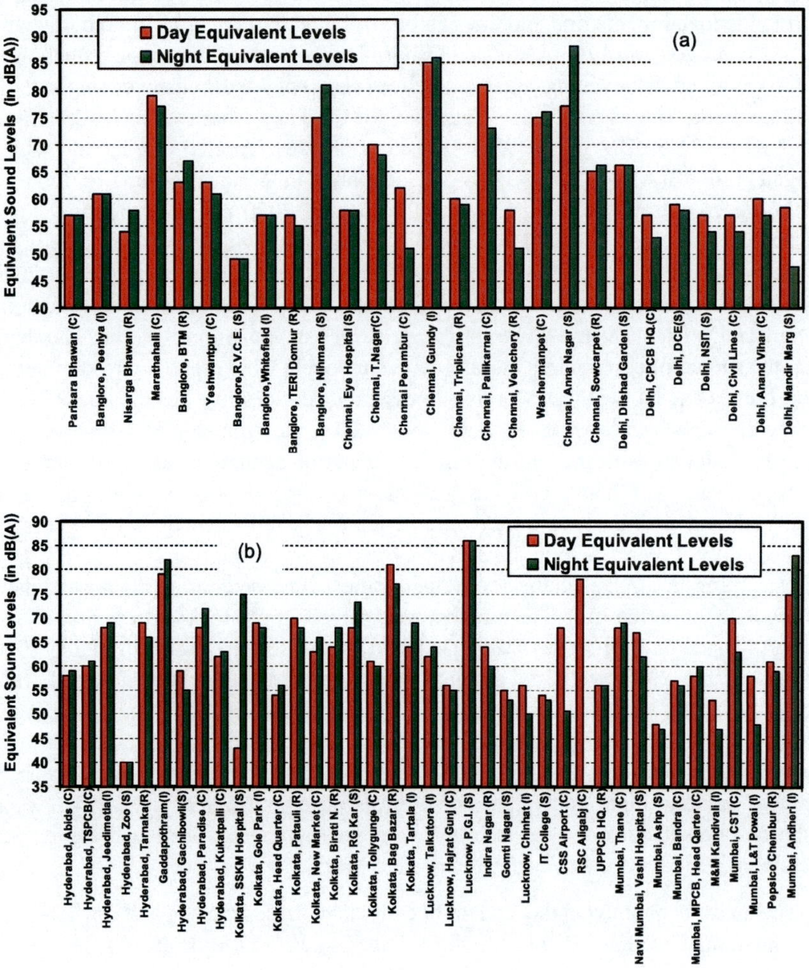

Fig. 5.16 (**a, b**) Day equivalent noise levels, L_{day} and night equivalent noise levels, L_{night} in dB (A) observed on Janta curfew day (22nd March 2020) for seven major cities

(47.1%) complied with day equivalent noise limits, while only 19 sites (27.1%) complied with the night equivalent noise limits for lockdown period.

- Significant increase was observed in the percentage of stations complying with norms in commercial areas. This can be attributed to the fact commercial activities such as malls, banquet halls, and restaurants are operated till late evening hours—and curtailment of the same resulted in increased compliance.
- Marginal increase in percentage of stations complying with noise standards during day time was observed in the residential areas. This can be attributed to the fact that noise in residential areas is due to household as well as vehicular

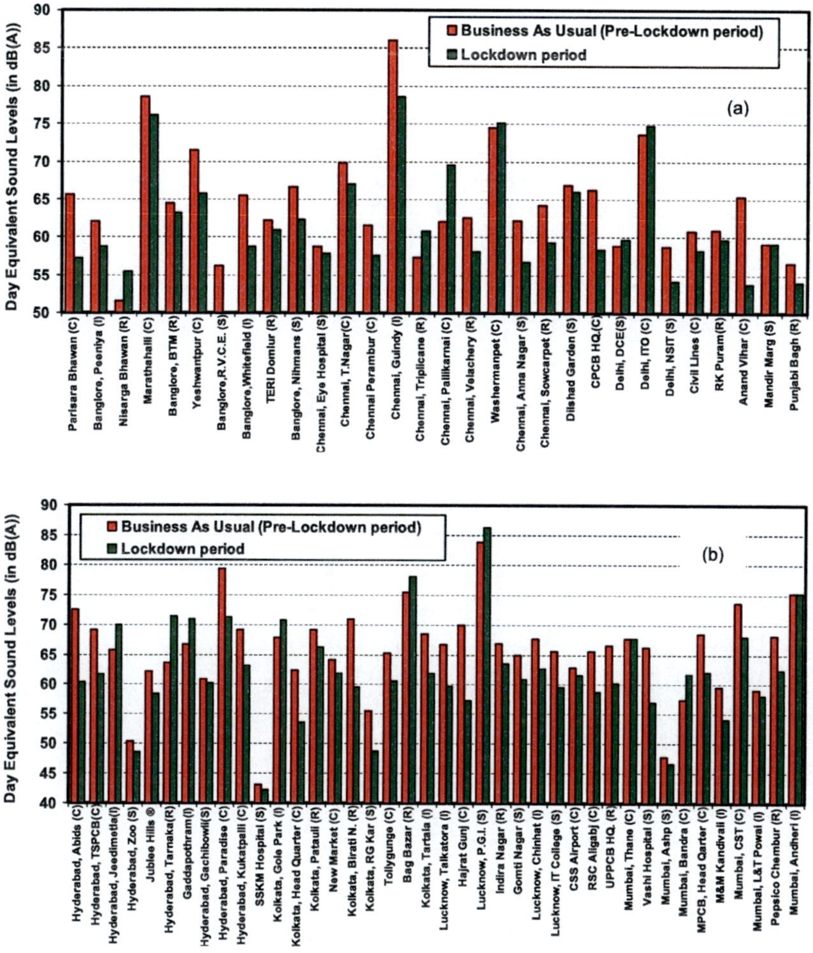

Fig. 5.17 (**a**, **b**) Average day equivalent noise levels, L_{day} in dB(A) observed during business as usual (pre-lockdown period) and lockdown period in year 2020

movement for transportation of people. During Janta curfew/lockdown, although the vehicular movement was minimal; the background sound due to household activities remained the same, and hence marginal impact of Janta curfew/lockdown observed in percentage of stations complying with noise standards.

- On the Janta curfew day, 31 sites (44.3%) complied with day equivalent noise limits, while only 14 sites (20%) complied with the night equivalent noise limits. Fourteen (nine industrial sites, four commercial sites, and one silence zone site) complied with the ambient noise limits on the day of Janta curfew day.

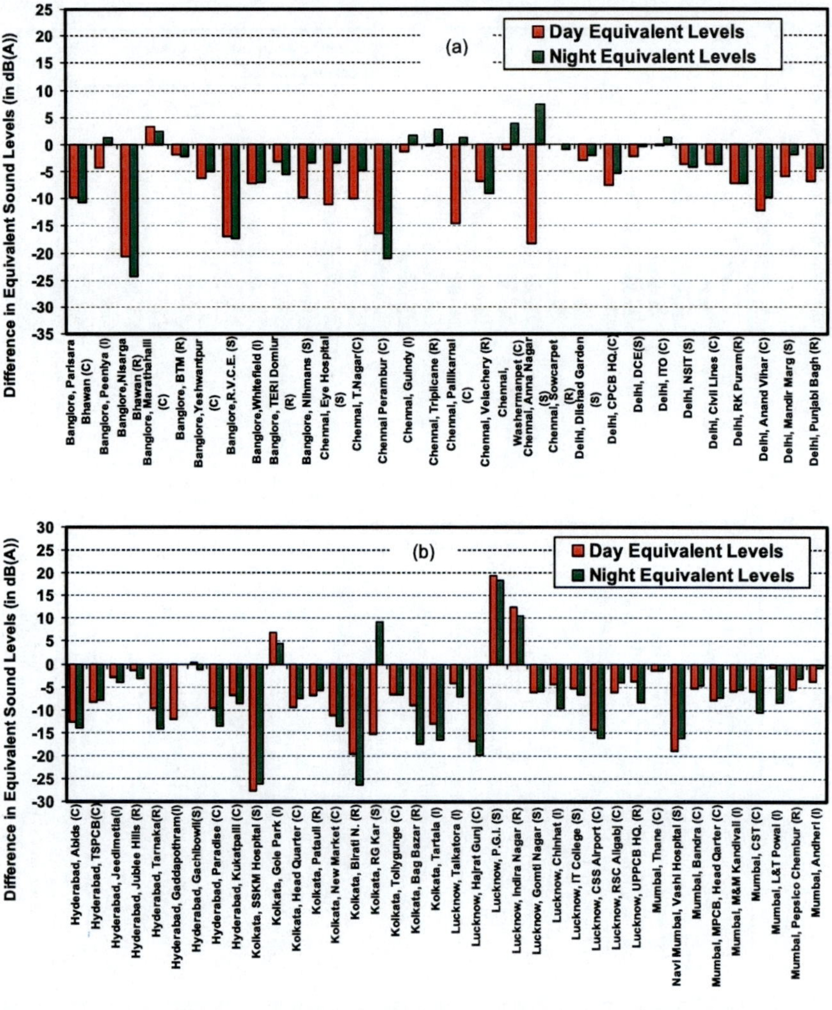

Fig. 5.18 (**a**, **b**) Difference in average day equivalent noise levels, L_{day} and night equivalent noise levels, L_{night} in dB(A) during the lockdown period and annual average ambient day and night equivalent noise levels observed in year 2019

5.11.1 *Impact of COVID-19 Lockdown on Environmental Noise*

A recent study reported by Mimani and Singh (2021) on the local anthropogenic noise level variation in Indian cities shows that all industrial, commercial, and residential zones experienced either a gradual or sudden decrease in noise levels, while the silence (S) zone was unaffected in lockdown period. The average maximum day- and night-time reduction with respect to the pre-lockdown period ranged from 4 to 13.8 dB(A) and 4 to 14.1 dB(A) in industrial, commercial and residential zones. The silence zone considered was not impacted throughout the lockdown and experienced a marginal noise reduction, while some of the sites registered an increment in noise levels attributed to an increased local traffic flux due to the patients visiting the hospitals due to COVID-19 [97]. The study reported by Garg et al., 2022 [64] also analyzed that only nineteen sites out of seventy sites under consideration comprising of nine industrial and ten commercial zone sites complied with the ambient noise limits in the lockdown period. Also, on the day of Janta curfew, fourteen sites including nine industrial sites, four commercial sites and one silence zone site met the noise limits. Evidently, no site lying in silence zone and residential zone complied with the ambient noise standards even during the lockdown period. These observations thus suggest that suitable control measures are essentially required in silence and residential zones in metropolitan cities for reducing the ambient noise levels and associated health hazards. The second wave of COVID-19 in April and May 2021 had been very serious in India particularly in the National Capital region of the country. In Delhi city, there had been many more than 25,000 COVID cases reported on some days. Consequently, a strict lockdown was imposed by government in various cities during the second wave also. Thus, a clear cut understanding of the noise scenario in the lockdown versus pre-lockdown period can be analyzed with the availability of more studies conducted across the different parts of the country in future. These studies shall be very helpful for consideration of imposing lockdowns, traffic control plans (odd–even rule) etc. for the ambient noise control perspective in the metropolitan cities of India.

5.12 Long-Term Versus Short-Term and Temporal Strategies

A continuous noise monitoring for a long-term is cumbersome and expensive especially for developing countries like India. Thus, short-term noise monitoring strategies and temporal sampling strategies should also be investigated whereby there is a trade-off between accuracy and costs incurred. The study by Morillas and Gajardo [98] based on 26 sampling points in Madrid revealed that it needs to take measurement for 9 days spread randomly throughout the year to obtain an estimate of L_{den} with a probability of success within a 90% confidence interval. If it requires a probability of 95%, the number of sampling days should be increased to 30–35. Gaja et al. [99] summarized 5 years of continuous noise measurements carried out in Valencia, Spain, and recommends that a random day strategy for sampling gives a more accurate representation than consecutive day strategy. Table 5.26 shows the Strengths, Weaknesses, Opportunities and Threats (*SWOT*) of noise monitoring strategies [100]. In Indian conditions, the reliability of temporal samplings is a critical issue as they may suffer from serious errors due to varied heterogeneous

Table 5.26 Strengths, Weaknesses, Opportunities and Threats (*SWOT*) of noise monitoring strategies (Garg et al. [100])

Noise Monitoring Strategy	Strengths	Weaknesses	Opportunities	Threats
Long-term (yearly)	Average yearly value is obtained with minimum uncertainty	Continuous long-term data acquisition and analysis is highly expensive and cumbersome	Useful for generating noise maps; Uncertainty calculation is easy	Dedicated equipment catering to needs of a single location requiring maintenance and calibration as per specific schedule
Short-term (random or continuous/ weekly or monthly)	Relatively less expensive; Trade-off between costs and accuracy of noise levels; Same equipment can be used for more sites, for example, a single instrumentation can be used to cover utmost 12 sites in a year if monthly strategy is followed	Moderate uncertainty levels; Don't exactly fulfill the requirements of Noise Directives	Best suited in Indian conditions wherein traffic density is not largely affected by weather or seasons;	May suffer from high uncertainty in areas where variability in traffic is high in different seasons or times; probability of levels within ± 1 dB(A) is less than 90%.
Temporal samplings (random or continuous/ days)	Economical and can be used to cover a larger part of city in less time; Selective random samplings based on scientific approach can provide a moderate uncertainty, for example, a 9 days random noise strategy can be helpful in covering utmost 40 sites in a year utilizing a single instrument	May suffer from serious errors, for example, areas with varied traffic density in different seasons, or noise emitted in festivals, or religious celebrations; parties, marriage processions, etc.; reliability is a critical issue in Indian context	Selective random strategy based on previous scientific evidence can be a good substitute for saving costs and time, for example, number of samplings days should be 30–35 for a probability of 95% (Morillas et al.); random week strategy better than continuous	Doubtful accuracy in some cases; Previous analysis of accuracy of short-term random samplings w.r.t to long-term average values is helpful for generating confidence interval and probability, for example, 14 consecutive days sampling gives a probability of 70% for levels within ± 1 dB (A), while 14 random days strategy give a probability of 88% (Gaja et al.)

traffic density, honking noise, noise emitted in festivals, or religious celebrations, parties, marriage processions, etc. Consequently, the short-term strategies are the best substitute not only for reliable accuracy of levels, but also for reducing the costs incurred on long-term noise monitoring. Thus, adopting a short-term strategy shall enable to cover the larger sites within a year rather than using a dedicated system for one site only. However, a recent study reported by Geraghty and Mahony (2016) recommended that long-term noise monitoring is necessary to accurately characterize long term indicators. The findings suggested that caution needs to be taken in the design of measurement experiments or in research that involves decisions to limit measurements to very short periods [101].

5.12.1 Short-Term Strategies

For short-term measurements with measurement time, $m << LT$, the error is calculated as [102]:

$$L_{\text{day,LT}} = L_{\text{day},m} \pm \varepsilon_d \tag{5.9}$$

$$L_{\text{night,LT}} = L_{\text{night},m} \pm \varepsilon_n \tag{5.10}$$

where ε_d and ε_n are the errors in short-term strategies for day and night levels for m measurement time. It may be noted here that value of $m = 30$ or 31 depending upon the month for short-term noise sampling considered. It may be noted here that *LT* is number of months in a year for which noise monitoring is conducted. It may be noted *LT* is equal to 365 days in case a single day value is utilized for calculating the annual representative value.

Figure 5.19 a-f shows the frequency histogram of calculated error from average yearly value for L_{day} and L_{night} for 1 and 2 months random and consecutive noise monitoring strategy. The noise monitored data for past 4 years (2011–2014) for 35 locations under study ($n = 35 \times 12 \times 4 = 1680$) was analyzed.

Table 5.27 shows the statistics for short-term monthly noise monitoring strategies. The results obtained indicate that it is possible to determine the L_{day} and L_{night} levels representative of the equivalent annual level by monthly samplings with a reasonable accuracy. An error of ±3 dB(A) from annual equivalent level is calculated with 95% probability for L_{day} and L_{night} value by one month noise monitoring strategy. For higher accuracy, the random 2 months strategy is recommended, whereby an error of ±2 dB(A) is calculated with a probability higher than 90%. These investigations were further extended to 3 months consecutive noise monitoring strategy and random 3 months strategy. Undoubtedly, higher is the measurement span, less is the error value. Figure 5.20a–d shows the frequency histogram of calculated error from average yearly value for L_{day} and L_{night} for 3 months random and 3 month consecutive noise monitoring strategy.

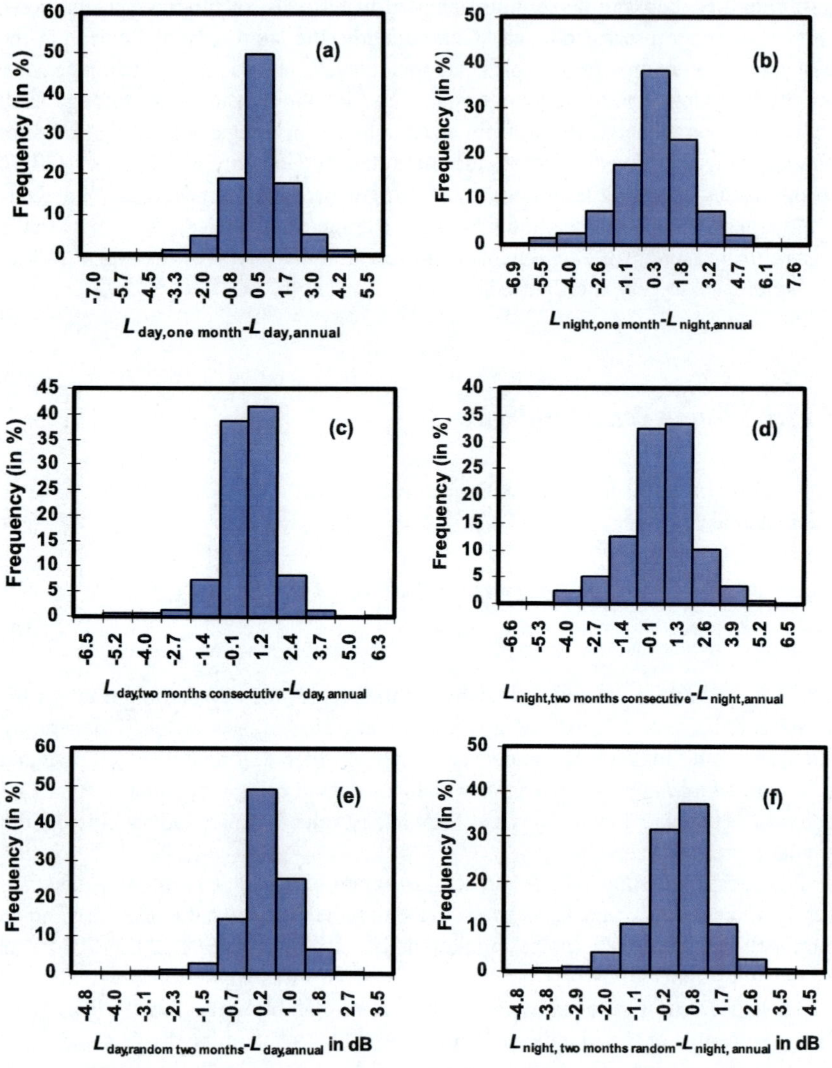

Fig. 5.19 (a–f) Frequency histogram of calculated error from average yearly value for L_{day} and L_{night} for 1 and 2 months random and consecutive noise monitoring strategy

In case of random three months strategy, the probability of error within ±2 dB is 98.8% for L_{day} and 96.9% for L_{night}. Interestingly, the probability of an error of ±1, ±2, and ± 3 dB in case of consecutive 3 months strategy is less than the probability calculated for 2 months random strategy, which implies that random two months strategy yields less error as compared to the consecutive 3 months strategy.

The probability calculated for the night equivalent noise levels are lesser than the day equivalent noise levels due to the fact that the standard deviation of monthly

Table 5.27 Statistics for short-term monthly noise monitoring strategies

Parameter	Mean dB(A)	Standard deviation in dB(A)	Probability [-1;1]	Probability [-2;2]	Probability [-3;3]	95 % Probability range in dB(A)	99 % Probability range in dB(A)
One month monitoring strategy							
L_{day}	-0.20	1.32	79.4 %	92.6 %	98.7 %	[-3.3;3.0]	[-4.5;4.2]
L_{night}	-0.41	1.86	65.9 %	83.8 %	96.8 %	[-4.0;3.2]	[-5.5;4.7]
Two consecutive months monitoring strategy							
L_{day}	-0.15	1.22	82.1 %	94.7 %	99.3 %	[-2.7;2.4]	[-4.0;3.7]
L_{night}	-0.22	1.68	69.0 %	86.9 %	97.1 %	[-2.7;2.6]	[-4.0;3.9]
Two random months monitoring strategy							
L_{day}	0.10	0.82	89.9 %	98.6 %	99.9 %	[-1.5;1.8]	[-2.3;2.7]
L_{night}	0.19	1.12	83.7 %	96.2 %	99.4 %	[-2.0;2.6]	[-2.9;3.5]
Three consecutive months monitoring strategy							
L_{day}	0.01	0.98	87.0 %	96.8 %	99.5 %	[-1.7;2.0]	[-2.6;2.9]
L_{night}	0.10	1.40	73.2 %	89.8 %	97.7 %	[-2.9;2.8]	[-3.8;3.7]
Three random months monitoring strategy							
L_{day}	0.10	0.77	92.5 %	98.8 %	99.9 %	[-1.1;1.3]	[-1.9;2.2]
L_{night}	0.10	1.10	84.2 %	96.9 %	98.8 %	[-1.8;1.6]	[-2.6;2.4]

night equivalent levels is higher than the day equivalent noise levels for many of the sites under consideration. The observations presented are consistent with previous studies of Gaja et al. 2003 and Morillas and Gajardo regarding higher accuracy of random samplings as compared to consecutive measurement strategy, but differs on the aspect of proposed random monthly samplings rather than temporal samplings (e.g., 9 random days). This is due to fact that the varied heterogeneous traffic density, varied weather conditions, honking noise, noise emitted due to public meetings, festival celebrations particularly some days before and after Diwali festival, religious ceremonies, etc. may result in doubtful accuracy of temporal samplings in Indian context. Consequently, short-term noise monitoring strategy followed can be a good trade-off between the accuracy and costs incurred and thus offer a Best Practicable and Economical Option (*BPEO*) for noise mapping of the larger parts of the country.

5.12.2 Temporal Sampling Strategies

The study of temporal sampling strategy has been focused by Morillas and Gajardo [98] and Gaja et al. [99] comprehensively. Gaja et al. [99] recommended a random

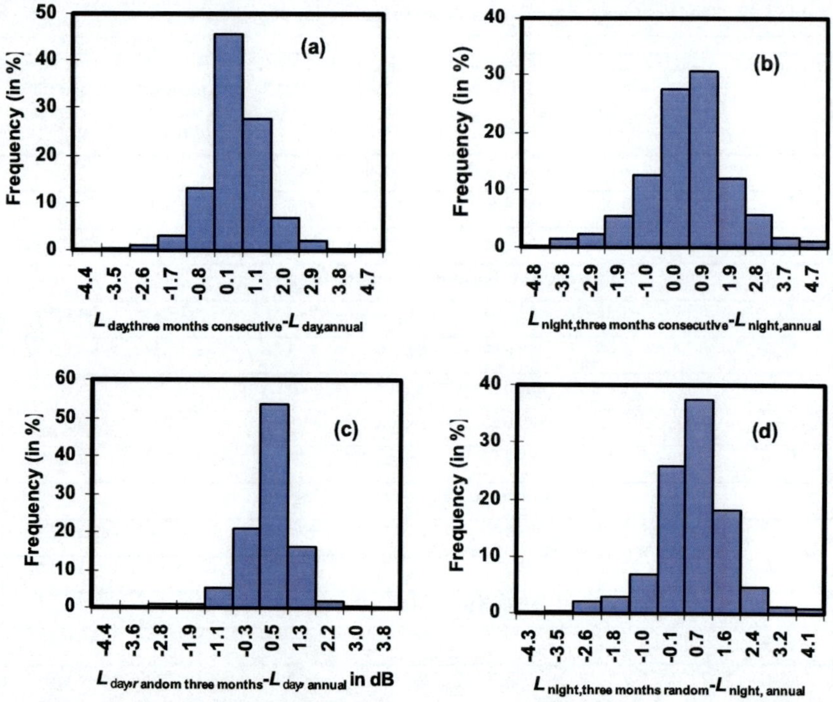

Fig. 5.20 (**a–d**) Frequency histogram of calculated error from average yearly value for L_{day} and L_{night} for 3 month random and 3 month consecutive noise monitoring strategy

9 days strategy for $L_{\text{Aeq,year}}$ levels assessment, whereby an error of ±3 dB is evaluated with a probability of 99%. The analysis of temporal sampling strategy was investigated from year 2017 noise monitored data for 365 days for 12 sites containing the yearly data of three sites each from commercial, residential, silence, and industrial zones. The error in temporal sampling strategy with respect to annual average value is calculated and corresponding probability for ±1, ±2, and ± 3 dB error is also evaluated. Table 5.28 shows the statistics for temporal sampling strategies for evaluation of annual average equivalent levels. One day approach seems to be of limited accuracy as the probability of an error of ±3 dB is 79% for day equivalent levels and 69% for night equivalent levels. The random 9 days strategy shows an error of ±3 dB with a probability of 90% for day equivalent levels, but less than 90% for night equivalent noise levels. It can be observed from Table 5.28 that random 3 weeks sampling strategy seems to be an optimized temporal sampling strategy as an error of ±3 dB is evaluated with a probability of success higher than 90% for both day and night equivalent levels. It is even better than 3 consecutive week's strategy. For higher probability of success with 95% confidence level, 30 consecutive days (or 1 month) strategy is the best approach for evaluation of the annual average day and night equivalent noise levels.

Table 5.28 Statistics for temporal sampling strategies for evaluation of annual average equivalent noise levels

	Parameter	Mean dB(A)	Standard deviation in dB(A)	Probability [-1;1]	Probability [-2;2]	Probability [-3;3]	90 % Probability range in dB(A)	95 % Probability range in dB(A)
1 day	L_{day}	-0.97	1.48	51.4 %	70.3 %	78.9 %	[-4.3;4.7]	[-5.1;4.9]
	L_{night}	-1.10	1.73	47.9 %	64.7 %	69.4 %	[-5.1;4.5]	[-5.5;5.3]
2 Consecutive days	L_{day}	-0.88	1.34	53.6 %	71.1 %	79.0 %	[-4.3;3.7]	[-4.8;4.6]
	L_{night}	-0.97	1.77	48.7 %	66.5 %	70.1 %	[-4.7;4.5]	[-5.0;4.8]
6 Consecutive days	L_{day}	-0.77	1.30	55.9 %	73.2 %	83.3 %	[-4.1;3.5]	[-4.5;3.9]
	L_{night}	-0.95	1.79	52.5 %	68.8 %	74.5 %	[-3.8; 4.2]	[-4.3;4.2]
9 Random days	L_{day}	-0.79	1.33	66.1 %	85.8 %	89.6 %	[-3.7;4.0]	[-3.9;4.3]
	L_{night}	-1.19	1.91	57.3 %	79.0 %	85.3 %	[-3.9;4.1]	[-3.7;4.7]
12 Consecutive days	L_{day}	-0.90	1.61	65.5 %	83.2 %	89.9 %	[-3.5;4.3]	[-3.7;4.2]
	L_{night}	-1.31	1.56	56.6 %	77.7 %	83.4 %	[-4.0;4.1]	[-4.3;3.8]
14 Random days	L_{day}	-0.89	1.58	74.1 %	89.2 %	91.1 %	[-3.4;3.5]	[-4.3;4.1]
	L_{night}	-1.23	1.88	62.9 %	79.8 %	88.8 %	[-3.3;3.7]	[-4.2;3.7]
3 Random weeks	L_{day}	-0.38	1.41	74.9 %	93.4 %	94.2 %	[-3.0;2.9]	[-3.5;3.3]
	L_{night}	-0.89	1.55	65.0 %	81.8 %	92.0 %	[-2.8;3.4]	[-4.1;3.5]
3 Consecutive weeks	L_{day}	-0.40	1.66	72.4 %	89.3 %	93.1 %	[-3.0;3.3]	[-3.8;3.6]
	L_{night}	-0.73	1.95	63.7 %	79.8 %	91.5 %	[-3.0;2.9]	[-4.4;3.6]
30 Consecutive days (One Month)	L_{day}	-0.20	1.32	79.4 %	92.6 %	98.7 %	[-2.1;2.3]	[-3.3;3.0]
	L_{night}	-0.41	1.86	65.9 %	83.8 %	96.8 %	[-2.6;2.4]	[-4.0;3.2]

The European Noise Directives 2002/49/EC recommends that noise measurements may be made over longer time such as 1 year. Most of the studies in Indian perspectives focus on a single day or two days sampling strategy (except that reported in *NANMN* project) at most to evaluate the day and night equivalent noise levels at a particular site, which can be helpful to understand the compliance with respect to ambient noise standards and take preventive measures for its control. Consequently, temporal sampling strategies although being cost-effective offer a limited accuracy as shown in Fig. 5.21a and b in estimating L_{day} and L_{night} long-term yearly value with a definite confidence levels. Random 3 weeks strategy is an optimized temporal strategy, whereby an error of ±3 dB is estimated with a probability of success higher than 90%. However, monthly strategy offers a more reliable and cost-effective approach in Indian perspectives for estimating L_{day} and L_{night} long-term yearly value with a definite confidence levels.

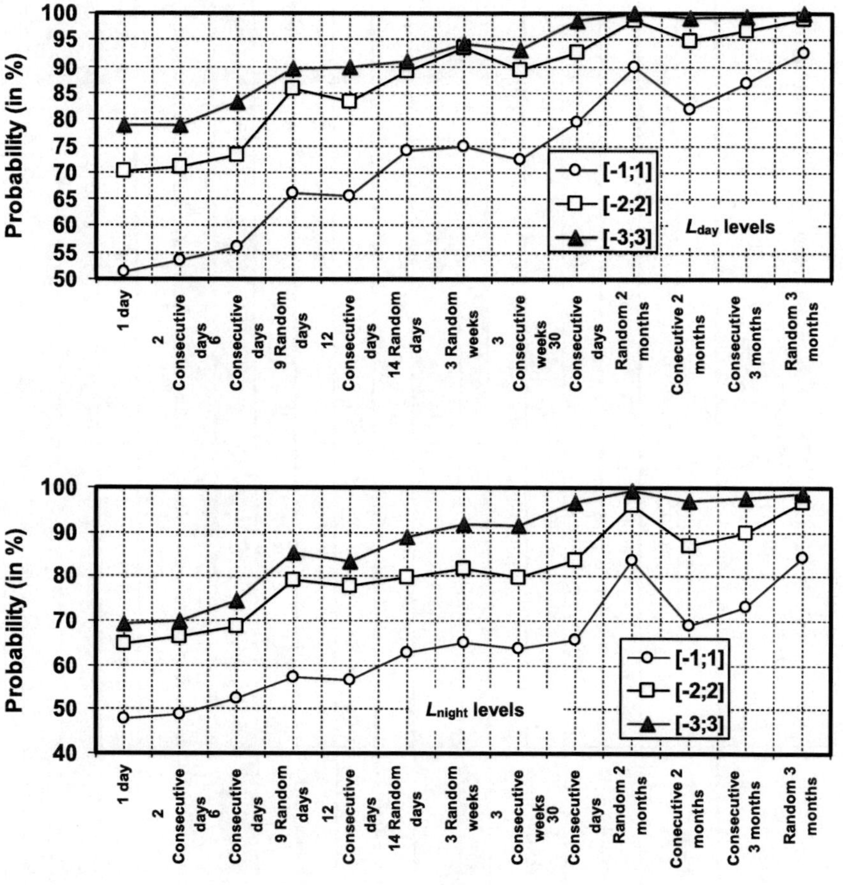

Fig. 5.21 (**a, b**) Comparison of short-term and temporal sampling strategies for L_{day} and L_{night} measurements for probability of success for an error of ±1, ±2, and ± 3 dB

5.13 Time-series Modeling of Ambient Noise Levels

A continuous long-term noise monitoring of ambient noise levels is cumbersome and quite expensive and in many cases the larger resources involved are not justified in terms of better accuracy than that achievable by cheaper and feasible temporal samplings [102]. The time-series analysis of ambient noise shall be thus instrumental in forecasting the future noise levels in addition to the continuous noise monitoring attributed to the stochastic nature of traffic noise. It can also serve as a suitable substitute to the continuous long-term noise monitoring provided the predicted data matches well with the actual measurement data. There had been some studies reported so far on application of this approach to various field including noise modeling as well as the short-term strategies [103–105]. Garg et al. [103] demonstrated the use of *ARIMA* approach for time series modeling of ambient noise levels. Consequently, both *ARIMA* and *ANN* approaches were compared for their suitability and forecasting efficiency [104]. Kumar et al. analyzed the short-term noise levels measured at 10 s intervals in the vicinity of a busy road carrying vehicular traffic using the *ARIMA* approach [105]. *ARIMA* approach has been widely used in air pollution monitoring and forecasting of air pollution levels.

5.13.1 ARIMA *Modeling Approach*

The *ARIMA* approach was first popularized by Box and Jenkins, and *ARIMA* models are often referred to as Box-Jenkins models [106]. In *ARIMA* modeling approach, a stationary time-series is made stationary by applying finite differencing of the data points. The mathematical formulation of the *ARIMA* (p, d, q) model using lag polynomials for a time-series data Y_t where t is an integer and Y_t are real numbers [107–109]:

$$\left(1 - \sum_{i=1}^{p} \phi_i L^i\right)(1 - L)^d Y_t = \left(1 + \sum_{i=1}^{q} \theta_i L^i\right)\varepsilon_t \tag{5.11}$$

where p, d, and q are integers greater than or equal to zero and refer to the order of autoregressive, integrated, and moving average parts of the model, respectively. The integer d controls the level of differencing. When $d = 0$, then it reduces to an *ARMA* (p,q) model. An *ARIMA* $(p,0,q)$ or ARMA (p,q) is a model for time-series that depends on p past values of itself and q past random error terms e_t. This method has form of Eq. (5.12) as [110]:

$$\begin{aligned} Y_t = \theta_0 + \phi_1 Y_{t-1} + \phi_2 Y_{t-2} + \ldots\ldots\ldots + \phi_p Y_{t-p} + \mu - \theta_1 e_{t-1} - \theta_2 e_{t-2} \\ - \ldots\ldots\ldots - \theta_q e_{t-q} + e_t \end{aligned} \tag{5.12}$$

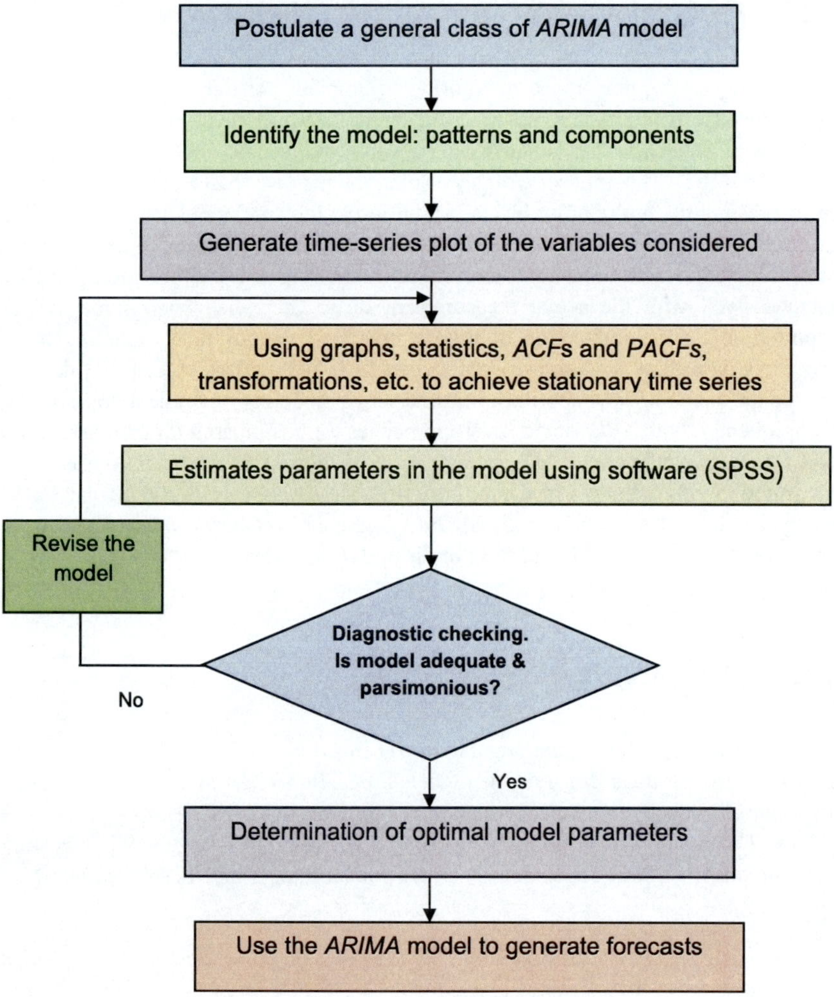

Fig. 5.22 Flowchart depicting the *ARIMA* methodology (Garg et al. [103])

where θ_1, θ_2. . . . . . . .θ_q are the finite weights, μ is the mean of series, Y_t is the forecasted output, Y_{t-p} is observation at time *t-p* and ϕ_1, ϕ_2, ...ϕ_p is a set of finite parameters determined by linear regression and e_t is error associated with the regression. Figure 5.22 shows the flowchart depicting the *ARIMA* methodology used in present study.

Although *ARIMA* models are quite flexible in handling the time-series data and modeling, their major limitation is the pre-assumed linear form of the model [111]. The time-series can be analyzed using many software packages such as SPSS software (Statistical Package for the Social Sciences). The characteristics of good *ARIMA* model are summarized as follows [112]:

- Stationary, that is, it has an *AR* term that satisfies some mathematic inequalities.
- Invertible, that is, it has *MA* coefficient that satisfies some mathematic inequalities.
- Parsimonious, that is, uses small number of coefficients to explain the available time-series data.
- Statistically independent residuals.
- Fits the available data sufficiently well.
- Root-mean squared error (*RMSE*) and mean-absolute percent error (*MAPE*) are acceptable, sufficiently small forecast errors.

5.13.2 ANN *Approach for Time-Series Modeling*

Artificial neural networks (*ANN*) are flexible computing frameworks for modeling a broad range of nonlinear problems. One of the most significant advantages of the *ANN* models over other classes of nonlinear models is that *ANN*s are universal operators that can approximate a large class of functions with a high degree of accuracy [113–115]. Single hidden layer feed forward network is the most widely used model for the time series modeling. The model is characterized by a network of three layers of simple processing units connected by acyclic links. The relationship between the output y_t and the inputs $(y_{t-1}\ldots\ldots\ldots y_{t-p})$ has the following mathematical representation [115]:

$$y_t = w_0 + \sum_{j=1}^{q} w_j g\left(w_{0,j} + \sum_{i=1}^{p} w_{i,j} y_{t-i}\right) + \varepsilon_t \tag{5.13}$$

where $w_{i,j}(i= 0,1,2,\ldots\ldots p, j = 1,2,\ldots\ldots q)$ and $w_j(j = 1,2,\ldots\ldots q)$ are the modal parameters called connection weights; p is the number of input nodes; q is the number of hidden nodes. *ANN* model performs a nonlinear functional mapping from the past observations to the future value y_t that is,

$$y_t = f\left(y_{t-1}\ldots\ldots\ldots y_{t-p}, w\right) + \varepsilon_t \tag{5.14}$$

where w is a vector of all parameters and $f()$ is a function determined by the network structure and connection weights. Thus, the neural network is equivalent to a nonlinear autoregressive model. The logistic function is often used as the hidden layer transfer function. In practice, simple network structure that has a smaller number of hidden nodes often works well. The choice of q, that is, number of hidden layer neurons, is data-dependent and there is no systematic rule in deciding this parameter. In addition to choosing an appropriate number of hidden nodes, another important task of *ANN* modeling of time-series is the selection of the number of lagged observations, p, the dimensions of the input vector [115]. The sound level data set can be judiciously divided into training data (70%), testing data (15%), and validation data (15%). However, there is no theory that can be used to guide the selection of p. The shot-gun (trial-and-error) methodology for specific problems is typically adopted which may be the primary reason for inconsistencies in literature [116]. Once a network structure (p, q) is specified, the network is ready for training a process of parameter estimation. The cost function is the objective to be fulfilled such that the mean squared error is minimum [115]:

$$E = \frac{1}{N} \sum_{n=1}^{N} e_i^2 \tag{5.15}$$

where e_i is the error between the values calculated by neural network for each input and expected response or target and N is the number of training samples. The back-propagation is the most widely used learning algorithm. The weights in a network are adjusted by comparing the actual response with the target response in such a way to minimize the mean squared error. *ANN* belong to the data-driven approach, that is, the analysis depends on the available data, with little a priori rationalization about relationships between variables and about the models [116]. Figure 5.23 shows the flowchart of methodology for the development of an *ANN* model. The ability of *ANNs* pertaining to adaptability, nonlinearity, and arbitrary function mapping thus make them functional for forecasting tasks [115, 116].

Table 5.29 shows the comparison of errors associated with time-series modeling of sound level parameters using *ARIMA* and *ANN* approach for a period of 1 year. The mean squared error, *MSE* observed for the training and validation data is quite low for both the *ARIMA* and *ANN* models so developed [104].

5.13.3 Other Modeling Strategies

It is thus evident that sound levels can be predicted with a reasonable accuracy using the statistical analysis of the present and historical time-series data sets obtained from continuous long-term noise monitoring. The *ARIMA* and *ANN* methodology can thus serve as a suitable substitute to the long-term noise monitoring and is thus indispensable for saving costs and time incurred on continuous noise monitoring. However, the *ARIMA* and *ANN* models so developed for sound level descriptors cannot be generalized for any specific site and as such these models can be developed afresh with different p, d, q values in case of *ARIMA* and different p and q values in case of *ANN* approach. Construction of an *ANN* model is time consuming and sometimes like a black-box wherein one cannot adjudge the weights and biases developed while training the network. However, in spite of these short-comings, *ANN* can serve as vital substitute to the cumbersome and expensive long-term traffic noise monitoring in metropolitan cities of India. Thus, future studies can further explore the potential of hybrid linear-nonlinear models in accurate forecasting of sound levels when applied to the dynamic time-series database [117].

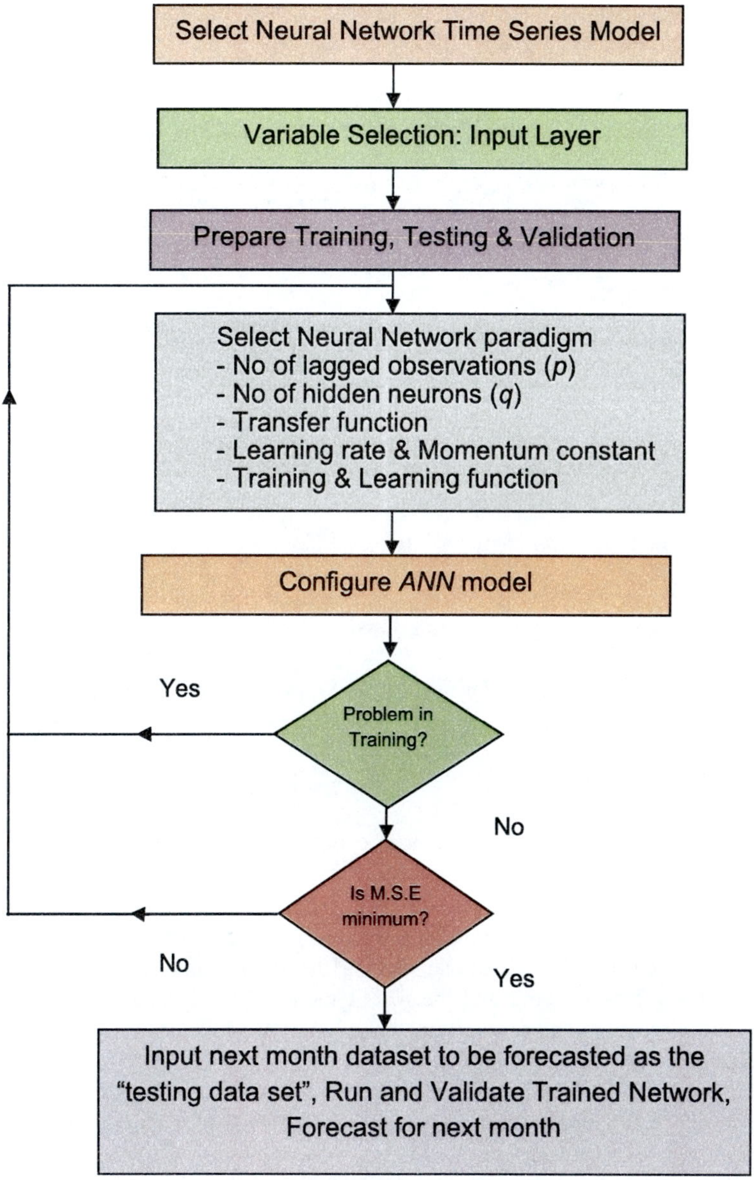

Fig. 5.23 Flowchart of methodology for development of an *ANN* Model (Garg et al., [104])

Recently, the support vector machine (*SVM*) method, which was suggested by Vapnik in 1995 [118], has been used in a range of applications such as in data mining, classification, regression, and time-series forecasting. The kernel functions and the optimal hyperparameters have the potential to improve the accuracy of *SVM*

Table 5.29 Statistical comparison of the errors associated with time-series modeling of sound level parameters using *ARIMA* and *ANN* approach (Garg et al. [104])

Statistical Parameter	L_{day} dB(A)		L_{night} dB(A)		L_{day} dB(C)		L_{night} dB(C)		L_{dn} dB(A)		$L_{eq,24h}$ dB(A)	
	ANN	*ARIMA*	*ANN*	*ARIMA*	*ANN*	*ARIMA*	*ANN*	*ARIMA*	*ANN*	*ARIMA*	*ANN*	*ARIMA*
Model Topology	**8-3-1**	**(0,0,14)**	**10-6-1**	**(0,0,7)**	**8-4-1**	**(0,0,14)**	**10-7-1**	**Simple**	**10-5-1**	**(1,0,7)**	**6-3-1**	**(0,1,9)**
***MSE* in dB(A)^2**	1.3	1.5	1.5	1.5	0.9	1.3	1.1	1.4	1.0	1.2	0.9	0.9
***RMSE* in dB(A)**	1.1	1.2	1.2	1.2	1.0	1.1	1.2	1.17	1.0	1.1	1.0	1.0
***MAE* in dB(A)**	0.0	0.1	0.1	0.1	0.0	0.1	0.1	0.1	0.1	0.1	0.0	0.0
Maximum Error in dB(A)	3.6	4.7	9.2	10.3	1.7	5.1	8.4	8.7	5.8	7.6	6.3	4.9
Minimum Error in dB(A)	-4.9	-4.7	-3.2	-2.9	-4.8	-5.1	-4.0	-3.9	-3.3	-3.5	-3.6	-2.9
R^2	0.4	0.2	0.3	0.2	0.4	0.3	0.2	0.1	0.5	0.1	0.7	0.8

forecasts. The research studies demonstrate that *SVM* method is a promising alternative approach for prediction of time-series data. The largest benefits of *SVM*s are the facts that a global solution exists. Furthermore, due to the specific optimization procedure it is assured that overtraining is avoided and the *SVM* solution is general [119]. A drawback of the *SVM*s is the fact that the training time can be much longer than for the Elman network and the *ARMA* model [120]. Figure 5.24 shows the flowchart of methodology for development of an *SVM* Model. The proper selection of hyperparameter combinations with advanced techniques such as grid search technique using 10 *K*-fold cross validation and heuristic techniques like Genetic algorithms can facilitate the development of an accurate forecasting model for time-series modeling and forecasting of ambient noise levels. Deep learning algorithms such as convolutional neural networks (CNN) and long short-term memory (LSTM) networks can also be explored for the time-series modeling, prediction, and forecasting of environmental noise levels.

Apart from time-series modeling, land use regression (LUR) models have been widely developed by researchers to explain the spatial variation of traffic noise. LUR modeling with accurate GIS source data can be a promising tool for noise exposure assessment with applications in epidemiological studies especially where noise maps

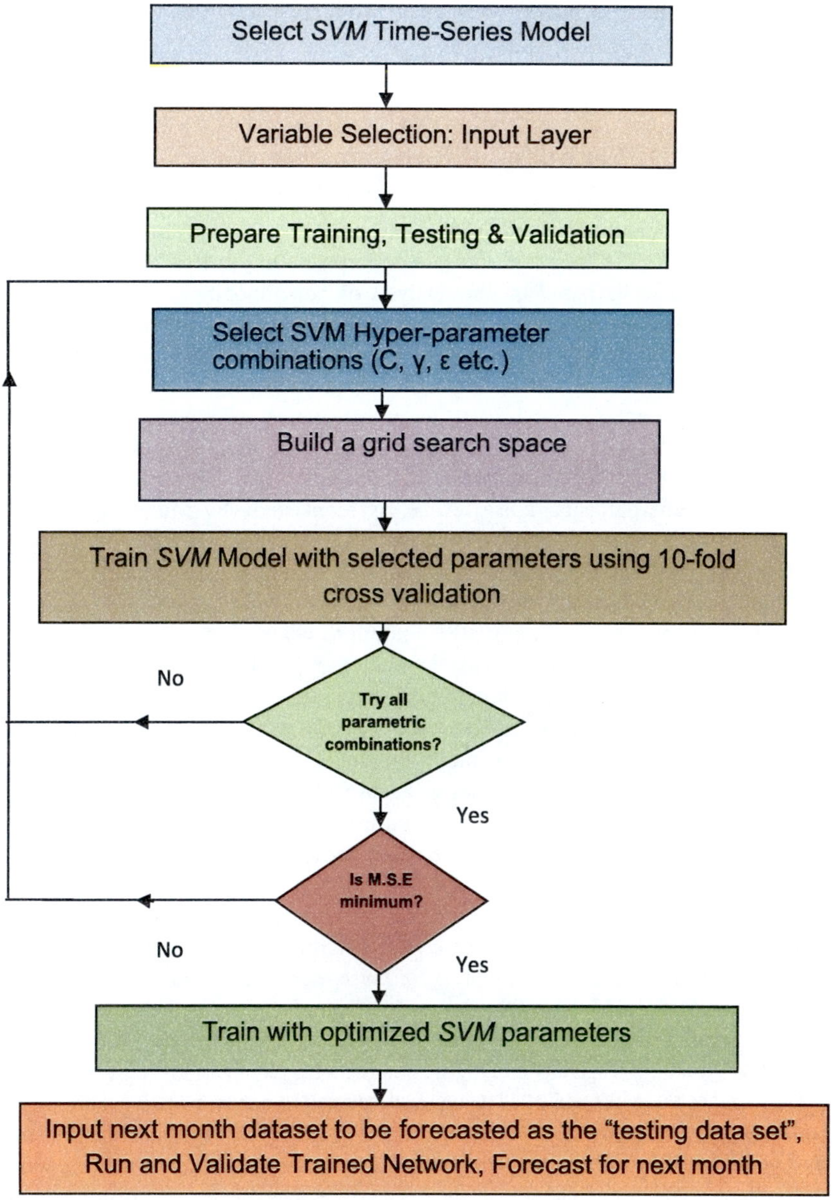

Fig. 5.24 Flow chart of methodology for development of an *SVM* model

are unavailable [121]. A study in Northwest China systematically examined the association between urban land-use and environmental noise change and utilizes LUR approach for developing a regional noise model based on GIS platform [122]. LUR models with good performance for $L_{eq,24h}$ and L_{night} in the full and octave-band frequencies had been developed in a study in Taiwan that can be applied for epidemiological studies [123].

5.14 Seasonal Variations of Environmental Noise

The seasonal variations of environmental noise levels had been always an unexplored factor. The long-term noise monitoring of the various sites thus facilitates such an analysis based on the analysis of noise monitored data in various seasons. The study by Geraghty and O'Mahony from a permanent noise monitoring network located in the city of Dublin, Ireland, showed a high degree of variability between months at all sites [101]. The EU Directive 2002/49/EC requires the value of noise descriptors: L_{den} and L_{night} as representative of a year. The seasonal variations thus may have implications on the use of the yearly average value recommended. There had been rarely some studies that reported the seasonal variations of environmental noise levels. A recent study by Munir et al. [124] focused on the analysis of temporal and seasonal variations of noise pollution in urban zones of Peshawar, Pakistan. The seasonal variation in L_{eq} revealed that the noise level in 70% of areas increased from winter to spring, 45% from spring to summer, 35% summer to autumn, 30% autumn to winter, 70% winter to summer, and 40% spring to autumn. The seasonal variations were observed to be attributed to the factors such as temperature, humidity, and wind speed and traffic was the main source of noise pollution identified in the area. The L_{eq} values observed in the sequence of summer > spring > winter > autumn. An Indian study focused on the assessment of day equivalent noise levels at curbside open-air microenvironment of Kolkata city under heterogeneous environmental conditions revealed that that prevailing traffic noise level of the microenvironment had weak negative (−0.21; $p < 0.01$) and very weak positive (0.19; $p < 0.01$) correlation with air temperature and relative humidity [125]. Thus, in order to understand the effect of seasonal variations on the environmental noise levels, a study for the ten locations in Delhi city was conducted based on the analysis of noise monitored data for year 2018 and 2019. Figure 5.25a and b shows the day and night equivalent sound levels in various seasons in year 2019 for nine locations in Delhi city. The difference of the annual average value and average day and night equivalent levels in various seasons defined as L_{diff} is reported in Table 5.30. The analysis seemed to be inconclusive, although it can be observed that maximal variations up to 6 dB(A) for day equivalent sound levels and 7 dB(A) for night equivalent sound levels had been observed for some locations. A generalized pattern on the variation of environmental noise levels in ascending/descending order for various seasons as observed by Munir et al. [124] could not be depicted from this analysis. However, the seasonal changes may bring a change in the traffic density as highlighted in one such study in Seferihisar, Turkey, wherein the noise levels in summer showed an increase both on weekdays and at weekends due to increased traffic [126]. Also, for the same noise levels, noise annoyance had been observed to be greater in warmer seasons as reported by

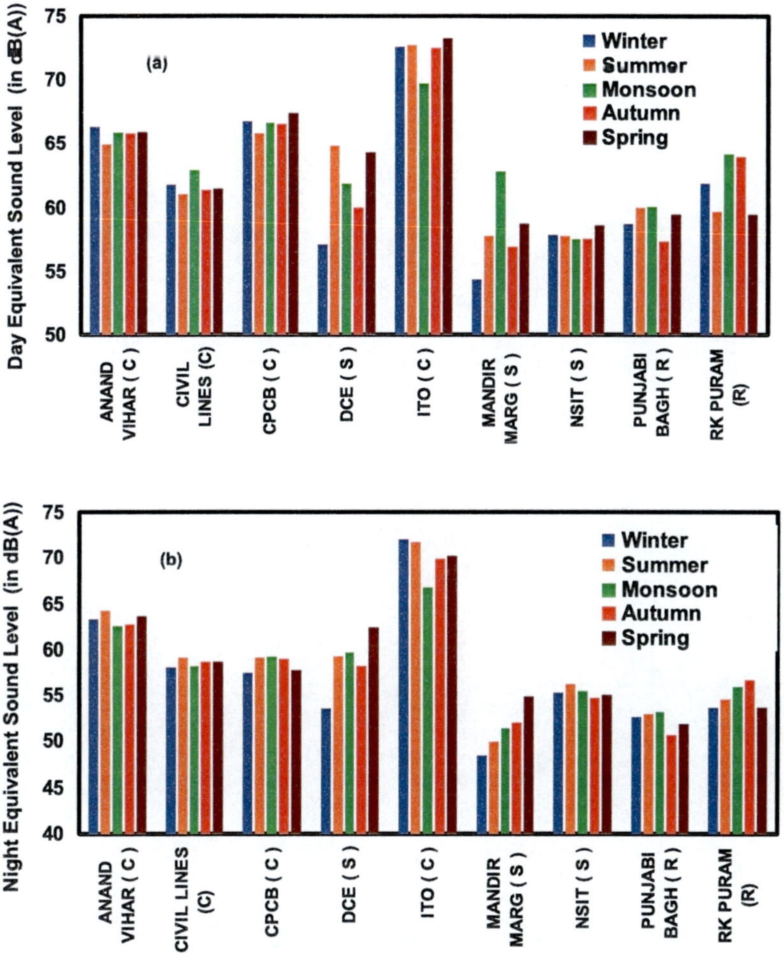

Fig. 5.25 (**a, b**) Day and night equivalent sound levels in various seasons in year 2019

Miedema et al. A 15 °C difference has about the same effect on noise annoyance as does a 1- to 3- dB difference in noise exposure [127].

Table 5.31 shows the correlation of the meteorological parameters with the noise descriptors for various seasons for the Delhi city for the year 2019. A positive correlation is observed between the temperature and noise, while a weak negative correlation is observed between humidity and noise. The recent study of Munir et al. [124] reported a strong correlation of temperature with noise levels for the Spring season when compared with other seasons. The present observations however show a strong correlation coefficient with ambient temperature for the summer and spring seasons in Delhi city. The day equivalent noise levels were observed to be the

Table 5.30 Seasonal analysis of day and night equivalent levels for ten sites in Delhi city for past 2 years

Parameters (in dB(A)) for day equivalent sound levels	Noise Monitoring sites in Delhi city																			
	CPCB, HQ.		DTU, Bawana		ITO		Dilshad Garden		NSIT, Dwarka		Civil Lines		R. K. Puram		Anand Vihar		Mandir Marg		Punjabi Bagh	
	2018	2019	2018	2019	2018	2019	2018	2019	2018	2019	2018	2019	2018	2019	2018	2019	2018	2019	2018	2019
Site Characteristics	Commercial		Silence		Commercial		Silence		Silence		Commercial		Residential		Commercial		Silence		Residential	
$L_{annual, avg, day}$	66.1	66.5	56.2	62.2	73.1	75.1	64.1	-	57.4	57.9	61.2	61.7	62.8	63.8	66.6	65.6	61.0	58.4	59.8	59.3
Standard Deviation	0.5	0.6	1.6	3.2	0.4	1.4	1.2	-	0.7	0.4	0.6	0.7	0.9	2.3	1.3	0.5	5.6	3.1	0.8	1.1
$L_{diff, Spring}$	0.6	-0.8	1.7	-2.1	0.3	1.9	-0.9	-	-1.0	-0.6	0.0	0.3	3.3	4.3	-0.7	-0.3	4.5	-0.3	0.4	-0.2
$L_{diff, Summer}$	0.3	0.8	-1.1	-2.6	0.6	2.3	0.9	-	0.2	0.2	0.8	0.7	2.5	4.1	-1.1	0.6	4.2	0.7	-0.4	-0.6
$L_{diff, Monsoon}$	-0.6	-0.1	-0.4	0.4	0.1	5.3	0.1	-	0.9	0.4	-0.9	-1.2	1.0	-0.4	-0.8	-0.3	-4.7	-4.4	-0.5	-0.8
$L_{diff, Autumn}$	-0.1	0	-1.6	2.3	-0.5	2.6	1.2	-	0.0	0.4	0.2	0.4	1.8	-0.2	2.0	-0.2	-6.3	1.4	1.4	2
$L_{diff, Winter}$	0.4	-0.2	1.7	5.1	-0.3	2.4	-1.7	-	-0.2	0	0.1	0	2.8	1.9	0.3	-0.7	5.3	4.1	0.9	0.5

Parameters (in dB(A)) for night equivalent sound levels	Noise Monitoring sites in Delhi city																			
	CPCB, HQ.		DTU, Bawana		ITO		Dilshad Garden		NSIT, Dwarka		Civil Lines		R. K. Puram		Anand Vihar		Mandir Marg		Punjabi Bagh	
	2018	2019	2018	2019	2018	2019	2018	2019	2018	2019	2018	2019	2018	2019	2018	2019	2018	2019	2018	2019
Site Characteristics	Commercial		Silence		Commercial		Silence		Silence		Commercial		Residential		Commercial		Silence		Residential	
$L_{annual,avg,night}$	58.0	58.6	52.3	58.9	69.9	73.5	57.0	-	55.6	59.2	58.8	58.7	60.3	-	65.9	62.9	-	51.8	53.4	52.7
Standard Deviation	1.0	0.9	2.2	3.2	0.6	2.1	0.6	-	0.6	0.5	0.6	0.4	1.6	-	1.4	0.6	-	2.5	2.1	1.0
$L_{diff, Spring}$	0.6	0.8	2.5	-3.5	0.7	3.3	-0.2	-	0.2	4.1	-0.4	0	6.5	-	1.7	-0.6	-	-3.1	1.9	0.9
$L_{diff, Summer}$	-0.2	-0.5	0.0	-0.3	0.3	1.8	0.0	-	0.6	3	-0.2	-0.4	4.8	-	1.2	-1.2	-	1.9	-2.9	-0.3
$L_{diff, Monsoon}$	-0.7	-0.7	1.4	-0.7	0.3	6.7	-0.4	-	-0.6	3.7	-1.1	0.4	3.0	-	0.5	0.4	-	0.4	0.0	-0.4
$L_{diff, Autumn}$	-0.1	-0.4	-1.6	0.7	-0.6	3.5	0.9	-	0.2	4.4	0.0	0.1	6.0	-	4.0	0.2	-	-0.1	1.8	2.1
$L_{diff, Winter}$	1.8	1.2	4.1	5.4	1.0	1.4	0.9	-	0.9	3.9	0.6	0.6	7.1	-	2.7	-0.3	-	3.5	1.9	0.1

Table 5.31 Correlation of the meteorological parameters with noise descriptors in various seasons for Delhi city

Seasons	Noise Descriptors	Relative Humidity (%)	Wind Speed (m/s)	Ambient Temperature (°C)
Winter	L_{day}	-0.302	-0.177	0.290
	L_{night}	-0.314	-0.079	0.502
	$L_{Aeq,24h}$	-0.329	-0.122	0.298
Spring	L_{day}	-0.213	0.234	0.332
	L_{night}	0.114	0.211	0.401
	$L_{Aeq,24h}$	-0.203	0.264	0.383
Summer	L_{day}	-0.411	0.088	0.512
	L_{night}	-0.110	0.192	0.321
	$L_{Aeq,24h}$	-0.431	0.511	0.533
Monsoon	L_{day}	-0.231	0.231	0.166
	L_{night}	0.115	0.298	0.344
	$L_{Aeq,24h}$	-0.157	0.233	0.267
Autumn	L_{day}	-0.267	-0.296	0.238
	L_{night}	0.115	-0.234	0.175
	$L_{Aeq,24h}$	-0.211	-0.322	0.221

highest in summer followed by monsoon, autumn, winter, and spring. The night equivalent noise levels were observed to be the highest in summer followed by autumn, monsoon, winter, and spring.

5.15 Uncertainty in Environmental Noise Measurements

Measurement uncertainty determination is very crucial in noise pollution assessments especially in order to quantify the probability of the success of a decision when a comparison between a measured level and a legislative (or technical) limit value is needed [128]. However, it is very difficult to postulate any probability distribution of the long-term environmental noise indicators. In noise mapping, each single value is a result of summing up the contribution of many sources. A careful and responsible analysis of uncertainties in noise prediction is complex and covers many influences [129]. ISO 1996-2:2017 [130] "Acoustics-Description, assessment and measurement of environmental noise-Determination of environmental noise levels" contains guidelines on assessing the uncertainties of the determined sound pressure levels. This depends on the sound source, measurement time interval, weather conditions, distance from the source, measurement method, and instrumentation [131]. The measurement uncertainty comprises of type A contribution calculated by means of statistical analysis for a series of individual measurement results and type B contribution calculated by means of methods other than analysis of series of results, for example, following some experimental characteristics of the measurement. The standard uncertainty (type-A) $u(\overline{X})$ is calculated for L_{day} levels as:

$$u(\overline{X}) = \sqrt{\frac{\sum\left(L_{\mathrm{day},i} - \overline{L_{\mathrm{day}}}\right)^2}{n(n-1)}} \tag{5.16}$$

where $\overline{L_{\mathrm{day}}}$ is average value for a measurement series, $L_{\mathrm{day},i}$ is monthly averaged L_{day} value, and n is number of months. The standard uncertainty for L_{night} can also be similarly calculated. Figure 5.26 shows the yearly standard deviation for 365 days in dB(A) for the day and night equivalent noise levels in year 2020. However, the estimation of type B involves various factors, such as calibration of the equipment, accuracy class of measuring device, wind speed, background noise, distance from road edge, meteorological conditions, etc., and pressure and temperature effects on the microphone pressure sensitivity, uncertainty in frequency response, uncertainty associated with microphone directional response, intrinsic sound level meter stability, etc. effect the measurement uncertainty.

ISO 1996-2:2017 describes the equation for determination of sound pressure levels as [130]

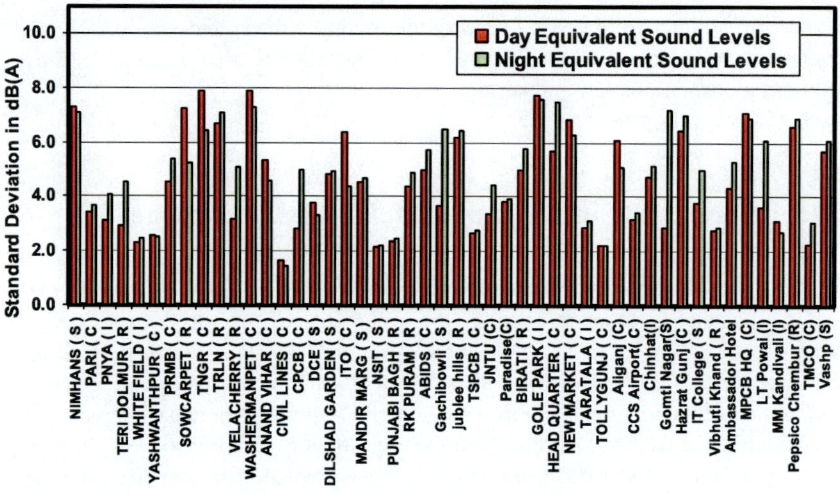

Fig. 5.26 Standard deviation in dB(A) for day and night equivalent sound levels

Table 5.32 Example of an uncertainty budget for a measured value (ISO 1996-2:2017)

Quantity	Estimate dB	Standard uncertainty, u_j dB	Magnitude of sensitivity coefficient, c_j c_j
$L' + \delta_{slm}$	L'	$u(L')$ 0.5[a]	$\frac{1}{1-10^{-0.1(L'-L_{res})}}$
δ_{sou}	0	u_{sou}	1
δ_{met}	0	u_{met}	1
δ_{loc}	0–6.0	u_{loc}	1
$L_{res} + \delta_{res}$	L_{res}	u_{res}	$\frac{10^{-0.1(L'-L_{res})}}{1-10^{-0.1(L'-L_{res})}}$

[a]0.5 dB refers to class 1 sound level meter; A class 2 sound level meter would have the standard uncertainty 1.5 dB

$$L = L' + 10\ \log\left(1 - 10^{-0.1((L'-L_{res})}\right) + \delta_{sou} + \delta_{met} + \delta_{loc} \tag{5.17}$$

where L is the estimated value during the specified condition for which a measured value is wanted, expressed in decibels (dB), L' is the measured value including residual sound, L_{res} is the residual sound, expressed in decibels (dB), δ_{sou} is an input quantity to allow for any uncertainty due to deviations from the expected operating conditions of the source, expressed in decibels (dB), δ_{met} is an input quantity to allow for any uncertainty due to metrological condition deviating from the assumed metrological conditions expressed in decibels (dB), δ_{loc} is an input quantity to allow for any uncertainty due to selection of receiver location, expressed in decibels (dB). The standard describes the uncertainty budget as shown in Table 5.32. The numbers given in Table 5.32 refers to A-weighted equivalent-continuous sound pressure levels only. Higher uncertainties are to be expected on maximum levels, frequency band levels, and levels of tonal components in noise. In many cases, the measured values shall be corrected to other source operating conditions not representing the measured cases but the yearly average. L' and L_{res} are both dependent on δ_{slm}, which is an input quantity to allow for any uncertainty of the measurement chain. In addition, L_{res} depends on δ_{res}, which is an input quantity to allow for any uncertainty due to residual sound.

The number of vehicle pass-bys needed to average the variation in individual vehicle noise emission depends on the required accuracy. If no better information is available, the standard uncertainty denoted $u_{sou}u_{sou}$ can be calculated by means of equation as:

$$u_{sou} = \frac{c}{\sqrt{n}}\ \text{dB} \tag{5.18}$$

where n is the number of pass-bys. For mixed traffic $C = 10$, for heavy vehicles only $C = 5$, and for passenger cars only $C = 2.5$. In each case, a more accurate standard uncertainty can be determined from the statistics of direct sound emission measurements of individual pass-bys either category by category or for a representative traffic mix. If only one or a few short-term measurements are carried out, they should be taken during favorable or very favorable propagation conditions. In that case, unless better information is available, the standard uncertainty for distances, $D \leq 400 \leq 400$ m is:

$$u_{\text{met,fav}} = 2 \text{ dB} \tag{5.19}$$

However, for distances D > 400 m, it is calculated as:

$$u_{\text{met,fav}} = \left(1 + \frac{D}{400m}\right) \text{dB} \tag{5.20}$$

This clause may not be entirely appropriate for aircraft noise as the direction of take-off and landing depends on the vector wind component along the runway. In some microphone locations, there may never be any favorable propagation. For the most common cases, default values for the standard uncertainties using different microphone location are given in Table 5.33 for traffic noise as mentioned in ISO 1996-2:2017. These values are to be used unless better information is available. For industrial noise and other locations, the uncertainties should be determined for each individual case. For stationary sources and low frequencies, it is not suitable to use the default correction of 3 dB in front of a facade instead, and it is recommended to use a flush-mounted microphone location with a 5.7 dB correction.

Table 5.34 shows an illustrative case of uncertainty budget for a single measurement under favorable propagation condition as recommended in ISO 1996-2:2017. The results from the road traffic noise exercises indicate that the magnitude of the uncertainty associated with a short-term measurement of $L_{\text{Aeq,1h}}$ calculated using the procedure presented in the ISO 1996-2:2017 was $\pm$ 4.4 dB with a confidence level of 95% as presented by Alves and Waddington [132]. ISO 1996-2 standard estimates a minimum uncertainty of 2 dB for measured noise levels, associated with factors such

Table 5.33 Standard uncertainty of corrections for reflections of different microphone locations relative to vertical reflecting surfaces

Microphone location	Standard uncertainty, u_{loc} u_{loc} dB
Traffic noise incident from all angles	
Reference location in a free field	0
Locations where there are no reflecting surfaces other than ground	0.5
Location using the correction 5.7 dB and meeting the requirements of B.4 of ISO 1996-2 standard	0.4
Location using the correction 3 dB and meeting the requirements of B.5 of ISO 1996-2 standard	0.4
Traffic noise with predominantly grazing incidence	
Reference location in a free field	0
Location using the correction 5.7 dB and meeting the requirements of B.4 of ISO 1996-2 standard	2.0
Location using the correction 3 dB and meeting the requirement of B.5 of ISO 1996-2 standard	1.0

Table 5.34 Uncertainty budget for a single measurement under favorable propagation condition as recommended in ISO 1996-2:2017 [130]

Quantity	Estimate	Standard uncertainty, u_i dB	Sensitivity coefficient, c_i	Uncertainty contribution $c_i u_i$ dB
$L_{eq,\,1\,h}$	$L' = 58$ dB	0.50	$\frac{1}{1-10^{-0.1(L'-L_{res})}} = 1.188$	0.59
δ_{slm}		0.5 (default)		
δ_{sou}	1000 vehicles	$\frac{10}{\sqrt{1000}} = 0.3$	1	0.30
δ_{met}	Favourable	2.00	1	2.00
δ_{loc}	+5.7 dB	0.40	1	0.40
δ_{res}	$L_{res} = 50$ dB	2.00	$\frac{10^{-0.1(L'-L_{res})}}{1-10^{-0.1(L'-L_{res})}} = 0.188$	0.38
$u(L_m) = \sqrt{\sum_{1}^{n}(c_j u_j)^2}$				2.18
Expanded Uncertainty (coverage prob. 95%)				**4.36**

as instrumentation, operating conditions (repeatability), weather and terrain conditions, and residual sound. Thus, for the developed computerized noise maps, the uncertainties due to the digital terrain model, the software used, etc. will also be added [133, 134]. Also, the accurate assessment of digital terrain model (DTM) is very crucial. Noise mapping with 5 m accuracy in elevation is insufficient and may drastically change the evaluation of the percentage of people affected by noise [134].

5.15.1 Bootstrap Method for Uncertainty Evaluation

The classical method of uncertainty evaluation has been always used for the evaluation and assessment of measurement uncertainty in environmental noise measurements and control. Also, with a major focus on noise mapping simulations carried out in different countries, the expression of uncertainty is very crucial [135]. However, there had been many studies reported on the applications of Bootstrap method for evaluation of uncertainty in environmental noise measurements due to its advantage of lack of requirement of assigning a probability distribution to the input quantities in the measurement model. It relies on the assumption that the sample is statistically representative of the complete population [136, 137]. Some studies such as that reported by Mateus et al. [136] conducted a systematic analysis for assessing the quality of sampling strategies for estimating the long-term environmental noise indicators at any given site. The study concluded that in order to reduce the influence of the sampling strategy on the overall uncertainty of the L_{Aeq}, it is much more worthwhile to increase the number of sampling episodes than the duration of the episodes [136]. The bootstrap method is based on the strategy of replicating the samples from the initial data referred as bootstrap samples. The replicated samples can be considered as the approximations of the true distributions of the estimations, and thus a good approximation of the distribution of relevant statistics such as mean and the standard deviation [138]. Figure 5.27 shows the algorithm of the bootstrap method [138].

The bootstrap estimation of the long-term noise indicators are [139]:

$$\overline{L^*_{\text{Aeq,LT}}} = \frac{1}{N}\sum_{k=1}^{N} L^*_{\text{Aeq,LT},k} \tag{5.21}$$

where N is the number of bootstrap replications, $L^*_{\text{Aeq,LT},k}$ is the level of the kth bootstrap estimate of index $L_{\text{Aeq,LT}}$. The bootstrap estimation of the type A uncertainty can be calculated as [139]:

$$s_{\text{B}}\left(\overline{L^*_{\text{Aeq,LT}}}\right) = \sqrt{\frac{\sum_{k=1}^{N}\left(L^*_{\text{Aeq,LT},k} - \overline{L^*_{\text{Aeq,LT}}}\right)^2}{N-1}} \tag{5.22}$$

In case the bootstrap distribution is asymmetrical, the bootstrap-t method can be utilized that employing the construction of the confidence intervals using t-student

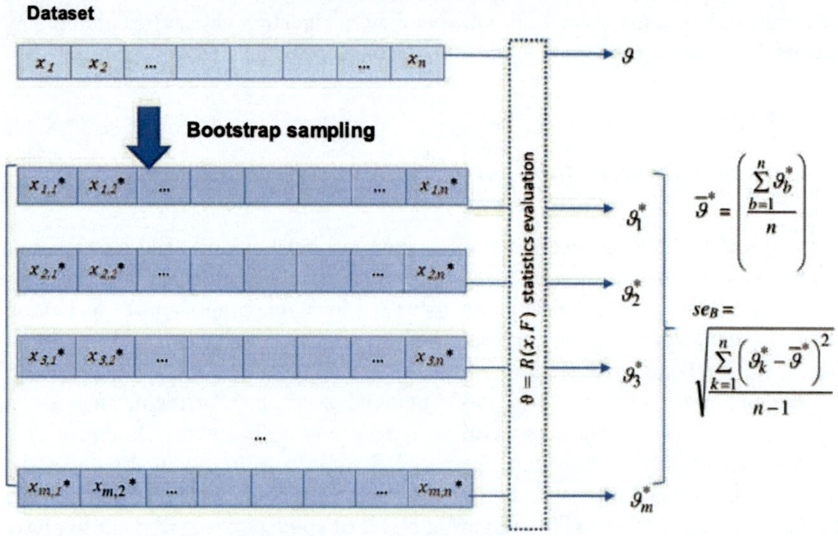

Fig. 5.27 Algorithm of the bootstrap method (Ruggiero et al. [138])

distribution [138]. The bootstrap method had been observed to estimate the equivalent sound pressure level with better precision when compared to the classical approach [140].

5.16 Conclusions and Outlook

The chapter shows the strategies and methodologies used for noise monitoring by various researchers across the globe. The ambitious *NANMN* project established in India for continuous round-the-clock noise monitoring is discussed. As a result, there had been some of the important findings and key conclusions derived which can be helpful for the policy makers for revision of ambient noise standards, adapting short-term noise monitoring and modeling techniques, understanding the noise scenario in festival seasons, and implementation of traffic reduction measures such as odd-even policy in Delhi city in the year 2016 and devising of noise control measures for reducing the noise pollutions levels in Indian cities. The following conclusions can be drawn from the analysis presented:

- The continuous monitoring since past decade shows that no site lying under residential and silence zone qualifies the ambient noise standards. Even in lockdown situation, no silence zone site or residential site met the ambient noise standards. Also, the analysis of noise monitored data during lockdown in Delhi in May 2021 revealed that only one residential site complied with day-time limit in the lockdown period. A 45 dB(A) L_{night} seems to be a very stringent limit. These observations suggest the need for a retrospective and prospective view on ambient noise limits reformulation of ambient noise standards especially for residential and silence zones. Also, the mixed category zone is available for some of these sites, whereby it is very difficult to exclusively classify them as either residential or silence zone or commercial or industrial zone. It is recommended that a mixed zone category should be specifically considered for noise abatement guidelines/ambient noise standards in future in Indian perspectives.
- The analysis of ($L_{\text{day}} - L_{\text{night}}$) for 70 sites in overall reveals that majority of observations show a difference between 5 and 10 dB(A) and less than or equal to 5 dB(A) in the past 10 years. These observations thus suggest that the night equivalent noise levels are comparable to the day equivalent levels for the majority of sites. In such a scenario, 10 dB nighttime adjustments in day-night average sound level are not appropriate particularly for the metropolitan cities considered in this chapter. On the similar analogy, these observations also suggest that 5 dB evening time correction in day-evening-night average sound level descriptor is not justified in Indian context as the evening time noise levels are as severe as the day equivalent noise levels. Thus, the 24-h equivalent continuous sound level, $L_{\text{Aeq,24h}}$, would rather be more suitable as it is a common way of expressing day-night average sound level without 10 dB night-time adjustment. It

is thus recommended that for developing an exposure-effect relationships and correlating noise annoyance and correlating the noise exposure with the health aspects; the single noise descriptor, 24-h equivalent continuous sound level, $L_{Aeq,24h}$, would rather be more suitable as compared to day-night average sound level or day-evening-night average sound level descriptors.

- The long-term noise monitoring showed that ambient noise levels were high when compared to the recommended standards for some sites and thus noise abatement measures are essentially required for controlling these levels. The interim target (IT) of 55 dB L_{night} as recommended in EU Night Noise Guidelines report was met by four sites in year 2020, seven sites in year 2019, and eight sites in year 2018. Overall, 94.3% of the observations exceeded the interim target recommended by Night Noise Guidelines (*NNG*) report, which indicates that 55 dB L_{night} is an ambitious target for all these sites under consideration.
- The analysis of noise limit exceedance factor (*NEF*) and average exceedance factor (*AEF*) shows that silence zone has the maximum *AEF* followed by residential zones and commercial zone. Thus, it suggests that suitable isolation and planned land-use pattern of commercial, residential, silence and industrial zones in a city shall be pivotal in reducing the ambient noise levels.
- The short-term noise monitoring strategy followed can be a good trade-off between the accuracy and costs incurred and thus offers a best practicable and economical option (*BPEO*) for noise mapping of larger parts of the country. An error of ±3 dB(A) from the annual equivalent level is calculated with 95% probability for L_{day} and L_{night} value by 1 month noise monitoring strategy. For higher accuracy, the random 2 months strategy is recommended, whereby an error of ±2 dB(A) is calculated with a probability higher than 90%.
- The accuracy and reliability of temporal samplings with respect to the annual average value is however limited in Indian perspectives. Although the temporal sampling strategy may not represent the annual average value for a particular site with a definite confidence level, yet they are widely employed for the assessment of noise scenario at a particular site in Indian scenario and for devising suitable measures for control. Random 3 weeks sampling strategy offers to be the best approach among all the temporal sampling strategies as an error of ±3 dB is evaluated with a probability of success higher than 90% for both day and night equivalent noise levels. However, monthly sampling strategies offer most cost-effective and reliable approach as compared to the temporal sampling strategies. In case of temporal sampling strategy of noise monitoring for one working day and one holiday (i.e., for 48 hours per site), the probability of accuracy of day equivalent noise level with respect to the annual average value of ±2 dB(A) is nearly 71% and that for night equivalent noise is nearly 67% as shown in Table 5.30 for the two consecutive days strategy.
- The *ARIMA*, *ANN* methodology, and other approaches such as support vector machines (*SVMs*) can serve as a suitable substitute to the long-term noise monitoring and is thus indispensable for saving costs and time incurred on continuous noise monitoring. These techniques are very useful in Indian perspectives as installing noise monitoring terminals at each and every corner of the cities

shall cause a lot of infrastructural and economic complications. Also, the use of smartphone technology in noise monitoring and mapping can be encouraged in Indian scenario provided the reliability of the developed apps w.r.t accuracy and precision of reported sound level measurements is ensured.

References

1. European Noise Directive (2002) Assessment and Management of Environmental Noise, 2002/49/EU. Official Journal of European Communities
2. WHO-JRC (2011) Burden of disease from environmental noise-Quantification of healthy life years last in Europe. European Centre for Environment and Health, JRC EU
3. Maibach M, Schreyer C, Sutter D, Van Essen HP, Boon BH, Smokers R, Schroten A, Doll C, Pawlowska B, Bak M (2007) Handbook on estimation of external costs in the transport sector. CE Delft Delft
4. Mizutani F, Suzuki Y, Sakai H (2011) Estimation of social costs of transport in Japan. Urban Stud 48(16):3537–3559
5. Nelson JP (1982) Highway noise and property values: a survey of recent evidence. JTEP 16(2):117–138
6. Theebe MAJ (2004) Planes, trains, and automobiles: the impact of traffic noise on house prices. J Real Estate Financ Econ 28(2–3):209–234
7. Bello JP (2018) SONYC: a system for the monitoring, analysis and mitigation of urban noise pollution. Communications of the ACM, New York
8. Steele C (2001) A critical review of some traffic noise prediction models. Appl Acoust 62(3): 271–287
9. Garg N, Maji S (2014) A critical review of principal traffic noise models: Strategies and implications. Environ Impact Assess Rev 46:68–81
10. The Noise Pollution (Regulation and Control) rules (2000) Ministry of Environment & Forests, India
11. Central Pollution Control Board (CPCB), Annual Report, 2011–12, 94–96
12. Worldwide Hearing Index, Mimi Hearing Technologies. https://www.mimi.io/en/blog/2017/3/8/worldwide-hearing-index-2017
13. World Economic Forum, Agenda, March (2017) These are the cities with worst noise pollution. https://www.weforum.org/agenda/2017/03/these-are-the-cities-with-the-worst-noise-pollution/
14. Arana M, Martin RS, Salinar JC (2014) People exposed to traffic noise in European agglomerations from noise maps. A critical review. Noise Mapp 1:40–49
15. Zannin P, Ferreira AMC, Szeremetta B (2006) Evaluation of noise pollution in urban parks. Environ Monit Assess 118:423–433
16. Murphy E, King EA, Rice HJ (2009) Estimating human exposure to transport noise in central Dublin, Ireland. Environ Int 35:298–302
17. Seong JC, Park TH, Ko JH, Chang SI, Kim M, Holt JB, Megdi MR (2011) Modeling of road traffic noise and estimated human exposure in Foulton County, Georgia, USA. Environ Int 37: 1336–1341
18. Skineer CJ, Grimwood CJ (2005) The UK noise climate 1990-2001: population exposure and attitudes to environmental noise. Appl Acoust 66(2):231–243
19. Piccolo A, Plutino D, Cannistraro G (2005) Evaluation and analysis of the environmental noise of Messina, Italy. Appl Acoust 66:447–465
20. Jamrah H, Omari AA, Sharabi R (2006) Evaluation of traffic noise pollution in Amman, Jordan. Environ Monit Assess 120:499–525

21. Yilmaz H, Ozer S (2005) Evaluation and analysis of environmental noise pollution in the city of Erzurum. Turkey Int J Environ Pollut 23(4):438–448
22. Li B, Tao S, Dawson RW (2002) Evaluation and analysis of traffic noise from the urban main roads in Beijing. Appl Acoust 63:1137–1142
23. Moriallas JMB, Escobar VG, Sierra JAM, Gomez RV, Carmona J (2002) A environmental noise study in the city of Caceres, Spain. Appl Acoust 63:1061–1070
24. Zannin PHT, Diniz FB, Barbosa WA (2002) Environmental noise pollution in the city of Curitiba, Brazil. Appl Acoust 63(4):351–358
25. Alberola J, Flindell H, Bullmore J (2005) Variability in road traffic noise levels. Appl Acoust 66:1180–1195
26. Olayinka OS (2012) Noise map: tool for abating noise pollution in urban areas 1:185. https://doi.org/10.4172/scientificreports.185
27. Björk EA (1994) Community noise in different seasons in Kuopio, Finland. Appl Acoust 42(2):137–150
28. Tsai K-T, Lin M-D, Chen Y-H (2009) Noise mapping in urban environments: a Taiwan study. Appl Acoust 70:964–972
29. Gaja E, Gimenez A, Sanchos S, Reig A (2003) Sampling techniques for the estimation of the annual equivalent noise level under urban traffic conditions. Appl Acoust 64:43–53
30. Scatolini F, Alves CJ (2016) Background noise analysis in urban airport surroundings of Brazilian cities, Congonhas Airport, São Paulo. Revista de Saúde Pública 50:69
31. Gunes M, Özyavuz M (2018) Noise mapping of Namik Kemal University Campus (Tekirdag – Turkey) by using geographic information systems. J Environ Prot Ecol 19(1):186–195
32. Mohammadi G (2009) An investigation of community respond to urban traffic noise. Iran J Environ Health Sci Eng 6(2):137–142
33. Leao S, Krezel A, Ong KL (2013) 2 Loud? Monitoring traffic noise with mobile phones. Paper presented at the 6th making cities liveable conference, Melbourne (VIC), 17–19 June 2013
34. Gozalo GR, Suárez E, Montenegro AL, Arenas JP, Morillas JMB, González DM (2020) Noise estimation using road and urban features. Sustainability 12:9217. https://doi.org/10.3390/su12219217
35. Schlatter F, Piquerez A, Habermacher M, Ragettli MS, Röösli M, Brink M, Cajochen C, Probst-Hensch N, Foraster M, Wunderli J-M (2017) Validation of large scale noise exposure modelling by long-term measurements. Noise Mapp 4(1):75–86
36. Mietlicki C, Mietlicki F, Ribeiro C, Gaudibert P (2014) The HARMONICA project, new tools to assess environmental noise and better inform the public. Forum Acusticum, 7–12 September 2014, Krakow
37. Zamora W, Calafate CT, Cano JC, Manzoni P (2017) Accurate ambient noise assessment using smartphones. Sensors (Basel) 17(4):917
38. Peckens C, Porter C, Rink T (2018) Wireless sensor networks for long-term monitoring of urban noise. Sensors 18(9):3161
39. Kumar R, Mukherjee A, Singh VP (2018) Traffic noise mapping of Indian roads through smartphone user community participation. Environ Monit Assess 189:17
40. Pődör A, Révész A (2017) Mapping and analyzing environmental noise using smartphones in small towns-a Hungarian case study. GI Forum 2:126–136
41. Kardous CA, Shaw PB (2016) Evaluation of smartphone sound measurement applications (apps) using external microphones-a follow-up study. J Acoust Soc Am 140(4):EL327
42. Grubeša S, Petošić A, Suhanek M, Đurek I (2018) Mobile crowdsensing accuracy for noise mapping in smart cities. Automatika 59:286–293
43. Maisonneuve N, Stevens M, Niessen ME, Steels L (2009) NoiseTube: measuring and mapping noise pollution with mobile phones. In: Athanasiadis et al (eds) Information technologies in environmental engineering. Environmental science and engineering. Springer, Berlin/Heidelberg, pp 215–228
44. Murphy E, King EA (2016) Smartphone-based noise mapping: integrating sound level meter app data into the strategic noise mapping process. Sci Total Environ 562:852–859

45. Kivelä I, Hakala I (2015) Area-based environmental noise measurements with a wireless sensor network. EuroNoise 2015, 31st May to 3rd June, Maastricht, 2351–56
46. Anachkova M, Domazetovska S, Petreski Z, Gavriloski V (2021) Design of low-cost wireless noise monitoring sensor unit based on IoT concept. J Vibroeng https://doi.org/10.21595/jve.2021.21709
47. European Commission report (2000) Position paper on EU noise indicators. ISBN 92-828-8953-X, pp 1–80
48. Can A, Aumond P, Michel S, De Coensel B, Ribeiro C, Botteldooren D, Lavandier C (2016) Comparison of noise indicators in an urban context. In: Proceedings internoise 2016, Hamburg
49. Prascevic M, Mihajlov D, Cvetkovic D (2015) The correlation between harmonica indices and noise indicators. ANUL XXII, NR. 2, ISSN 1453-7397. anale-ing.uem.ro/2015/230.pdf. Accessed on 17.05.2021
50. Guski R (1997) Psychological methods for evaluating sound quality and assessing acoustic information. Acta Acust United Acust 85(3):765–774
51. Kang J (2017) From dBA to soundscape indices: managing our sound environment. Front Eng Mng 4(2):184–192
52. Zwicker E, Fastl H (2007) Psychoacoustics –facts and models. Springer, Berlin
53. Aletta F, Kang J, Axelsson Ö (2016) Soundscape descriptors and a conceptual framework for developing predictive soundscape models. Landsc Urban Plan 149:65–74
54. Ricciardi P, Delaitre P, Lavandier C, Torchia F, Aumond P (2015) Sound quality indicators for urban places in Paris cross-validated by Milan data. J Acoust Soc Am 138(4):2337–2348
55. Axelsson Ö, Nilsson ME, Berglund B (2010) A principal components model of soundscape perception. J Acoust Soc Am 128(5):2836–2846
56. Schomer PD (2005) Criteria for assessment of noise annoyance. Noise Control Eng 53(4): 132–144
57. Fujii K, Atagi J, Ando Y (2002) Temporal and spatial factors of traffic noise and its annoyance. J Temp Des Arch Environ 2(1):33–41
58. Berglund B, Hassmén P, Soames Job RF (1996) Sources and effects of low-frequency noise. J Acoust Soc Am 99(5):2985–3002
59. Rey Gozalo G, Trujillo Carmona J, Barrigón Morillas JM, Vílchez-Gómez R, Gómez Escobar V (2015) Relationship between objective acoustic indices and subjective assessments for the quality of soundscapes. Appl Acoust 97:1–10
60. Kang J (2007) Urban sound environment. Taylor and Francis Incorporating Spon, London
61. Asensio C, Aumond P, Can A, Gascó L, Lercher P, Wunderli JM, Lavandier C, de Arcas G, Ribeiro C, Muñoz P, Licitra G (2020) A Taxonomy proposal for the assessment of the changes in soundscape resulting from the COVID-19 lockdown. Int J Environ Res Public Health 17: 4205. https://doi.org/10.3390/ijerph17124205
62. Héritier H, Vienneau D, Foraster M, Eze IC, Schaffner E, Thiesse L, Ruzdik F, Habermacher M, Köpfli M, Pieren R, Schmidt-Trucksäss A, Brink M, Cajochen C, Wunderli JM, Probst-Hensch N, Röösli M, SNC Study Group (2018) Diurnal variability of transportation noise exposure and cardiovascular mortality: a nationwide cohort study from Switzerland. Int J Hyg Environ Health 221(3):556–563
63. Can A, Gauvreau B (2015) Describing and classifying urban sound environments with a relevant set of physical indicators. J Acoust Soc Am 137(1):208–218
64. Garg N, Gandhi V, Gupta N K (2022) Impact of COVID-19 lockdown on ambient noise levels in seven metropolitan cities of India. Appl Acoust 188:108582
65. Garg N, Sinha AK, Gandhi V, Bhardwaj RM, Akolkar AB (2016) A pilot study on establishment of ambient noise monitoring network across the major cities of India. Appl Acoust 103: 20–29
66. Garg N, Sinha AK, Sharma MK, Gandhi V, Bhardwaj RM, Akolkar AB, Singh RK (2017) Study on the establishment of a diversified national ambient noise monitoring network in seven major cities of India. Curr Sci 113(7):1367–1383

67. Garg N, Sinha AK, Gandhi V, Bhardwaj RM, Akolkar AB (2017) Evaluation and Analysis of Environmental Noise Pollution in Seven Major Cities of India. Arch Acoust 42(2):175–188
68. SGS Weather and Environmental Systems Pvt. Ltd, New Delhi, India. http://www.sgsweather.com
69. Central Pollution Control Board (2015) Status of Ambient noise level in Indian in 2011-2014. CPCB report, NANMN/02/2015–16. https://cpcb.nic.in
70. Central Pollution Control Board (2015) Status of Ambient noise level in Indian in 2015. CPCB report, NANMN/03/2015–16. https://cpcb.nic.in
71. Central Pollution Control Board (2016) Status of Ambient noise level in Indian in 2016. CPCB report, NANMN/04/2016. https://cpcb.nic.in
72. Central Pollution Control Board (2018) Status of Ambient noise level in Indian in 2017. CPCB report, NANMN/05/2017, August, 2018. https://cpcb.nic.in
73. Babisch W (2002) The noise/stress concept, risk assessment and research needs. Noise Health 4(16):1–11
74. World Health Organization (2009) Night noise guidelines for Europe. Copenhagen, WHO Regional Office for Europe
75. Garg N (2019) On Suitability of day-night average sound level descriptor in Indian scenario. Arch Acoust 44(2):385–392
76. Garg N, Kumar A, Saini PK, Maji S (2015) A retrospective view of ambient noise standards in India: status and proposed revision. Noise Control Eng J 63(3):266–278
77. Kundu Chowdhury A, Debsarkar A, Chakrabarty S (2016) Assessment of seasonal variations of average traffic pollution levels in curbside open-air microenvironments in Kolkata, India. Health Scope 5(2):e33081. https://doi.org/10.17795/jhealthscope-33081
78. Kalawapudi K, Singh T, Dey J, Vijay R, Kumar R (2020) Noise pollution in Mumbai Metropolitan Region (MMR): an emerging environmental threat. Environ Monit Assess 192:152. https://doi.org/10.1007/s10661-020-8121-9
79. Delhi Pollution Control Committee website. http://182.75.174.13/Public/Default.aspx?site=fc490e5f-5cd7-496d-a75a-621e02a6829
80. Garg N, Sinha AK, Gandhi V, Bhardwaj RM, Akolkar AB (2017) Impact of Diwali celebration on environmental noise pollution in India. Acoust Aust J 45:101–117
81. Sharma O, Mohanan V, Singh M (1999) Characterization of sound pressure levels produced by crackers. Appl Acoust 58(4):443–449
82. Lokhande S K, Garg N, Jain M C, Rayalu S (2022) Evaluation and analysis of firecrackers noise: Measurement Uncertainty, legal noise regulations and noise induced hearing loss. Appl Acous 186: 108462
83. Central Pollution Control Board report (2017) State of Pollution Abatement measures: firecrackers (Patakas), Control of Urban Pollution Series. CUPS/88/2017–18, October, 2017. http://www.indiaenvironmentportal.org.in/files/file/Firecracker%20report,%20Oct.%202017.pdf. Accessed on 17.05.2021
84. Central Pollution Control Board report (2019) Report on ambient air quality and noise on Deepawali 2019, 1–6. www.cpcbenvis.nic.in/airpollution/Deepawali-2019.pdf. Accessed on 17.05.2021
85. Central Pollution Control Board report (2020) Report on Ambient Air Quality and Noise on Deepawali 2020, 1–6. https://cpcb.nic.in/uploads/Diwali-Report-2020.pdf. Accessed on 17.05.2021
86. Maharashtra Pollution Control Board report (2019) Report on ambient noise monitoring during Ganesh Festival, 2019. http://www.mpcb.gov.in/. Accessed on 17.05.2021
87. Kalawapudi K, Singh T, Vijay R, Goyal N, Kumar R (2021) Effects of COVID-19 pandemic on festival celebrations and noise pollution levels. Noise Mapp 8:89–93
88. Central Pollution Control Board, India (2017) Delhi's ambient noise levels influenced by traffic flow – case studies, Control of Urban Pollution Series. CUPS/87/2017–18, September, 2017
89. Hindustan Times (2016) New Delhi, Friday, January, 04, 2016, 3

90. Garg N, Sinha AK, Gandhi V, Bhardwaj RM, Akolkar AB (2017) Effect of odd-even vehicular restrictions on ambient noise levels at ten sites in Delhi city. Indian J Pure Appl Phys 55:687–692
91. Goel R, Tiwari G, Mohan D (2016) Evaluation of the effects of the 15-day Odd-Even scheme in Delhi: a preliminary report, transport research and injury prevention programme. Indian Institute of Technology, Delhi
92. Hindustan Times (2016) New Delhi, April, 22, 2016, 3
93. Munoz P, Vincent B, Domergue C, Gissinger V, Guillot S, Halbwachs Y, Janillon V (2020) Lockdown during COVID-19 pandemic: impact on road traffic noise and on the perception of sound environment in France. Noise Mapp 7:287–302
94. Basu B, Murphy E, Molter A, Basu AS, Sannigrahi S, Belmonte M, Pilla F (2021) Investigating changes in noise pollution due to the COVID-19 lockdown: the case of Dublin, Ireland. Sustain Cities Soc 65:102597
95. Aletta F, Osborn D (2020) The COVID-19 global challenge and its implications for the environment-what we are learning. UCL Open Environ 1:5. https://doi.org/10.14324/111.444/ucloe.000008
96. Asensio C, Pavón I, de Arcas G (2020) Changes in noise levels in the city of Madrid during COVID-19 lockdown in 2020. J Acoust Soc Am 148(3):1748
97. Mimani A and Singh R (2021) Anthropogenic noise variation in Indian cities due to the COVID-19 lockdown during March-to-May 2020. J Acoust Soc Am.150(5): 3216–3227.
98. Morillas JMB, Gajardo CP (2014) Uncertainty evaluation of continuous noise sampling. Appl Acoust 75:27–36
99. Gaja E, Gimenez A, Sancho S, Reig A (2013) Sampling techniques for the estimation of the annual equivalent noise level under urban traffic conditions. Appl Acoust 64:43–53
100. Garg N, Saxena TK, Maji S (2015) Long-term versus short-term noise monitoring: strategies and Implications in India. Noise Control Eng J 63(1):26–35
101. Geraghty D, O'Mahony M (2016) Investigating the temporal variability of noise in an urban environment. Int J Sustain Built Environ 5(1):34–45
102. Brambilla G, Lo Castro F, Cerniglia A, Verardi P (2007) Accuracy of temporal samplings of environmental noise to estimate the long-term L_{den} value, In: Proceedings internoise 2007, Istanbul, Turkey, 1–8
103. Garg N, Soni K, Saxena TK, Maji S (2015) Applications of autoregressive integrated moving average (ARIMA) approach in time-series prediction of traffic noise pollution. Noise Control Eng J 63(2):182–194
104. Garg N, Sharma MK, Parmar KS, Soni K, Singh RK, Maji S (2016) Comparison of ARIMA and ANN approaches in time-series predictions of traffic noise. Noise Control Eng J 64(4): 522–531
105. Kumar K, Jain VK (1999) Autoregressive integrated moving averages (ARIMA) modeling of a traffic noise time-series. Appl Acoust 58:283–294
106. Box GEP, Jenkins GM (1970) Time-series analysis, forecasting and control. San Francisco, CA
107. Adhikari R, Agrawal RK (2013) An introductory study on time-series modeling and forecasting. Lambert Academic Publishing
108. Cai XH (2008) Time series analysis of air pollution CO in California South Coast area with seasonal ARIMA model and VAR model. M.S thesis, University of California
109. Stevenson S (2007) A comparison of the forecasting ability of ARIMA models. J Prop Invest Finance 25:223–240
110. Díaz-Robles LA, Ortega JC, Fu JS, Reed GD, Chow JC, Watson JG, Moncada-Herrera JA (2008) A hybrid ARIMA and artificial neural networks model to forecast particulate matter in urban areas: the case of Temuco, Chile. Atmos Environ 42:8331–8340
111. Zhang GP (2003) Time series forecasting using a hybrid ARIMA and neural network model. Neurocomputing 50:159–175

112. Paul JC, Hogue MS, Rahman MM (2013) Selection of best ARIMA model for forecasting average daily share price index of Pharmaceutical companies in Bangladesh: a case study on Square Pharmaceutical Ltd. Glob J Manage Bus Res Finance XIII:15–26
113. Chen A, Leung MT, Hazem D (2003) Application of neural network to an emerging financial market: forecasting and trading the Taiwan Stock Index. Comput Oper Res 30:901–923
114. Zhang GP, Qi GM (2005) Neural network forecasting for seasonal and trend time series. Eur J Oper Res 160:501–514
115. Khasei M, Bijari M (2010) An artificial neural network (p,d,q) model for time series forecasting. Expert Syst Appl 37:479–489
116. Zhang G, Patuwo BE, Hu MY (1998) Forecasting with artificial neural networks: the state of the art. Int J Forecast 14:35–62
117. Garg N, Maji S (2017) Investigations on modeling transportation noise and its abatement. Lambert Academic Publishing House, pp 1–368
118. Vapnik VN (1995) The nature of statistical learning theory, 1st edn. Springer, New York
119. Samsudin R, Shabri A, Saad P (2010) A comparison of time-series forecasting using support vector machine and ANN model. J Appl Sci 10(11):950–958
120. Thissen U, van Brakel R, de Weijer AP, Melssen WJ, Buydens LMC (2003) Using Support Vector machines for time-series prediction. Chemom Intell Lab Syst 69:35–49
121. Aguilera I, Foraster M, Basagaña X, Corradi E, Deltell A, Morelli X, Phuleria HC, Ragettli MS, Rivera M, Thomasson A, Slama R, Künzli N (2015) Application of land use regression modelling to assess the spatial distribution of road traffic noise in three European cities. J Expo Sci Environ Epidemiol 25:97–105
122. Xie D, Liu Y, Chen J (2011) Mapping urban environmental noise: a land use regression method. Environ Sci Technol 45(17):7358–7364
123. Chang TY, Liang CH, Wu CF, Chang LT (2019) Application of land-use regression models to estimate sound pressure levels and frequency components of road traffic noise in Taichung. Taiwan Environ Int 131:104959
124. Munir S, Khan S, Nazneen S, Ahmad SS (2021) Temporal and seasonal variations of noise pollution in urban zones: a case study in Pakistan. Environ Sci Pollut Res. https://doi.org/10.1007/s11356-021-12738-8
125. Chowdhury AK, Debsarkar A, Chakrabarty S (2015) Critical assessment of day time traffic noise level at curbside open-air microenvironment of Kolkata City, India. J Environ Health Sci Eng 13:65
126. Say N, Ökten SÖ, Aysu A, Yalçinkaya NM (2017) Seasonal change and spatial distribution of traffic noise in Seferihisar. Turkey Arch Acoust 42(4):631–642
127. Miedema HME, Fields JM, Vos H (2005) Effect of season and meteorological conditions on community noise annoyance. J Acoust Soc Am 117(5):2853–2865
128. Ruggiero A, Russo D, Sommella P (2016) Determining environmental noise measurement uncertainty in the context of the Italian legislative framework. Measurement 93:74–79
129. Probst W (2005) Uncertainties in the prediction of environmental noise and in noise mapping. Acoustique Technique, 34–39. http://www.bruit.fr/revues/78_11099.PDF
130. ISO 1996-2:2017, Description, measurement and assessment of environmental noise. Part 2: determination of environmental noise levels
131. Manvell D, Aflalo E (2005) Uncertainties in environmental noise assessments – ISO 1996, Effects of instrument class and residual sound. Forum Acusticum, 2005, Budapest
132. Alves S, Waddington D (2014) Estimation of uncertainty using revised draft ISO 1996-2. In: Proceedings of forum acusticum, 7–12 September, 2014, Krakow
133. Arana M, San Martín R, Nagore I, Pérez D (2009) Using noise mapping to evaluate the percentage of people affected by noise. Acta Acust 95:550–554
134. Arana M, San Martín R, Nagore I, Pérez D (2011) What precision in the digital terrain model is required for noise mapping? Appl Acoust 72:522–526
135. Gómez DMM, Jaramillo AM, Ochoa J (2020) Analysis of the measurement uncertainty and its effects on noise mapping validations. J Environ Manage 266:110606

136. Mateus M, Carrilho JAD, da Silva MCG (2015) Assessing the influence of the sampling strategy on the uncertainty of environmental noise measurements through the bootstrap method. Appl Acoust 89:159–163
137. JCGM 100 (2008) Evaluation of measurement data – guide to the expression of uncertainty in measurement. Joint Committee for Guides in Metrology, Paris (JCGM/WG 1)
138. Ruggiero A, Russo D, Sommella P (2016) Determining environmental noise measurement uncertainty in the context of the Italian legislative framework. Measurement 93:74–79
139. Batko W, Stępień B (2010) Application of the bootstrap estimator for uncertainty analysis of the long-term noise indicators. Acta Phys Pol A 118:11–16
140. Liguori C, Pietrosanto A, Ruggiero A, Russo D, Sommella P (2017) Adopting bootstrap for the uncertainty estimation of road traffic noise measurement. In: 7th IMEKO TC19 symposium on environmental instrumentation and measurements at: aguascalientes, Mexico, August 3–4, 2017

Chapter 6
Noise Impact Assessment Studies

Abstract This chapter presents an overview of noise impact assessment studies for road, railways, aircraft, and construction noise. The various traffic noise models developed in recent years are discussed in details. A comparison of principal traffic noise models adopted in recent years on the basis of technical attributes including source modeling and sound propagation algorithms is presented. The chapter also describes the Federal Transit Administration (FTA) Noise Impact Assessment criteria. Some case studies of noise impact assessment of transit trains at elevated and curved track and that for high-speed trains is presented. The various aircraft noise models developed in recent times is discussed. The chapter also discusses the acoustic induced vibration in structures and correlates it with low-frequency rumbling noise and building damage criteria.

6.1 Introduction

Environmental impact assessment (EIA) can be defined as a process for identifying, predicting, evaluating, and mitigating the biophysical, biological, hydrological, ecological, meteorological, noise, economic, social, and other relevant effects prior to development, then using the conclusions as a tool for planning and decision-making. In other words, impact assessment, simply defined, is the process of identifying the future consequences of a current or proposed action [1]. Noise is one of the major environmental pollutants in recent times apart from air and water pollution. Thus, noise impact assessment is equally important and imperative for new projects so that future noise scenario can be predicted and cost-effective solutions can be devised and implemented for abatement and control. The noise impact assessment should cover the following: understanding the existing noise scenario and establishing the baseline condition; prediction of the noise impact on the residential and other sensitive developments; identifying and establishing the need for mitigation measures to meet the legal requirements; and verification of the effectiveness of the mitigation measures if mitigation measures are needed. Baseline noise refers to the noise environment in an area that is affected by the land traffic noise sources, for example, expressways/major arterial roads/mass rapid transit

N. Garg, *Environmental Noise Control*,
https://doi.org/10.1007/978-3-030-87828-3_6

(MRT) tracks [2]. Control of environmental impacts at the planning stage is a very effective strategy. A small change in the layout or design of the proposal can avoid or minimize a noise impact that may otherwise need substantial and often additional costly retrospective measures for noise mitigation. Selection of the most appropriate options may require consideration of the other environmental effects [3]. The increase in traffic volume, heavy vehicles, and ultimately the traffic noise and its overall impact to the nearby community shall be very helpful for ascertaining the noise impact due to a particular project and also for planning and devising suitable measures for noise abatement and control so as to comply with the legal limits or accomplishment of noise goals. The development of validated prediction models in this regard shall be indispensable for conducting noise impact assessment studies and devising suitable abatement and control measures.

6.2 Principal Traffic Noise Models

In recent times, there have been various models developed in many countries that are validated and reliable for traffic noise predictions [4–20]. Garg et al. (2014) [4] presented a comprehensive review of the various principal traffic noise models developed in recent times. Although the characterization of source in conjunction with the advanced numerical methods for sound propagation has significantly reduced the uncertainty in traffic noise predictions, the approach followed is quite complex and requires specialized mathematical skills for predictions which is sometimes quite cumbersome to be understood by the town planners. The Harmonised Accurate and Reliable Methods for the EU Directive on the Assessment and Management Of Environmental Noise (HARMONOISE) model has clearly segregated the rolling and propulsion noise of the vehicles as [9, 10]:

$$L_{WR}(f) = a_R(f) + b_R(f)\ \log\left(\frac{v}{v_{ref}}\right) \tag{6.1}$$

$$L_{WP}(f) = a_P(f) + b_P(f)\ \log\left(\frac{v - v_{ref}}{v_{ref}}\right) \tag{6.2}$$

where v_{ref} is 70 km/h. The coefficients a_R, b_R, a_P, and b_P are given in one-third octave bands in frequency range 25–10 kHz. In July 2014, the Member States of European Union had adopted the Common Noise Assessment Methods in Europe (CNOSSOS-EU) Model. The model has superseded HARMONOISE model with an objective of enhancing reliability and comparability of noise exposure data across the EU Member States [18]. The CNOSSOS-EU method is valid for determining noise in frequency range from 125 Hz to 4 kHz for road traffic noise. The vehicles are grouped into four separate categories with regard to their characterization of noise emission, while the fifth category is foreseen as an open class for new vehicles that may be developed in future. The coefficients are given in octave bands for each vehicle category and for a reference speed $v_{ref} = 70$ km/h and virtual reference road surface consisting of an average of dense asphalt concrete and stone mastic asphalt. This framework provides a coherent and harmonized approach in modeling the various noise sources such as road, rail, aircraft, and industries [19]. The model is, however, undergoing validation in various stages for its implementation in strategic noise mapping in Europe as directed by Commission Directive 2015/996/EC of May 2015. As of first January 2019, all EU Member States are required to use this method for developing noise maps as per the European Environmental Noise Guidelines. In order to have a unified and harmonized calculation approach, the EU working group analyzed the various issues faced by this method on some of the sound propagation algorithms used and means to resolve them [20].

6.3 Comparison of Models

The principal traffic noise models developed in recent years have been equipped with all the source and propagation characteristics for traffic noise predictions and developing noise maps. The sound propagation algorithms have been dealt differently in some of these models utilizing the faster computational techniques for predictions. In order to account for the extension of the reflecting surface element the concept of Fresnel-zones is applied [21]. Numerical propagation methods, viz., parabolic eq. (PE), boundary element method (BEM), and straight ray approach (RAY), are employed for computing the effects of atmosphere, ground surface and obstacles on sound waves. The French model NMPB-Routes 2008 [15] defines two classes of meteorological conditions, either homogeneous or downward-refraction. NMPB-Routes-2008 exhibits a more moderate trend to the overestimation of noise level in downward refraction conditions. ISO 9613-2 also specifies an engineering method for calculating the attenuation of sound during propagation outdoors as [22]:

$$L_{ft}\,(\mathrm{DW}) = L_{\mathrm{W}} + D_C - A \tag{6.3}$$

where L_{W} is octave band sound power level in dB(A) radiated by sound source in reference to sound power of 1 pW, D_{C} is directivity correction in dB in direction from center of source to receiver and A is combined attenuation due to geometrical divergence (A_{div}), atmospheric absorption (A_{atm}), ground effect (A_{gr}), barrier attenuation (A_{bar}), and miscellaneous effects, viz., foliage, industrial sites, housing, etc. It is imperative to compare these models using actual measurements conducted at various sites of a city or an area for ascertaining the suitability and validity of these models. However, there had been only a few studies carried out so far for analyzing the comparison of these models. For instance, in one such study reported by Lisle (2016), the modeling was undertaken with neutral weather conditions to allow comparative assessment [23]. It is thus evident that these models have been validated in the respective countries and thus are to be accordingly calibrated in

Indian conditions to make them suitable for reliable road traffic noise predictions. Table 6.1 shows the revised comparison of all these principal traffic noise models based on the various technical attributes as presented earlier by Garg et al. (2014) [4]. Thus, these models are to be accordingly modified in Indian conditions for distances less than 10 m due to spatial limitations. A validated model developed in terms of acoustical characterization of vehicles including the sound propagation algorithms and with GIS integration for developing noise maps shall meet these objectives. Figure 6.1 shows a typical illustration of the noise map generated from an Indian study using the CRTN model (or CoRTN model, UK) for Dhanbad township area, in Jharkhand, India, which served as a decision support tool for devising noise control action plans [24].

6.4 Traffic Noise Models in Indian Scenario

In Indian scenario, various traffic noise models had been developed in different parts of the country in recent times as shown in Table 6.2.

The majority of the models developed in Indian scenario are area specific and there is no validated and comprehensive model developed so far for traffic noise predictions and forecasting. Also, honking correction is very crucial parameter to be considered for validation of these models. Honking is a frequent phenomenon in Indian roads context therefore it was observed that honking has significant impact on traffic noise besides traffic volume and vehicular speed. Garg et al. (2015) [30] recommended some empirical formulations based on multiple linear regression (*MLR*) approach and genetic algorithms (*GA*) for predicting single noise descriptors, L_{Aeq} and L_{10} in terms of total traffic flow (Q), average speed of vehicles (V), and percentage of heavy vehicles (p) moving on the road.

$$L_{\mathrm{Aeq}} = 55.29 + 3.98\ \log\ Q + 4.34\ \log\ V + 0.15p; R^2 = 0.54 \tag{6.4}$$

$$L_{10} = 56.45 + 3.55\ \log\ Q + 6.34\ \log\ V + 0.07\,p; R^2 = 0.44 \tag{6.5}$$

$$\begin{aligned} L_{\mathrm{Aeq}} = {} & 2.56 \times \log_{10}(LMV) + 0.798 \times \log_{10}(MCV) + 0.093 \\ & \times \log_{10}(HV) + 21.825 \times \log_{10}\left(V_{avg}\right) + 71.636 \times \left(V_{avg}\right)^{-0.226} \end{aligned} \tag{6.6}$$

$$\begin{aligned} L_{10} = {} & 3.198 \times \log_{10}(\mathrm{LMV}) + 0.512 \times \log_{10}(\mathrm{MCV}) + 0.577 \\ & \times \log_{10}(\mathrm{HV}) + 26.161 \times \log_{10}\left(\mathrm{V_{avg}}\right) + 55.592 \times \left(\mathrm{V_{avg}}\right)^{-0.214} \end{aligned} \tag{6.7}$$

where $\mathrm{L_{Aeq}}$ is the predicted equivalent continuous sound pressure level, LMV is the Light Motor vehicles comprising of Cars, three-wheelers and two-wheelers, MCV is the medium commercial vehicles including vans, SUVs, etc. and HV is the heavy vehicles comprising of buses and trucks and $\mathrm{V_{avg}}$ is the average traffic flow speed.

Table 6.1 Comparison of principal traffic noise models (TNMs) based on various technical attributes

Technical attributes	FHWA TNM model	CoRTN model	RLS 90 model	ASJ RTN-model 2018	HARMONOISE model	sonROAD 18 model	NMPB –Routes – 2008	CNOSSOS-EU
References	FHWA traffic Noise model (TNM), 1998; TNM 3.0, 2019	Givargis et al. 2008; Steele 2011	Steele 2011; Probst et al. 2010.	Sakamoto S (2020)	Defrance et al. 2007; Jonasson et al. 2007; Watts 2005; Jónsson 2008; JRC report 2010.	Heutschi et al. 2018	Dutilleux G et al. 2010; Sétra 2009; Kephalopoulos 2012.	Kephalopoulos et al. 2012; 2014 [18, 19]
Government users	USA Federal Highway Administration's Traffic Noise Model	UK Calculation of Road Traffic Noise	Germany *Richtlinien für den Lärmschutzan Straben* (Guidelines for Noise Control on Streets)	Japan	Proposed for EU Member States **H**armonised **A**ccurate and **R**eliable **M**ethods for the EU Directive on the Assessment and Management **O**f Environmental **Noise**	Switzerland	France (Nouvelle Méthode de Prévision du Bruit des Routes)	EU Member States (CommonNOise aSSessment MethOdS)
Applications	Highway, road networks	Highway, single traffic stream	Highways, car parks, Simple streams only	Highway/constant speed/in different traffic conditions	Road and railway traffic	Road traffic noise	Highway, road networks	Road traffic, railway, aircraft and industrial noise source
Predicts traffic volumes	No	No	Yes	No	No	No	No	No
Traffic conditions	Constant speed, acceleration, decelerate	Constant speed, grades	Constant speed, grades, quasi-intersections, interruptions	Constant speed, Acceleration/ Deceleration mode,	Constant speed, Acceleration/ Deceleration mode. Corrections	Constant speed	Steady speed, acceleration, deceleration	Steady speed, acceleration, deceleration

(continued)

Table 6.1 (continued)

Technical attributes	FHWA TNM model	CoRTN model	RLS 90 model	ASJ RTN-model 2018	HARMONOISE model	sonROAD 18 model	NMPB –Routes – 2008	CNOSSOS-EU
				Junctions, Signalized intersection, road tunnels, depressed and semi underground roads, Flat/overhead roads & double-deck via ducts	for slip & acceleration/ deceleration defined			
Vehicle types	Automobiles, Medium Trucks, Heavy Trucks, Buses and Motorcycles	Light vehicles/ Heavy vehicles	Light vehicles/ Heavy vehicles/Car parks	Light vehicles (Passenger cars and small sized vehicles) and Heavy vehicles (Medium sized and large sized) & Motorcycle	Light vehicles, Medium Heavy, Heavy, other heavy vehicles & two wheelers	Ten vehicle categories acc to SWISS-10 scheme and CNOSSOS classes. Future electric vehicles shall be included	Light vehicles <3.5 tonnes and Heavy Goods vehicles, 3.5 tonnes or higher	Four Categories: Light motor vehicles; Medium heavy vehicles; Heavy vehicles and Powered two-wheelers. Open category is also defined for future needs
Propagation	Energy type Propagation in one-third octave band is modeled considering atmospheric absorption,	Energy type	Energy type The calculation is made starting from an average level $L_{m,E}$ measurable at a distance of 25 m from the center of	Energy type The model is developed based on geometrical acoustics and it contains effects of shielding by	Energy type Reference model employs three propagation models viz., parabolic Eq. (PE) model,	Energy type son ROAD 18 is based on the fundamental assumptions and empirical formulations of CNOSSOS. It will	Energy type Propagation model includes calculation of probability of occurrence of downward refraction conditions for each	Energy type The method helps to determine the equivalent continuous sound pressure level at a receiver

	divergence, acoustical characterization and topography of intervening ground, walls, berms and their combinations, intervening rows of buildings and intervening areas of heavy vegetation		the road lane. It includes corrections due to presence of obstacles, vegetation, air absorption, reflection and diffraction, ground absorption, etc.	barriers or buildings, ground surface, air absorption and meteorological condition. The procedures of application to general roads (flat, bank, cut, and viaduct) and special road sections (interchanges, junctions, signalized intersections, road tunnels, depressed/semi-underground roads, flat roads with overhead viaducts, and double-deck viaducts)	straight-RAY (RAY) & boundary element method (BEM). Atmospheric refraction is taken into account by PE. In the region outside the source region, a PE model is used. For a flat ground surface, the crank-Nicholson PE (CNPE) model or the Green's function (GFPE) model is used. For a ground surface with smooth hills, the generalized-terrain PE model (GTPE) is used. BEM and RAY are used for obstacles with complex shapes, while CNPE & GFPE for rectangular obstacles. Point-to-point propagation is	be coupled with an appropriate propagation model such as ISO 9613-2. Prediction uncertainty of ±1.5 dB primarily due to pavement performance is evaluated	direction, search for propagation trajectories between each source and receiver, calculation of attenuation in downward refraction conditions & homogeneous conditions. Sound levels are weighted by average occurrence p_i of downward refraction conditions and homogeneous conditions	point in downward-refraction: Propagation conditions (positive vertical gradient of effective sound celerity) from the source to the receiver; homogeneous atmospheric conditions (null vertical gradient of effective sound celerity) over the entire area of propagation. The method provides results per octave band, from 63 Hz to 4 kHz, made for each of the center frequencies. The limit of validity is distance of 800 m for a normal distance to the road

(continued)

Table 6.1 (continued)

Technical attributes	FHWA TNM model	CoRTN model	RLS 90 model	ASJ RTN-model 2018	HARMONOISE model	sonROAD 18 model	NMPB –Routes – 2008	CNOSSOS-EU
					dependent on speed-sound gradient calculated from the wind speed, wind direction and temperature profile			
Basic model	It computes vertical sub-source vehicle emissions depending upon vehicle type, pavement type, and throttle conditions. Seventeen constants defined depending upon variables for converting A-weighted noise level emissions to one-third octave band spectra. Emission based on vehicles pass by at 15 m over	L_{10} in terms of total hourly flow is calculated at a reference distance of 10 m and reference hourly mean traffic speed of 75 kmph	Average level L_{mE} at a distance of 25 m from center of road lane and is function of amount of vehicles per hour and of % of heavy trucks	Sound Power level is defined as function of vehicle speed with change in noise generated due to pavement type, road gradient and noise directivity considered in correction terms. Sound Power level defined for steady and non-steady traffic flow. For signalized intersections and junctions, the coefficient of	Each vehicle category is represented by two point sources each having a specified sound power contribution from rolling and propulsion noise. For calculating the sound power from whole vehicle, the sound power between lowest and the highest source is distributed. The effect of speed & acceleration is taken into account in formulation of	Sound Power level derived from maximum pass by level of single vehicle at a distance of 7.5 m and at a height of 1.2 m above ground. Reference speed of 70 km/h is taken, road pavement property, temperature effects, and correction for road gradient is accounted	Noise emission of traffic lane and characterized by its sound Power level per meter and per vehicle $L_{w/m/veh}$ which is sum of power unit noise component and rolling noise component. Rolling noise component defined for three road surfaces *R1*, *R2*, and *R3*	For each road vehicle, the emission model consists of a set of mathematical formulations for rolling noise and propulsion noise. The sound power level emitted by one of the sources (rolling or propulsion) is a function of the vehicle speed (20 km/h to 130 km/h).For two-wheelers, only propulsion noise is considered. Model

	flat absorptive ground			deceleration and steady flow are same, while coefficients for acceleration and non-steady flow are the same	source strength for rolling and propulsion noise			calculates noise in the frequency range from 125 Hz to 4 kHz for road traffic noise, from 125 Hz to 4 kHz for railway noise, from 63 Hz to 4 kHz for industrial noise, and from 50 Hz to 10 kHz for aircraft noise
Geometrical divergence	Adjustment for distance A_d from the elemental roadway segment to receiver defined $A_d = 10\log\left(\frac{15}{d}\right)\left(\frac{\alpha}{180}\right)$ is angle (in °) subtended by elemental roadway segment	Δ_d is distance adjustment defined	$D_{S,m}$ is attenuation to distance and air absorption defined in model $D_s = 11.2–20 \times \log(s)-(s/200)$ Where s is distance between emission and imission point	$L_A = L_{WA}-8-20\log(r) + \Delta L_{cor}$ Where L_{WA} is A-weighted sound Power level of single running vehicle and ΔL_{cor} accounts for corrections for diffraction, ground effect, and atmospheric absorption	Point source: $\Delta L_d = 10\log(4\pi d^2/d_0^2)$ d = propagation distance, $d_0 = 1$ m Line source: $\Delta L = 10\log(\Delta\theta/4\pi d)$ $\Delta\theta$ = angle of view from receiver to segment	$A_{div,f} = 20\log\times(d) + 11$	$20\log_{10}(d) + 11$	For a point sound source in free field, $A_{div,f} = 20\log\times(d) + 11$
Source characteristics and height of source	Simple stream Energy apportioned to two source heights: One at pavement	Single stream Noise levels are obtained at a reference distance of 10 m	Single stream The starting point of the calculation is L_{mE} measurable at a distance of 25 m	Simple straight stream	Harmonoise distributes 80% of tire/road noise on a source 0.01 m above the ground	Single vehicle with microphone position at a distance of 7.5 m and at height of 1.2 m.	Each source line is broken down into a set of sound point sources, placed 0.05 m above	Two source heights are recommended by the method at

(continued)

Table 6.1 (continued)

Technical attributes	FHWA TNM model	CoRTN model	RLS 90 model	ASJ RTN-model 2018	HARMONOISE model	sonROAD 18 model	NMPB –Routes – 2008	CNOSSOS-EU
	level and one at 1.5 m above the pavement except for heavy trucks, where the upper height is 3.66 m above the pavement	from the nearest carriageway edge of highway.	from center of road lane. The model is also able to evaluate the sound emission of the parking lot		and 20% either on 0.30 m or 0.75 depending on type of vehicle. For propulsion noise, it is the other way round. For heavy construction equipment, additional point source for exhaust noise at 3.5 is used		roadway. GdBN08 describes the pass-by maximum levels in dB(A) measured at 7.5 m horizontal distance and 1.2 m height above the ground surface	0.5 m (source A), 4.0 m (source B).
Input data	Traffic type, flow, speed, road and emission data, local characteristics	% heavy vehicles, flow, speed, road, and environmental data, gradient	Traffic type, flow, park or road data	Traffic type, speed, barrier geometry, road surface and gradient, flow (steady/non steady) distance from source to prediction point, mean wind speed, and density of buildings	Traffic speed, composition, intensity (flow), flow characteristics, viz., acceleration/deceleration	Vehicle type, average speed, pavement type, air temperature, slope of road	Average hourly flow rate for each category of vehicle, speed and traffic flow type of each vehicle category, road platform surface category, road gradient	Average hourly flow rate for each category of vehicle, speed and traffic flow type of each vehicle category, road platform surface category, road gradient
Noise descriptor	One hour L_{Aeq}, day-night average sound level, day-evening-night average sound level	$L_{10\ (1\text{-hour})}$ and $L_{10\ (18\text{-hour})}$	L_{eq}, $L_{m,E}$, L_m (mean level for each lane)	L_{WA}, L_{AN}, $L_{Aeq,T}$	$L_{Aeq,\ T}$, L_{den} & L_{night}	A-weighted sound Power level, $L_{eq,A}$	L_{Aeq}, $L_{Aeq,LT}$	$L_{Aeq,T}$, L_{day}, $L_{evening}$, L_{night} and L_{den}

Type of mapping	Multiple dual points→grid	Line→point	Line→point	Line→point	Incoherent line source→point	Line→point	Line→point	Noise emission is represented by a source line in terms of spectral sound power per meter
Gradient effect	Model computes adjusted speeds based on user input speeds, roadway grade, and traffic control devices. TNM reduces input speeds depending upon steepness and length of upgrades	Gradient correction: $\Delta_G = 0.3\ G$ dBA	Gradient correction: $R_{RS} = 0.6\ \lvert g \rvert - 3$ for $g > 5\%$ $R_{RS} = 0$ for $g \leq 5\%$	Gradient correction: $\Delta L_{grad} = 0.14\ i_{grad} + 0.05\ i_{grad}^2$ Where i_{grad} is gradient of road (%). It is applied only to heavy vehicles ascending inclined roads	The effect of gradient is described as $a = a_1 + g \sin(\alpha)$ where a_1 is acceleration of vehicle	Gradient correction for road gradient accounted	Correction term ΔL_m defined for uphill, downhill, and horizontal pavements. Three potential gradient defined: Horizontal (gradient less than 2%), upwards (gradient of 2% to 6%), and downwards (gradient of 2% to 6%)	Correction term $\Delta L_{WP,grad}$ which is a function of the slope s (in %), the vehicle speed (in km/h), and the vehicle class
Directivity	Subsource-split ratio for vehicle emission, r_i defined in terms of five constants	Angle of view adjustment defined	Not mentioned	Directivity function defined	Directivity functions defined for rolling and propulsion noise	Vertical directivity accounted	Not mentioned	Horizontal directivity effects are implicitly taken into account; vertical directivity is neglected
Ground effect	TNM model for reflection coefficients based on approach of Chessell incorporating the single-parameter	Not defined	Level difference caused by ground absorption and meteorological influences in free field, D_{BM} defined in model	Correction for ground effect ΔL_{grad} for excess attenuation defined. The coefficient for the excess	Analytical formula established by Chien and Soroka. Additional correction factor of coherence due to	ISO 9613-2 calculates ground effect of +3 dB independently of octave band up to distance of 7.5 m. Further away, ground type	Attenuation due to ground in downward refraction conditions $A_{ground,F}$ and in homogeneous conditions $A_{ground,H}$ in third octave band	Attenuation due to the ground in favorable conditions and homogeneous conditions is defined

(continued)

Table 6.1 (continued)

Technical attributes	FHWA TNM model	CoRTN model	RLS 90 model	ASJ RTN-model 2018	HARMONOISE model	sonROAD 18 model	NMPB –Routes – 2008	CNOSSOS-EU
	ground impedance model			Attenuation due to the ith ground surface as a function of the average height of the Propagation path is defined	presence of turbulent eddies near ground surface defined. Each impedance discontinuity is modeled through a Fresnel weighting approach	has a significant effect	defined. Formula is asymptotic approx. Of Chein and Soroka's formulation	
Atmospheric absorption and rarefaction	Atmospheric absorption defined in terms of ambient air temperature, reference air temperature 20 ° C, and oxygen relaxation frequency f_{r0}	No mention	$D_{S,m}$ attenuation due to distance and air absorption defined	Correction term ΔL_{air} is calculated considering standard state of atmosphere (20 ° C, 60% RH, and 101.325 kPa) as a function of distance from source to prediction point	Uses straight rays and curves the ground to simulate refraction; radius of curvature determined from maximum height of curve. Effect of air absorption is calculated with ISO 9613-1.Curved ground-analogy is adopted by inverse curving of the terrain rather than curving sound rays	Air absorption in third octave band *f* according to ISO 9613-1 for temperature + 8 °C & RH 76%	Atmospheric rarefaction in downward conditions taken into account by means of height correction terms. Turbulence also taken into account. Attenuation due to atmospheric absorption A_{atm} defined	Atmospheric absorption A_{atm} during propagation over a distance *d* is calculated as $A_{atm} = (\alpha_{atm} \cdot d/1000)$. The values of the α_{atm} are given at 15 °C, a relative humidity of 70% and an atmospheric pressure of 101.325 pa

Meteorological effects	TNM does not account for atmospheric effects such as varying wind speed or direction or temperature gradient. TNM assumes neutral atmospheric conditions	Not mentioned	D_{BM} is attenuation due to ground and atmospheric effect	Change in L_{Aeq} due to effect of wind defined $\Delta L_{m,line}$	$A_{excess,j}$ is excess attenuation represents the effect of ground, meteorology, barrier, and air absorption. Standard P2P model for homogeneous atmosphere is used. Sound speed profile is approximated by means of Lin-log function	ISO 9613 is considered, K_{met} is accounted for night periods	Two classes of meteorological conditions: Homogeneous and downward propagation defined	Classes of meteorological conditions: Homogeneous atmospheric conditions, atmospheric downward refraction conditions
Correction for road surfaces	TNM defines energy average emission levels depending upon road: Dense-graded asphaltic concrete, Portland cement concrete, open-graded asphaltic concrete and an "average" composite pavement type	Correction for concrete and bituminous surface; impervious bituminous and pervious road surface	Correction for road surface D_{Stro} defined	Frequency characteristics of road vehicle noise on dense asphalt, porous asphalt, and Koukinou II (porous asphalt formed in layers with the top layer having a maximum chipping size of 13 mm and the bottom layer of stone mastic asphalt)	Correction for road temperature, tires with and without studs, road surface wetness and aging defined	Pavement correction defined	Road pavement influence addressed by grouping pavement into three categories (*R*1, *R*2, and *R*3), correction for ageing effect. Correction for air temperature included	The emission coefficients are valid for the reference road surface, that is, a virtual road surface corresponding to an average of dense asphalt concrete and stone mastic asphalt in a representative maintenance condition. For porous road surface, a correction is given

(continued)

Table 6.1 (continued)

Technical attributes	FHWA TNM model	CoRTN model	RLS 90 model	ASJ RTN-model 2018	HARMONOISE model	sonROAD 18 model	NMPB –Routes – 2008	CNOSSOS-EU
Noise at intersections and roundabouts	Not mentioned	Not mentioned	Correction term for increased effect of traffic light controlled intersections	Two calculation methods defined, viz., practical and simplified method defined	Micro simulation based correction factor C_s applied to emission level. C_s is evaluated as average of correction function $C(x)$ over length of segment estimated by simulated noise emission profiles. Curve $C(x)$ fitted to noise emission profiles using least squares method	Constant vehicle speed is assumed	Not mentioned	Coefficients $C_{R,m,k}$ and $C_{P,m,k}$ depend on the kind of junction (for a crossing with traffic lights; roundabout)
Impedance effect	TNM allows users to enter various ground types based on effective flow resistivity (cgs rayls) measured by Embelton. The ground type and associated effective flow	Not mentioned	Not mentioned	Complex values of specific acoustic impedance are considered in wave-based numerical analysis	For porous road surface: Hamet model; ISO road surface: One parameter model with flow resistivity 2 MPas/m^2, surface with cluster: One parameter model with		Acoustic absorption of ground is represented by a frequency independent dimensionless coefficient G between 0 and 1. *Gtrajet* is defined as the fraction of absorbent ground present in the whole of the path	Acoustic absorption of a ground is represented by a dimensionless coefficient G, between 0 and 1, independent of the frequency. G_{path} is defined as the fraction of absorbent

	resistivity (EFR) is defined. TNM averages the ground impedance in vicinity of reflection point using Boulanger's approach				flow resistivity 200 MPas/m^2		covered. $G = \min\left[\left(\frac{300}{\sigma}\right)^{0.57}, 1\right]$ σ in cgs rayls	ground present over the entire path covered
Diffraction effect	Multiple reflections between parallel barriers computed in two dimensions; double diffraction included; TNM chooses most effective pair of barriers based on their input heights in accordance with "Foss selection algorithm"	Reflection adjustment defined	D_B is attenuation due to topography and building dimension defined in model. D_E is correction for absorption characteristic of building surfaces	The fundamental correction term for diffraction is calculated as a function of path length. Empirical formulation for simple barrier, finite length barrier, thick barrier, multiple barrier with overhang, edge modified, low height, and transmission through barrier defined	BEM and RAY can be used for obstacles of complex shapes. CNPE and GFPE can also be used for rectangular obstacles; GTPE is used for propagation over smooth hills. Degouts approximation of Fresnel integrals gives attenuation as a function of path length difference and wavelength. Reflections from the faces of wedges/thick barriers are taken into account as ground effects	As per ISO 9613 standard	Correction term defined in model in terms of Fresnel number (N) and corrective term C_h. Barrier diffractions is calculated using Maekawa's approximated formulation considering barrier as a hard surface.	For pure diffraction with no ground effect, a mathematical formulation in given. For diffraction with ground effects, attenuation due to the diffraction between the source and the receiver and attenuation due to the ground effect on the source side and receiver side is defined

(continued)

Table 6.1 (continued)

Technical attributes	FHWA TNM model	CoRTN model	RLS 90 model	ASJ RTN-model 2018	HARMONOISE model	sonROAD 18 model	NMPB –Routes – 2008	CNOSSOS-EU
Tire type corrections	No	No	No	No	Yes	No	No	Correction for studded tires defined
Bridges, tunnels, viaducts, defined	No Rows of buildings as optional element included	No	No RLS considers parking lot emissions	Yes	No GTPE is used for ground surface with smooth hills	–	No Special elements like trenches, tunnels, and partial covers included	Correction for structural radiation (bridges and viaducts) for railway noise recommended
Vegetation effect	Attenuation through dense foliage and tree zones included in propagation path. Berms can be selected with user selectable heights, top widths, and side slopes	Not mentioned	D_B is attenuation coefficient due to topography and building dimensions	Not mentioned	Rough terrain with vegetation can be described by terrain roughness and ground impedance. Diffraction effects of earth mounds taken into account by Deygout approx. The attenuation through tress $A_{scat,i}$ defined	Correction $A_{fol,f}$ according to ISO 9613-2	Roughness parameter defined for sparse habitat (farms, villages, trees, and hedges)	Not mentioned Diffraction on thin screens, thick screens, buildings, earth berms (natural or artificial), and by the edges of embankments, cuttings, and viaducts defined

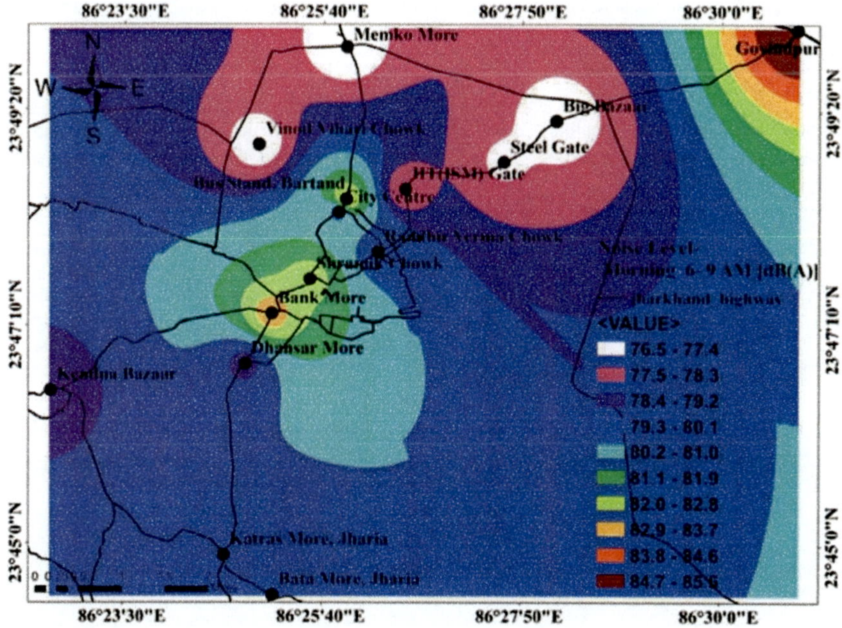

Fig. 6.1 Noise map of Dhanbad city during morning house (6.00 a.m. to 9.00 a.m.) using CRTN model (Debnath and Singh 2018, [24])

6.4.1 Honking Correction

In one of the study focused on traffic noise assessment in the urban agglomeration of Nagpur, India; it was observed that honking induced an additional 2 to 5 dB(A) noise over and above the traffic noise [36]. One such study [37] showed that restriction on horns, trucks, and noisy buses reduced the equivalent traffic noise between 6 and 10.2 dB(A) in residential areas. A considerable reduction of 7.6 dB(A) was observed during a ban on horns together with limiting the commercial vehicles to 10%. The recent study of Kalaiselvi and Ramachandraiah (2016) for six different locations in Chennai city showed that noise levels increase by 0.5–13 dB in the presence of horns depending on the traffic flow [29]. Understanding the real traffic noise characteristics requires the quantification of some of the basic traffic flow characteristics such as speed, flow, level of service (LOS), and density. In a given roadway, the noise level changes with density and LOS on the road. A new factor for horn correction is introduced with respect of levels of service (LOS) as per IRC 106-1990 [38] as shown in Table 6.3. The increase in number of honking per minute is correlated with increase in L_{eq} (in dB(A)) as [29]:

$$y = 4.4847 \ln (x) - 3.0346; R^2 = 0.92 \tag{6.8}$$

Table 6.2 Recommended traffic noise models in different parts of India for traffic noise predictions [24–35]

Author	Area of study	Empirical formulation recommended
Debnath and Singh 2018	Dhanbad	CRTN model was used for traffic noise modeling and noise maps were developed, CRTN model was validated with experimental observations
Sonaviya and Tandel 2020	Surat	A new honking correction in conjunction with German RLS-90 model is used for heterogeneous traffic conditions
Thakre et al. 2020	Nagpur	The study revealed an increment of 5–6 dB(A) in noise level, 4–6 times in honking, and 1.7 times in traffic volume for a minor road when compared for year 2012 and 2019
Lokhande et al. 2016	Nagpur	Noise modeling and prediction are performed using a modified Federal Highway Administration (FHWA) model
Singh D et al. 2016	Patiala, Punjab	An alternative data analysis technique, based on algebraic graph theoretic approach, has been presented for the assessment of road traffic noise
Kalaiselvi and Ramachandraiah 2016	Chennai	A new factor for horn correction is introduced with respect of level of service (LOS). In heterogeneous traffic conditions, horn events increase noise level (L_{den}) by 0.5–13 dB(A) as compared to homogeneous traffic conditions
Garg et al. 2015	Delhi	$L_{Aeq} = 55.29 + 3.98 \log Q + 4.34 \log V + 0.15p;$ $R^2 = 0.54$ $L_{10} = 56.45 + 3.55 \log Q + 6.34 \log V + 0.07\ p;$ $R^2 = 0.44$ where Q is the total traffic flow, V *is* average speed of vehicles, and p is the percentage of heavy vehicles
Sharma et al. 2014	Nagpur, Maharashtra	$L_{eq} = 59.44 + 4.461 \log Q_e + 0.1983S_e - 0.3771d + 0.582\ h$ for $S_e > 30$ km/h and honking >0, where Q_e is the equivalent traffic flow, S_e is equivalent vehicle speed, d is the distance, and h is total honking per minute
Mishra et al. 2014	Delhi	Reported noise levels along for two sites in Delhi city and along bus rapid transit system corridor in Delhi city
Kumar et al. 2011	NH 58 highway	$L_{eq} = 19.92224 \log [Q(1 + 0.1VP)] + 12.59764$ where VP is percentage of heavy vehicles
Agarwal et al. 2011	Jaipur	$L_{eq}\ (LV) = 8.43 \log (x) + 39.522$, LV is light vehicles and HV is heavy vehicles. $L_{eq}\ (HV) = 12.67 \log (x) + 27.579$, where x is equivalent no of vehicles
Rajakumara et al. 2009	Bangalore	Acceleration lane model: $L_{eq} = 59.21 + 0.043\ S_E + 5.71 \log Q_E - 0.0197\ R$ Deceleration lane model: $L_{eq} = 65.12 + 0.061\ S_E + 2.14 \log Q_E + 0.923 \log L - 0.041\ R$, where R is position of sound level meter and L is queue of waiting vehicles on deceleration lane

Table 6.3 Horn correction values as per levels of service (LOS) as per IRC 106–1990. (Kalaiselvi and Ramachandraiah 2016, [29])

LOS	Description	Speed (km/h)	Flow (vehicles/h/lane)	% of time occupied by horn sound	Increase in L_{eq} dB(A)
A	Traffic flows at or above the posted speed limit and all motorists have complete mobility between lanes	>80	<700	0	0
B	Slightly congested, with some impingement of maneuverability	65–80	700–1100	1	0
C	Ability to pass or change lanes is not assured. Most experienced drivers are comfortable, and posted speed is maintained, but roads are close to capacity	55–65	1100–1550	3–6	2
D	Speeds are somewhat reduced, motorists are hemmed in by other vehicles	45–55	1550–1850	6–13	6
E	Flow becomes irregular and speed varies rapidly, a road above its designed capacity	25–45	1850–2000	13–26	10
F	Flow is forced, every vehicle moves in lockstep with the vehicle in front of it	<25	Unstable	>26	12

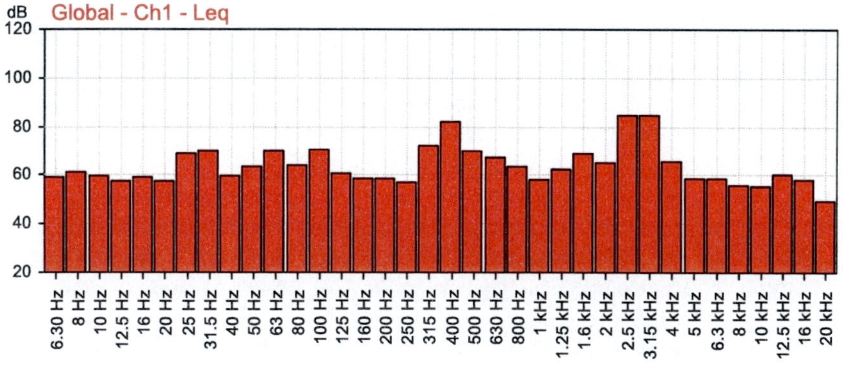

Fig. 6.2 Frequency spectrum of a car horn blared (Y axis shows the sound pressure level in dB, X axis show the frequency in one-third octave band)

where x is number of honkings per minute and y is increase in L_{eq} dB(A). The levels of service (LOS) as per IRC 106–1990 is correlated with increase in L_{eq} (in dB(A)).

The spectrum of honking noise as observed in field studies in some of the areas where honking noise was dominant in Delhi city was observed to be dominated in octaves of 2.5 kHz, 3.15 kHz, and 4 kHz as shown in Fig. 6.2 [39]. Thus, it is imperative that for areas having high traffic noise, the sound insulation in octaves of 2.5–4 kHz should be higher, which necessitates that the coincidence dip observed in sound transmission loss characteristics of facades, windows, and doors of dwellings has to be controlled in this frequency range. The field traffic noise measurements for some sites in Delhi region by Garg et al. (2015) [30] and in subsequent experimental investigations had revealed that honking noise contributes to an additional noise of 1–7 dB(A) over and above the traffic noise based on the intensity and frequency of events, which suggests that it must be accounted as a correction in Indian models for traffic noise predictions and forecasting.

6.5 Implications of TNMs, Validation, and Prediction Uncertainty

The recent traffic noise models (TNMs) are based on scientific principles appending numerical methods as a consequence of increasing computer skills and facilitates inclusion of separate source and propagation part in order to allow independent updating of source and propagation models. It is very difficult to weigh the pros and cons of all different approaches specifically when all these models have been validated in their respective countries. Also, the concept of an "ideal model" so proposed by Steele in 2001 [9] is to a larger extent compensated by the recent models covering each technical attributes exclusively based on scientific principles and mathematical computations. However, the application of these models widely is still uncertain as the source model is best suited for that particular country only. There are various commercial software packages available for traffic noise predictions like CadnaA (Datakustik, Germany); Sound Plan (U.K); IMMI (Wölfel); Mithra-SIG (CSTB); NovaPoint (Vianova Systems); Predictor-LIMA (B&K, Denmark). The validity of noise predictions have to be however ascertained by comparison with actual measurements for various sites for their suitability in respective area/state/country. Accuracy, precision, high computation speed, flexibility, integration with GIS and its suitability for long-term, and 24 h predictions are the desired model characteristics that need to be accomplished for getting recognition and usefulness in public domain by Pollution Control boards or agencies of the particular area or country. The recent models include meteorological effects whereby a number of 25 classes have been recommended that causes the determination of long-term average sound levels with an error of utmost 2 dB at 1000 m range. Comparison of noise maps calculated with different meteorological conditions show only negligible differences [40]. There are various technical aspects pertaining to the development of a model which can be helpful for enhancing the accuracy, usability and reliability of the traffic noise model as follows:

- Source specific empirical models, effect of pavement type, that is, road surface corrections.
- Effect of vehicle acceleration and deceleration.
- Emission characteristics of all sources including the electric vehicles in future, parking lots, honking corrections.
- Effect of varied source heights.
- Gradient effect.

- Signalized intersections or roundabouts.
- Effect of multiple lanes.
- Harmonized approach for sound propagation algorithms, horizontal and vertical distribution, urban street canyons effect.
- Development of linear sound emission law for propulsion and rolling noise radiated by a vehicle.
- development of linear sound emission law for the interrupted flow conditions,
- Validation of the developed model from actual measurements particularly for the roads with heavy, medium, and low traffic density and highways and GIS interfacing.

Figure 6.3 summarizes the implications of recently developed noise modeling projects for traffic noise predictions. The European good practice [41] recommends the use of speed limit or accurate average speed measurements as an input for noise emission models to generate noise maps. Traffic noise prediction models that aim to be accurate in the vicinity of interrupted traffic flows will however have to model the

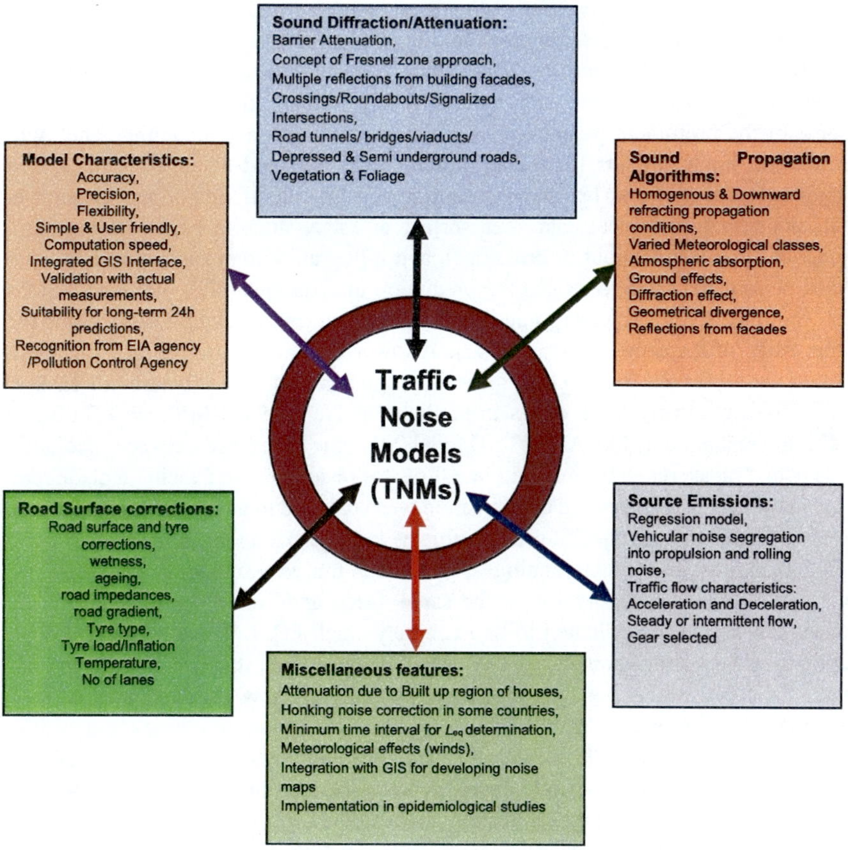

Fig. 6.3 Implications of recently developed noise modeling projects for traffic noise modeling and predictions

temporal and spatial evaluations of vehicles speeds and accelerations and thus micro-simulation modeling is essential to include these dynamic effects [42]. Donato (2007) observations showed that it is possible to determine the hourly L_{eq} with an associated imprecision of $\Delta L_{eq} = 2.5$ dB(A) with measurement lasting less than 15 min for hourly traffic flow greater than 60 vehicles per hour. These times are further shortened to value less than 5 min for $N > 170$ vehicles per hour [43]. Maruyama et al. (2013) investigations showed that if the number of vehicle passing during measurement time interval exceeds 170, ΔL_{AeqT} is thoroughly within the range of ± 1 dB [44]. ISO 9613-2 [22] specifies an estimated accuracy for broadband L_{AT} (DW) for downward propagation to be ± 3 dB for source and receiver height $0 < h < 5$ m and $0 < d < 100$ m where d is distance between source and receiver. However, it has been pointed out in some studies, for example, Brittain et al. (2008) and Parzych et al. (2004) [45, 46], that empirical formulations recommended in ISO 9613-2 used to predict the outdoor propagation, including ground effects imposes limitations which should be rectified.

The accuracy of the input data is very crucial for the traffic noise predictions with lowest uncertainty. The input data: traffic flow, vehicle type, speed limits, spatial layout of road network, presence of roundabouts or crossings on a road segment, land cover types over the study area (buildings, grassland, water bodies, etc.), height and location of building, meteorological data, prevailing wind direction, road surface type and age, road gradient, and topography (elevation model of the study area) are some of the prominent parameters required for precise noise modeling. The high-resolution input datasets for traffic flow, composition and average speed, spatial layout of road network, land-cover (buildings or grassland), height and location of buildings, meteorological data, road surface and age, slope of each road segment, junction type (roundabout or crossings), and topography (elevation model of area) shall be indispensable to reduce the prediction uncertainties [47]. The CNOSSOS-EU framework itself is undergoing further validation, testing, and revision during the next stage of implementation [19, 47]. A lower resolution noise model had been tested and observed for its performance for exposure ranking (Spearman's rank = 0.75) with relatively large errors in predicted noise levels (RMSE = 4.5 dB(A)) [47]. In some models like ASJ RTN-Model 2018, the difference between prediction and actual measurements were within ± 3 dB for residential area facing a plane road and cut road [5]. The vehicle classification was analyzed to account for an error less than ± 1 dB if ratio of medium sized vehicles to large sized ranges from 10 to 80%. The variations in vehicle running speed under the assumption that all vehicles classified in each category run at the same speed and sound power level on the change in L_{Aeq} was evaluated to be extremely small [5]. In some of the TNMs, a correction factor for geometrical divergence is considered that accounts for propagation distance only, but does not include angle of view of a receptor to road segment. For instance, CoRTN model accounts for an angle of view correction, whereby view angles of 2° and 90° can account for correction of −20 dB (A) and − 3 dB(A), respectively [47]. Relative building heights, line-of-sight between noise sources and receptors can be obtained from the high-resolution LiDAR digital terrain model (DTM) [47]. The software used and feeding of the accurate input parameters is a big challenge for prediction accuracy. For instance, the

FHWA TNM with two software versions 2.5 and 3.0 had been evaluated to have some difference in the predicted results owing to the correction to the computation of Fresnel ellipses for accounting the ground impedance between the source and the receiver [7]. Also, the noise model so developed requires a comprehensive validation in many stages site-by-site basis so that legal bodies and pollution control boards can use it for reliable predictions. In recent years, CadnaA, Sound PLAN, IMMI, B&K Predictor LIMA, etc. are some of the widely used software in national and international scenario for noise mapping. Results from a study by Karantonis et al. (2010) [48] confirmed that the CadnaA noise modeling package is accurate and effective for modeling road traffic noise in Australia. However, in Indian scenario the various source models used in these software's have to be validated first with actual experimental observations for generating reliable noise maps with least uncertainty levels. In Indian perspectives, Director General of Mine Safety (DGMS) has circulated recommendations of tenth National Conference on Safety in Mines (2007) where it has been stipulated that noise mapping should be made mandatory of various places in the mines premises based on the various machines being used in concerned mines along with personal noise dosimetry of individual workmen exposed to noise level above 85 dB(A) [49]. Thus, development of a noise model including the validated source and propagation algorithms, interfaced with Geographical Information Systems, validated at many stages and following a harmonized approach for source and propagation algorithms, input parameters, and software employed shall be an indispensable initiative in this regard for noise mapping and noise impact assessment for metropolitan cities of India. More recently, the Indian Institute of Technology, Roorkee and Indian Institute of Technology (Banaras Hindu University), Varanasi of India had been working towards developing a propagation and mitigation model of mixed traffic noise for planning of mid-sized Indian cities focusing Kanpur, Gorakhpur and Varanasi cities in Uttar Pradesh state of India. Also, CSIR-National Physical Laboratory, New Delhi; CSIR-Central Road Research Institute, New Delhi and CSIR-National Environmental Engineering Research Institute, Nagpur are also some premier institutes in India working towards developing a validated noise model. Thus, such initiatives focused on district/state level can be extended to other states for developing a country based harmonized noise model. Evidently, a validated and harmonized model integrated with GIS interface can serve as a substitute to the long-term noise monitoring particularly for the big metropolitan cities and shall be indispensable for developing noise maps and ascertaining the efficacy of noise action plans.

6.6 Noise Impact Criteria

The Federal Transit Administration (FTA), USA, in 1995 published impact assessment procedures to be used for mass transit projects. These criteria are based upon the comparison of the existing outdoor ambient noise to future outdoor sound levels from the proposed project. The noise impact criteria (Fig. 6.4) are defined by two curves, which allow increasing project noise levels as ambient noise increases up to a point beyond which the impact is determined based on the project noise alone. The land use category 1 relates to areas where quietness is an essential element in their intended purpose and outdoor L_{eq} metric is used. The category 2 relates to residences and buildings where people normally sleep and outdoor L_{dn} metric is prescribed. Category 3 includes institutional land uses with primarily daytime and evening use outdoor L_{eq} metric is used. For category 1 and 2 land uses, the maximum project noise level to be considered to cause no impact is 65 dB(A) ($L_{eq(1hr)}$ or L_{dn}) regardless of the existing noise. The upper limit of the severe impact range is 75 dB(A) for category 1 and 2 land uses. The upper limit of 75 dB(A) is associated with an unacceptable living environment. Project noise between the two curves is considered to have moderate impact on the community. Category 3 land uses are considered less noise sensitive than category 1 land uses. The noise metrics L_{dn} is used for category 2 land uses, where night time sensitivity is a factor as shown in Table 6.4 [50].

Figures 6.4 and 6.5 show the noise impact criteria and increase in cumulative noise levels allowed by the criteria. For example, if the existing noise dominated by trains was measured to be an L_{dn} of 63 dB(A) at a particular location, and the new

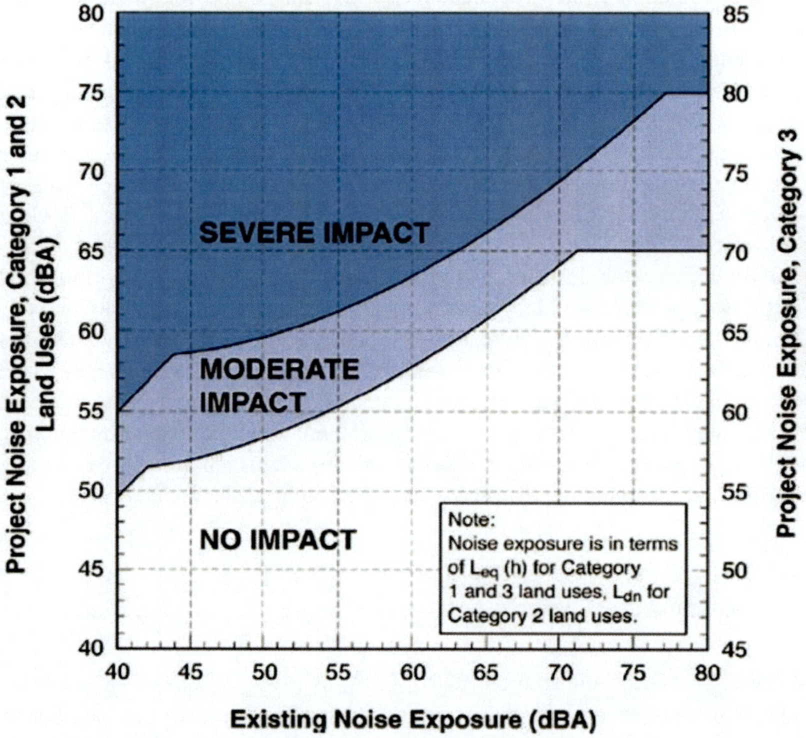

Fig. 6.4 Noise impact criteria for transit projects. (Federal Transit Administration 2018, [50])

Table 6.4 Land use categories and metrics for transit noise impact criteria. (Federal Transit Administration 2018, [50])

Land use category	Land use type	Noise metric, dB (A)	Description of land use category
1	High sensitivity	Outdoor $L_{eq(1hr)}$ [a]	Land where quiet is an essential element of its intended purpose. Example land uses include preserved land for serenity and quiet, outdoor amphitheaters and concert pavilions, and national historic landmarks with considerable outdoor use. Recording studios and concert halls are also included in this category.
2	Residential	Outdoor L_{dn}	This category is applicable all residential land use and buildings where people normally sleep, such as hotels and hospitals.
3	Institutional	Outdoor $L_{eq(1hr)}$ [a]	This category is applicable to institutional land uses with primarily daytime and evening use. Example land uses include schools, libraries, theaters, and churches where it is important to avoid interference with such activities as speech, meditation, and concentration on reading material. Places for meditation or study associated with cemeteries, monuments, museums, campgrounds, and recreational facilities are also included in this category.

[a] $L_{eq(1hr)}$ for the loudest hour of project-related activity during hours of noise sensitivity

combination of diesel and electric trains is projected to be an L_{dn} of 65 dB(A), the change in the noise exposure due to the project would be 2 dB. Referring to Fig. 6.5, a 2 dB increase with an existing noise exposure of 63 dB(A) would be rated as a moderate impact. The criterion for moderate impact is a noise exposure increase of 10 dB for an existing noise exposure level of 42 dB(A) or less, but only a 1 dB increase when the existing noise exposure is 70 dB(A).

The *severe* impact threshold defines the noise limits above which a significant percentage of the population would be highly annoyed by project noise levels, while the *moderate* impact threshold defines areas where the change in noise is noticeable, but may not be sufficient to cause a strong and adverse community reaction. Table 6.5 shows that as the existing noise exposure increases from 45 dB(A) to 75 dB(A), the allowed transit noise exposure increases from 51 dB(A) to 65 dB(A). However, the allowed increase in the cumulative noise level decreases from 7 dB (A) to 0 dB(A).

The justification for this is that people already exposed to high levels of noise should be expected to tolerate only a small increase in the amount of noise in their community.

Federal Transit Administration, USA, recommended the equations for the noise impact criteria methodology using spreadsheets, computer programs, or other analysis tools as shown in Table 6.6.

It may be noted here that due to type of land-use category included in category 3, the criteria allow the project noise for category 3 sites to be 5 dB greater than category 1 and category 2 sites. Each curve represents a threshold of noise impact,

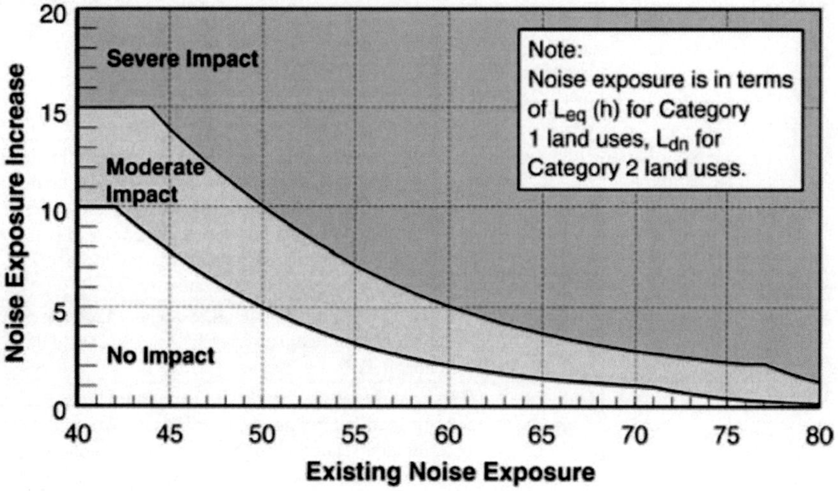

Fig. 6.5 Increase in cumulative noise levels allowed by criteria. (Federal Transit Administration 2018, [50])

Table 6.5 Noise impact criteria: effect on cumulative noise exposure [50]

Existing noise exposure L_{eq} dB (A)	Allowable project noise exposure L_{eq} dB(A)	Allowable combined noise exposure dB(A)	Allowable noise exposure increase dB (A)
45	51	52	7
50	53	55	5
55	55	58	3
60	57	62	2
65	60	66	1
70	64	71	1
75	65	75	0

with impact indicated for points on or above the curve. For land-use category 1 and 2, the absolute limit is 65 dB(A) for moderate impact and 70 dB(A) for severe impact. For land-use category 3, the absolute limit is 75 dB(A) for moderate impact and 80 dB(A) for severe impact.

6.6.1 Noise Impact Assessment: Case Study

The steps in general noise impact assessment recommended are as follows [50]:

- Identify the noise sensitive receivers and existing noise exposure.
- Determine project noise source reference levels.
- Estimate project noise exposure by distance.

Table 6.6 Threshold of moderate and severe impacts. (Federal Transit Administration 2018, [50]).

Threshold of Moderate Impact		
Category 1 and 2		
	$11.450 + 0.953\, L_E,$	$L_E < 42$
	$L_p = 71.662 - 1.164\, L_E + 0.018\, L_E^2 - 4.088 \times 10^{-5}\, L_E^3,$	$42 \leq L_E \leq 71$
	$65,$	$L_E > 71$
Category 3		
	$16.450 + 0.953\, L_E,$	$L_E < 42$
	$L_p = 76.662 - 1.164\, L_E + 0.018\, L_E^2 - 4.088 \times 10^{-5}\, L_E^3,$	$42 \leq L_E \leq 71$
	$70,$	$L_E > 71$
Threshold of severe Impact		
Category 1 and 2		
	$17.322 + 0.940\, L_E,$	$L_E < 44$
	$L_p = 96.725 - 1.992\, L_E + 3.02 \times 10^{-2}\, L_E^2 - 1.043 \times 10^{-4}\, L_E^3,$	$44 \leq L_E \leq 77$
	$75,$	$L_E > 77$
Category 3		
	$23.322 + 0.940\, L_E,$	$L_E < 44$
	$L_p = 101.725 - 1.992\, L_E + 3.02 \times 10^{-2}\, L_E^2 - 1.043 \times 10^{-4}\, L_E^3,$	$44 \leq L_E \leq 77$
	$80,$	$L_E > 77$
L_E = the existing noise exposure in terms of L_{dn} or $L_{eq\,(1hr)}$		
L_p = the project noise exposure which determines impact in term of L_{dn} or $L_{eq\,(1hr)}$		

- Combine noise exposure by all sources.
- Inventory impacts—inventory of noise-sensitive receivers that fall within the moderate and severe noise contours.
- Determine the noise mitigation needs.

For instance, for evaluating the noise impact of train running on an elevated corridor in a community, the input data corresponding to number of train pass bys during the day (6 a.m. to 10 p.m.) and night (10 p.m. to 6 a.m.), maximum number of train pass-bys during hours, number of vehicles per train for each time period, maximum number of vehicles per train during hours, train speed, guideway configuration, location of highway, and street grade crossings, if any, and noise barrier location if any has to be conserved for impact study. Thus, the $L_{eq(1h)}$ for each source and due to the combined sources is to be evaluated. If the project noise will affect any residential receivers, calculate the L_{dn} using the combined $L_{eq(1hr)}$ for both the daytime and nighttime periods separately at a distance of 15 m (or 50 ft. specified in FTA Manual [50]). Now, the project noise exposure for locations beyond the reference distance such as for noise-sensitive land uses can be predicted using geometrical divergence relationship considering transit trains as a line source. The noise-sensitive land uses within the specific contours using general assumptions for shielding attenuation from rows of buildings can be then tabulated. Finally, the estimates of the noise reduction from mitigation in the community areas can be applied and tabulation of noise impacts can be repeated for assessment and further devising the control measures.

The general rules of thumb to determine the effect of shielding from intervening rows of buildings can be: to assign −4.5 dB of shielding attenuation for the first row of intervening buildings only, and to assign −1.5 dB of shielding attenuation for each subsequent row, up to a maximum total attenuation of 10 dB. The noise level thresholds at which the project noise would cause moderate and severe impacts using the estimated existing noise exposure can be determined. All the noise-sensitive land use buildings and sites that lie between the impact contours and the project boundary can be thus tabulated. For residential buildings, an estimate of the number of dwelling units is satisfactory. Figure 6.6 shows the procedure for general noise impact assessment of mass transit projects. A case study of a transit train running at elevated corridor in Delhi city is considered for illustration of application of the noise impact assessment strategy. The overall at-grade noise of metro train measured at maximum speed at a distance of 5 m from track was 75 ± 2.3 dB(A) in at-grade and elevated corridor and at 15 m is 69 ± 4.7 dB(A). The L_{eq} (1 hr) due to metro train noise at a distance of 15 m can be also estimated using the following equations [50, 51]:

$$L_{eq}\ (1\mathrm{hr}) = SEL_{ref} + 10\ \log\ (V) + C_s\ \log\ (S/50) - 35.6 \tag{6.9}$$

where SEL_{ref} is the reference sound exposure level for source, V is hourly volume of vehicle per hour, S is average vehicle speed in mph, and C_s is the speed constant. C_s is recommended as 15 for diesel buses, 28 for electric buses, 21 for hybrid buses, and 30 for automobile and van pools. Additionally, the L_{eq} (1 hr) due to locomotive horns at a distance of 15 m can be calculated as:

$$L_{eq}\ (1\mathrm{hr}) = SEL_{ref,horn} + 10\ \log\ (V) - 35.6 \tag{6.10}$$

The daytime L_d and nighttime, L_n can be calculated as:

$$L_d = L_{eq}\ (1\ \mathrm{hr})\ \text{where}\ V = V_d \tag{6.11}$$

$$L_n = L_{eq}\ (1\ \mathrm{hr})\ \text{where}\ V = V_n \tag{6.12}$$

where V_d is the number of trains activity in day time and V_N is number of trains activity in night time, N_D is the average hourly number of events that occur during

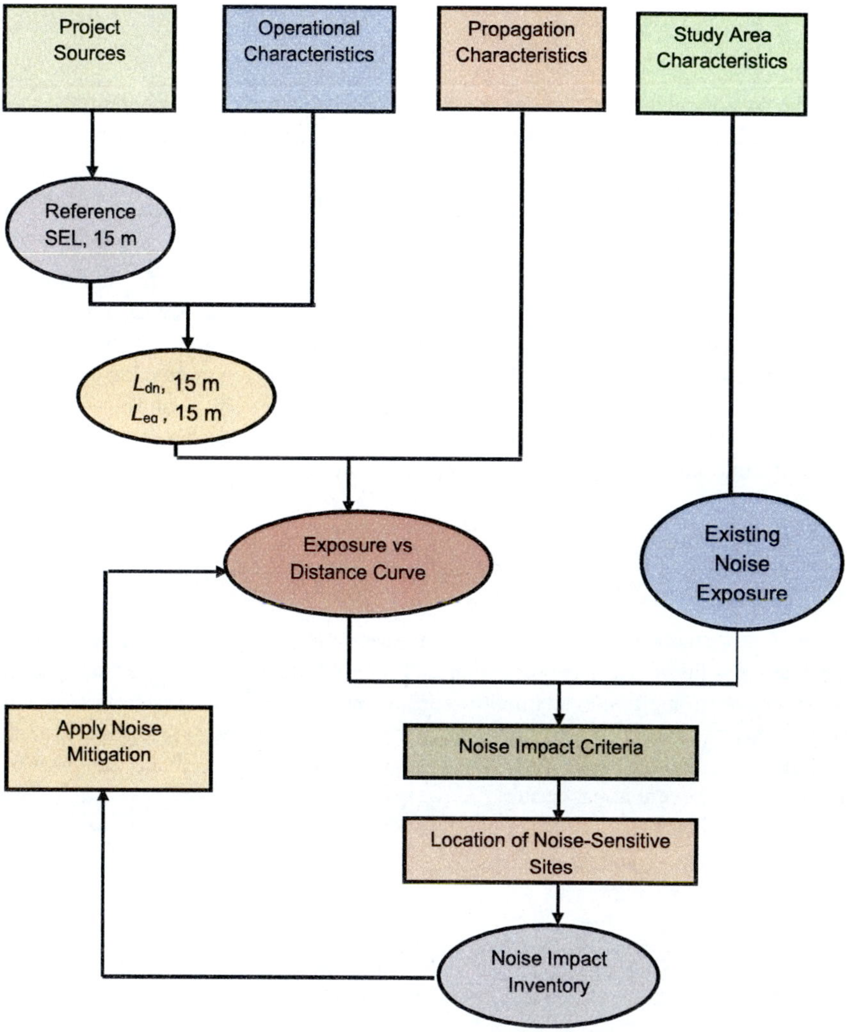

Fig. 6.6 Procedure for general noise impact assessment of mass transit projects. (Federal Transit Administration 2018, [50])

day time, and N_N is the average hourly number of events that occur during the night time.

$$V_d = \left[\left(\sum_{6a.m}^{10p.m} N_D \right) / 16 \right] \text{ and } V_n = \left[\left(\sum_{10p.m}^{16a.m} N_N \right) / 8 \right] \tag{6.13}$$

$$L_{eq,total\ (1hr)} = 10 \ \log \left(\sum\nolimits_{allsources} 10^{L_{eq}/10} \right) \tag{6.14}$$

The 24 h A-weighted L_{eq} at 15 m can be thus calculated as:

$$L_{eq} = 10\ \log\left(\left(\frac{1}{24}\right)\sum 10^{\left(L_{eq}(1hr)/10\right)}\right) \tag{6.15}$$

The 24-hour day-night average sound level (in Indian scenario, whereby the day levels are considered from 6.00 a.m. to 10.00 p.m.) at 15 m is calculated as:

$$L_{dn} = 10\ \log\left[16 \times 10^{\mathrm{L_d}/10} + 8 \times 10^{((\mathrm{L+n_{10}})/10)}\right] - 13.8 \tag{6.16}$$

where L_{dn} is day-night average sound level; L_{d} is average hourly L_{eq} noise level during the daytime (6 a.m. to 10 p.m.); L_{n} being the average hourly L_{eq} noise level during the night time (10 p.m. to 6 a.m.); and 13.8 is the adjustment for the number of hours in a day 10 × log(24). The barrier adjustment is recommended as −5 dB (A) for noise barrier. It may be noted there that the long-term sound level can be calculated either in $L_{\mathrm{Aeq,24h}}$ or L_{dn}. The distance correction factor recommended for 15 m for line source (fixed-guideway and highway) is:

$$C_{\mathrm{distance}} = -15\ \log\ (d/15) \tag{6.17}$$

$$L_{\mathrm{eq}}\ (\mathrm{h})\ \mathrm{receiver} = L_{\mathrm{eq}}\ (\mathrm{h}) - 15\ \log_{10}\ (d/15) \tag{6.18}$$

where L_{eq}(h) receiver is the source noise at receptor level at the reference distance of 15 m and d is the source-receptor distance. Table 6.7 shows the predicted increase in noise exposure due to elevated metro corridor at different distances from elevated metro corridor. Thus, using eqs. 6.19 and 6.20 for moderate and severe impact for an area having ambient level, L_{E} of 72.5 dB(A) can be used to evaluate the onset of moderate and severe impact criteria:

$$L_{\mathrm{p}} = 71.662 - 1.164\ L_{\mathrm{E}} + 0.018\ L_{\mathrm{E}}^2 - 4.088 \times 10^{-5}\ L_{\mathrm{E}}^3 \text{ for } 42 < \mathrm{L_E} < 71 \text{ } \textit{moderate} \text{ impact} \tag{6.19}$$

$$L_{\mathrm{p}} = 96.725 - 1.992\ L_{\mathrm{E}} + 3.02 \times 10^{-2} L_{\mathrm{E}}^2 - 1.043 \times 10^{-4}\ L_{\mathrm{E}}^3 \text{ for } \textit{severe} \text{ impact} \tag{6.20}$$

It is observed that the accentuated noise levels amount to be 1.6 dB(A) depending upon the ambient noise conditions at the site. Thus, it can be observed from the

Table 6.7 Predicted increase in noise exposure due to elevated metro corridor. (Garg et al. 2011 [51])

Distance from metro corridor up to (m)	Existing Noise L_{eq} dB (A)	Project L_{eq} dB (A)	Combined noise exposure dB(A)	Increase in noise exposure dB(A)	Project L_{eq} criteria, dB(A)		
					No impact	Moderate impact	Severe impact
15	72.5	69	74.1	1.6	<66	66–71	>71
30	69.5	64.5	70.7	1.2	<64	64–69	>69
60	67.6	60	68.3	0.7	<63	63–68	>68
100	63.1	56.6	64.0	0.9	<60	60–65	>65

Table 6.8 Onset of moderate and severe impact criteria

Existing Noise L_{eq} dB(A)	Project L_{eq} dB(A)	Onset of moderate impact	Onset of severe impact
72.5	69	66	71
		66–69 = −3	71–69 = 2
		Onset of moderate impact distance	**Onset of severe impact distance**
		24 m	11 m
	Erection of noise barrier (−5 dB(A))		
	64	66–64 = 2	71–64 = 7
		No impact	No impact

Table 6.7 considering the project L_{eq} criteria: the noise impact is moderate up to 30 m, while there is no impact for distances greater than 60 m for land-use categories 1 and 2.

The noise levels drop off approximately 3 dB per doubling of distance for line sources and approximately 6 dB per doubling of distance for point sources over. This reduction in noise will occur over a shorter distance in areas containing obstacles blocking the path of sound propagation. Table 6.8 shows the onset of moderate and severe impact criteria.

6.6.2 *Noise Impact Study of Transit Trains at Curvature*

Transit trains noise at curves is an area of continuous research and development as the induced noise and vibration not only brings discomfort to the passengers, but also to the residents in close proximity. So, it is imperative to have noise and vibration reduction either at the source side or at the path to minimize the impact to the local community. Curve tracks may cause squeal noise, excessive rail wear, and corrugation. Both top-of-rail squeal and flanging noise are associated with curves, particular the sharp curves ($R < 500$ m), whereas rolling noise is generally associated with tangent track [52]. Lubrication on tight curves is a simple and effective approach to control the friction as high coefficient of friction (COF) tends to cause squeal and corrugation [53]. In one such study reported by Garg et al. (2011) [54], the gauge face lubrication didn't have a significant effect on the A-weighted noise generated by the train transit system while passing through a curved track. Figure 6.7 shows the difference in spectrum of the sound pressure level monitored on the far way side for normal and lubricated track at a distance of 10 m from the track.

Too low COF or too high COF leads to wheels or rail troubles such as skid at braking, overrun at a station, wheel/rail wear, and corrugation, for example [53]. The measurements revealed increased wheel/rail interaction and squealing noise on lubricated track causing an increase in sound pressure level in range 160 Hz to

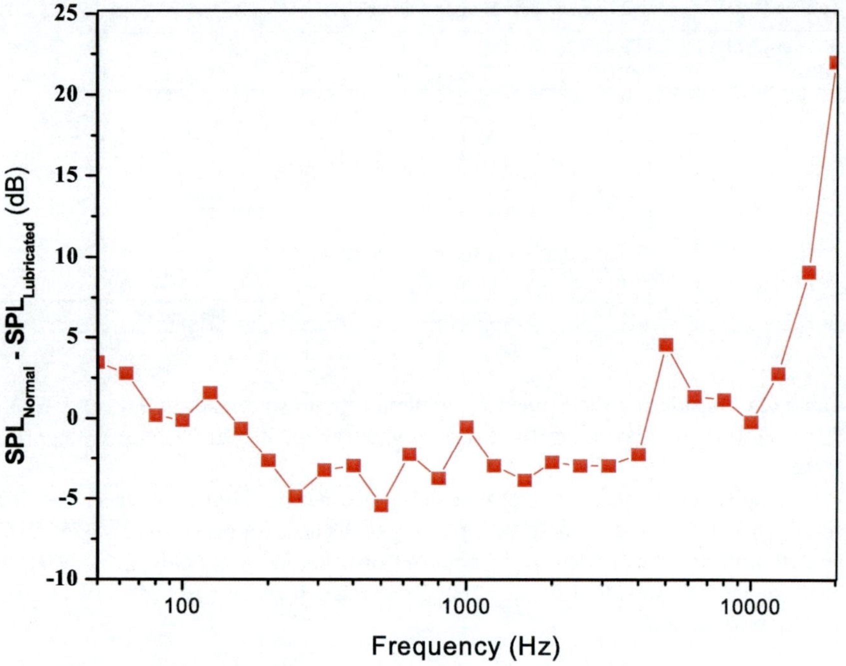

Fig. 6.7 Difference in sound pressure level monitored on normal and lubricated track on far way side. (Garg and Sharma 2010, [54])

4 kHz; the maximum increase is observed in the frequency range of 200 Hz to 800 Hz. However, when both the near side and far side measurements were taken into account, it was observed that the increase in sound pressure level is dominated in the frequency range of 250 Hz to 400 Hz. The negation of lubrication effect in controlling the overall A-weighted noise emission is attributed to an additional reaction force component, which accentuates the wheel/rail interaction comparatively more than the reduced high-frequency squeal and flanging noise for the lubricated track. This additional force component is generated due to longitudinal and spin creepage on the flange while encountering a curved track and is observed to cause an increase in noise emission at the frequency range of 250 Hz to 400 Hz, where the rail/wheel interaction is dominant; in both way side as well as interior noise levels. This enhanced rail/wheel interaction accentuated the sound pressure level in the bands where rail/wheel interaction dominates as is evident from Fig. 6.8 showing the average spectra for both dry and lubricated (wet) tracks. This study also concluded that parameters such as A-weighted noise level, *SEL*, or L_A are not sufficient to determine the effectiveness of noise reduction measurement due to lubrication. The noise spectrum can only reveal the true picture of any noise mitigation program at rail track curvature [54].

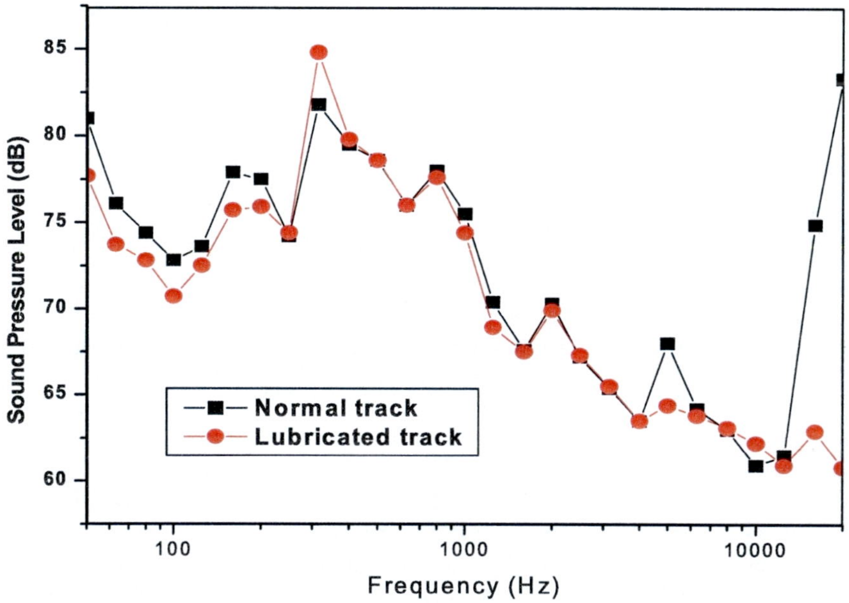

Fig. 6.8 Comparison of average sound pressure level monitored on way side. (Garg and Sharma 2010, [54])

6.6.3 *Noise Impact Study of High-Speed Trains*

High-speed trains can cause severe environmental noise impact in the residential areas. Aerodynamic noise is the dominant source at higher speeds and thus the noise impacts may extend out to long distances on either side of the tracks in otherwise quiet residential areas. At low speeds traction noise is predominant, rolling noise becomes dominant up to speeds around 200 km/h, then aerodynamic noise prevails (typically around 250 km/h or more). Pamanikabud and Paoprayoon (2003) developed a noise prediction model for elevated mass rapid transit (MRT) platforms as [55]:

$$L_{Aeq}\,(1\mathrm{h}) = 69.894 + 0.847\,Acc_n + 1.0222\,Dec_n + 0.429\,Acc_f + 0.661\,Dec_f, R^2 = 0.84 \quad (6.21)$$

where Acc_n is train acceleration rate while departing a station on near side, where Acc_f is train acceleration rate while departing a station on far side, Dec_n is train deceleration rate while entering a station on near side and Dec_f is train deceleration rate while departing a station on far side. Such a model can be used for noise prediction of the transit trains on elevated corridors in commercial lines. A study reported by Chen et al. (2007) [56] for evaluating the noise impact on residents due to high-speed maglev train uses L_{ASmax} to correlate it to L_{Aeq} under reference conditions based on Rathe's model [56]:

$$L_{Aeq} = L_{ASmax} + 10\ \log\left(\frac{len}{v}\right) - 10\ \log\ (2\alpha) + 10\ \log\ (V) - 32.3 \tag{6.22}$$

where L_{ASmax} is reference L_{ASmax} under reference conditions, *len* is reference source length, V is reference train speed, kph, α is $\tan^{-1}\left(\frac{len}{2y}\right)$ radians, y is reference observer distance from track centerline, and V is hourly volume of train traffic in trains per hour. For the German Magnetic Levitation Noise Standards, 59 dB (A) $L_{\text{Aeq,1h}}$ is the upper limit allowed near residential land for day time and 49 dB (A) $L_{\text{Aeq,1h}}$ for the night time. It is recommended that about 50 m should be used as the noise limit boundary for residential land when the maglev operates at 430 kph speed. The American Public Transit Association (APTA) guidelines for maximum airborne noise from train operations near general types of buildings recommended maximum pass-by sound levels of 75 dB(A) for high- density residential areas and 80 dB(A) for commercial and industrial areas for single-family dwellings. The guidelines recommended 80 dB(A) for multi-family dwellings and 85 dB(A) for commercial buildings for high density residential areas [57]. The Japanese Shinkansen maximum L_{ASmax} value recommended is 70 dB(A) for residential area and 75 dB(A) for nonresidential area. The relationship between percentage highly annoyed and L_{Aeq} is derived for Shanghai Maglev Train (SMT) in China as [56]:

$$\%\text{HA} = 7.13\ L_{\text{Aeq}} + 28.06, R^2 = 0.88 \tag{6.23}$$

$$\%\text{HA} = -108.31 + 1.94\ L_{\text{ASmax}}, R^2 = 0.94 \tag{6.24}$$

It was suggested that at least 240 m should be used as the noise limit boundary when SMT operates at 430 kph; however, there is 25% highly annoyed at 300 m away from the track where the train was operating at maximum speed suggesting a stricter limit boundary for SMT than the Japanese standard. It was observed that % HA increases by 2% when L_{ASmax} rises by 1 dB(A). The study concluded that noise nuisance factors based upon maximum speeds rather than average measures are more relevant measure for noise control policies for the high-speed trains.

6.7 Aircraft Noise Model

The European vision "Flightpath 2050" provides a roadmap and goals for the future of airborne transport systems. One of the major objectives is the reduction of the aircraft noise impact by 65% [58]. Thus, noise modeling, assessment, and control are the key features to be focused for aircraft noise across the globe especially for the busiest airports and community nearby. The Federal Aviation Administration (FAA) standard methodology for noise assessments has been the Integrated Noise Model

(INM) since year 1978. INM is a computer program used by over 1000 organizations in over 65 countries [59]. The guidance and underlying database/noise calculation methodology for INM is mentioned in European Civil Aviation Conference (ECAC) Doc 29 [60]. The different ways of approaching along-path integration divide noise modeling algorithms into the three different types: closest point of approach (CPA), Segmentation and Simulation. The CPA algorithm is the simplest and fastest. Principally, it divides a ground track into straight segments and circular arcs. The vertical flight profile is represented by contiguous straight segments. The distance from the observer to the closest point of approach and coordinate geometry facilitates the calculation of L_{max}. To allow for the curvature of the flight path and aircraft speed variations, empirical adjustments are made to the infinite segment L_{E}. Segmentation algorithms are developments of CPA algorithms which calculate the separate contributions to L_{E} from all noise-significant flight path segments. The level-time-history at any specific observer location is then developed by calculating the sound radiated towards it from each flight path point. The aircraft event level (L_{max} or L_{E}) is calculated in terms of components as shown in Fig. 6.9 as: aircrafts operation, ground tracks, vertical profile of height, speed and engine power, noise emission profile, ground roll, reverse thrust, free-air attenuation, lateral directivity, lateral attenuation, receiver factors, noise metrics, longitudinal directivity, and integration along the flight path [60].

The different approaches used for aircraft noise modeling [61] are: fully numerical, where the source and propagation are simulated in time-dependent computational fluid dynamics (CFD) and computational aeroacoustics (CAA) run; CFD methodology in conjunction with acoustic analogy; fully analytical wherein the acoustic field is derived based on databases containing the measured acoustic data, either from component wind-tunnel tests or from full-scale aircraft and for varying operational conditions. INM contains an acoustic database of noise versus power (or operational mode for helicopters) versus distance (NPD) values, augmented by a database of spectral characteristics. The noise levels in the NPD data (for all aircraft types) have been adjusted for time-varying aircraft speed atmospheric absorption, distance-duration effects (if the simplified adjustment process is used for exposure-based noise levels), and spherical divergence. Noise certification tests are specified by International Civil Aviation Organization (ICAO) and adopted by all aircraft manufacturing states [62]. Single event noise levels are determined at three reference points: approach, under the descent path 2000 m before landing threshold, lateral (or sideline), at the point where noise is greatest on a line 450 m to the side of the initial climb after lift-off, and flyover, under the departure climb path, 6500 m from start of roll (brakes are released and the aircraft starts its takeoff) as shown in Fig. 6.10.

The information services needed to perform noise mapping include: noise monitoring terminals located at appropriate locations around the airport, airport radars and interface facilities, identification of specific aircraft, for example, links to airport flight information systems, other information relevant to sound propagation such as meteorological data, and elevation and GIS-data. INM, FLULA, ANoPP, ANoPP2, SOPRANO, IESTA, SonAIR, PANAM, AzB, STAPES, Propnoise, Piano, and UPACS-LES are some of the aircraft noise modeling tools developed by various researchers and used in different countries (Table 6.9) [60–62]. Aircraft noise

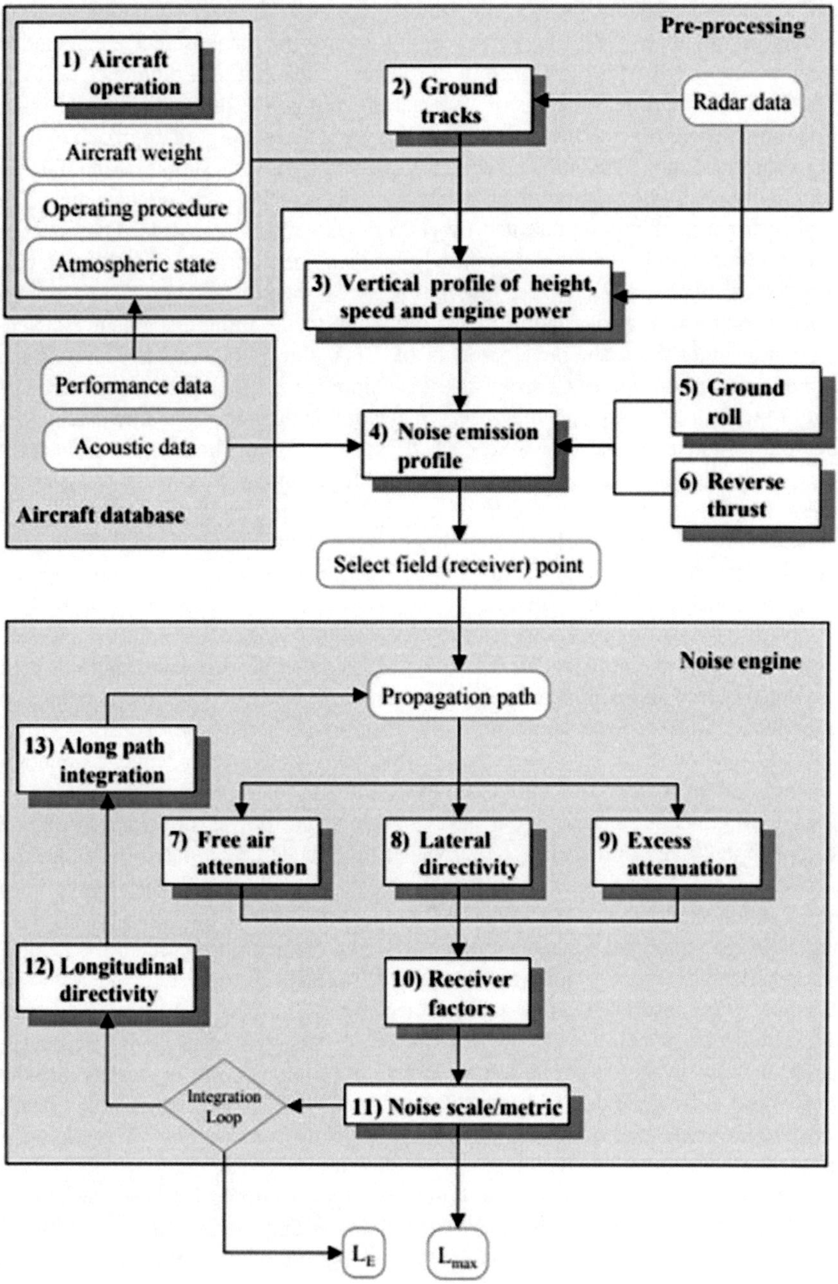

Fig. 6.9 Determination of aircraft noise event level showing the elements of a noise modeling system (ECAC Doc. 29, [60])

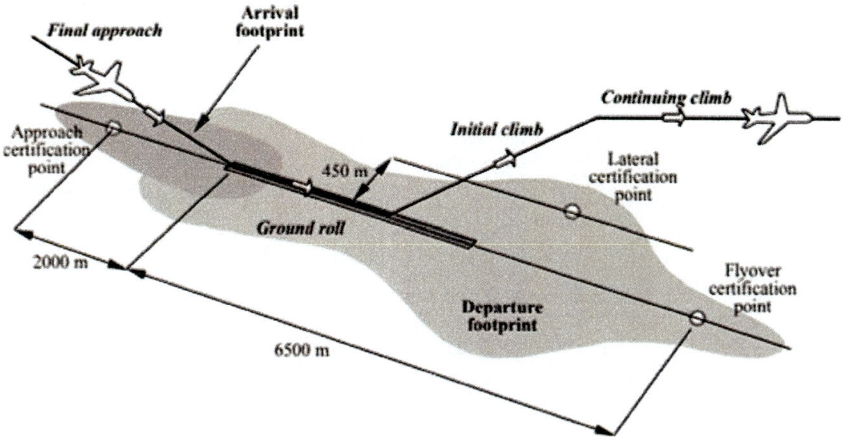

Fig. 6.10 Certification reference point for aircraft noise (ECAC Doc. 29, [60])

assessment and mitigation can be accomplished on different levels of complexity and noise action plans (NAP) such as introducing a new land-use management scheme as reported by Vogiatzis (2012) for Larnaka International Airport, Cyprus [63]. Thus some of the important measures as suggested in various studies [63] like restrictions of all new education uses, religious operations, theatres, cultural, and sensitive building constructions; allowing residential constructions with a prerequisite of implementing desired sound insulation for high acoustic comfort and strengthening the sound insulation of existing residences, introduction of public complaint management system and dissemination of environmental information to general public can be very instrumental for noise control.

Recently, a simplified model (Rapid Aviation Noise Evaluator (RANE)) had been reported for the calculation of aviation noise that relies upon the concept of noise cylinders around defined flight-tracks with the noise radius determined from publicly available noise-power-distance curves. However, the current version is limited to single run ways and the approach based on noise cylinders requires further development for the assessment of more complex situations and multiple flight paths [64].

6.7.1 *Aircraft Noise Contours*

Noise contour is a line on a map that represents equal levels of noise exposure. Different models and tools have been developed across the globe for developing noise contours and maps for aircraft noise. The numbers of annual flights, aircraft types, take-off and landing procedure of various aircraft types and flight tracks, coordinates and elevation of runways ends, weather, and terrain data are critical inputs to the contour model. Furthermore, some of these models like INM contain the noise-power-distance (NPD) tables for different aircraft types that relate the

Table 6.9 Aircraft noise modeling in ECAC Member States. (ECAC.CEAC Doc 29, [60])

Country	Noise model	Noise index	Noise limits
France	Integrated noise model	$L_{den,}$ the daytime is 06 h00–18 h00, the evening-time is 18 h00–22 h00 and the nighttime is 22 h00–06 h00	The noise exposure plan defines four land-use planning zones (A, B, C, D) around the aerodrome delineated by L_{den} noise contours The upper limit is not defined but lower limit for A zone is 70 dB(A) L_{den}
Germany	AzB noise model	A-weighted equivalent sound pressure level $L_{pAeq,day}$ for the night period: $L_{pAeq,night}$ and a "number above threshold (NAT)" contour is used	Maximum limit is 65 dB (A) for day time and 55 dB for nighttime and 6 × 57 dB $NAT_{indoors}$
The Netherlands	L_{Amax} noise-power-distance relationship (NPD) tables	L_{den} and L_{night} noise indices are used which are based on the sound exposure level (SEL)	L_{den} and L_{night} contours calculated over a rectangular grid at 500 m spacing
Norway	Norwegian model for aircraft noise calculation, NORTIM	Equivalent aircraft noise (EFN)	Four noise zones defined, IVth zones: EFN > 70 dB or EFN > 105 dB during daytime (07–22) or EFN > 100 dB during nighttime (22–07)
Switzerland	Swiss noise modeling system FLULA2	A weighted L_{eq}	The typical noise limit is 60 dB(A) (Immission limit for residential areas). Noise limits for the one-hour L_{eq} at night are typically 10 dB lower than day limits
United Kingdom	ANCON 2 noise modeling system	Daytime (0700–2300 hrs) and nighttime noise (2300–0700 hrs)	L_{eq} values are calculated over a rectangular grid at spacings between 50 and 500 m over an area large enough to contain the lowest contour level of interest

distance between aircraft and observer to L_{max} of noise levels for a range of thrust settings. The radar tracks for the different flight routes or from the Aeronautical Information Publication (AIP) can be used for precise noise computations. In Indian scenario, the ambient air quality standards with respect to noise in airport noise zone specify the daytime limit to be 70 dB (A) and nighttime to be 65 dB(A) for busy airports having more than 50,000 aircraft movements per year. Noise mapping for all airports should be carried out as per the requirements specified in the Director General Civil Aviation's requirements by the airport operators. The noise contour areas, determination of the number of people exposed to particular noise levels are the final conclusions that can be derived from the noise mapping. Thus, the noise

Table 6.10 Airport noise contours analysis: an illustration

L_{eq} in dB(A)	Area (in km^2)	Population	Household
>50	106	15,000	6000
>53	55	7000	3000
>56	29	2500	1000
>59	17	1100	500
>62	8	500	110
>65	4	200	50
>68	3	45	14
>71	1.5	20	6
>74	0.6	3	1

levels due to multiple movements can be studied using the different noise metrics like L_{Aeq}, L_{den}, L_{Amax}, L_{day}, $L_{evening}$, and L_{night} and also the number of flights with noise levels higher than the guidelines can be analyzed. Table 6.10 shows a typical example of detailed analysis of noise contours available through noise mapping as an illustration. According to the European legislations, the map shall be updated every 5 years. In special cases, the competent authority may issue an order to update the existing maps. The map shall include noise levels down to 55 L_{den} and night noise levels down to 50 L_{night}. Strategic noise mapping and action plans are vital tools for defining the main strategies to reduce noise exposure of residents and introduce and preserve "quite zones." A recent study presented by Vogiatzis and Remy (2019) explicitly presented the benefits associated with strategic noise mapping for noise reduction for an international airport in Greece [65].

The usual practice in aircraft noise assessment is to compare the *before* and *after* situations, the numbers of people resident between noise exposure contours depicting low, moderate, and high levels of exposure and thus analyze the changes in the extent of annoyance in the community. Figure 6.11 shows the illustrative aircraft noise contours for a hypothetical two-runway airport [60].

Figure 6.12 shows the typical illustration of an airport noise map developed for Ankara Esenboğa Airport, Turkey, showing the noise contours for L_{den} and L_{night} noise levels. The maps of cumulative areas exposed to 55, 65, 75 dB(A) during both 24 h and nighttime noise indicators for the year 2012 were shown describing the specific areas and population exposed to various noise levels. For instance, the calculated population exposure for Ankara Esenboğa Airport, Turkey, was observed to be about 0.014% (200 out of 1,433,961) of the population exposed to 65 dB(A) or higher noise levels. Thus, such an analysis can also be helpful for predicting future noise scenario based on capacity expansion of the airport and devising suitable noise action plans for reducing noise impact [66].

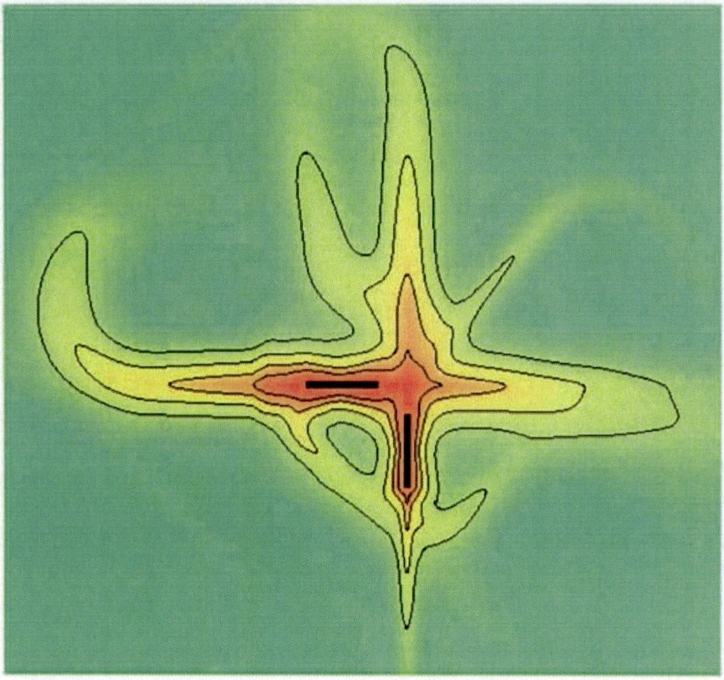

Fig. 6.11 Illustrative aircraft noise contours for a hypothetical two-runway airport (ECAC.CEAC Doc 29 2016, [60])

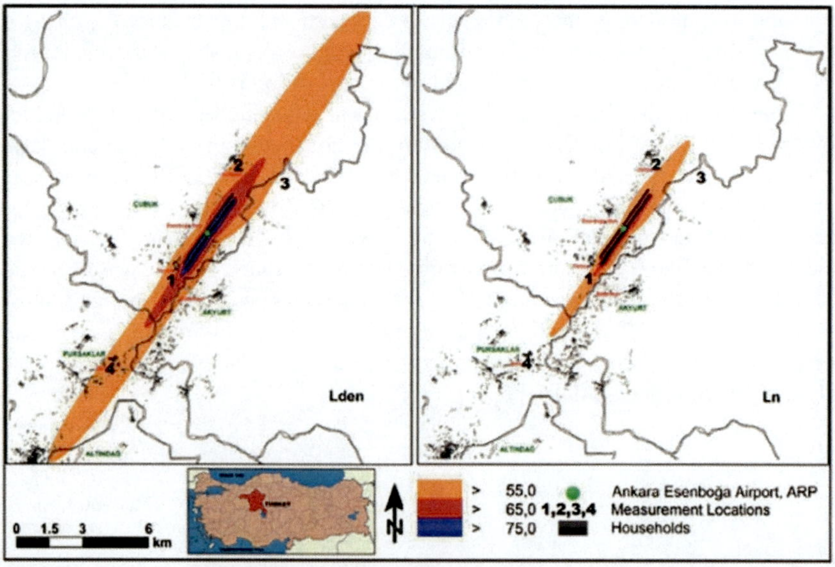

Fig. 6.12 Airport noise map developed for Ankara Esenboğa Airport, Turkey (Ozkurt N 2014 [66]) showing areas exposed to 55, 65, 75 dB(A) for L_{den} and L_{n} descriptors

6.8 Acoustic-Induced Vibration

Structural vibrations result from both air borne and ground borne excitation. The induced vibration in buildings results from various external sources, viz., traffic, blasts, construction activities, sonic boom, low-frequency noise from aircrafts fly-over, impulsive impacts, or human activities. The acoustic waves exert fluctuating forces over the building elements, causing them to vibrate which may be enhanced by resonances in case the frequency of sound waves interacting lie within the domains of natural frequency of structure. Thus, studies on acoustic-induced vibration in structures shall be helpful in ascertaining the structural integrity when subjected to high acoustic loads [67].

6.8.1 Diffuse Field Studies

An investigation had been reported earlier in a diffuse field conditions to measure the vibrations levels generated due to diffuse sound field excitation by Garg and Maji (2013). The measurements had been reported in Reverberation Chambers at CSIR-National Physical Laboratory, New Delhi, India, of dimensions 6 m × 6.5 m × 7 m. A pink noise generated the acoustic excitation through a dodecahedral loudspeaker system coupled with an amplifier. The sound field was generated in a varying range from 50 dB to 120 dB and the corresponding vibration levels on floor and walls of the chambers were measured. The magnitude of vibration level was measured at different points on the floor and walls of reverberation chamber. The standard deviation of magnitude was observed to be ±2.88 mm/s^2 for walls and ± 3.67 mm/s^2 for floor vibration. The linear relationship of vibration levels induced due to acoustic excitation was observed, whereby the acceleration response increased generally as the sound pressure levels increased and followed a straight line relationship based on the assumption of linear behavior of the structure. The induced vibration of the floor ($L_{a\ (floor)\ \mathrm{rms}}$) in frequency domain (Fig. 6.13) generated due to the random acoustic excitation was also correlated by a simple regression fit as [67]:

$$L_{\mathrm{a\ (floor)}} = L_{\mathrm{p}} - 10\ \log\ (f) - 10\ \mathrm{dB}; f \geq LLF \qquad R^2 = 0.74 \tag{6.25}$$

where L_{a} is 20log (a/a_{ref}) wherein a_{ref} is 10^{-5} m/s^2. Thus the above empirical relation can be used to predict the floor vibration induced to acoustic excitation in a diffuse field at different frequencies. The reverberation chamber has a lower limiting frequency (*LLF*) for good diffusion of 125 Hz. The relation between random and the lower limiting frequency was observed to be [67]:

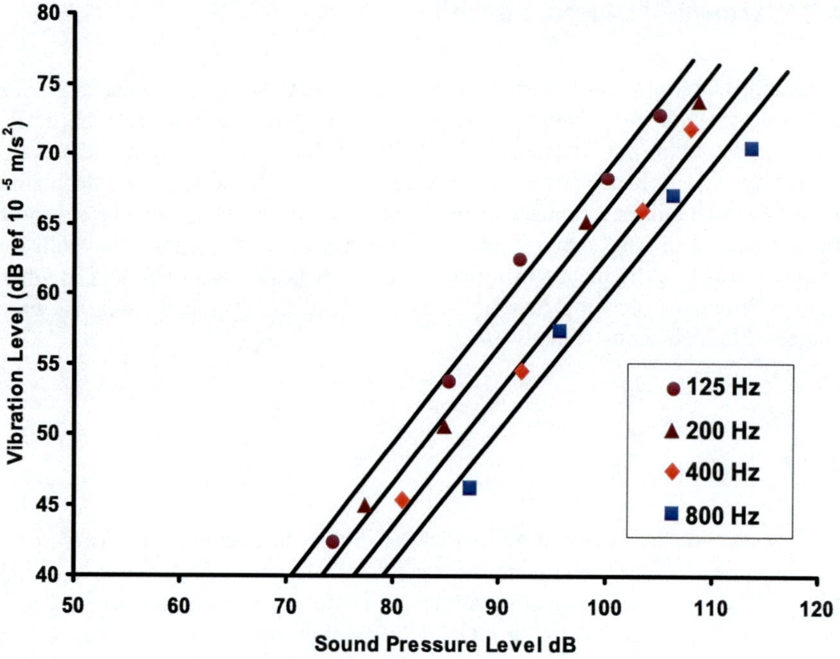

Fig. 6.13 Induced floor vibration due to acoustic excitation (L_{in}) of filtered band at different center frequency (Garg and Maji 2013, [67])

$$L_{\mathrm{a\ (floor)}}\ (random) = L_{\mathrm{a\ (floor)}}\ (LLF) + 3\ \mathrm{dB} \tag{6.26}$$

The diffuse field conditions are however difficult to achieve in a normal build up areas and thus there may exist deviations from the results predicted due to empirical formulation. The cross-over frequencyf_s can also be empirically calculated as [68]:

$$f_s = 2000\sqrt{\frac{T_{60}}{V}} \tag{6.27}$$

where T_{60} is the reverberation time of the room in seconds and V is the Volume of the room in m^3. In the normal rooms, the frequency f of the eq. (6.25) becomes equivalent to f_s of eq. (6.27). For sound pressure level of 1 Pa (or 94 dB), the eq. (6.25) reduces to [67]:

$$L_{a\ (floor)} = 84 - 10\ \log\ (f)\ \mathrm{dB} \tag{6.28}$$

The overall response of the floor was observed as 0.0002 g/Pa (20.4 mm/s^2/Pa) and for the wall as 0.0015 g/Pa (13.2 mm/s^2/Pa).

6.8.2 Field Studies

In most of the practical situations, the interaction of sound waves with structure in free-field conditions is quite cumbersome to model as various uncertainties are involved in acoustic-elastic coupling. Table 6.11 shows an illustration of the sound and vibration levels registered during the various incidents/physical phenomena measured experimentally [69–71].

The noise level within a building element due to structural vibration is dependent on vertical floor vibration as proposed by Kurzweil [72]. The floor vibration level L_a (room) can be estimated from [72]:

$$L_a\ (\text{room}) = L_a\ (\text{tunnel}) - C_g - C_{gb} - C_b\ \left(\text{dB re } 10^{-6}\text{ g (rms)}\right) \tag{6.29}$$

where L_a (tunnel) is octave band acceleration on the wall of a subway tunnel during a train pass by, C_g is vibration attenuation due to propagation throughout the ground, C_{gb} is coupling loss between ground and building, and C_b is vibration attenuation due to propagation in the building. Ground-borne noise is a low-frequency rumbling sound inside buildings, caused by vibrations of floors, walls, and ceilings. The sound pressure level in the room L_p (room) can be thus determined in terms of octave band center frequency f as [72]:

$$L_p\ (\text{room}) = L_a\ (\text{room}) - 20\ \log_{10} f + 37\ \text{dB} \tag{6.30}$$

Kurzweil's equation simplifies to $L_p = L_v - 27$ dB, where L_v is floor vertical velocity level in dB re 1 nanometer/second and L_p is the sound pressure level in the room in dB re 20 μPa [72, 73]. In some cases, it is measured as L_{pASmax}, in decibels, where L_p denotes the sound pressure level, A denotes the A-weighting, that represents the sensitivity of the human ear, *S* denotes slow time response over 1 s and max denotes the maximum value during a single train pass-by. However, for ground-borne sound, there are no relevant national or international standards setting criteria for the acceptable levels. There is an overlap in which an average human is more sensitive to audible noise than to perceptible vibrations. Vibrations in the magnitude of 0.01–0.03 mm/s (velocity levels of 80–90 dB) lie just below our perception level. Vibrations are noticeable around 0.3–1 mm/s (velocity levels of 110–120 dB) for most humans. Very strong noticeable vibrations for most humans have a level of about 1–10 mm/s (velocity levels of 120–140 dB). Below 1 Hz, humans are sensitive to vibrations because this can cause seasickness [74]. The analysis of measurements conducted for train induced ground vibration for a specific site revealed that the average ground vibration at 4 m distance from the center line was 130 mm/s^2 in acceleration mode and 336 ± 73 μm/s for passage of trains at a speed of 80 km/h. Thus, it can be inferred that there exists a safe margin of 15 dB for damage to occur,

Table 6.11 Sound and vibration levels registered during various incidents/phenomena

Location/area/ incident/ phenomena	Sound level generated in dB(A)	Vibration level generated in the structure	Remarks
Ceremonial gun firing	125 dB(A)	6×10^{-3} g (rms)	The floor modes of 25 Hz, 42 Hz, 69 Hz, and 100 Hz get amplified during the excitation
Overflying aircrafts landing and take-off over an ancient monument	120 dB(A)	3×10^{-3} g (rms)	Major resonant modes excited in structure lying in frequency range 10 Hz to 100 Hz
Blasting, explosions, and for sonic booms	120–130 dB (A)	0.3 and 3.0 g	Damage to houses may occur at peak acceleration values between about 0.3 and 3.0 g in the frequency range of 10 to 100 Hz, respectively
An historical monument site during sound and light show	67.3 dB(A)	The floor vibration was 9.1 ± 2.0 μm/s and 2.74 mm/s^2 during the show.	The residual floor vibration in velocity mode was 1.7 ± 2.0 μm/s and in acceleration mode at site was 350 μm/s^2
Floor vibration levels in metro train while running	–	At grade	54–171 mm/s^2
		Elevated	5.1–26.8 mm/s; 29–198 mm/s^2
		Underground	100–1000 mm/s^2
Train at grade with speed of 100 km/h	94.3 dB at 1 m	153 mm/s^2 and 392 μm/s at 4 m	Average ground vibration at 4 m from the center line was measured as 130 mm/s^2 and 336 ± 73 μm/s for passage of trains at a speed of 80 km/h

if a peak particle velocity (ppv) of 3 mm/s is considered to be onset of damage. Similarly, the passage of metro trains in an underground station at a speed of 50 km/h registered a vibration level of 20 mm/s^2 and an rms velocity of 47.8 μm/s at a distance of 5 m, which presented a larger safe margin for damage to occur [75]. The German standard, DIN 4150–3:1999 [76] includes guidelines for residential buildings together with criteria for both commercial/ industrial buildings and high sensitivity structures.

Vibration levels required to produce damage in buildings are generally much higher than those would be tolerated by humans. The threshold of architectural damage to building occurs at peak particle velocity of 5 mm/s. For old historical buildings or buildings sensitive to vibration the recommendation is 3–8 mm/s peak particle velocity for the frequency range of 10–50 Hz and 3 mm/s for frequency below 10 Hz as described in DIN 4150 standards (Table 6.12). Rudder (1978) [77]

Table 6.12 Guideline values of vibration velocity for evaluating the effects of short-term vibration, DIN4150-3:1999 [76]

Type of structure	Guideline values for vibration velocity in mm/s			
	Foundation frequency			Plane of floor of uppermost story
	1–10 Hz	10–50 Hz	50–100 Hz	
Buildings used for commercial purposes, industrial buildings and buildings of similar design	20	20–40	40–50	40
Dwellings and buildings of similar design and/or use	5	5–15	15–20	15
Structures that, because of their sensitivity to vibration, do not correspond to those listed in lines 1 and 2 and are of great intrinsic value (e.g., buildings that are under a preservation order)	3	3–8	8–10	8

Table 6.13 Vibration limits relating to minor or cosmetic damage to buildings from piling operations (from BS 5228: Part 4: 1992)

Building classification	Intermittent vibration (ppv, mm/s)	Continuous vibration (ppv, mm/s)
Residential in generally good repair	10	5
Residential where preliminary survey reveals significant defects	5	2.5
Industrial/commercial-light and flexible structure	20	15
Industrial/commercial-heavy and stiff structure	30	15

suggested the threshold level for building damage of 2.5 mm/s corresponding to −22 dB re 1 g at 50 Hz. It was also proposed that additional safety factor that the maximum permitted acceleration to be −35 dB re 1 g or 0.18 m/s^2 (0.02 g). A precise guidance on acceptable vibration levels in structures is provided in BS 5228: Part 4: 1992 [78] Code of practice for noise and vibration control applicable to piling operations as shown in Table 6.13. This standard recommends that conservative threshold for minor or cosmetic damage should be taken as peak particle velocity of 10 mm/s for intermittent vibration and 5 mm/s for continuous vibrations to determine whether there is any risk of building damage, particularly from construction works involving piling. This criterion applies across all the frequency bands. The transient vibration guide values for cosmetic damage recommended in BS 7385: Part 2:1993 [79] recommends a peak particle velocity in frequency range of 4–15 Hz as 50 mm/s for reinforced or framed structures, industrial, and heavy commercial buildings. The peak component particle velocity for un-reinforced or light framed structures, residential or light commercial type buildings in frequency range of 4–15 Hz is recommended between 15 and 20 mm/s increasing to 50 mm/s at 40 Hz and above.

In practical situations, as the diffuse field conditions don't prevail, the interaction of sound waves with structural modes gets reduced resulting in weak coupling and thus lower vibration levels are registered. However, eq. (6.25) quantifies the maximum vibration levels induced in a diffuse field setup for adjudging the severity of vibration levels induced due to high acoustic loads and correlating with structural integrity as recommended by DIN 4150 and BS 7385 standards. Vibration levels in one-third octave bands are acceptable if they are less than the following: 0.48 mm/s at 1 Hz, reducing to 0.18 mm/s at 2 Hz, then falling to 0.1 mm/s at 8 Hz and above to 80 Hz [80].

6.9 Construction Noise Impact Assessment

Construction activities generate noise levels that have the potential impact upon nearby noise-sensitive receptors especially in residential zones. There is currently no statutory guidance pertaining to the maximum permissible noise level for a project's construction phase. BS 5228 standard [81] recommends a method called "ABC method" for setting construction noise limits for residential dwellings, but it does not provide guidance for commercial or office buildings. The method designates noise sensitive locations into a specific category (A, B, or C) based on existing ambient noise levels in the absence of construction noise. A significant effect is deemed to occur if the total L_{Aeq} noise level, including construction noise, exceeds the threshold level for the category as summarized in Table 6.14. This strategy applies to residen-

Table 6.14 BS5228 [81] construction noise assessment (ABC Method)

	Threshold values, in decibels (dB)		
Assessment category and threshold value period (L_{Aeq})	Category A	Category B	Category C
Nighttime (23:00–07:00)	45	50	55
Evenings and weekends	55	60	65
Daytime (07:00–19:00) and Saturdays (07:00–13:00)	65	70	75

Note A: Category A: threshold values to use when ambient noise levels (when rounded to the nearest 5 dB) are less than these values

Note B: Category B: threshold values to use when ambient noise levels (when rounded to the nearest 5 dB) are the same as category A values

Note C: Category C: threshold values to use when ambient noise levels (when rounded to the nearest 5 dB) are higher than category A values

Note D: 19:00–23:00 weekdays, 13:00–23:00 Saturday, and 07:00–23:00 Sundays tial receptors only. The residential receptor is accordingly categorized based on the rounded ambient noise levels less than, equal to, or higher than the values in Category A, respectively. For instance, for a site, where the rounded daytime ambient noise level is below 65 dB(A), then it is a Category A receptor and the threshold for potential significance is 65 dB(A). If the rounded daytime ambient noise level is equal to 65 dB(A), then the receptor is assigned to Category B and the threshold level is 70 dB(A).

BS 5228 also suggests another method, called as 5 dB(A) change method. This states that "Noise levels generated by site activities are deemed to be potentially significant if the total noise (pre-construction ambient plus site noise) exceeds the pre-construction ambient noise by 5 dB $L_{Aeq,T}$ or more, subject to lower cut-off values of 65 dB, 55 dB and 45 dB $L_{Aeq,T}$ from site noise alone, for the day-time, evening and night-time periods, respectively; and a duration of one month or more, unless works of a shorter duration are likely to result in significant effect." These assessment criteria can be judiciously applied until there are any harmonized specific criteria developed for construction noise impact assessment.

6.10 Noise Impact Assessment of Industrial and/or Commercial Nature

The significance of sound of industrial nature at noise sensitive residential receptors can be ascertained by BS 4142:2014 standard [82] that primarily provides a numerical method for evaluation of such impact. The subjective response of sound is affected by various factors apart from sound levels. The significance of sound of an industrial and/or commercial nature depends upon both the margin by which the rating level of the specific sound source exceeds the background sound level. Typically, the greater this difference, the greater the magnitude of the impact. A difference of around +10 dB or more is likely to be an indication of a significant adverse impact, while a difference of around +5 dB indicates an adverse impact. Table 6.15 shows an illustration of noise impact assessment study.

Table 6.15 shows a case study for an industrial plant generating continuous sound levels whereby the subjective assessment identified a mild to prominent tone. The ambient noise levels, residual noise levels, and background levels were measured from which the significance of impact (low, adverse, or severe) was evaluated. It may be noted that sound with prominent impulses are more annoying than continuous types with the same equivalent sound pressure level. Also, adverse impact will always not be a cause of complaint and not every complaint is an evidence of an adverse impact [82].

Table 6.15 Case study of noise impact assessment as per BS 4142:2014 [82]

Result		Relevant Clause of BS 4142	Comments
Measured ambient sound level	L_{Aeq}(15 min) = 54 dB	7.1	
Residual sound level	L_{Aeq}(15 min) = 40 dB	Clause 7.3.3, $L_s = 10\lg(10^{La/10} - 10^{Lr/10})$; L_s is the specific sound level; L_a is the ambient sound level; and L_r is the residual sound level.	
Background sound level (daytime)	L_{A90}(15 min) = 37 dB	8.1, 8.1.3, 8.1.4, 8.3	
Specific sound level	L_{Aeq}(15 min) = 54 dB	7.3.15	
Acoustic feature correction	+4 dB	Clause 9.2; A penalty of 2 dB for a tone which is just perceptible, 4 dB where it is clearly perceptible and 6 dB where it is highly perceptible. A penalty of 3 dB for impulsivity which is just perceptible, 6 dB for clearly perceptible and 9 dB for highly perceptible. A penalty of 3 dB can be applied where tonal/impulsive characteristics is not there, though otherwise is readily distinctive.	
Rating level	(54 + 4) dB = 58 dB	9.3.4	
Background sound level	L_{A90}(15 min) = 37 dB	8.1, 8.1.3, 8.1.4, 8.3	
Excess of rating over background sound level	(58–37) dB = 21 dB	11	
Assessment of noise impact	Adverse	11	The excess of the rating level over the background sound level is huge and the uncertainty doesn't have any influence.
Uncertainty of assessment		10	

6.11 Noise Mitigation

Mitigation is one of the important aspects of noise impact assessment apart from baseline studies, noise prediction, assessment, and reporting. The types of mitigation which might be employed may be classified as: avoidance; reduction; and compensation. Avoidance can also be achieved by: controlling the hours of operation; limiting the duration of operation; limiting the number of events; or limiting the number of different sources operating concurrently. Reduction includes various measures like erection of barriers, enclosures, etc. Compensation may include measures applied outside of the development area, such as fitting double/secondary glazing to the affected premises. In some cases, financial compensation is also offered as a provision of alternative to the nearby residents. The noise impact assessment should be reviewed by the appropriate authority and the local planning authority should ensure that all mitigation measures are executed to reduce the

Table 6.17 Classification of short-term and long-term noise impact. (Design Manual for Roads and Bridges 2011 [84])

Long-term impact classification	Short-term impact classification	Sound level change (in dB) L_{pAeqT} T = either 16 hr day or 8 hr night
Negligible	Negligible	≥0 dB and < 1 dB
	Minor	≥1 dB and < 3 dB
Minor	Moderate	≥3 dB and < 5 dB
Moderate	Major	≥5 dB and < 10 dB
Major		≥10 dB

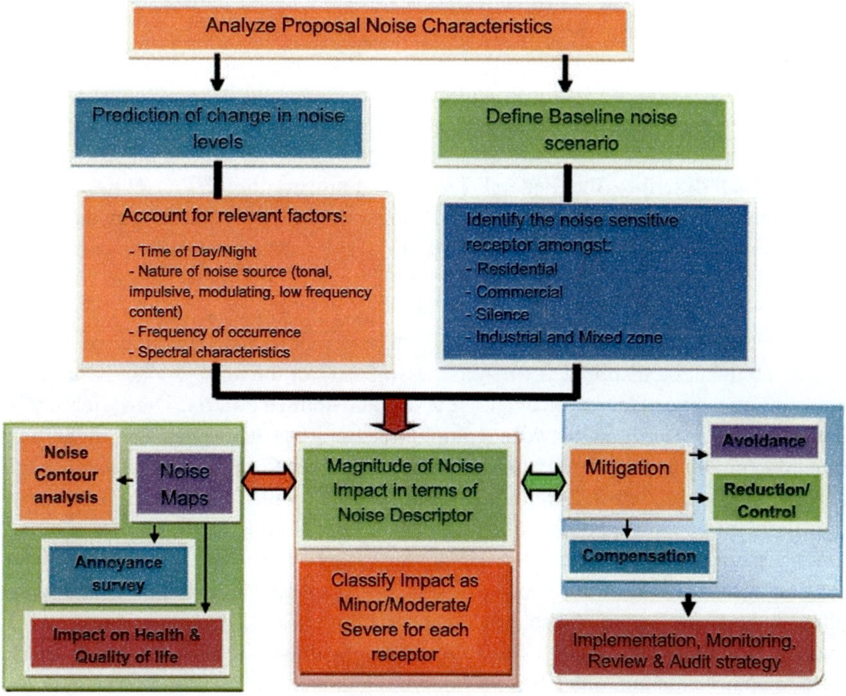

Fig. 6.14 Summary of noise impact assessment and mitigation

significant adverse effects. Also, consideration should be given for the need for any additional mitigation or monitoring measures and the lowest observed adverse effect level (*LOAEL*) and the significant observed adverse effect level (*SOAEL*) that would occur due to a project [83]. Table 6.17 describes the classification of short-term and long-term noise impact [84]. Thus, identification of a significant effect will depend on: the magnitude of the impact (classification); the number of dwellings that experience the impact and grouping of the dwellings based on the impact [85].

Figure 6.14 summarizes the methodology of noise impact assessment and mitigation with some modifications to that proposed in report [83]. It is imperative in recent times that apart from annoyance induced due to noise, adverse effects on health and quality of life should be considered for impact assessment and finally mitigation. Noise mapping can prove to be indispensable in such perspectives to understand the *before* and *after* noise scenario, number of people annoyed, and initiation and execution of noise action plans. Monitoring, review, and audit strategy can be followed for bringing synergy in the process and also for surveillance and controlling the progress.

6.12 Conclusions

The chapter presents the methodology of conducting noise impact assessment studies due to traffic, train transit, and aircraft systems in urban agglomerations. A comparison of principal traffic noise models on the basis of technical attributes including source modeling and sound propagation algorithms is presented. The chapter also describes the Federal Transit Administration (FTA) Noise Impact Assessment Criteria. The various aircraft noise models developed in recent times are discussed. Also, the acoustic induced vibration in structures and its correlation with maximum vibration level generated in diffuse field conditions is discussed. The I-INCE report (2011) recommended that the process of the noise impact assessment should aim to achieve the best negotiated outcome for all those involved and should not be a bureaucratic approach where the assessment of the noise impact is simply a comparison between the predicted noise levels and defined criteria. The variability in the reactions within and between communities is vital to be acknowledged and flexibility in the application of such criteria be considered in the impact assessment [3]. It is evident that a harmonized methodology with a simple approach for development of a traffic noise model shall be more suitable for town planners and urban authorities for traffic noise predictions, environmental impact assessment studies, identification of hot spots, assessment of percentage annoyance, and planning and execution of noise abatement measures in Indian perspectives. A national initiative involving various research institutes, experts, town planners, traffic departments and other stake holders such as that followed in other countries (e.g. ASJ RTN-Model 2018) towards developing a validated noise model integrated with GIS interface can be very helpful for conducting noise impact assessment and control especially for the proposed *Smart Cities* and high speed rail corridor in India. Noise contours may prove to be beneficial in showing the spread of predicted noise levels in the area. The practical, socioeconomic, and other implications of such impact assessment should be precisely considered prior and shared in a form suitable for public consultation, possibly by way of displays or other easily accessible information. Thus, it is imperative that all the future planning, policies, legal framework, and action plans should aim to reduce and control the environmental noise and mandatorily conduct noise impact assessment studies for the new establishments and projects such as operation of high-speed Bullet trains in future in India in order to avoid the adverse impacts on human health and quality of life.

References

1. Sahzabi HY (2004) Application of GIS in the environmental impact assessment of sabalan geothermal field, NW-IRAN. United Nations University, Geothermal Training Programme, Iceland, Report 19
2. Technical Guidelines for Land Traffic Noise Impact Assessment (2016) National Environmental Agency (NEA), Singapore, July, 2016. https://www.nea.gov.sg/docs/default-source/our-services/technical-guidelines-for-noise-impact-assessment-.pdf. Accessed on 20.02.2021
3. I-INCE Publication No. 11-1 (2011) Guidelines for community noise impact assessment and mitigation, Final report of I-INCE technical study group on Community noise: Environmental Noise Impact Assessment and Mitigation (TSG 6), March, 2011
4. Garg N, Maji S (2014) A critical review of principal traffic noise models: strategies and implications. Environ Impact Assessment Rev 46:68–81
5. Sakamoto S (2020) Road traffic noise prediction model "ASJ RTN-model 2018": report of the research committee on road traffic noise. Acoust Sci Tech 41(3):529–589
6. Heutschi K, Locher B, Gerber M (2018) sonROAD18: Swiss implementation of the CNOSSOS-EU road traffic noise emission model. Acta Acust United Acust 104:697–706
7. Federal Highway Administration (2019) Technical Manual: Traffic Noise Model 3.0. FHWA-HEP-20-012. U.S Department of Transportation, December, 2019
8. Department of Transport (1998) Calculation of Road Traffic Noise. Department of Transport, Welsh Office, HMSO, U.K.
9. Steele C (2001) A critical review of some traffic noise prediction models. Appl Acoust 62(3): 271–287
10. Watts G (2005) Harmonoise prediction model for road traffic noise, PPR 034. 2005
11. Jonasson HG (2007) Acoustical source modeling of road vehicles. Acta Acust United Acust 93: 173–184
12. Defrance J, Salomons E, Noordhoek I, Heimann D, Plovsing B, Watts G, Jonasson H, Zhang X, Premat E, Schmich I, Aballea F, Baulac M, de Roo F (2007) Outdoor sound propagation reference model developed in the European Harmonoise project. Acta Acust United Acust 93: 213–227
13. Kragh J, Svein Å, Jonasson HG (2002) Nordic environmental noise prediction methods. Nord 2000, Summary report, DELTA, Denmark
14. Kragh J, Jessen BB (2001) Noise emission data for vehicles driving at constant speed in road traffic measured in 1999–2000 by Delta. Delta Acoustics and Vibration report, AV 1018/01, Lyngby
15. Dutilleux G, Defrance J, Ecotière D, Gauvreau B, Bèrengier M, Besnard F, Le DE (2010) NMPB-routes 2008: the revision of the French method for road traffic noise prediction. Acta Acust United Acust 96:452–462
16. Road noise prediction (2009) 1- Calculating sound emission from road traffic, Road noise prediction, 2-Noise propagation computation method including meteorological effects (NMPB 2008), Sétra
17. JRC Draft report (2010) Workshop on selection of Common noise assessment methods in EU, JRC.I.02.Form.CAT.039-Version, Brussels
18. Kephalopoulos S, Paviotti MF, Anfosso-Lèdèe F (2012) Common noise assessment methods in Europe (CNOSSOS-EU) 1-180
19. Kephalopoulos S, Paviotti M, Anfosso-Lédée F, Maercke DV, Shilton S, Jones N (2014) Advances in the development of common noise assessment methods in Europe: the CNOSSOS-EU framework for strategic environmental noise mapping. Sci Environ 482-483: 400–410
20. Kok A, van Beek A (2019) Amendments for CNOSSOS-EU: description of issues and proposed solutions, RIVM Letter Report No 2019-0023, National Institute for Public Health and the Environment (Netherlands)

21. Hothersall DC, Harriott JNB (1995) Approximate models for sound propagation above multi-impedance boundaries. J Acoust Soc Am 97:918–926
22. ISO 9613-2 (1996) Acoustics-Attenuation of sound during propagation outdoors-Part 2: General method of calculation
23. Lisle S de (2016) Comparison of road traffic noise prediction models: CoRTN, TNM, NMPB, ASJ RTN, Acoust Aus 44(3):409–413
24. Debnath A, Singh PK (2018) Environmental traffic noise modeling of Dhanbad township area, a mathematical approach. Appl Acoust 129:161–172
25. Sonaviya DR, Tandel BN (2020) Integrated road traffic noise mapping in urban Indian context. Noise Mapping 7:99–113
26. Thakre C, Laxmi V, Vijay R, Killedar DJ, Kumar R (2020) Traffic noise prediction model of an Indian road: an increased scenario of vehicles and honking. Environ Sci Pollut Res 27:38311–38320
27. Lokhande SK, Pathak SS, Kokate PK, Bodhe GL (2016) Evaluation and forecasting of highway traffic noise in the city of Nagpur, India. Int J Earth Sci Eng 9(4):1529–1535
28. Singh D, Nigam SP, Agrawal VP, Kumar M (2016) Modeling and analysis of urban traffic noise system using algebraic graph theoretic approach. Acoust Aus 44(2):249–261
29. Kalaiselvi R, Ramachandraiah A (2016) Honking noise corrections for traffic noise prediction models in heterogeneous traffic conditions like India. Appl Acoust 111:25–38
30. Garg N, Mangal SK, Saini PK, Dhiman P, Maji S (2015) Comparison of ANN and analytical models in traffic noise modeling and predictions. Acoust Aus 43(2):179–189
31. Sharma A, Bodhe GL, Schimak G (2014) Development of a traffic noise prediction model for an urban environment. Noise Health 16:63–67
32. Mishra RK, Shukla A, Parida M, Rangnekar S (2014) EIA based comparative urban traffic noise analysis between operational and under construction phase public transport corridor. Int J Traffic Trans Eng 4(3):352–362
33. Kumar K, Katiyar VK, Parida M, Rawat K (2011) Mathematical modeling of road traffic noise prediction. Int J Appl Maths Mechan 7:21–28
34. Agarwal S, Swami SL (2011) Comprehensive approach for the development of traffic noise prediction model for Jaipur City. Environ Monitoring Assessment 172:113–120
35. Rajakumara HN, Gowda RMM (2009) Road traffic noise pollution model under interrupted traffic flow condition. Environ Monitoring Assessment 14:251–257
36. Vijay R, Sharma A, Chakrabarti T, Gupta R (2015) Assessment of honking impact on traffic noise in urban traffic environment of Nagpur, India. J Environ Health Sci Eng 13:10
37. Ali SA, Tamura A (2003) Road traffic noise levels, restrictions and annoyance in greater Cairo. Egypt Appl Acoust 64(8):815–823
38. IRC-106 (1990) Guidelines for capacity of urban roads in plain areas, Indian Road Congress
39. Garg N, Maji S (2016) A retrospective view of noise pollution control policy in India: status, proposed revisions and control measures. Curr Sci 111(1):29–38
40. Probst W (2010) New Techniques in noise prediction. Proceedings of 20th International Congress on Acoustics, ICA, 23-27th August, Sydney
41. European Commission Working Group (2006) Assessment of exposure to noise. Good practice guide for strategic noise mapping and the production of associated data on noise exposure, *WG-AEN* report, Version 2
42. Coensel BD, Botteldoren D, Vanhove F, Logghe S (2007) Microsimulation based corrections on the road traffic noise mission near intersections. Acta Acustica United Acustica 93:241–252
43. Donato S. R. de (2007) Estimation of the minimum integration time for determining the equivalent continuous sound level with a given level of uncertainty considering some statistical hypothesis for road traffic. Noise Control Engg J 55:526–531
44. Maruyama M, Kuna K, Sone T (2013) The minimum measurement time for estimating L_{AeqT} of road traffic noise from the number of vehicles pass-bys. Appl Acoust 74:317–324
45. Brittain FH, Hale ME (2008) Some limitations of ray-tracing software for predicting community noise from industrial facilities. Proceedings of NOISE-CON 2008, Michigan

46. Paryzch D (2004) Handling of barriers in ISO 9613-2, Special session on ISO 9613-2. Proceedings of NOISE-CON, Baltimore
47. Morley DW, de Hoogh K, Fecht D, Fabbri F, Bell M, Goodman PS, Elliot P, Hodgson S, Hansell A, Gulliver J (2015) International scale implementation of the CNOSSOS-EU road traffic noise prediction model for epidemiological studies. Environ Pollut 206:332–341
48. Karantonis P, Gowen T, Simon M (2010) Further comparison of traffic noise predictions using the CadnaA and SoundPLAN Noise prediction models. Proceedings of 20th International Congress on Acoustics, ICA 2010, 23–27 August 2010, Sydney, Australia
49. Manwar VD, Mandal BB, Pal AK (2016) Environmental propagation of noise in mines and nearby villages: a study through noise mapping. Noise Health 18(83):185–193
50. Federal Transit Administration (2018) Transit Noise and Vibration Impact Assessment Manual. FTA Report No. 0123, September, 2018, U.S Department of Transportation
51. Garg N, Sharma O, Maji S (2011) Noise impact assessment of mass rapid transit systems in Delhi city. Indian J Pure Appl Phys 49:257–262
52. Eadie DT, Santoro M (2006) Top-of-rail friction control for curve noise mitigation and corrugation rate reduction. J Sound Vib 293(3–5):747–757
53. Tomeoka M, Kabe N, Tanimoto M, Miyauchi E, Nakata M (2002) Friction control between wheel and rail by means of on-board lubrication. Wear 253(1–2):124–129
54. Garg N, Sharma O (2010) Noise emissions of transit trains at curvature due to track lubrication. Indian J Pure Appl Phys 48:881–885
55. Pamanikabud P, Paoprayoon S (2003) Predicting mass rapid transit noise levels on an elevated station. J Environ Manag 67:353–362
56. Chen X, Tang F, Huang Z, Wang G (2007) High-speed maglev noise impacts on residents: a case study in Shanghai. Trans Res Part D 12:437–448
57. Staiano M (2003) Noise-impact estimates per FTA and APTA criteria. J Sound Vib 267(3): 407–418
58. Flightpath 2050 Europe's Vision for Aviation (2011) Report of the High Level Group on Aviation Research. European Commission. Directorate-General for Research and Innovation
59. Integrated Noise Model (INM) version 7.0 Technical Manual (2008) Federal Aviation Administration, U.S Department of Transportation
60. ECAC.CEAC Doc 29 (2016) Report on Standard Method of Computing Noise Contours around Civil Airports. Volume 1: Applications Guide, 4th Ed., December, 2016. https://www.ecac-ceac.org/. Accessed on 20.02.2021
61. Bertsch L, Simons D G, Snellen M (2015) Aircraft Noise: The major sources, modelling capabilities and reduction possibilities. IB 224–2015 A 110, March 2015, DLR (German Aerospace Center), ELIB-Eintrag-2015-Bertsch-95939-Bericht.pdf (dlr.de). Accessed on 20.02.2021
62. International Civil Aviation Organization: International Standards and Recommended Practices (2005) Environmental Protection. Annex 16, Volume I, Aircraft Noise, 4th Ed., July 2005
63. Vogiatzis K (2012) Airport environmental noise mapping and land use management as an environmental protection action policy tool. The case of the Larnaka international airport (Cyprus). Sci Total Environ 424:162–173
64. Torija AJ, Self RH, Flindell IH (2017) A model for the rapid assessment of the impact of aviation noise near airports. J Acoust Soc Am 141(2):981–995
65. Vogiatzis K, Remy N (2014) Strategic noise mapping of Herakleion: the aircraft noise impact as a factor of the international airport relocation. Noise Mapping 1:15–31
66. Ozkurt N (2014) Current assessment and future projections of noise pollution at Ankara Esenboğa airport, Turkey. Trans Res Part D 32:120–128
67. Garg N, Maji S (2013) Vibration induced due to acoustic excitation in diffuse field conditions. Acoust Australia 41(3):219–224
68. Schroeder MR (1996) The "Schroeder frequency" revisited. J Acoust Soc Am 99(5):3240–3241
69. Sharma O, Mohanan V, Singh M (1997) Damage criteria for induced vibration in buildings. J Acoust Soc India 25:III 3.1–III 3.5

70. Mohanan V, Sharma O (1997) Induced structural vibration on Khajurao temples due to overflying aircrafts. J Acoust Soc India XXVII, 1–4:123–127
71. Sharma O, Singh M, Garg N (2010) Investigations on induced vibrations due to acoustic excitation from Sound and Light show at Sri Brihadiswara temple, Thanjavur, Tamil Nadu, India. NPL Technical Report No. AC.C.10–1.F
72. Kurzweil LG (1979) Ground-borne noise and vibration from underground rail systems. J Sound Vib 66(3):363–370
73. Taylor RT (2016) The relationship between floor vibration from an underground source and airborne sound pressure level in the room. Int J Rail Trans 4:247–255
74. van Leeuwen HJA, van Zwienen AC (2016) The determination of railway vibrations levels in practice. Internoise 2016, Hamburg
75. Garg N, Sharma O (2010) Investigations on transportation induced ground vibrations, Proceedings of 20th International Congress on Acoustics, ICA 2010 23-27th August 2010, Sydney, Australia
76. DIN 4150-3 (1999) Structural vibration, Part 3: Effects of vibration on structures
77. Rudder FF (1978) Engineering guidelines for analysis of traffic induced vibration, US Dept. of Transportation Report No. FHWA RD-78-166
78. BS 5228, Part 4 (1992) Noise Control on Construction and Open Sites Part 4: Code of Practice for Noise and Vibration Control Applicable to Piling Operations. British Standards Institution
79. BS 7385-2 (1993) Evaluation and measurement for vibration in buildings. Guide to damage levels from ground borne vibration. British Standards Institution
80. AS2670.2 (1990) Evaluation of Human Exposure to Whole–body Vibration-Continuous and Shock-induced Vibration in Buildings (1 to 80 Hz)
81. British Standards Institution (2014) British Standard 5228–1:2009+A1:2014. Code of practice for noise and vibration control on construction and open sites – Part 1: Noise. British Standards Institution
82. British Standards Institution (2014) British Standard 4142:2014. Methods for rating and assessing industrial and commercial sound. British Standards Institution
83. Guidelines for Environmental noise impact assessment (2014) The institute of environmental management and assessment (IEMA), November, 2014. http://programmeofficers.co.uk/Preston/CoreDocuments/LCC393.pdf. Accessed on 08.06.2021
84. Design Manual for Roads and Bridges (DMRB) (2011) Environmental assessment. Section 3 Environmental Assessment Techniques. Volume 11, section 3, Part 7: Noise and Vibration, HD 213/11, The Department for Regional Development, Northern Ireland. https://programmeofficers.co.uk/Preston/CoreDocuments/LCC026.pdf. Accessed on 08.06.2021
85. High Speed Rail (West Midlands – Crewe) Environmental Statement (2017) Volume 5: Technical appendices. Sound, noise and vibration methodology, assumptions and assessment (SV-001-000). High Speed Two (HS2) Limited, Birmingham

Chapter 7
Environmental Noise Control Strategies

Abstract Environmental noise should always be assessed at the planning phase. This chapter presents the various aspects of noise control and discusses the active and passive noise control strategies. The main areas of environmental noise mitigation or abatement are transportation noise control, architectural design, use of passive control measures, and urban planning. Roadway noise and aircraft noise are the most pervasive sources of environmental noise. The chapter reviews the various passive noise control strategies for enhancing the sound insulation of building elements. The cost/benefit analysis of noise abatement measures and proposal of a Noise Abatement Programme and Noise Abatement Goal in Indian scenario are also presented. A comprehensive analysis and suitability of noise control strategies for Indian perspectives and soundscape approach for noise management and control are discussed.

7.1 Introduction

Transportation noise has emerged as a serious problem globally. The alarming increase in vehicular population and excessive use of horns has become havoc for the society causing serious annoyance among the community. Residents near the vicinity of roads and airports or dwellings under directly the funnel zone of aircraft are the major sufferers, and thus proper measures have to be undertaken to prevent the exterior noise entering inside the dwellings. The complex and pervasive problem of noise pollution has no single solution, rather it requires a combination of short-, medium-, and long-term approaches and careful consideration of the nature of the noise source. There are many sources of noise pollution, each requiring tailored abatement measures [1]. The biggest source of environmental noise is road traffic, exposure to which far exceeds rail and aircraft sources combined [2]. In urban areas, road traffic accounts for 80% of all noise pollution [3]. Although fewer people are exposed to air traffic noise than that from road or rail, it is reported to cause greater annoyance [4, 5]. In the European Union, more than 100 million people are exposed to road traffic noise levels exceeding 55 dB(A) L_{den} inside the big agglomerations and along major road infrastructure [6]. This value is above the onset of negative

N. Garg, *Environmental Noise Control*,
https://doi.org/10.1007/978-3-030-87828-3_7

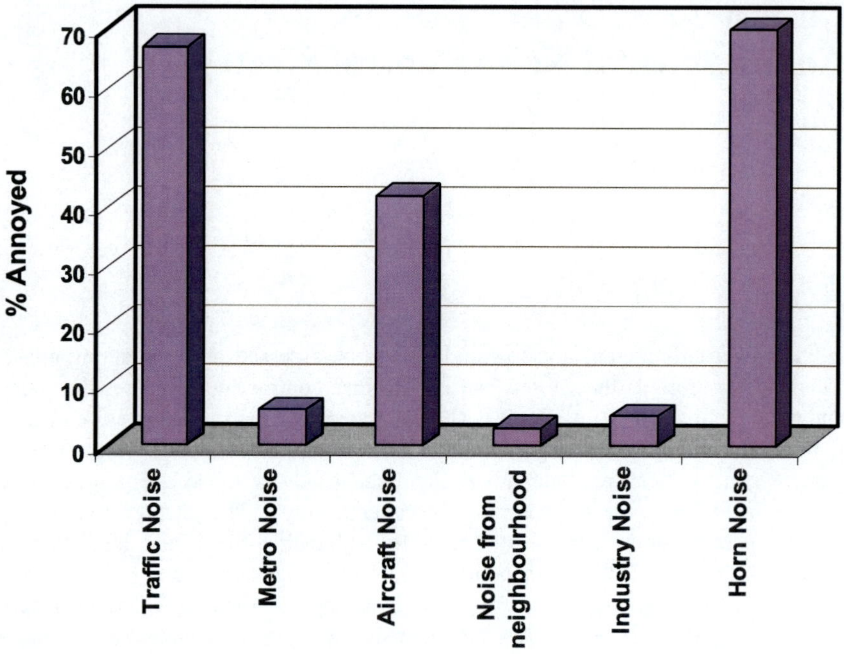

Fig. 7.1 Noise annoyance due to various noise sources reported in socio-acoustic survey in Delhi city

health effects following the World Health Organization. As per the World Economic Forum 2017 survey, Delhi city was the second worst city for noise pollution followed by Cairo, Mumbai, Istanbul, and Beijing. The study found that average city dweller of Delhi, India, has the greatest hearing loss equivalent to 20 years older than their actual age [7, 8]. Guangzhou in China had the worst noise pollution, while Zurich in Switzerland had the least. The top five quietest cities were all in Europe: Zurich, Vienna, Oslo, Munich, and Stockholm [8]. A socio-acoustic survey (details in Appendix V of book) conducted for assessment of impact of noise on community in Delhi city revealed that more than 60% people are affected from road traffic and horn noise and around 40% are annoyed due to aircraft noise [9]. The noise from metro train and industry is not so serious as compared to response given for the traffic noise as shown in Fig. 7.1. The operation of metro trains had been assessed to cause an increase in ambient noise level by 2–3 dB (A) in medium and high traffic density areas and thus causes a relatively less impact on the community as compared to the road traffic noise [10].

The reaction toward road traffic noise was also found to be serious as revealed from the survey results from around 700 individuals in Delhi city of India as shown in Fig. 7.2. Twenty-five percent people felt stressed and 28% felt mentally tired when exposed to traffic noise, while 17% felt irritated and angry, 1% had sleep disturbances induced, and 2% feel headache when exposed to long-term traffic

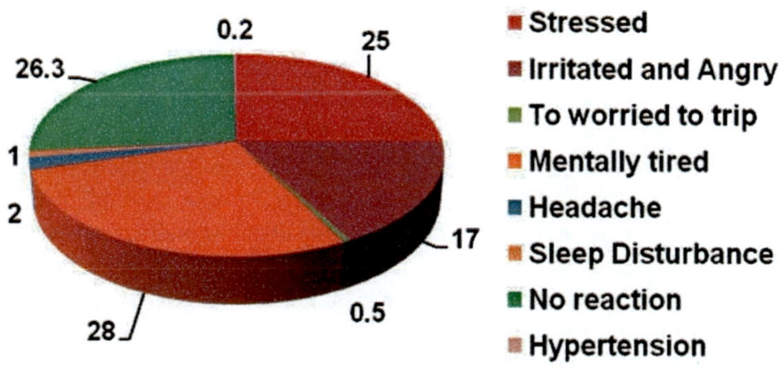

Fig. 7.2 Reaction toward road traffic noise reported in a small socio-acoustic survey in Delhi city

noise. The lifestyle of people becoming habitual toward noise pollution is reflected from the fact that around 26.3% people have no reaction. Only a small percentage of people were too worried for trip and very little reported hypertension problem associated with noise.

With economic expansion and infrastructural growth, the concern toward the harmful effects of noise has considerably increased and created an inevitable need for devising strategic noise abatement plans for existing and future roads. This chapter discusses the passive noise control strategies for noise abatement and control. It is emphasized that a proper control at source, propagation path, and increased sound insulation of dwellings is very essential for achieving fruitful results. Proper planning with construction of low noise pavements, erection of environment friendly noise barriers, and efficient traffic management can play a key role in fighting against the traffic noise pollution. An important decision-support tool is the cost-benefit analysis, which can help to prioritize different noise abatement options and ensure that limited funds are spent to greatest effect [11]. Traffic management measures are some of the most affordable measures. Cost varies widely between local noise abatement measures. Noise barriers are considered the least cost-effective approach, despite their significant noise abatement ability [4]. For facade insulation, costs are generally high compared to other measures, but comparatively little when implemented in new buildings with high thermal insulation standards [11]. Facade insulation may be reserved for the sites wherein the environmental noise limits are rigid and mandatory [12] and have an estimated cost of €3.5 per m^2 [13]. Low-noise tires are considered particularly cost-effective, due to their significant noise abatement but minimal side effects. Quiet tires can reduce noise by around 4 dB at no additional cost [13].

7.2 Active and Passive Noise Control

Noise control is either an active or passive means of reducing the sound emissions. Passive noise control is sound reduction through the use of sound insulation materials, sound-absorbing tiles, or a mufflers. Active noise control (ANC) also known as noise cancellation, or active noise reduction (ANR), is a method for reducing unwanted sound by an additional sound specifically designed to cancel the first. Active noise canceling is best suited for low frequencies. Active noise control (ANC) is the application of the principle of the superposition of waves to noise attenuation problems. Much progress has been made toward applying ANC to narrow-band, low-frequency noise in confined spaces. ANC was believed to utilize the mechanism of phase cancellation to achieve the desired attenuation [14]. Figure 7.3 shows the comparison between the two noise control methods. The ANC system is very effective for attenuating low-frequency noise, where passive methods are expensive and ineffective. Most practical ANC systems use adaptive filters to automatically track variations of unknown noise characteristics and environment. The most commonly used adaptive filter is the finite impulse response (FIR) filter with the filtered-X least mean-square (FXLMS) algorithm [15, 16]. The commercially available digital signal processors [17] and simplified real-time experimentation of ANC algorithms, thus promoting development and application of ANC

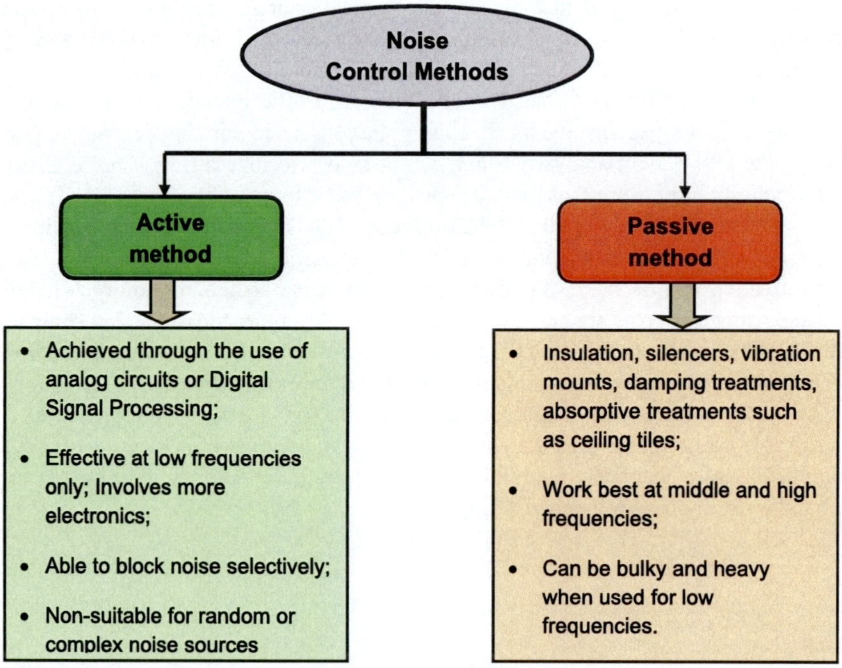

Fig. 7.3 Active and passive noise control strategies

systems. The relationships among the noise reduction performance of ANC systems, the changed system transfer functions caused by the speed and direction of the moving sources, and the corresponding frequency shifting due to the Doppler effect have not yet been investigated so far due to which the use of ANC is difficult for traffic noise reduction with moving noise sources [18]. The effectiveness of applying active noise control system to improve the sound insertion loss of the noise barrier has been demonstrated in some studies [19]. The insertion loss of noise barriers can be significantly increased by introducing a multichannel active noise control system in which sensors and actuators are optimally arranged. However, active noise control system is more useful in low-frequency range rather than passive noise control system. It has been observed in some studies [19] that the control system significantly increases the insertion loss of the noise barrier at low frequencies when the ground on both sides of the barrier is totally nonreflective. In practice, a nearby reflective ground may deteriorate the performance of the active noise-control system. The effects of ground reflection on the effectiveness of the active-noise barrier are considered. Active-control system can be effective at improving the performance of the noise barrier for noise with multiple-frequency components [19].

The application of active noise control (ANC) methods can offer a viable solution in case of double partition panel as the performance of practical double panel structures rapidly deteriorates toward the lower frequencies. It has been shown that a considerable improvement of the insertion loss of the double partition panel can be obtained by using small loudspeakers as secondary sources inside the cavity between two panels, provided that the sound field inside the cavity of the partition is substantially attenuated [20]. Active noise control (ANC) is currently being explored as a potential solution for controlling noise that propagates through an open window, whereby the ANC system is installed at the opening of the sliding window [21]. The performance of the active control system for a window with different source and receiver configurations are compared in one such study, and the best result is achieved when the secondary sources are placed in the center at the bottom of the staggered window [22]. Existing studies on ANC of open windows have shown that noise attenuation (up to 10 dB) is attainable under certain scenarios and the global control of noise is possible with ANC of open windows at low frequencies, regardless of noise incidence angles [23]. ANC can be applied to windows by controlling the sound pressure within the cavity of a double glazed window system. This is known as cavity control and tends to show a higher increase in sound transmission loss but tend to compromise thermal insulation due to the larger inter-pane spacing required for loudspeaker control sources [24, 25]. The active control of the sound field in a double-glazed window cavity by means of loudspeakers and microphones inside the cavity had been shown to be applicable to reduce the transmission of sound, whereby the total sound pressure level can be reduced by nearly 8 dB and somewhat more than 5 dB with feed forward and feedback control, respectively [26].

Passive noise control measures, however, when used for lower frequencies tend to get bulky and heavy as the size and mass of passive treatments usually depend on the acoustic wavelength, making them thicker and more massive for lower frequencies [27]. Passive treatments become more effective at higher frequencies and often provide an adequate solution without the need for active control [28].

7.3 Source-Path-Receiver Concept of Noise Control

The environmental noise problem can be resolved in terms of its three basic elements: the source, the conveying medium, and the receiver as follows:

- To modify the source to reduce its noise output
- To alter or control the transmission path and the environment so as to reduce noise level reaching the recipient
- To provide the receiver with personal protective device

Controlling noise pollution at the source should be the primary consideration as preventing or reducing the noise emissions in the first place itself is the most effective way. The primary sources causing the increase in ambient noise levels can be due to the transportation systems: vehicular, vehicular honking, railway, aircrafts flyover noise, construction or residential noise, loudspeaker noise or noise emissions during the festivals or marriage processions, etc. The sound wave transmission path can be also controlled so as to dissipate sound energy before it reaches the receiver by putting in place a structure to absorb or deflect the sound such as noise barriers. Lastly, the receiving end also has to be protected in case the measures implemented at the source and transmission path cannot effectively control noise pollution. Studying and making decisions about the technical alternatives is a complex issue involving many factors that may themselves come into conflict or may depend upon uncertain situation [29]. Measures taken at the noise source are in general the most effective measures and result in the best reduction in noise exposed people. Stricter noise limits for vehicles and low noise road surfaces are cost-effective measures, while noise barriers are the most expensive measure. Noise barrier and earth berms appear to have greatest potential for implementation and can achieve noise reductions in the range of 5–15 dB(A). Noise barriers have a relatively high cost in range of 2.1 million dollars per linear mile; however, these options still appear to be the most feasible and reasonable mode of noise mitigation available for both the existing and future roadway projects. Earth berms are often preferred alternatives to noise barriers; however, the implementation of berms can gave engineering concerns, due to the large horizontal foot print typically required to achieve adequate berm height [30]. However, low-noise pavements are a cost-effective option to reduce traffic noise as they can reduce the investments in noise abatement measures compared to expensive noise barriers especially for intra-urban roads. A Norwegian study of estimated costs and effects of different traffic noise abatement measures reported that priority should be given to actions in central urban areas: stricter regulations of vehicles, area traffic management, low-noise road surfaces, and insulation of most exposed dwellings. Building insulation is the most effective measure in relation to costs when priority is given to the most heavily exposed houses. The costs naturally depend on the ambitions of noise abatement goals. Extensive use of screens is not very cost-effective. Building of new major roads, bypasses, and tunnels in urban areas proved to have little cost-effectiveness. Low noise road surfaces seem to be more cost-effective than noise barriers [31]. The measures taken at the noise source in general are the most cost-effective measures and result in best reduction in respect of noise exposed people and noise annoyance [32]. Land use planning is also a valuable tool for noise abatement. It is wiser to plan the land uses to avoid as much as possible placing noise sensitive areas near to noise producing land uses. Also, when locating areas for new industrial estates, there should be buffer zones between these land uses and the residential areas.

7.4 Source Noise Control

Control at the source, that is, propulsion and rolling noise radiated by vehicles, should be of prime concern while reducing the source noise levels. The tire width, tire hardness, treads patterns, groove depth, and road surface are the pivotal factors affecting the tire noise. An optimal combination of tire and road surface is instrumental in reducing the road traffic noise. The European studies [33] revealed the difference in noise level between quietest and noisiest tire/road surface combination was 9 dB(A) for cars and vans and 7 dB(A) for trucks. Low noise road surfaces can be divided into thin dense surfaces, porous surfaces, and poro-elastic pavements. Porous asphalt has the potential for reduction of vehicle noise by 5–8 dB, while the poro-elastic pavements that combine the porous structure with a soft aggregate showed reductions of 10–12 dB for normal tires and 13–15 dB for studded tires [34]. Sandberg [35] investigations revealed that the tire and road construction result in concentration of major noise generating and influencing mechanisms in range 700–1300 Hz. A softer tire/road contact with softer tire treads as well as road surface that increase belt stiffness but decrease contact stiffness and porous road surfaces designed for favorable sound absorption in 800–1250 Hz could be possible ways of dealing with the multi-coincidence peak. Apart from rolling noise, key technologies for propulsion noise, viz., common rail injection system, pilot injection, exhaust gas recirculation, electronic power train management, multiple muffler system for exhaust noise control acoustic shielding of power trains, application of resonators, and active noise control, could be instrumental in bringing the noise reduction at the source itself. Traffic calming can cause up to 4 dB L_{Aeq} reduction and 7 dB L_{Amax} reduction depending upon traffic congestion, speed, road layout, etc. Junctions and roundabouts can cause 1–4 dB L_{Aeq} reduction as compared with the intersections. Narrow cushions, round-top humps, and correct distancing between such measures to avoid acceleration and deceleration is preferable. A 20% reduction in traffic results in 1 dB noise reduction provided other factors such as speed and composition are unchanged [36]. Pavement surface characteristics listed in Table 7.1 are the major attributes to tire pavement interactions and are considered as cost-effective options to mitigate traffic noise [37]. A recent study reported by Ögren et al. (2018)

Table 7.1 Effect of single pavement characteristics on noise levels (Liao et al. 2014, [37])

Pavement surface characteristics	Main noise generation mechanism	Main frequency range	"Crossover" frequency	Correlation with noise	Main effect on noise levels	
					Lower frequencies	Higher frequencies
Texture (MPD—mean profile depth)	Tread impact	400–1000 Hz	1600	Good/fair	Increase	Decrease
Porosity (V_A)	Air pumping	1000–10,000 Hz	–	Fair/poor	Decrease	Decrease or increase
Stiffness (E^*)	–	–	–	Fair/poor	Increase	Increase
Roughness (IRI)	–	–	–	Poor	–	–
Aggregate size (NMAS—nominal maximum aggregate size)	Tread impact	400–1000 Hz	–	Good/fair	–	–

[38] suggested that measures taken at the tire/road source, such as low-noise tires or low noise pavements, are most effective for reducing the noise exposure. Reducing the speed is also relatively efficient. Reducing the rolling noise by introducing low-noise emission tires or low-noise pavements would have a significant effect on road traffic noise exposure. Strategies aimed only at separating new residential areas from roads with dense traffic or introducing electric vehicles would cause only a limited improvement [38]. Rolling noise is the most important source for higher speeds, but propulsion noise is also important at lower speeds especially for heavy vehicles [39]. Use of electric vehicles will reduce the propulsion noise, which will reduce the total noise emission significantly at low speeds [38]. The total noise emitted by electrical light vehicles is 2.7 dB lower at 30 km/h, but only 0.4 dB lower at 110 km/h. For heavy vehicles, a reduction of 10 dB at 30 km/h and 1.5 dB at 90 km/h is achieved [40]. Low noise tires are readily available in different forms, and reductions of 3–5 dB can be achieved [41, 42]. Low noise pavements can be either elastic such as rubberized asphalt or porous, porous or drain asphalt, or both (poro-elastic pavements). Reduction of the order of 3–10 dB can be achieved as compared to the standard pavements and even higher reductions can be achieved from poro-elastic surfaces [38, 43–45]. Studies have concluded that porosity has a significant influence on the sound absorption characteristics of pavements, while the layer thickness has insignificant impact. The acoustical performance of pavement diminishes with aging especially for porous surfaces [44]. From a practical point of view, three key factors associated to noise abatement should be considered: pavement type, traffic speed, and traffic density [46]. Porosity within asphalt pavements decreases the noise levels at every frequency (except at some frequency). Thus, to reduce the noise levels more efficiently, the factors such as pore-distribution, interconnected content, pore shape, and pore size should be taken into consideration. Pavement surface texture increases the noise levels below 1600 Hz, especially on impervious asphalt pavements [37].

7.5 Sound Insulation of Building Elements

The control strategies focused on strengthening the sound insulation of receiver, that is, dwellings so as to reduce the noise exposure and provide acoustic comfort to the residents, are the last resort when there is no control neither at the source nor any treatment can be done on the pathway of sound propagation. The main considerations for design of sound insulative material are the building elements, viz., the walls, windows, roof, ceilings, and exterior facade. Thus, a proper treatment of the building elements would considerably reduce the outside noise and protect the residents from hazards of noise pollution. Sound transmission loss (*TL*) is a performance of sound insulation measured in reverberation chambers. There are varied single-number ratings, viz., sound transmission class (*STC*), weighted sound reduction index (R_w), outdoor-indoor transmission class (*OITC*), and spectrum adaptation terms C, C_{tr} used for describing the sound insulation properties of partition wall

Table 7.2 *STC* rating required for building elements (Metropolitan Council report, Minnesota 2006 [48])

Specified noise level reduction	Required *STC* rating needed for compliance			
dB(A)	Roof-ceiling	Walls	Windows	Doors
20	40	40	30	20
25	45	45	35	25
30	50	50	40	30
35	55	55	45	35
40	60	60	50	40

panels used in dwellings, offices, exterior facades, etc. The *STC* value is defined as *TL* value where the *STC* contour intersects the 500 Hz line (ASTM E413-87 1999 [47]). The better the *STC* of materials, higher the sound insulation it provides. R_w is used to facilitate the comparison of sound insulation performance of different materials in European continent. The sound reduction required through building elements depends upon the existing ambient noise levels in that area and expected noise level reductions for acoustic comfort of residents. Table 7.2 gives the sound transmission class (*STC*) requirements for various building elements on basis of specified noise level reduction as recommended by Metropolitan Council, Minnesota [48].

In order to differentiate between the field and laboratory performance, it was recommended to select a wall or floor/ceiling system rated at 5 *STC* points above the level of sound attenuation required based on Dunn's [49] observations related to comparison of dB (A) reductions for aircraft and traffic noise with *STC* rating for 104 different building elements. It was concluded that aircraft noise and traffic noise on an average is attenuated by 4.6 and 6 dB less than numerical value of appropriate *STC* rating. Dunn's average values of 5 and 6 dB were used as corrections to table of *STC* values in Australian standards.

7.6 Sound Insulation of Window Glazing

Windows are considered to be the weakest parts of building facades. Thus, the sound insulation of windows is very pivotal in achieving acoustic comfort in the dwellings. Window properties in addition to glass thickness, inter-pane spacing, and treatment of cavity either with absorptive materials or simply air can appreciably affect the sound transmission loss. The double glazing has been the prominent solution for controlling the traffic noise, although they suffer from decrement on sound insulation at low frequencies attributed to mass-air-mass resonance. The various aspects affecting the sound insulation as summarized from the previous studies [50–56] are as follows:

- Air-gap plays a vital role in controlling the sound transmission of double glazing. The percentage contribution is 63% for R_w value, while it is 45.6% for $R_w + C_{tr}$

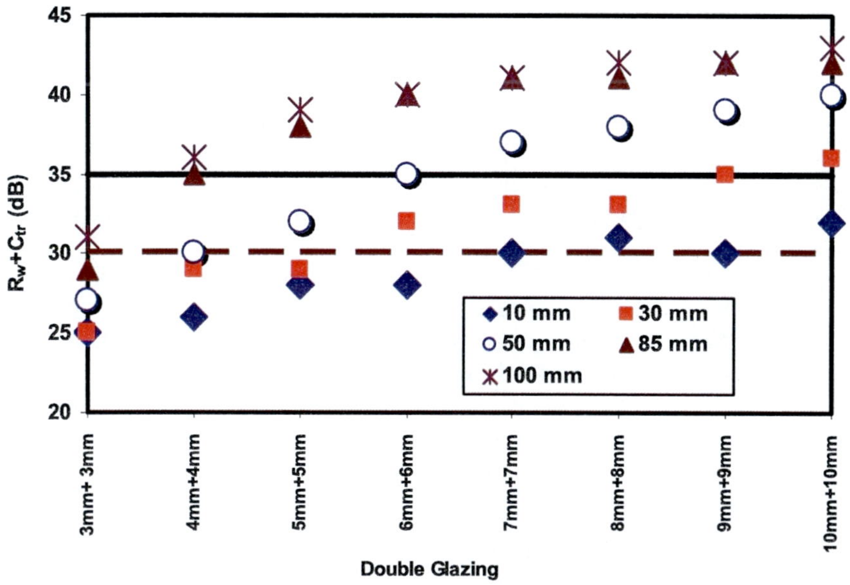

Fig. 7.4. ($R_w + C_{tr}$) value of double glazing with identical glass each side (Garg et al. 2016, [53]).

value. It is observed experimentally that a mass-air-mass resonance of the windows is the main culprit behind the decrement of transmission loss at low frequencies particularly. A double window provides enhanced sound insulation as compared to a single glazing and thus design considerations should be focused on choosing the thickness and depth of airspace to bring the mass air mass resonance frequency below 100 Hz. There has to be a trade-off in choosing the air gap as an air gap of less than 30 mm will lead to pronounced mass-air-mass resonance dip, while increasing the air space above 100 mm increases sound insulation at low frequencies more than at high frequencies as mass-air-mass resonance is lowered, but the standing wave resonances affects the high-frequency insulation.

- The back pane thickness has higher percentage in determination of single-number rating R_w, $R_w + C$ and *STC* value as compared to front pane thickness, while front pane thickness has slightly higher contribution in case of single number quantity $R_w + C_{tr}$. These investigations do not contradict with the reciprocity law as back pane thickness had been experimentally observed to be more sensitive in controlling the co-incidence dip.
- The type of lamination has a very little role (<3%) which is statistically insignificant. An optimum design of the front and back pane thickness in conjunction with proper selection of air-gap can be instrumental in enhancing the sound insulation as shown in Fig. 7.4.

The results analyzed on *Insul SW* software leads to some important aspects (Fig. 7.4) regarding the fulfillment of objective of $R_w + C_{tr} \geq 35$ dB for a double glazing as:

- A double glazing with identical 6 mm glass on both sides if air gap is ≥50 mm.
- A double glazing with identical 4 mm glass on both sides for an air-gap ≥85 mm.
- For glazing having different glass on each side for controlling the co-incidence dip and an average air-gap of 50, 5, and 10 mm combination can achieve the desired value.
- For an average air gap of 85, 3 and 6 mm glazing can achieve $R_w + C_{tr} \geq 35$.
- For window glazing, the coincidence dip is controlled by increasing the back pane thickness, as it shifts to a lower frequency with the increase in back pane thickness, but is independent of inter-pane distance [51].
- The regression for the single-number rating and variables air-gap (d) and thickness (t) of identical glass in a laminated (PVB 0.76 mm) double glazing is modeled as [56]:

$$R_w = 1.146t + 0.092d + 30.169; R^2 = 0.92 \tag{7.1}$$

$$R_w + C_{tr} = 1.878t + 0.121d + 16.827; R^2 = 0.92 \tag{7.2}$$

$$STC = 1.182t + 0.101d + 29.57; R^2 = 0.83 \tag{7.3}$$

It may be noted that these empirical formulations are valid for range 30 mm $\leq d \leq$ 100 mm and 3 mm $\leq t \leq$ 10 mm.

Experimental studies on the abatement of traffic noise with a dominant horn noise component revealed that the sound insulation in octaves of 2.5 to 4 kHz should be higher. Thus, the coincidence dip has to be controlled in this region. Significant increase in ($R_w + C_{tr}$) value is observed in sandwich window constructions with either one double glazing or both are double, and thus, these are best for areas having high levels of traffic noise with dominant honking component as shown in Fig. 7.5 [55].

7.7 Sound Insulation of Multilayered Building Constructions

The sound transmission through masonry constructions has always been a gray area of research for its applications in facades and walls in dwellings for outside noise abatement. A dry wall is a high performance light weight partition system consisting of GI steel frame, encased with gypsum plasterboards on either side attached through self-drilling drywall screws [57]. The dry wall technology has many advantages compared to the masonry constructions as: speed of installation is three to four times faster than masonry constructions; lighter in weight than masonry constructions,

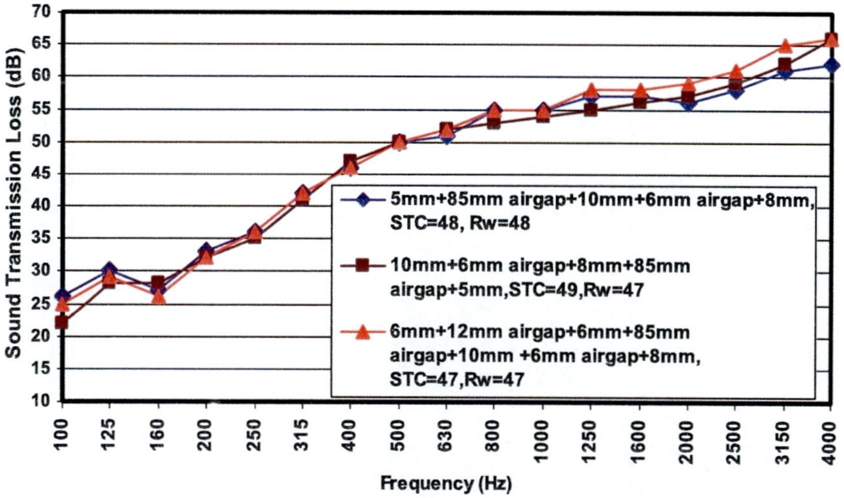

Fig. 7.5 Sound transmission loss of improved sandwich constructions with either one double glazing or both is double (Garg et al. 2012 [55])

higher sound insulation and fire resistance, and versatility, recyclable, and less heat convection (0.16 W/m K) as compared to the brick wall (0.81 W/mK). The sound insulation provided by the drywall constructions can however be significantly enhanced by combination with masonry constructions for its suitability for building facades. Figure 7.6 shows the modest improvements in *STC* in dry wall constructions comparable to conventional masonry structures [58, 59] as presented by Garg et al. 2011 [60]. It can be observed that the modest improvements in *STC* is attributed to addition of glass fiber batts, resilient furring channels and further addition of gypsum boards on each side. A similar laboratory experimentation [59] for wood studs at 600 mm o.c with 65 mm glass fiber batts and 12.5 mm gypsum board on each side revealed 11 points increase in *STC* value (Vth configuration). In case (VIth configuration) of a 31 × 92 mm, 25 gauge (0.5 mm) non-load-bearing steel studs at 400 mm o.c with 89 mm glass fiber batt and 12.5 mm gypsum board on each side, the *STC* value increases up to 10 points with further addition of Gypsum boards [59]. The use of Load bearing steel studs of varying thickness results in increase in *STC* by two points. The air-layer in the composite structure brings impedance mismatch between the different medium, while the sound absorption of porous materials brings further attenuation leading to improved insulation of multilayered structures [61]. Structurally decoupling the drywall panels from each other (by using resilient channel, steel studs, a staggered-stud wall, or a double stud wall) can yield an *STC* with good low-frequency transmission loss as well.

Figure 7.7 shows a similar analysis in terms of the increments in ($R_w + C_{tr}$) value (in dB) in masonry construction along with gypsum board with addition of various options using data from Warnock report [58, 62] and prediction from validated *Insul SW* software (version 7.0.4). The I block shows the sound insulation characteristics

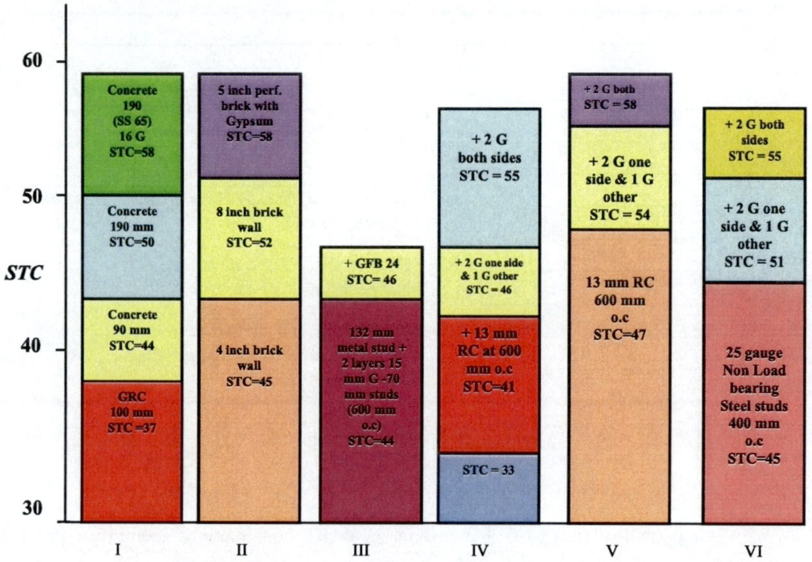

Fig. 7.6 Increments in *STC* (sound transmission class)[a] value in masonry and drywall constructions upon addition of various options (resilient furring channels RC 13, 2 gypsum board on one side and one on other, 2 gypsum board on both sides, non-load-bearing steel studs, addition of glass fiber batt) (Garg et al. 2011, [60])
[a]Analysis presented in Fig. 7.6 is based on the results published by Warnock [58] and Quirt [59] in *STC* rating. *STC* rating has not been assigned any units

of 90 mm and 140 mm concrete block having (R_w + C_{tr}) value less than 50. An unexpected decrement (II block) in the value of (R_w + C_{tr}) for 16 mm gypsum drywall mounted on 13 mm resilient metal channels applied to either side of concrete block 190 mm thick with 19 mm glass fiber batt compressed in between. The weighted sound reduction index is 56 and spectrum adaptation terms come out to be (−4, −11). The sandwich double wall constructions with 90 mm concrete block (BLK 90) on both sides appreciably accentuates the (R_w + C_{tr}) value as expected (III block), yet these constructions find their limited applications due to practical and economic constraints. The IVth block shows an unexpected decrement in both the configurations, viz., 16 mm dry wall attached to 190 mm concrete block and 16 mm dry wall attached on each side to 190 mm concrete block owing to a decrement in value of C_{tr} to −7 in each case, while for bare 190 mm concrete block it is calculated to be −4.7. The analytical predictions for brick wall constructions (using *Insul SW* software) show that the sound transmission can be significantly enhanced by addition of two gypsum boards attached via steel studs and sound absorptive material in cavity.

To ensure mass-air-mass resonance (*m.a.m*) is below 80 Hz, the air space between single layer of wall board and concrete block should be at least 60 mm; for a double layer of wall board, the space may be as small as 35 mm [63]. Guillen et al. [64]

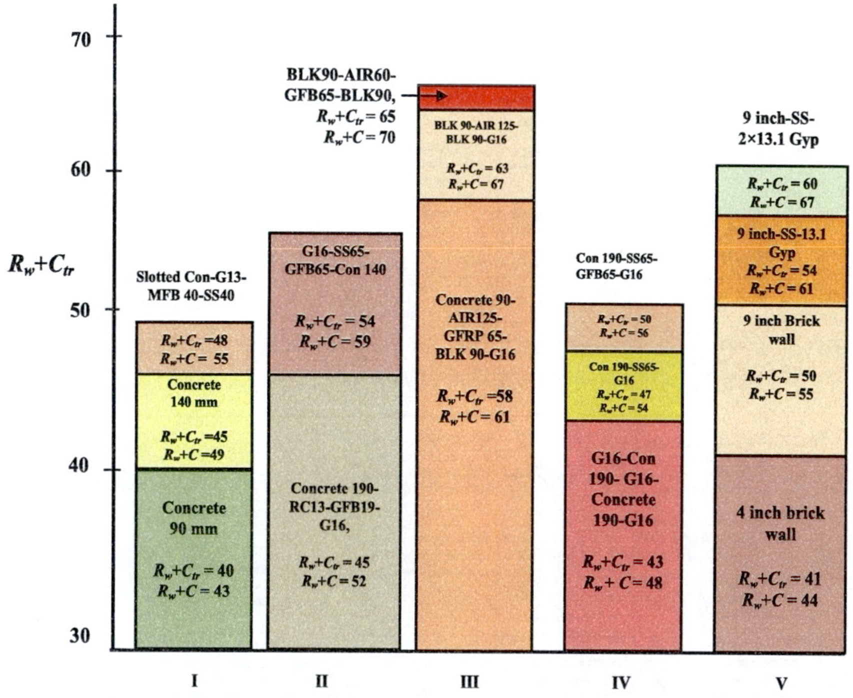

Fig. 7.7 Increments in (R_w + C_{tr}) value (in dB) in masonry construction along with gypsum board with addition of various options, viz., resilient furring channels RC 13, gypsum board attached either on one side or on both, steel studs, and addition of glass fiber batt [58, 62]

study showed that that masonry-air cavity-gypsum walls had higher sound reduction index than masonry-air cavity-brick ones. The importance of strengthening the facades is evident from Norwegian study wherein for noise reduction of 7 dB inside the dwelling, the percentage highly annoyed respondents dropped from 42% to 16% [65]. The drywall technology alone suffers from poor low-frequency sound insulation characteristics. Thus, a sandwich multilayered massive construction with drywall attached is a good substitute although practically cumbersome and expensive. The following conclusions have been derived from various parametric studies done [66] as follows:

- The type of stud frame and the number of gypsum layers attached play a significant role in affecting the R_w + C value of sandwich gypsum constructions.
- Double studs are the best preferred attachment followed by staggered studs, steel studs with resilient railings. Double stud wall systems perform the best as they have independent leaves. The next best performance is obtained with staggered stud walls. The stud spacing has also been analyzed to be noncritical in controlling the overall sound insulation characteristics especially when structural breaks are provided in the walls consistent with Quirt (1985) [76] observations. An

Fig. 7.8 Cause-and-effect analysis for improving sound transmission loss of multilayered building elements (Garg et al. 2016 [53])

optimum depth of 82 mm in case of cavity filled with sound absorptive material is at least required for $m.a.m < 50$ Hz.

- The density and thickness of the sound absorbing material has no major role in deciding the sound insulation characteristics, although it is evident that inclusion of sound absorbing material will shift the low-frequency *m.a.m* and also interrupt the standing wave resonances creeping in the cavity.

All these factors influencing the sound insulation characteristics of multilayered building elements especially as reported in terms of widely used sound transmission class (*STC*) rating and R_w have been shown in a cause-and-effect analysis diagram

(Fig. 7.8) presented earlier by Garg et al. 2016 [53] based on exhaustive literature survey [58–80].

7.8 Airborne Sound Insulation Criteria

Legal sound insulation requirements are very vital tool to fight against the outdoor noise and the ever-increasing traffic noise. In view of acoustic comfort, there has been a considerable research particularly in Europe for better sound insulation criteria to regulate the airborne and impact sound insulation between dwellings and airborne sound insulation of facades. Rasmussen [81, 82] conducted an extensive survey describing the legal sound insulation requirements between dwellings in Europe that revealed considerable differences not only in level of requirements, but also in terms of descriptors and frequency range applied. Table 7.3 shows the common indices used to describe the airborne and impact sound insulation [83].

The introduction of impact rating $L_{n,w} + C_I$ ($L_{n,w}$ is weighted normalized impact noise level and C_I is spectrum adaptation term) has been established for impact noise. C_I (spectrum adaptation term) is the frequency spectrum adaptation term used to describe the impact noise. In general, a lower number of $L_{n,w} + C_I$ means the better the impact sound insulation of the construction element (the sound pressure level is lower). The rating is derived from data that has been adjusted (normalized) to receiving room absorption of 10 m^2 to account for the influence a receiving room has on sound pressure level. Low frequencies have a considerable impact while evaluating impact sound in lightweight buildings [84]. The facade insulation is expressed in terms of weighted standardized sound level difference of facades ($D_{2m,nT,w}$), which is the difference between the outdoor sound pressure level at 2 m in front of the facade and the sound pressure level in the receiving room, corresponding to a reference value of the reverberation time (T = 0.5 s for

Table 7.3 Common indices used to describe airborne and impact sound insulation (National Building Codes of India 2016 [83])

S no.	Airborne (*A*) impact (*I*)	Lab (*L*) field (*F*)	Measured values		Single number quantity	
			Name	Symbol	Name	Symbol
(i)	*A*	*F*	Standardized level difference	D_{nT}	Weighted standardized level	$D_{nT,w}$
(ii)	*A*	*L*	Sound reduction index	R	Weighted sound reduction index	R_w
(iii)	*I*	*F*	Standardized impact sound pressure level	L'_{nT}	Weighted standardized impact sound pressure level	$L'_{nT,w}$
(iv)	*I*	*L*	Normalized impact sound pressure level	L'_n	Weighted normalized impact sound pressure level	$L'_{n,w}$

Table 7.4 Sound insulation requirements for multi-family housing (Machimbarrena and Rasmussen 2017 [85])

Country	Airborne requirement (dB)	Impact requirement (dB)	Facade sound insulation or indoor level Requirement (dB)
Argentina	$R'_{w} \geq 50$	$L'_{n,w} \leq 53$	$D_{2m,\,n,\,T,\,w} \geq 53$ $D_{2m,\,n,\,T,\,w} \geq 36$
Brazil	$D_{nT,\,w} \geq 45$	$L'_{nT,w} \leq 80$	$D_{2m,\,n,\,T,\,w} \geq$ 20–30 depending upon surrounding noise
Chile	$R_A/R'_A \geq 45$	$L_{n,w} \leq 75$	$D_{nT,\,w} + C_{tr} \geq X$ X depending on outdoor noise for L_{eq} (h) < 60 dB(A) $\Rightarrow X = 20$
Austria	$D_{nT,\,w} \geq 55$	$L'_{nT,w} \leq 48$	$R'_{res,\,w} \geq X$; X depends on outdoor level for $61 < L_d < 65$ and $51 < L_n < 55$ dB; $X = 43$
Belgium	$D_{nT,\,w} \geq 54$	$L'_{nT,w} \leq 58$	Goal indoor level during peak hours: $L_{Aeq} \leq 34$ dB, always: $D_{nA,\,tr} \geq 26$
Czech Republic	$R'_{w} \geq 53$	$L'_{n,w} \leq 55$	Indoor level: $L_{Aeq} \leq 40$ (day) $L_{A\,eq} \leq 30$ (day)
Denmark	$R'_{w} \geq 55$	$L'_{n,w} \leq 53$	Indoor level: $L_{den} \leq 33$
España (Spain)	$D_{nT,\,A} \approx D_{nT,\,w} + C_{100\text{–}5000} \geq 50$	$L'_{nT,w} \leq 65$	$D_{2m,\,nT,\,Atr\,(100\text{ to }5000)} \approx D_{2m,\,nT,\,w} + C_{tr} \geq X$ ($X = 30$ for $L_{den} < 60$ dB(A))

dwellings). Meeting the legal requirements doesn't ensure sufficient acoustic comfort and thus several countries in Europe have adopted classification schemes in a way of increasing order of better acoustic comfort which are further higher criteria than the legal requirements with an objective to provide the enhanced acoustic comfort. Table 7.4 shows the sound insulation requirements for multifamily housing in different countries [85].

The traditional frequency range for legal sound insulation was in the range of 100–4 kHz in one-third octave bands. However, in some countries, with the adaptation of light weight building practices, the need to include the low frequencies below 100 Hz becomes pertinent and thus frequency bands down to 50 Hz are applied in the regulatory requirements in Sweden [86]. The recent studies have recommended standardized field level difference ($D_{nT,w}$) as the suitable single-number rating for describing the in situ airborne sound insulation between dwellings [82, 87]. A study by Rasmussen (2010) on sound insulation descriptors used in European continent reported the following criteria for airborne sound insulation between dwellings as [82]:

$$D_{nT,w} + C_{50-3150} \geq 55 \text{ dB} \tag{7.4}$$

The desire for improved acoustic comfort class intends to offer real acoustic comfort so that 90% of inhabitants are satisfied with the sound insulation. The condition $D_{nT,w} \geq 54$ dB leads to an estimated 70% of the inhabitants satisfied with sound insulation [88], where $D_{nT,w}$ is weighted standardized level difference. There have been many studies reported that focus on correlating the sound descriptors with subjective perception. The improved acoustic comfort class may be achieved when more than 90% of the inhabitants are satisfied with the sound insulation [89]. Rindel [90] investigations suggested that a change of 1 dB in sound insulation rating corresponds to a change of about 4% in the percentage of people disturbed by intruding noise.

7.8.1 Sound Insulation Criteria: National and International Perspectives

7.8.1.1 National Building Codes of India

In Indian perspectives, limited information prevails pertaining to the legal sound insulation criteria and use of sound insulation descriptors. Sound transmission class (*STC*) is the widely used single-number rating among manufacturers and testing laboratories in India. The National Building Code (NBC) is a national instrument providing guidelines for regulating the building construction activities across the country [83]. The weighted sound reduction index, R_w, is recommended to be 50 dB between the living room in one house or flat and living room and bed-room in another or elsewhere between houses or flats is recommended as 45 dB. R_w of 35 dB is specified between one room and another in the same house or flat. The floor of a room immediately above the bed room or living room shall result in an impact sound pressure level, $L'_{nT,w}$ not greater than 60 dB. Typically a 150 mm thick concrete floor with thick carpet (12 mm) covering would satisfy the requirement. The reverberation time in schools have been specified as 1–1.25 s for assembly halls, 0.75–1.25 s for music teaching rooms, and 0.6 s for classrooms. The reverberation time should not exceed 1 seconds in all general offices. For purpose of the sound insulation, rooms in educational buildings are classified from Class A to E as shown in Table 7.5. The recommended minimum sound reduction (D_w) between rooms of same class is 25 dB for Class A, 35 dB for Class C or D, and 45 dB for Class B or E. In case of schools, the concrete floor of the room immediately above the teaching rooms shall provide an impact sound pressure level, $L'_{nT,w}$ not greater than 70 dB. Sensitive areas such as operation theatres, intensive care units (ICU), a sound insulation (D_w) of 45 dB between the rooms shall be provided. The recent studies reported in European continent focused on revising the sound regulation requirements and recommendations on sound regulation criteria for acoustic comfort in terms of single-number ratings necessitates a similar retrospective analysis and legal framework in Indian scenario also.

Table 7.5 Classification A to E recommended in national building codes of India for rooms in educational buildings (National Building Codes of India 2016 [83])

Class A	Noise producing	(a) Workshops
		(b) Kitchens
		(c) Dining rooms
		(d) Gymnasiums
		(e) Indoor swimming pools
Class B	Noise producing but needing quiet at times	(a) Assembly halls
		(b) Lecture halls
		(c) Music rooms
		(d) Typing rooms
Class C	Average	(a) General class rooms
		(b) Practical rooms
		(c) Laboratories
		(d) Offices
Class D	Rooms needing quiet	(a) Libraries
		(b) Studies
Class E	Rooms needing privacy	(a) Medical rooms
		(b) Staff rooms

7.8.1.2 International Scenario

Acoustic classification schemes (ACS) label a number of quality classes for different levels of acoustic comfort and protection with best class first, that is, in descending order. Table 7.6 lists the descriptors used for sound insulation classes in some of European countries summarized by Rasmussen (2019) based on exhaustive survey [91]. There exist different schemes in different countries indicating the diversity of class intervals, ranges, and denotations reflecting different levels of acoustic comfort and protection. For instance, some countries have no low-frequency descriptors in any classes, while some countries applied $R'_{w} + C_{tr}$ descriptor in the upper two classes. Rasmussen work [92] reported potential for implementation of acoustic classification scheme as: new housing—Class B; new multi-storey housing—Class C and renovated housing—for example, Class D.

The regulations specify the minimum requirements, while qualification scheme have different quality classes reflecting different levels of acoustic comfort and protection. The ISO WG29 task for developing an ISO ACS based on COST TU0901 proposal has opted for a compromise solution whereby for the upper classes where high protection against noise is desired, the low frequency down to 50 Hz limits are included and alternate limits are indicated without low frequencies included. The low-frequency limits were not included as some countries objected to their inclusion due to high uncertainty and other reasons [93]. Table 7.7 shows the class limits for airborne sound insulation and impact sound pressure level [93]. It defines six classes (A to F) with 4 dB steps that correspond to a substantial step in subjective assessment between the classes; the total range is 20 dB plus an additional

Table 7.6 Airborne sound insulation between dwellings. Descriptors in acoustic classification schemes in Europe (Rasmussen 2019 [91])

Airborne sound insulation between dwellings—main class criteria in dB						
Country	Class A NL, IT:I DE/PL:III/AQ-4	Class B NL, IT:II DE/PL:II/AQ-3	Class C NL, IT:III DE/PL:I/AQ-2	Class D NL, IT: IV PL:AQ-1	Class E NL: V PL: AQ-0	Class F
DK	$R'_{w} + C_{50\text{–}3150}$	$R'_{w} + C_{50\text{–}3150}$	R'_{w}	R'_{w}	R'_{w}	R'_{w}
FI	$R'_{w} + C_{50\text{–}3150}$	$R'_{w} + C_{50\text{–}3150}$	R'_{w}	R'_{w}	N/A	N/A
IS	$R'_{w} + C_{50\text{–}3150}$	$R'_{w} + C_{50\text{–}3150}$	R'_{w} (5)	R'_{w}	N/A	N/A
NO	$R'_{w} + C_{50\text{–}3150}$	$R'_{w} + C_{50\text{–}3150}$	R'_{w} (5)	R'_{w}	N/A	N/A
SE	$D_{nT,w} + C_{50\text{–}3150}$	$D_{nT,w} + C_{50\text{–}3150}$	$D_{nT,w} + C_{50\text{–}3150}$	$D_{nT,w} + C$	N/A	N/A
LT	$R'_{w} + C_{50\text{–}3150}$ or $D_{nT,w} + C_{50\text{–}3150}$	$R'_{w} + C_{50\text{–}3150}$ or $D_{nT,w} + C_{50\text{–}3150}$	R'_{w} or $D_{nT,w}$ (5)	R'_{w} or $D_{nT,w}$	R'_{w} or $D_{nT,w}$	N/A
IT	R'_{w}	R'_{w}	R'_{w}	R'_{w}	N/A	N/A
DE	$D_{nT,w}$	$D_{nT,w}$	$D_{nT,w}$	N/A	N/A	N/A
DEGA	R'_{w}	R'_{w}	R'_{w}	R'_{w}	R'_{w}	npd
AT	$D_{nT,w} + C_{50\text{–}3150}$	$D_{nT,w} + C_{50\text{–}3150}$	$D_{nT,w}$	$D_{nT,w}$	npd	N/A
NL	$D_{nT,w} + C$	$D_{nT,w} + C$	$D_{nT,w} + C$	$D_{nT,w} + C$	$D_{nT,w} + C$	N/A
PL	$R'_{w} + C_{tr}$	$R'_{w} + C_{tr}$	$R'_{w} + C$	$R'_{w} + C$	$R'_{w} + C$	N/A
TR	$D_{nT,w} + C$	$D_{nT,w} + C$	$D_{nT,w} + C$	$D_{nT,w} + C$	$D_{nT,w} + C$	$D_{nT,w} + C$
ISO/ FDIS	$D_{nT,w} + C_{50\text{–}3150}$	$D_{nT,w} + C_{50\text{–}3150}$	$D_{nT,w} + C$	$D_{nT,w} + C$	$D_{nT,w} + C$	$D_{nT,w} + C$

(1) *DK* Denmark, *FI* Finland, *IS* Iceland, *NO* Norway, *SE* Sweden, *LT* Lithuania, *IT* Italy, *DE* Germany, *AT* Austria, *NL* Netherlands, *PL* Poland, *TR* Turkey

(2) Classes indicated in descending order, that is, the best class first, class A and B denote high acoustic comfort

(3) The descriptors indicated are from ISO/FDIS 19488 (Aug. 2018), the same as in ISO/NP TS 19488 (Jan. 2019)

(4) In addition, there is another scheme, DEGA-Empfehlung 103 with six classes A*-E and class F = npd, descriptor R'_{w} applied

(5) Use of $C_{50\text{–}3150/5000}$ is recommended also in Class C

(6) For row housing, class C has a special 5 dB stricter criterion to match the building regulations

denotation NPD (no performance determined). Many countries preferred their national regulations to be in Class C, which is difficult to match for all of them [93]. In England and Wales, $D_{nT,w} + C_{tr}$ is the descriptor used for airborne sound insulation between dwellings. An analysis of requirements in 35 European countries pertaining to the airborne sound insulation between the dwellings reported by Rasmussen (2019) showed that $D_{nT,w}$ values range from 48 to 56 among these countries; R'_{w} ranges from 50 to 55. Some countries like Spain have requirements in terms of $D_{nT,A}$ ($D_{nT,w} + C_{100\text{–}5000}$) ≥ 50, Sweden: $D_{nT,w} + C_{50\text{–}3150} \geq 52$, Poland: $R'_{w} + C \geq 50$, while for England and Wales $D_{nT,w} + C_{tr} \geq 45$ [92]. Table 7.8 shows

Table 7.7 Class limits for airborne sound insulation and impact sound pressure level in dwellings (Rasmussen and Machimbarrena 2019 [93])

Type of space	Class A dB	Class B dB	Class C dB	Class D dB	Class E dB	Class F dB
Airborne sound insulation between dwellings and other rooms—class limits						
Between habitable rooms in a dwelling and rooms outside the dwelling in all directions	$D_{nT,50} \geq 58$	$D_{nT,50} \geq 54$	$D_{nT,A} \geq 52$	$D_{nT,A} \geq 48$	$D_{nT,A} \geq 44$	$D_{nT,A} \geq 40$
From common stairwells or access areas into habitable rooms in dwellings, where there is an entrance door in the separating wall	$D_{nT,A} \geq 46$	$D_{nT,A} \geq 42$	$D_{nT,A} \geq 38$	$D_{nT,A} \geq 34$	$D_{nT,A} \geq 30$	$D_{nT,A} \geq 26$
From premises with noisy activities into habitable rooms in dwellings	$D_{nT,50} \geq 64$	$D_{nT,50} \geq 60$	$D_{nT,A} \geq 58$	$D_{nT,A} \geq 54$	$D_{nT,A} \geq 50$	$D_{nT,A} \geq 46$
Different descriptors are applied to reflect the use of different frequency range and weightings. Instead of $D_{nT,A}$ ($D_{nT,w}$ + C), $D_{nT,w}$ may be applied if 2 dB is added to the limit value. If $D_{nT,A}$ is applied instead of $D_{nT,50}$, 2 dB should be added to the limit value of $D_{nT,50}$						
Impact sound pressure level in dwellings—class limits						
In habitable rooms in dwellings from other dwellings in all directions	$L'_{nT,w} \leq 46$ and $L'_{nT,50} \leq 50$	$L'_{nT,w} \leq 50$ and $L'_{nT,50} \leq 54$	$L'_{nT,w} \leq 54$	$L'_{nT,w} \leq 58$	$L'_{nT,w} \leq 62$	$L'_{nT,w} \leq 66$
In habitable rooms in dwellings from: common stairwells and access areas, balconies or terraces or bath rooms not belonging to own dwelling	$L'_{nT,w} \leq 50$	$L'_{nT,w} \leq 54$	$L'_{nT,w} \leq 58$	$L'_{nT,w} \leq 62$	$L'_{nT,w} \leq 66$	$L'_{nT,w} \leq 70$
In habitable rooms in dwellings from premises with noisy activities	$L'_{nT,w} \leq 40$ and $L'_{nT,50} \leq 44$	$L'_{nT,w} \leq 44$ and $L'_{nT,50} \leq 48$	$L'_{nT,w} \leq 48$	$L'_{nT,w} \leq 52$	$L'_{nT,w} \leq 56$	$L'_{nT,w} \leq 60$
Experience has shown that when applying the low-frequency rating potentially disturbing high-frequency sounds are not rated appropriately, and for this reason, two descriptors are applied in order to account for both hard floor impact sounds as well as low-frequency footsteps sounds. The limit values for $L'_{nT,w}$ are 4 dB lower than those specified for $L'_{nT,50}$ Impact sound from small balconies and rooms (area less than 4 m^2) are not included, for example, toilets and utility rooms Premises with noisy activities are rooms for shared services like laundries, central boiler house, joint/commercial kitchens, or commercial premises like shops, workshops, or cafés. However, in each case, noise levels shall be estimated and the sound insulation designed accordingly, for example, for party rooms, discotheques etc.						

the in situ airborne sound insulation data as an illustrative example of current typical separating wall constructions (typical solutions and performance) in different countries [94]. A common acoustic classification scheme with a number of quality classes was proposed in COST Action TU0901 European project [94], wherein 32 countries participated. The COST Action TU0901 Acoustic Classification Scheme (ACS)

Table 7.8 Current typical separating wall constructions in different countries (Fausti et al. 2014 [94])

Country	Wall description	Approx. total width (mm)	Typical airborne sound insulation		
			Country descriptor (100–3150 Hz)	$D_{nT,w}$ (100–3150 Hz)	D_{nTA} (or $D_{nT,w} + C$) (50–5000 Hz)
Australia	Two leaves of 110 mm clay brick masonry with 50 mm cavity and 50 mm thick glass wool/polyester insulation and 13 mm plaster board fixed to studs both sides	300	$D_{nT,w} + C_{tr}$ 45 dB	49 dB	49 dB
Denmark	200 mm concrete	200	R'_w 55 dB	53 dB	53 dB
France	180 mm concrete	180–300	$D_{nT,w}$ + C 53 dB	53 dB	53 dB
Germany	240 mm brick (density equal or greater than 2000 kg/m^3), plaster board on both sides +15 mm plaster	270	R'_w 53 dB	52 dB	51 dB
Netherlands	120 mm limestone – 60 mm cavity – 120 mm limestone	250–300	R'_w + C 54 dB	55 dB	54 dB
Slovenia	200 mm reinforced concrete wall with 50 mm aerated autoclaved concrete and 10 mm plaster finish	260	R'_w 57 dB	57 dB	55 dB
Turkey	Gypsum plaster both sides + gypsum board + metal profile +50 mm glass wool +100 mm concrete hollow block	180	R'_w 49 dB	48 dB	47 dB
Austria	250 mm cellular clay block, lined with 70 mm mineral wool backed gypsum board on both sides and plaster on one side	340	$D_{nT,w}$ 61 dB	61 dB	59 dB
UK	100 mm solid aggregate (1500 kg/m^3) block with 75 mm cavity, 8 mm render coat with 12.5 mm gypsum board	320	$D_{nT,w} + C_{tr}$ 53 dB	60 dB	56 dB
Poland	240 mm calcium silicate wall plastered on both sides with 10–15 mm thick gypsum plaster	260	R'_w + C 51 dB	53 dB	52 dB
Czech Republic	15 mm lime plaster, 250–300 mm hollow bricks, 15 mm lime plaster	280–300	R'_w 53 dB	52 dB	51 dB
Croatia	20 mm plaster +201 mm reinforced concrete +20 mm plaster	250	R'_w 53 dB	52 dB	51 dB
Sweden	200 mm homogeneous on-site cast concrete	200	R'_w + C 57 dB	56 dB	54 dB

Table 7.9 Translation equations between descriptors (Monteiro et al. 2017 [95])

(x)	(y) $D_{nT,50}$	$D_{nT,100}$
R'_w	$y = 0.63x + 20.23$	$y = 0.83x + 9.60$
$R'_w + C$	$y = 0.71x + 16.89$	$y = 0.90x + 7.40$
$R'_w + C_{(50\text{–}3150\ \text{Hz})}$	$y = 0.89x + 7.77$	$y = 0.85x + 12.30$
$D_{nT,w}$	$y = 0.71x + 14.77$	$y = 0.92x + 3.63$
$D_{nT,w} + C_{tr}$	$y = 0.85x + 12.02$	$y = x + 5.83$
$D_{nT,w} + C$	$y = 0.80x + 11.02$	$y = x + 1.0$
$D_{nT,A\ (100\text{–}5\ \text{kHz})}$	$y = 0.79x + 10.64$	$y = x + 0.23$

proposal recommends airborne sound insulation between a dwelling and premises with noise activities as $D_{nT,50} \geq 68$ as class A limit. These limits are further reduced by 4 dB for class limits B to F ($D_{nT,50} \geq 48$). A recent study by Monteiro et al. (2017) [95] translated the existing airborne sound insulation national requirements in the 32 countries to $D_{nT,50}$ using the developed translation equations. The equivalent requirements of $D_{nT,50}$ varied from 50 ($D_{nT,50} > 50$ for countries like Portugal and Spain) to 55 ($D_{nT,50} > 55$ for countries like Denmark and Switzerland). Table 7.9 describes the translation equations between descriptors [95]. The inclusion of low-frequency performance is a subject of consideration, although for the impact sound as well there seems to be a consistent international attention to the low-frequency aspect [96]. However, harmonization of these descriptors pertaining to the acoustic regulations and classification schemes shall facilitate exchange of knowledge about constructions having enhanced sound insulation and optimizing the acoustic comfort criteria [97].

A study by Garg et al. 2013 [98] recommended the airborne sound insulation both as prescriptive and verification criteria. The Class C proposed referred to the minimum criteria that must be followed as chosen in other countries, while Class B represented the normal acoustic comfort (NAC) criteria and is defined for the acoustic comfort having an average noise level reduction of the order of 35 dB. Class A represented the enhanced acoustic comfort (EAC) criteria, whereby high acoustic comfort levels achieved through a noise level reduction of 40 dB. The $R_w + C_{tr,50\text{–}3150} \geq 50$ dB criteria is chosen for Class A, when the outdoor noise level is within 70 dB(A) to 75 dB(A). The criteria are enhanced by 5 dB for an outdoor noise level more than 75 dB(A). Clarity in description of legal requirements for minimum insulation required can be very helpful for the selection of suitable building materials. Administrative and legal aspects particularly related with the enforcement of sound regulation requirements (either prescriptive or verification [98]) and inculcating knowledge about strengthening the building facades through training and awareness programs shall be a vital step in controlling the outside traffic noise.

7.9 Noise Barriers

Noise barriers are one of the most effective ways to mitigate road traffic noise although being the least cost-effective. Noise barriers can reduce the A-weighted noise levels depending on their design and height. If the barrier surface density exceeds 20 kg/m^2, a reduction of 5 dB can be achieved by having a barrier tall enough to break the line of sight from the road to the receiver and an additional 1.5 dB reduction can be achieved for each additional meter of height [99]. The attenuation effect of a barrier is sensitive to the atmospheric conditions because of diffraction and refraction. The influence of these phenomena varies with the weather. This makes it difficult to calculate how much reduction a noise barrier can actually bring [100]. Noise barriers derive their performance by blocking the line of sight, thus creating a sound shadow [101], as shown in Fig. 7.9. The profiled noise barriers are very useful for screening transportation noise in situations where the maximum height of the barrier needs to be limited and extra screening is required.

Noise barriers are generally classified into two basic categories depending on its structure: (a) ground-mounted, and (b) structure-mounted. There are three basic types of ground-mounted noise barriers systems. These are: (a) noise berm, (b) the noise wall, and (c) combination of noise berm and noise wall. Structure-mounted noise barriers systems can be classified in two categories. These are: (a) noise wall on bridges, and (b) noise wall on retaining wall [102]. The major innovative diffraction-edge modifications include T-shaped barriers, multiple-edge barriers, Y-shaped barriers, tubular capping, and phase interference devices. Multiple edge barriers give consistent improvements in insertion loss over a wide area compared with plane reflective barriers. Apart from attenuation, economical, structural, aesthetic constraints, compatibility with environment, installation, maintenance, and safety aspects equally play an important role. For instance, T-shaped barriers with an

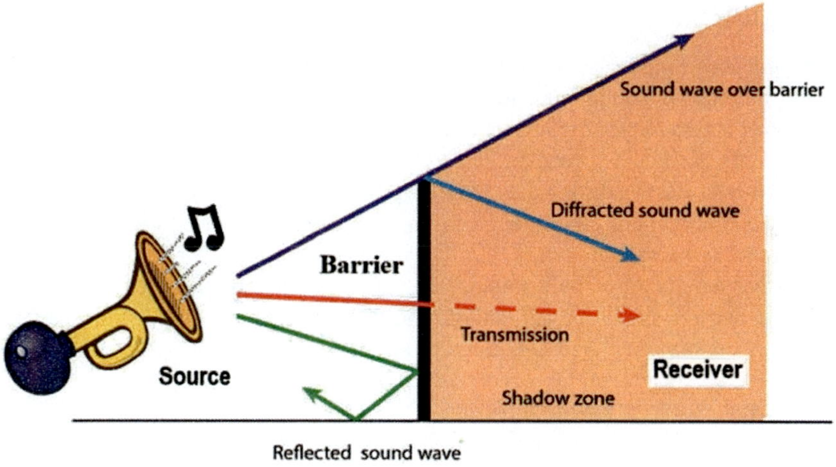

Fig. 7.9 Diffraction effect of a noise barrier

absorptive material have a significant potential of reducing noise levels by 2–3 dB, which could reduce the barrier height by an average 5 ft; however, they suffer from the disadvantage of persistence maintenance and replacement of absorbing material after sometime. Thus, it is critical for designer to consider all the worthy factors and develop a decision matrix allocating suitable weights to various factors. The evaluation matrix could be used to score the best practicable option based on acoustic performance as well as economic, constructability, maintenance, and aesthetic considerations. In one such study of Arizona Department of Transportation (ADOT) on scoring the barrier options T-top design with absorptive material is recommended over the conventional barrier of concrete or masonry construction [103].The European EN 1793 standards refers to the intrinsic acoustic performance of noise barriers or Noise Reducing Devices (NRDs) and for in-situ performance assessment. A single-number index of the sound absorption evaluation DL_α and airborne sound insulation DL_R is recommended. Similarly, a single-number rating of sound reflection performance weighted to the normalized traffic spectrum, DL_{RI} and that for airborne sound insulation performance weighted to the normalized traffic spectrum, DL_{SI} are also recommended. The EN 1794 standards deal with non-acoustic performance and EN 14389 standards deal with long-term performance. The details of in-situ performance evaluation of noise barriers is discussed in Appendix part of the book.

7.9.1 Noise Barrier Design Morphology

The efficiency of noise barrier largely depends upon its geometry. Besides the height of the barrier and its top element form, the cross section of the barrier contributes to its performance well. The boundary element method is often used a numerical tool for simulating the behavior of proposed barrier shapes, both in 2D and 3D spaces [104].

- Sound absorbing material is often used on the source side of the barrier to reduce the buildup of sound pressure level [101]. An absorptive treatment may be appropriate to reduce the impact of reflected noise.
- The "line of sight" between the source and receiver must be cut off completely by the barrier. Barrier design also play a major role where effective noise barriers should be high and long enough to break the line-of-sight between the sound source and the receiver.
- The barrier should be built as close as possible to either the source or receiver. Noise barriers should be tall enough and long enough to avoid degradations in barrier performance. Barriers do not necessarily have to be of constant height as varying the height of the barrier may also help to alleviate the monotonous appearance of long length of barriers.
- Barriers should be constructed of solid, non-porous material. Minimum density of material should be 20 kg/m^2. The sound transmission loss of barrier material must be at least 10 dB higher than the barrier attenuation [101]. The materials used for fabrications of noise barrier can be: steel (painted, galvanized, stainless),

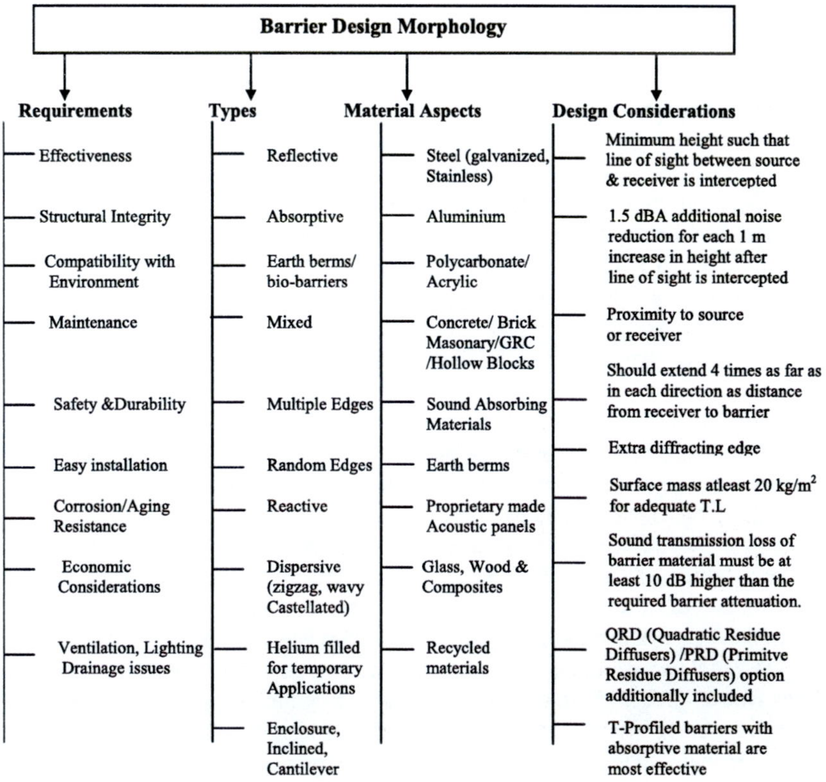

Fig. 7.10 Design morphology for noise barriers (Garg et al. 2012, [9])

aluminum, polycarbonate or acrylic sheets concrete, brick or glass fiber-reinforced concrete (GRC), proprietary-made acoustic panels and landscaped earth berms.

- Capped barriers of same height as conventional barriers could potentially provide greater noise reductions than conventional barriers. A specified noise reduction could potentially be provided by a capped barrier of lower height than a conventional barrier [105]. If high quality absorptive materials are available, the relative screening performance of cylinder-profile and mushroom-profile caps can be improved, by up to 8 dB(A) relative to a plane screen of the same overall height. However, the same can be said for T-profile configurations when the uppermost surfaces are treated with the same materials. Double cylinder-profile caps were observed to be more efficient than single cylinder-profile caps. When absorptive these offered similar average screening performance to T-profile configurations with the same overall height [106]. A 3 m high T-shaped barrier provides the same performance as a 10 m high plain barrier [107].

The potential benefits of strategic planting may also be considered so as to reduce the monotonous appearance of long length of barriers. Planting greatly assists in reducing the dominance of noise walls. Figure 7.10 shows the design morphology for noise barriers [9].

In general, the higher the barrier, the greater the level of noise reduction. In the USA, barrier heights of 6 and 7 m are very common. In Japan, normal barrier heights for highways is 3 m and measured barrier noise reduction ranges between 15 and 23 dB, but barriers of height 8 m are used in suburbs. Barrier heights of 5 to 6 m are commonly used in Australia. Measured insertion losses of barriers of height ranging from 3 to 6 m are typically between 5 and 12 dB. Typically, if a barrier thickness is more than 3 m, the barrier is considered as thick [108]. A sound barrier eventually will become the part of the surrounding landscape and could be a cause of impact for ecosystems, the road uses and those who live alongside the road. In one such study reported by Arenas [99], although most residents felt that sleeping conditions improved after the barrier was built, most negative reactions were the loss of sunlight and visual impact. Attention should be paid for selection of materials used in the construction of barriers, particularly for the areas experiencing extreme weather conditions. A good design should also account for minimal maintenance, other than cleaning or damage repair for many years [99]. Earth berm noise barrier types are generally the most sustainable. The relative sustainability of road noise barrier types using the three multi-criteria analysis (MCA) techniques, that is, simple additive weighting (SAW), Preference Ranking Organization Method for Enrichment Evaluations (PROMETHEE), and Elimination et Choice Translating Reality (ELECTRE III), has been presented [109]. Slutsky and Bertoni [110] showed that only a 3° angle is enough to remove the effect of multiple reflections between screens (for a 45 m wide road). For narrower roads, the angle should be between 10° and 15°. Since absorbing materials are quite expensive, it might be a good and more cost-effective alternative to tilt the noise barriers instead [100]. Hothersall et al. compared different types of noise barrier tops: T-, Y-, and arrow-shaped. The Y- and arrow profiles performed less efficiently than the T-profile [111]. For both reflective and absorptive treatments, the relative insertion loss of a T-top increases with increasing the top width, but the benefit to be obtained for widths in excess of 1.6 m may not be an economic proposition. The absorptive treatment applied to T-top should have a high value of absorption coefficient for frequencies center around 1 kHz. Significant attenuation can be obtained with an absorptive top for highways [112]. However, there are practical disadvantages of using an absorbing surface as it might be damaged by moisture, water, and dust and lose their properties over time. Recent work has tended to be concentrated on more complex devices such as the use of quadratic residue diffusers (QRD), on the optimization of multi-edge barriers and T-tops and also the basic shapes of noise barriers. The best shape of barriers for utilizing QRD among the tested models is the T-shape profile barrier. It is found that reducing the design frequency of QRD shifts the performance improvement toward lower frequency, and therefore the most efficient model for traffic noise is a barrier covered with a QRD tuned to around 400 Hz. The amount of improvement in mean insertion loss with a 400 Hz tuned QRD T-shape barrier compared to an equivalent absorbent T-shape barrier was found to be 0.9 dB(A). Reducing the well width of utilized QRD on T-shape profile barriers reduces the overall A-weighted insertion loss of the barriers [113]. Reducing the well width in quadratic residue diffuser increases the absorption coefficient of diffusers because of viscous and thermal effects in the narrower wells [114]. Studies have shown that the efficiency of T-shaped barriers with reactive top surfaces is highly frequency dependent and the global efficiency of such barriers increases with number of wells considered on the top surface [115]. The improvement in global efficiency compared with a flat top surface T-shaped barrier can be up to 2–3 dB(A) for 5–9 wells. Noise barriers and embankments should have a robust appearance and harmonize with the character of the road and its surroundings. The use of different materials, colors, and of vegetation can contribute significantly to breaking the monotony of long and high noise barriers.

7.9.2 Insertion Loss Calculation

The effectiveness of noise barrier is expressed by "insertion loss" descriptor and determined from an in situ measurement according to ISO 10847 standard that cover only typical in situ situations [116]. Insertion loss is sometimes also termed as attenuation and is calculated by the difference between the acoustic level in absence (L_{dir}) and in presence of the barrier (L_{screen}) as:

$$IL = L_{\mathrm{dir}} - L_{\mathrm{screen}} \tag{7.5}$$

ISO 10847:1997 establishes a method for determining the insertion loss of noise barriers, assuming the existence of only one insertion loss for the same outdoor noise barrier. Since during the day distinct vertical gradients of sound speed may occur, different curved path radius of the sound rays can be formed, which leads to a different insertion loss throughout the day, for the same noise barrier [117]. The insertion loss is calculated in terms of path length difference. Path length difference is defined as the difference between the diffracted path from the source over the top

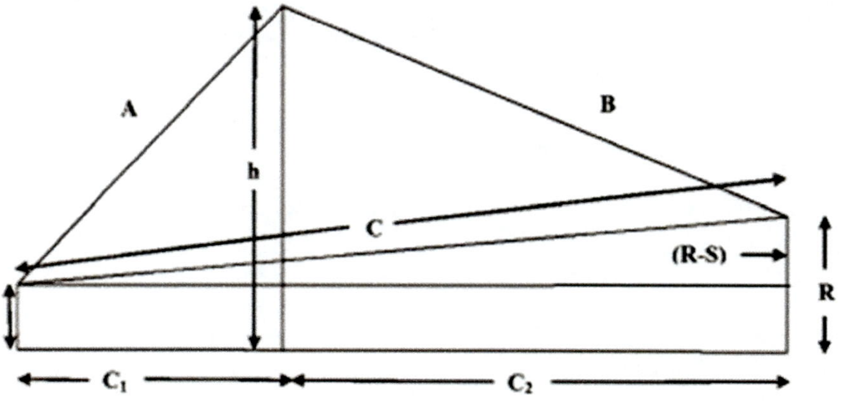

Fig. 7.11 Path length difference (δ)

of barrier to the receiver and the direct path from source to the receiver as if barrier were not present [118]. Path length difference, δ (Fig. 7.11) can be calculated as: $\delta = A + B - C$, where A is the distance of source from barrier top, B is the distance of receptor from barrier top, and C_1 is the distance of source from barrier and C_2 is the distance of receiver from the barrier, h is the height of barrier, R is the height of receiver above the roadway, and S is the source height of vehicle above the roadway. Fresnel number (N) is a dimensionless value used to predict the attenuation provided by a noise barrier, defined as $N = \frac{2\delta}{\lambda}$.

$$A = \sqrt{C_1^2 + (h - S)^2} \tag{7.6}$$

$$B = \sqrt{C_1^2 + (h - R)^2} \tag{7.7}$$

$$C = \sqrt{(C_1 + C_2)^2 + (R - S)^2} \tag{7.8}$$

Maekawa's work is well established for predicting the insertion loss behind the rigid, thin barriers [119]. A mathematical description of this curve, which involves simple calculations and is therefore often used instead of the Maekawa curve, was suggested by Kurze and Anderson [120]. The Maekawa chart and the corresponding Kurze–Anderson formula have found extensive use. Their main advantages are their simplicity and the accuracy of their predictions. The Maekawa curve can be described mathematically by the following formula suggested by Kurze and Anderson having only a small discrepancy of 1.5 dB from the curve for $N < 1$. Maekawa's curve and the exact solution are in agreement for a broad range of cases. The discrepancies are less than 1.5 dB. Discrepancies greater than that can be found in the following two cases: (a) source or receiver is very close to or at the barrier; (b) receiver is very close to or at the shadow boundary [121].

7.9.2.1 Kurze Anderson Equation

Insertion loss can be estimated by using the model proposed by Kurze and Anderson [120] given by the following equation:

$$IL = 5 + 20\,\log\left(\frac{\sqrt{2\pi \mathrm{N}}}{\tanh\sqrt{2\pi \mathrm{N}}}\right)\mathrm{dB},\ \text{up to } N = 12.5 \tag{7.9}$$

$$IL = 20\ \mathrm{dB}\ \text{for}\ N > 12.5,\ N\ \text{is the Fresnel number} \tag{7.10}$$

Maekawa's curve is one of the most established methods for predicting the insertion loss behind the barriers. For the simple case of a barrier modeled as a half plane, the insertion loss (IL) is given versus a single parameter, the Fresnel number (N_1). Maekawa's curve gives predictions that deviate largely from experimental data and analytical solutions when the source or receiver is very close to or at the barrier surface, or when the receiver is very close to or at the shadow boundary.

The first Fresnel number (N_1) is the conventional Fresnel number and is associated with the relative position of the source to the barrier and the receiver. The second Fresnel number (N_2) is defined similarly to N_1 and represents the relative position of the image source to the barrier and the receiver. It is shown that if a second Fresnel number (N_2) is appropriately defined, the *IL* obtained by the existing analytical solutions can be expressed versus N_1 and N_2 for several types of incident radiation (plane, cylindrical, and spherical). Accordingly, the single curve in Maekawa's chart can be replaced by a family of curves. Each curve corresponds to a different N_2 and provides the IL versus N_1. The corrections to Kurze and Anderson expression were provided by Menounou (2001) to make it more suitable for locations of the source or receiver close to the barrier or for the receiver close to the boundary of bright and shadow zones. The analytical expression for insertion loss is [121]:

$$IL = IL_s + IL_b + IL_{sb} + IL_{sp} \tag{7.11}$$

where

$$IL_s = 20\log_{10}\frac{\sqrt{2\pi N_1}}{\tanh\sqrt{2\pi N_1}} - 1 \tag{7.12}$$

$$IL_b = 20\log_{10}\left[1 + \tanh\left(0.6\log_{10}\frac{N_2}{N_1}\right)\right] \tag{7.13}$$

$$IL_{sb} = \left(6\tanh\sqrt{N_2} - 2 - IL_b\right)\left(1 - \tanh\sqrt{10N_1}\right) \tag{7.14}$$

$$IL_{sp} = \begin{cases} 3\ \text{dB} & \text{for plane waves} \\ 10\log_{10}((1 + (A+B)/d) & \text{for cylindrical waves} \\ 10\log_{10}\left((A+B)^2/d^2 + (A+B)/d\right) & \text{for spherical waves} \end{cases} \tag{7.15}$$

The quantity N_2 is the Fresnel number calculated for a wave traveling from the image source to the receiver where the image source is generated by reflection from the barrier. The second term is a function of N_2/N_1 which is the measure of the proximity of either the source or the receiver to the barrier surface. The third term is appreciable and needs to be computed only when N_1 is very small, which in turn is a measure of the proximity of the receiver to the shadow boundary [121, 122]. A Matlab code for the design of noise barriers is included in Appendix VI of book.

7.9.2.2 ISO 9613 Formula

The diffraction effect is calculated as [123]:

$$D_z = 10 \log \left[3 + \frac{C_2}{\lambda} C_3 z k_{met}\right] \text{ dB} \tag{7.16}$$

where C_2 is 20 and include the effect of ground reflection, C_3 is 1 for single diffraction, $C_3 = \left[\frac{1+\left(\frac{5\lambda}{e}\right)^2}{\frac{1}{3}+\left(\frac{5\lambda}{e}\right)^2}\right]$ for double diffraction, where λ is wavelength of sound at the nominal mid band frequency of octave, z is difference between path lengths of direct and diffracted sound, k_{met} is correction factor for meteorological effects, and e is distance between two diffraction edges in case of double diffraction. For single diffraction, the path length difference z is calculated by

$$z = \sqrt{(d_{ss} + d_{sr})^2 + a^2} - d \tag{7.17}$$

where d_{ss} is distance from source to first diffraction edge in meters, d_{sr} is distance from diffraction edge in meters and a is component distance parallel to barrier edge between source and receiver in meters. For double diffraction, the path length difference z is calculated as:

$$z = \sqrt{(d_{ss} + d_{sr} + e)^2 + a^2} - d \tag{7.18}$$

The correction factor k_{met} for meteorological condition is calculated as:

$$k_{met} = \exp\left(\frac{-1}{2000}\sqrt{\frac{d_{ss} d_{sr} d}{2z}}\right) \text{ else } k_{met} = 1 \text{ for } z \le 0 \tag{7.19}$$

7.9.2.3 Harmonoise Model

The Harmonoise model calculates the insertion loss caused by a screening object using Deygout's solution based on Fresnel number, N_f as follows [124]:

$$N_f = \text{sign}\,(h_{eff}) \frac{2(R_S + R_R - R)}{\lambda} \tag{7.20}$$

where h_{eff} is effective barrier height, R_S is path length between source and top edge of barrier, R_R is path length between type of barriers and receiver and sign (h_{eff}) is zero for $h_{eff} = 0$, sign $= -1$ for $h_{eff} < 0$ and sign $= 1$ for $h_{eff} > 0$. The attenuation due to barrier is given by:

$$A_{diff} = 0 \text{ for } N_f \le -0.25 \tag{7.21}$$

$$A_{\text{diff}} = 6 - 12\sqrt{-N_{\text{f}}} \text{ for } -0.25 \leq N_{\text{f}} < 0 \quad (7.22)$$

$$A_{\text{diff}} = 6 + 12\sqrt{N_{\text{f}}} \text{ for } 0 \leq N_{\text{f}} < 0.25 \quad (7.23)$$

$$A_{\text{diff}} = 8 + 8\sqrt{N_{\text{f}}} \text{ for } 0.25 \leq N_{\text{f}} < 1 \quad (7.24)$$

$$A_{\text{diff}} = 16 + 10 \log (N_{\text{f}}) \text{ for } N_{\text{f}} \geq 1 \quad (7.25)$$

7.9.2.4 Optimization in Noise Barrier Design

The optimization in barrier design is to reduce the costs by decreasing the barrier dimensions, such as heights and lengths. The function must be minimized as [125]:

$$\text{Minimize}\, E(X) = \sum_{i=1}^{k} H_i r_i \quad (7.26)$$

where $X = \{H_1. \ldots .H_2. \ldots\ldots, H_{k,} r_1. \ldots .r_k\}$ and H_i and r_i are barrier height and length. The constraints on overall sound pressure level are followed by:

$$L_0 - D_z \leq L_b \text{ for target sound pressure level}, L_b, \text{ and} \quad (7.27)$$

$\lambda < l_l + l_r$, for the wavelength condition (horizontal dimension that is normal to the source-receiver line is larger than the acoustic wavelength, λ at nominal mid-band frequency band). L_0 is the overall sound pressure level without barrier, and D_z is the barrier attenuation that should not be greater than 20 dB in the case of single diffraction and 25 dB in the case of double diffraction. A robust method to design noise barrier using the global optimization of simulated annealing algorithm had been proposed by Mun et al. (2009) [125] that yields cost reductions. Such an analysis utilizing the heuristic algorithms can be used for optimization of costs associated with noise barrier design.

7.9.3 Noise Barrier Applications

Innovative combinations such as perforated metal sheeting on one side with inclusion of absorptive material within; corrugated profile adds more value to the insertion loss provided by barrier. However, the conventional concrete or masonry structure serves to be the best practicable option for highway barriers considering the long-term stability and maintenance free services as there are some limitations associated with transparent and polycarbonate barriers. Laminated glass requires frequent cleaning due to accumulation of dirt, while polycarbonate sheets become opalescent over time as it can absorb water. Plastics are prone to damage from fire and vandalism and some, for example, polyethylene, become brittle after prolonged exposure to sunlight. Application of multiple edges noise barriers or the so-called profiled noise barriers having better acoustical performance than conventional concrete structures can be very effective for specific applications wherein a single source has to be screened. A case study of noise barrier designed by CSIR-National Physical Laboratory, New Delhi, for Delhi Metro Rail Corporation (DMRC), India, that consists of half Y shaped structure that includes a 2 mm thick aluminum alloy as the backplate, 1 mm thick aluminum alloy sheet (30% open) as facing surface, 50 mm thick glasswool of density 40 kg/m^3 wrapped in tissue bags and polythene cover and the whole assembly will be fixed in GI framework serves as an excellent illustration of specific source noise abatement [126]. "T" is the simplest crowning that can be added to the top of barriers and gains of the order of 2–3 dB (A) can be obtained depending on the barrier height, the cap width, the surface impedance, and the source-receiver configuration [115]. Monazzam studies [113, 114] revealed that a QRD (quadratic residue diffuser) barrier could increase the performance of rigid barrier within certain frequency domain depending on frequency design of the QRD on the top. The results showed that the best shape of barrier for using QRD among the tested models is T-shaped profile barrier. The improvement of mean insertion loss with a 400 Hz tuned QRD T-shaped barrier compared to a T-shaped barrier with flat absorbent top surface was found to be around 1 dB(A). The improvement in global efficiency compared with a flat top surface T-shaped barrier can be up to 2–3 dB(A) for 5–9 wells [115]. Also special bio-barriers can be designed to cater individual requirements without affecting the aesthetics of the surroundings. The use of noise barriers is gaining importance in recent times in Indian perspectives also. In one such study of Rominder et al. (2013) [127], a noise barrier was designed for busy Britannia Chowk flyover in Delhi city of India. The noise barrier of height 2.67 m made of glass fiber-reinforced plastic sheet had been effective in reducing noise level to acceptable standard both in day time and night time. The noise barrier has been effective in reducing noise level by 19 dB (A) during the day time and the night time.

Internationally, there have been many innovative solutions designed and implemented pertaining to noise barrier applications. One example is a photovoltaic noise barrier (PVNB), as shown in Fig. 7.12, where a noise barrier located along a highway or railway is used as substructure for PV modules. A PVNB can be an example of a public-private partnership where three parties government, residents, and private investors can benefit from [128]. A PVNB located near Töging, Germany, one of the largest PVNBs in the world, is approximately 1 km long and 6000 m^2 in area, and has nearly a 2 MWp capacity. Generally, the degree to which PVNBs attenuate noise levels hinges on the proportion of the solar panels' glass surfaces to the noise barrier's other materials, as the glass surface of a PV module can only be applied for sound reflection [129]. PVNB barriers find good opportunities of electricity production. Recent years have seen a growing interest in the potential for use of sonic crystals also as noise barriers. Sonic crystals take their name from the analogous effects of photonic crystals on light. Rigid scatterers like wood or aluminum have been used to study the behavior of sonic crystals reflecting

Fig. 7.12 LHS: Picture of a photovoltaic noise barrier (Source: KOHLHAUER GmbH, Germany); RHS: sonic crystal noise barrier consisting of three rows of microperforated cylindrical shells 3 m high with diameter 16 cm distributed in a square lattice with lattice constant 22 cm (Garcìa-Chocano et al. 2012, [134])

the acoustic waves. They consist of periodic arrays of circular cylinders and are known to give high attenuation at selective frequencies as a consequence of multiple scattering. The frequencies of highest attenuation can be found by assuming that an integer number of half wavelengths fits the distance between the scatterers. Additional attenuation bands may be achieved if the array is composed of resonant elements instead of solid cylinders. The main advantage of sonic crystals (SCs) is that by varying the distance between the cylinders and the design of the resonant elements it is possible to attain peaks of attenuation in a certain range of frequencies. In principle, if SC barriers are used as highway noise barriers on both sides of a road, there should be less reverberation than between solid barriers [130]. The acoustic barriers based on SC are usually called sonic crystals acoustic screens (SCAS) are very light and easily built, and on the other, by varying the constant lattice and the filling fraction it is possible to adjust the attenuation bands to the sound spectrum to be reduced [131]. Sonic crystal barrier consist of periodic arrays of circular cylinders that are most commonly in a vertical rather than horizontal alignment. They offer high attenuation at selected frequencies based on spacing of the cylinders and design of cylinders. Sound attenuation up to 20 dB and 25 dB was obtained [131, 132] with arrangements of metallic cylinders in air and it has been concluded that sonic crystal can compete with mass law based sound screens with the advantage of less volume and weight. A systematic study of noise barriers based on sonic crystals made of cylinders that use recycled materials like absorbing component was reported by Sánchez-Dehesa J et al. (2011) [133]. The barriers consisted of only three rows of perforated metal shells filled with rubber crumb possess attenuation properties resulting from a combination of sound absorption by the rubber crumb and reflection by the periodic distribution of scatterers. The results suggested that the proposed structures (sonic crystal consisting of microperforated cylindrical shells) are suitable to attenuate sound in severe noise environments [134]. Figure 7.12 shows the pictorial view of a sonic crystal noise barrier [134].

7.9.4 Effect of Vegetation

Vegetation can be used in urban areas for transportation noise control as relatively inexpensive and aesthetically pleasing noise barriers. The visual shielding of a noise source has considerable psychological advantage as well. A minimum of 2–5 m wide dense shrubs followed by medium height trees are effective in offering 2–5 dB (A) noise reduction from traffic noise in residential areas [135]. Earth mounds are characterized by their distinct advantages over other noise barriers on account of their natural appearance, less cost, easiness to maintain, and unlimited life span. Depending upon the availability of space and other statutory requirements one can work out suitable noise control packages consisting of earth berms, plantations, bushes, shrubs, small noise barriers, etc. keeping in view of economy, logistics, aesthetics of the surroundings, etc. into consideration [136]. Acoustical landscaping is defined as the optimization of the relief (shaping and ground cover) to increase the noise shielding of a source. Two common examples are depressing the road and raising the landscape near the road. Acoustical landscaping belongs to the group of so-called environmental methods for the purpose of (surface transport) noise reduction [137]. Natural environments are highly preferred by the people [138], leading to improved noise perception [139]. Tyagi et al. (2013) [140] results indicated that sufficiently dense vegetation belts across the roadsides may prove as effective noise barriers and significant attenuation may be achieved over the critical middle frequencies (1–4 kHz). Excess attenuation of 15 dB or more is observed at 1 kHz for a number of sites. Attenuation performance of vegetation belts in the low-frequency range (200–500 Hz) and over the high-frequency range (8–12.5 kHz) is found to be even better. Huddart [141] reported a variation of 5.4 L_{10} dB(A) to 9.3 L_{10} dB(A) in attenuation per doubling of distance over seven different sites, while Tyagi et al. observed an attenuation of 5.1 L_{10} dB(A) to 10.9 L_{10} dB(A) over 11 different sites. Vegetation is a complex medium consisting of air, soil, stems, trunks, branches, and foliage. Thus, when used as a barrier against noise, the effects vary extensively. Van Renterghem et al. (2012) [142] predictions suggested that practically achievable vegetation belts can compete to the noise reducing performance of a classical thin noise barrier (on grassland) with a height of 1–1.5 m (in a nonrefracting atmosphere). Significant noise reduction is predicted to occur for a tree spacing of less than 3 m and a tree stem diameter of more than 0.11 m. The effect of the presence of tree stems, shrubs, and tree crowns is predicted to be approximately additive. The vegetation belt has additional non-acoustic advantages such as being CO_2-neutral or positive, having a much higher aesthetic value, and its potential to improve local air quality, for example, by capturing road traffic-originated fine particles. The study utilized full 3D finite-difference-time domain calculations and revealed that height was not a critical parameter as traffic noise usually occurred at low heights, close to the road surface. Pathak et al.'s (2008) studies [143] showed that the maximum reduction occurs in between frequency range 400 Hz and 4 kHz. Scattering by vegetation is a high-frequency phenomenon [142], while the peak attenuation at low frequency is due to the ground effect [144]. Aylor (1972) suggested that attenuation predicted by the scattering theory agreed well with field observations when [145]:

$$\frac{2\pi a}{\lambda} \geq 1 \tag{7.28}$$

where a is the radius of trunk and λ is the wavelength of sound. This implies that a trunk/branch diameter of ~ 11 cm or more is required for scattering of sound at 1000 Hz. For frequencies lower than 1 kHz, ground attenuation is considered to be the prominent factor in absorption of sound due to formation of a highly porous humus layer. For higher frequencies (above 1 kHz), the attenuation is caused by vegetation. Trunks and branches will mainly reduce the sound by scattering. Leafs on the other hand will scatter and absorb the sound due to viscous dissipation and energy losses caused by heat conduction through the boundary layer of the leaves [145, 146]. The important scientific conclusions realized from various studies in recent years pertaining to the effect of vegetation in noise control are as follows:

- Significant noise reduction is predicted to occur for a tree spacing of less than 3 m and a tree stem diameter of more than 0.11 m [142].
- The effect of the presence of tree stems, shrubs and tree crowns is predicted to be approximately additive [142].
- With increasing tree stem diameter and decreasing spacing, traffic noise insertion loss is predicted to be more pronounced [142].
- The effectiveness of vegetation in attenuating traffic noise is best paired with a setback made up of soft ground, which becomes more effective as the distance from the highway increases. A belt of vegetation of 5 m depth is ideal to achieve traffic noise reduction [147].
- A medium to sparse vegetative barrier gave rise to highest potential of noise abatement. Dense, interlocking vegetation may not be an absolute for noise reduction [147].
- Greater trunk sizes that generally contribute to denser planting intensity was shown to have a slight advantage over smaller trunk sizes [147].
- Tree belts need not necessarily be densely planted as mean noise insertion loss of 6.4 dB(A) for dense planting and 5.2 dB(A) was found for sparse to medium planting amounting to a difference of 1.2 dB(A) [147].
- One such study [148] suggested an inverse relationship between aesthetics and perception of how noise barrier would perform. The transparent and deciduous vegetative noise barrier, judged most aesthetically pleasing were judged as least effective in attenuating noise.
- Width of the flora belt is a significant noise reducing factor [149]. A greater attenuation occurs when the source and receiver height are lower than flora belt owing to the existence of shadow zone behind the barrier.

Figure 7.13 shows the vegetative noise barriers for urban noise abatement [150]. The major disadvantages of using vegetation for screening is that the benefit is not immediate and it may take many years for the plants to attain sufficient height

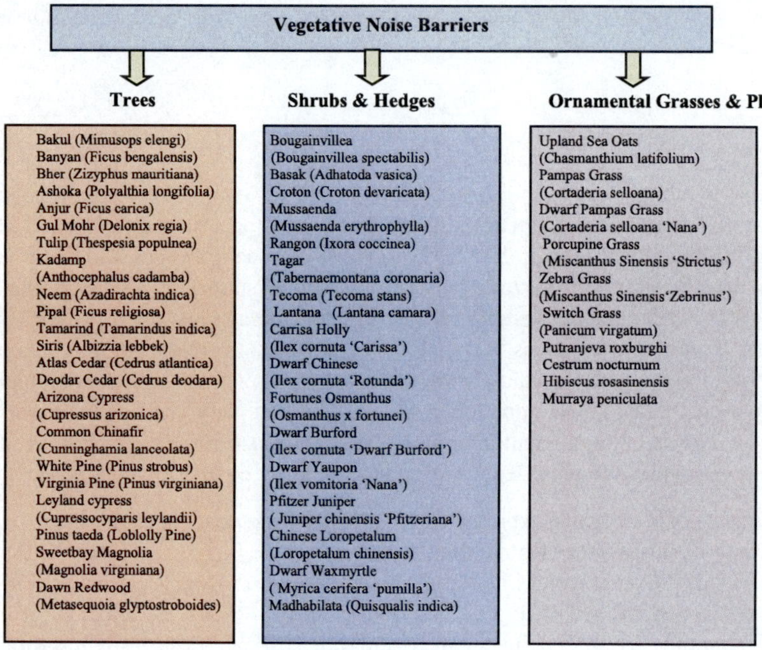

Fig. 7.13 Vegetative noise barriers for urban noise abatement (Garg et al. 2012, [9] and Georgia Forestry Commission report, [150])

and belt width. Further a degree of maintenance is required to achieve optimum performance throughout the life. The vegetation effect is considered as a mid- and high-frequency absorption per meter. Fujiwara et al. observed that the belt of trees with a 30 m width resulted in 8 dB more attenuation than same width of land without trees [151]. Pal et al. (2000) study recommended the minimum thickness of green belt of 40 m along the road for different locations in coal mining complexes. Excess attenuation for higher frequencies was higher than for the lower frequencies [152].

Garg et al. (2018) [153] employed a multiple regression approach to evaluate the noise reduction of vegetation along the road side in Delhi city. The dependent factor was the relative attenuation that includes attenuation due to geometrical divergence, that is, distance effect, and the independent factors were width, height, and distance from the line source (<15 m). Table 7.10 shows the multiple regression model for traffic noise reduction effects of vegetation for distance less than 15 m from the line source. The regression model suggests that width and height are the most prominent factors for overall reduction in traffic noise. However, some of the factors such as visibility, average density, average canopy branch, average diameter of trunk, and height as discussed in Pal et al. (2008) [152] may be considered to develop a more comprehensive model that can exclusively identify the effect of each parameter on the overall noise attenuation. The model described in Table 7.9 can be used for

Table 7.10 Multiple regression model for traffic noise reduction effects of vegetation (Garg et al. 2018 [153])

Variable	Unstandardized coefficient (*B*)	Standardized coefficient (β)	*t*-value
Belt width	3.19	2.11	3.95*
Height	2.57	0.91	3.44*
Receiver height	−0.63	−0.21	−4.11*
Source height	0.24	0.11	1.13*
Distance	−0.37	−0.57	−9.21*
Constant	1.87		1.11

$*P \leq 0.001$, $R^2 = 0.77$, $F = 51.20$

ascertaining the overall attenuation due to traffic noise source along the road side for distance less than 15 m.

7.10 Noise Abatement Program and Noise Goal

The alarming increase in vehicular population has created an urgency of devising strategic action plans for managing noise issues and effects, including noise abatement particularly for the dwellings near the vicinity of roads. The success of noise abatement measures essentially requires a strategic noise abatement planning with formulation of noise abatement goal complimentary with the financial year plans and special budgetary allocation for accomplishing the noise goal in next two decades. Figure 7.14 shows the proposed strategy of such a plan [154]. The strict implementation of a noise control program utilizing an inventory of noise control methods is must to combat the high ambient noise levels. Table 7.11 summarizes the noise abatement approaches and their cost effectiveness core as recently summarized by the European Commission report [1] and also by Münzel et al. (2018) [155]. Noise barriers, building insulation, traffic management, and quiet road surfaces are some of the important measures that can be undertaken to control the traffic noise. The options of bypass, roundabouts, flyovers, etc. for reducing the traffic density is an indispensable step. Earth berms and vegetative barriers prove to be cost-effective solution in such cases. Thus, a task force or an expert committee setup such as the National Committee for Noise Pollution Control (NCNPC), India for looking into such initiatives in coordination with the transport authority or traffic management department, development authority, and pollution control board can periodically review the progress for accomplishment of desired noise reduction targets. Setting of noise abatement goal such as that by Dutch National Traffic and Transportation plan with a goal of decreasing the number of houses exposed to a noise level of >70 dB (A) by 100% serves as a good illustration in Indian perspectives also. Continuous noise monitoring for detection of noisy hot spots and implementing the best practicable and economical options (BPEO) is essentially required for accomplishing the

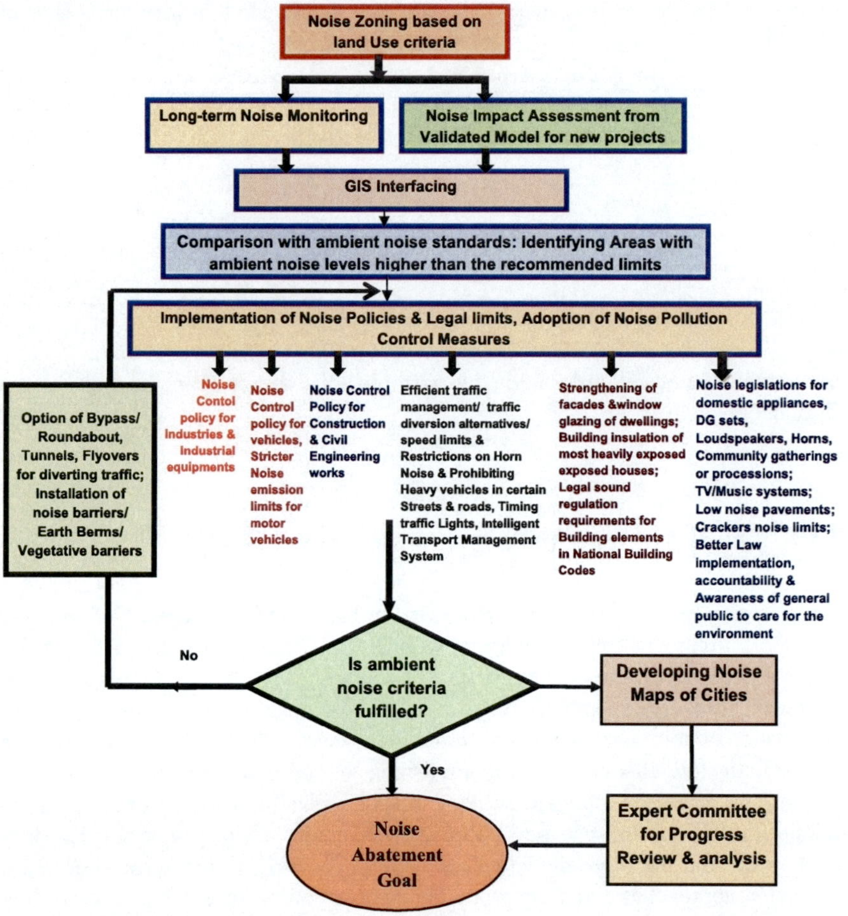

Fig. 7.14 Flow chart of a noise pollution control strategy for reducing the ambient noise levels in Indian cities (Garg et al. 2016 [154])

desired objectives. Chapter 10 discusses about the evaluation of Best Practicable and Economical options for traffic noise management and control.

New technologies and improvements are important contributions in reducing noise levels for all transportation noise sources (e.g., development of quieter engines, low noise tires for vehicles, and low-noise brake blocks for trains, electric cars). Thus, an inventory of noise control package focused on source, path, and the receiver shall meet the desired noise reduction targets. Figure 7.15 shows the proposed framework for community noise assessment and abatement in India [154].

Efficient traffic management is one of the vital reforms in this regard. Restrictions in use of horns especially near sensitive spaces like hospitals, schools, etc., reducing speed limits, timing traffic lights to achieve smooth traffic flow, smooth driving avoiding frequent acceleration and deceleration and prohibiting heavy vehicles in certain road and streets can be very fruitful in achieving the desired objectives.

Table 7.11 Noise abatement approaches (Münzel et al. 2018 [155])

Abatement procedures	Reduction in noise, dB	Cost-effectiveness score (range: 1–5; lowest score, 1; highest score, 5)
Noise barriers	3–20	2
Brake blocks for trains	8–10	4
Building insulation	5–10	1
Building design	2–15	3
Changing driving styles	5–7	3
Quiet road surfaces	3–7	5
Low-noise tires	3–4	3
Land-use planning and design	Unknown	4
Electric cars	1	1
Traffic management	3	3

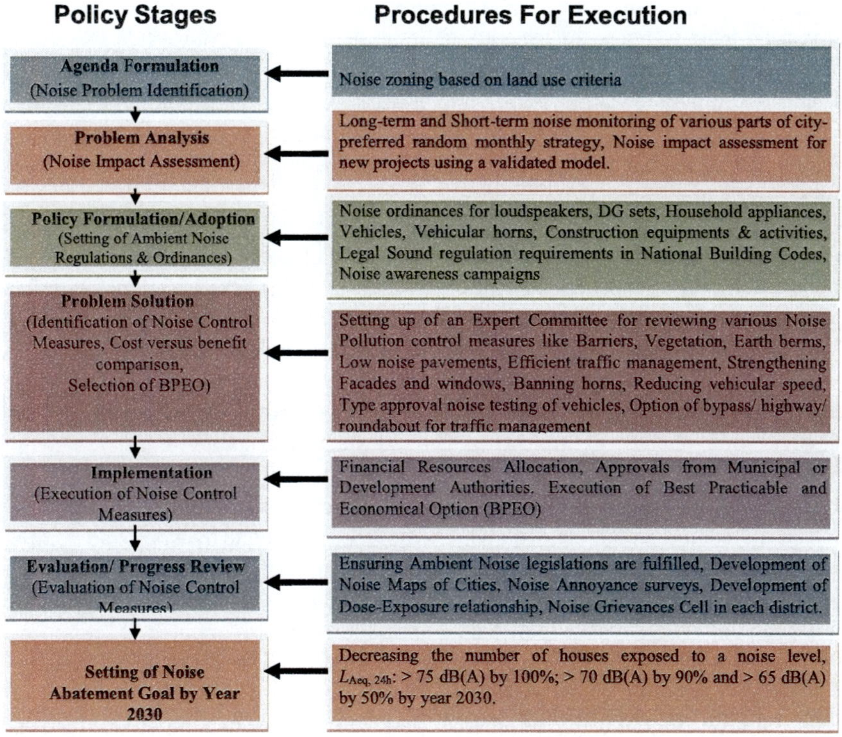

Fig. 7.15 Proposed framework for community noise assessment and abatement in India (Garg et al. 2016, [154])

Fig. 7.16 Illustrative noise ordinance signs for inculcating public awareness toward noise pollution control [156, 157]

Honking is a frequent phenomenon in Indian road context, and thus, setting of "No Honking" zones in urban agglomerations especially in residential and silence zones is must in Indian perspectives.

Illustrative noise ordinances sign shown in Fig. 7.16 as suggested by Chanaud (2014) can be a very effective step for educating about the existence and enforcement of noise ordinances especially for heavy traffic density areas [156, 157]. It is envisaged that awareness of general public in maintaining a "noise free society" is must. Social websites like *shutheshor.com* and *awaaz.org* can be very instrumental in generating mass awareness and voice against growing noise pollution in Indian cities. Recently, the Government of Delhi, India, has started a website *www.ngms.delhi.gov.in* and helpline no. 155271 for lodging complaints if someone violates noise rules and creates excessive noise.

Organizing "**Noise Awareness Campaigns**" as a part of the ongoing "**Swatch Bharat Mission**" introduced by the Government of India in year 2014 shall be indispensable in controlling the noise pollution in India. CSIR-National Physical Laboratory, New Delhi organized a one day International workshop on Noise Pollution Monitoring, Mapping and Control in December, 2021 followed by a Noise Awareness Week to inculcate the awareness towards noise pollution in the country. The proposals in Master Plan of Delhi (MPD 2021) target on "traffic de-congestion" include the following: Unified Metro Transport Authority; synergy between land-use and transport; and integrated multi-modal public transport system for reducing the dependence on personalized vehicles. Road- and rail-based mass transport system is recommended to be a major mode of public transport and optimal use of existing road network and development of missing links. National Transport Oriented Development (TOD) policy can serve as guidelines and play a catalytic role in formulating state/city level policies for promotion of transit oriented development. Decongestion plans for busy road junctions, special drives for "No Honking" (Fig. 7.17), promotion of carpooling policy for reducing road traffic, launching of odd-even traffic measures for some time, launching car free day initiatives, installation of synchronized traffic signaling, constructing flyovers in Delhi city for deflecting the traffic and easing the traffic density, restricted entry of heavy vehicles, and installation of noise barriers at hotspots are some of the major administrative issues that should be realized for reducing noise pollution levels in metropolitan cities of India [7].

Fig. 7.17 Illustrative signs for inculcating public awareness toward restricting the honking and promotion of carpooling for reducing noise (CPCB report, [7])

7.11 Noise Compatible Land-Use Planning and Management

Land-use planning is one of the four pillars recommended in balanced approach to noise control by ICAO. Noise compatible land-use planning is a community planning strategy for reducing the traffic noise along highways and along the airport. It considers land-use options and noise issues more effectively by encouraging the location of less noise-sensitive land uses next to highways, for example, *by using adjacent land for activities, services, or businesses that are not disrupted by noise.* Effective noise compatible land use planning can eliminate the need for noise barriers in highway programs [158]. The International Civil Aviation Organization (2008) guidance on balanced approach to aircraft noise and management categorizes land-use planning in three categories: planning, mitigation, and financial instruments. Figure 7.18 summarizes the noise compatible land-use planning and management measures (ICAO 2008 [159]). This methodology concentrates on the determination of the compatibility of various land uses with transportation noise and then noise zoning the areas accordingly to the type of development. For instance, airport noise defines the boundaries wherein residential development can be possible or not. Vogiatzis (2012) study ascertained the land use management strategy as an effective tool for protection from aircraft noise for Larnaka International Airport in Cyprus. The proposed land use action plan restricted the noise sensitive receptors, allowed residential construction with external shell having weighted sound reduction index $R_w \geq 40$ dB; increasing the sound insulation of houses exposed to $L_{night} > 60$ dB(A) and allowing the industrial use of limited nuisance [160]. The widely used land use controls are: zoning, comprehensive plan, land acquisition, easement purchase, development rights (purchase and transfer), and land banking. In case of land banking, a public agency purchases land for future use and development. Purchase and transfer development rights ensure long-term assurances that land uses will remain compatible. The master plan is a pivotal approach to ensure long-term development and compatible uses. The master plan is the policy document guiding land-use, whereas zoning is the means of implementing the policy [161].

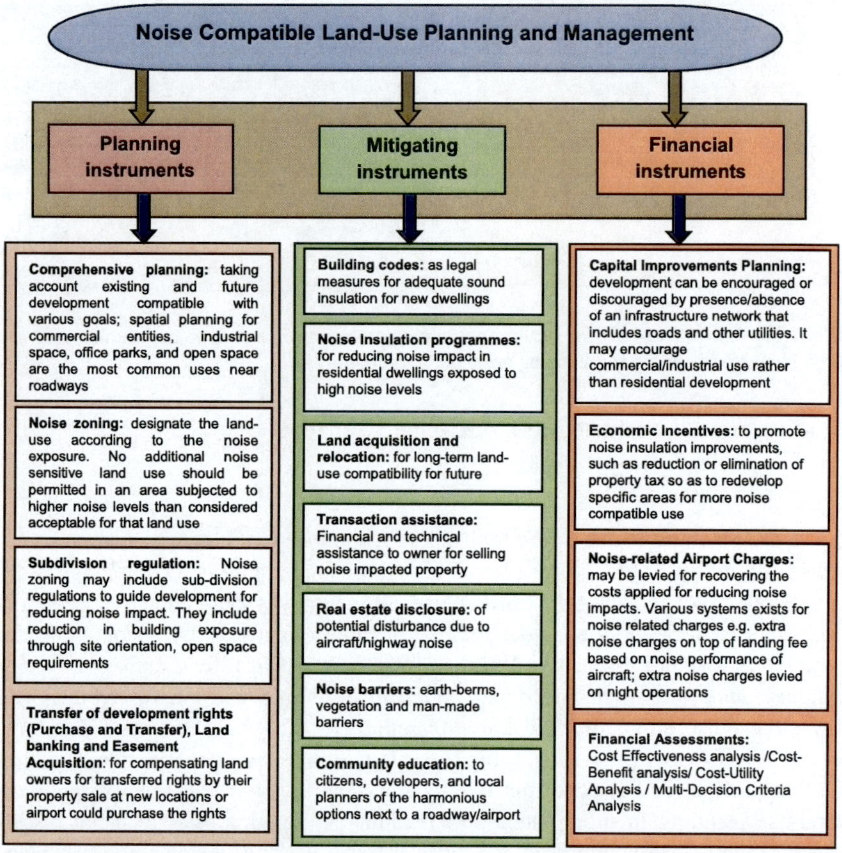

Fig. 7.18 Noise compatible land-use planning and management measures (Based on ICAO 2008 [159])

7.12 Economic Considerations on Noise Control

Noise abatement is a balance between the costs for prevention and reduction of noise and the benefits of lower environmental noise levels for society. The noise abatement incurs direct costs (investment costs for installment of application of the noise measure, for example, the construction costs for a noise barrier; increased costs for maintenance and preservation of the noise measure, for example, higher maintenance costs and lower durability for porous road surfaces, or acoustic grinding to obtain and preserve a lower rail roughness) and the indirect costs (e.g., longer flying distances for noise-optimized take-off and approach routes) [162]. Figure 7.19 shows the noise reduction versus the life cycle costs per road meter for various source measures, porous road surfaces and noise barriers; noise reductions for a typical highway situation with mixed traffic (data from [163, 164]). Cost values are

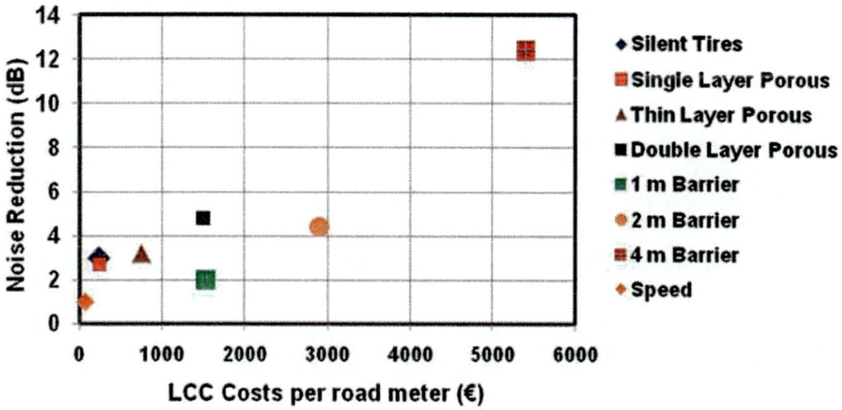

Fig. 7.19 Noise reduction versus the costs per road meter for various source measures, porous road surfaces and noise barriers; noise reductions for a typical highway situation with mixed traffic (Peeters and van Blolkand 2018 [162], data from [163] and [164])

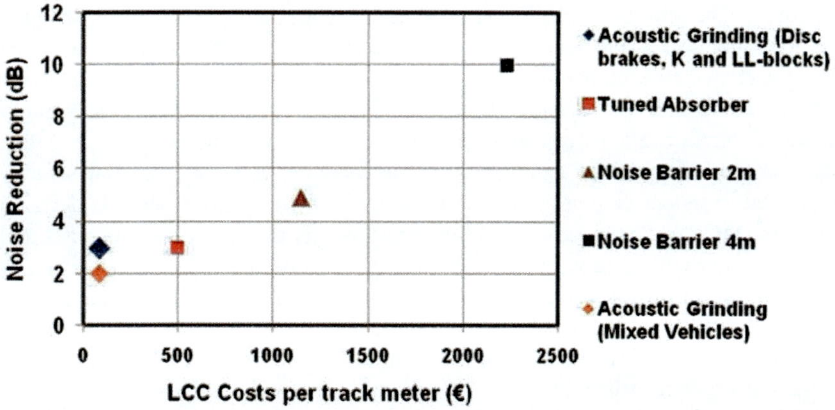

Fig. 7.20 Noise reduction versus costs per track meter for various measures; costs are LCC costs over a period of 30 years; noise reductions based on Dutch legislation (Peeters and van Blolkand 2018 [162], adapted from Kuijpers A et al. 2009, [165])

the life cycle costs, investment plus 30 years of maintenance costs, for a 20 m wide road, taken from [162].

Similarly for the rail traffic noise abatement, Fig. 7.20 shows the noise reduction versus the life cycle costs (LCC) per road meter for various abatement measures [162, 165]. The LCC costs used depend highly on the method and assumptions used to determine the costs. Differences in assumptions lead to quite different costs, for example, for noise barriers near roads (Fig. 7.18) and near railways (Fig. 7.19). Cost estimates for these noise abatement show that retrofitting K- and LL-blocks is the most cost-effective option. The major priority is given to measures at the source and specifically to the retrofitting of in-use rolling stock with silent braking systems (LL-Blocks and K-Blocks); K-Block are more effective than LL-Blocks in noise reduction, but require adapting braking system causing additional costs and more maintenance activities and are more expensive than LL-Blocks [166].

The various decision-making methods for noise abatement measures are: cost minimization (CM); cost-effectiveness analysis (CEA); cost-utility analysis (CUA); cost-benefit analysis (CBA); and multi-criteria decision analysis (MCDA). The cost-minimization analysis primarily considers the only variable costs parameter with the goal of finding the cheapest alternative that fulfills the legal noise limits. In a cost-effectiveness analysis (CEA), the goal is to optimize the balance between costs on one side and a particular measurable output quantity on the other side. In addition to the cost-minimization, the objective is not just to find the cheapest solution, but to find an optimal solution, based on the incremental costs versus incremental returns. Cost-utility analysis (CUA) is a method similar to cost-effectiveness. In a cost-utility analysis, the output parameter is not a single measurable health quantity, such as the life expectancy, but a "virtual" quantity that combines separate health-related quantities in a single unit. Units commonly used are DALY (Disability Adjusted Life Years) or QALY (Quality Adjusted Life Years), both of which aim to estimate the number of "healthy life years." The cost-benefit analysis (CBA) involves not only the costs, but also all the benefits are expressed in monetary units. Some benefits are directly related to a monetary value, such as reduced hospital costs due to lower occurrence of diseases. Benefits that are time-related, such as effects on mobility (traveling times, traffic jams) can quite easily be expressed in money. Other benefits, such as the reduction of annoyance or the increase of safety, are not directly expressed in monetary units and need to be "monetarized." Multi-criteria decision analysis (MCDA) is a form of multi-criteria analysis that combines the costs and other decision criteria (e.g., annoyance, maintainability, roadside aesthetics, social security) into a decision evaluation process, which includes attributing a score, on an arbitrary scale, to each of the criteria and assigning weighting factors [167]. Table 7.12 describes the SWOT analysis of different decision methods for noise abatement measures [162].

The prime techniques for measuring the economic benefits of noise reduction: cost of abatement, cost of illness (health expenditures due to noise exposure), contingent valuation method (CVM), and hedonic price method (HPM). Contingent valuation method is based on stated preferences of a representative sample of the public, who are to be affected by the new noise standard. The benefit of noise reduction can be estimated by asking people how much they are willing to pay for a given amount of noise reduction. Hedonic price method is the most popular approach for estimating the true value of noise reduction. Cost-benefit analysis is applied when the effects of the measures are defined in monetary terms as are the costs. The cost utility analysis uses a health-related quantity as yield of measure or policy. Usually, this is the DALY (Disability Adjusted Life Years) and the QALY (Quality Adjusted Life Years). The World Health Organization has published a method to interpret the health effects in loss of DALY's [168]. All the methods are used in practice, but one can distinguish the situations where one method will

Table 7.12 Strengths, weaknesses, opportunities and threats analysis of different decision methods for noise abatement measures (EPA Network Interest Group on Traffic Noise Abatement 2018, [162])

Method	Benefit/other criteria	Strengths	Weaknesses	Opportunities	Threats
Cost-benefit analysis (CBA)	Multiple criteria expressed in monetary units, summed to a single value; Every benefit is translated to monetary units	Direct comparison between costs and benefits in the same units; reliable and transparent	Requires monetization: criteria that cannot be monetarized cannot be taken into account; may be inaccurate in some cases	Easy to understand an explain to the public; people naturally understand monetary units	Monetization is difficult for some criteria; objections against monetization of health and environmental impacts sometimes
Multi-decision criteria analysis (MCDA)	Multiple criteria each on different arbitrary scale, every benefit and costs are scored and then combined and compared	Any criteria can be included and quantified; sensitivity analysis is relatively easy; allows for evaluation of the accuracy of results	Requires development and tailoring of weighting factors for each project	Allows for public participation and democratic decision-making	Weighting factors are subjective and ambiguous in some situations
Cost minimization (CM)	Output is fixed for the required noise reduction	Simple and fast; only one right answer	Benefits are disregarded; may lead to inefficient use of available budget	Easy to explain; Inexpensive to execute	Low public involvement; focus on costs alone may lead to lower acceptance
Cost-effectiveness analysis (CEA)	Single parameter; output parameter is usually the noise reduction (in dB × persons)	Relatively simple; accurate	Selection of "most effective measure" may be subjective (depending on implementation); no consideration of additional benefits	Optimal distribution of available budget; well-fit for legislation (limited complexity vs. high effectiveness)	Low public involvement
Cost-utility analysis (CUA)	Single parameter; health impact is expressed in DALY or QALY units	Method is well described; Relatively easy to execute	Limited accuracy, mainly of disability weights; selection of 'best measure' may be subjective	Focus on health/ annoyance leads to higher acceptance; well-known and internationally accepted.	For small-scale projects, health effects may show to be limited ("milli-DALY")

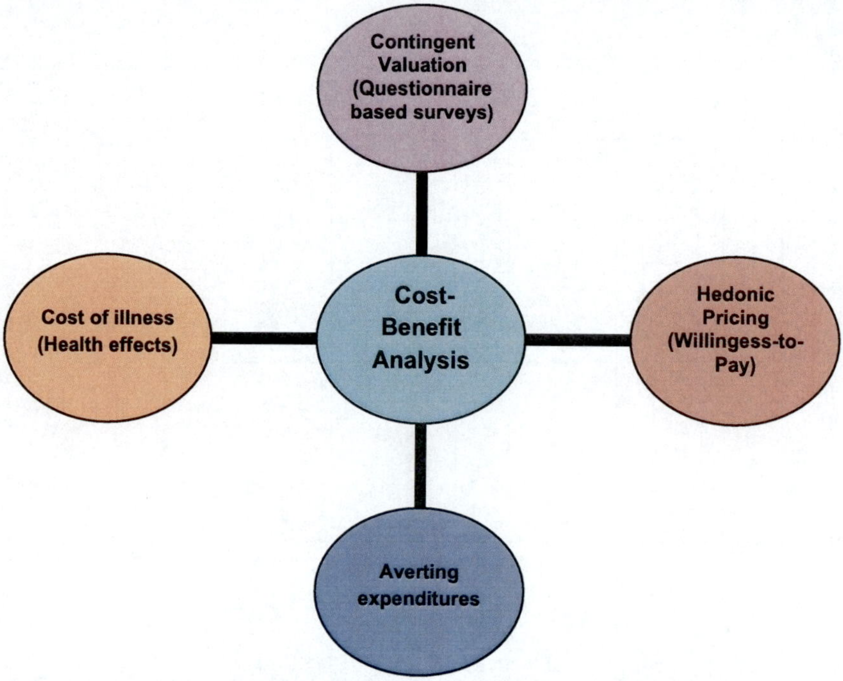

Fig. 7.21 Prime considerations in cost-benefit analysis (CBA) for noise control

perform better than other. However, the utility part is difficult to explain as one needs to explain such terms like DALY's [169]. In one such study reported by Klæboe et al. (2011) [12], cost-effectiveness analysis showed that facade sound insulation was the least expensive noise control alternative per dwelling. However, the cost-benefit analysis showed that benefit were less than 20% of the costs. A mixed noise abatement policy employing low-noise asphaltic road in addition to facade insulation was therefore considered. The economic assessment of noise effects is challenging and no broad consensus exists about how this should be done. Cost-benefit analysis (CBA) takes a more holistic approach and is more demanding than the cost-effectiveness approach as it considers all the impacts of a measure or project. The prime concern of CBA is to accomplish best overall performance in monetary terms versus the cost of a measure or a project. Figure 7.21 shows the prime considerations in cost-benefit analysis (CBA) for noise control. Cost-effectiveness analysis (CEA) method is best suited to prioritize interventions to reduce noise [170]. Economists present the state of the practice in noise effect valuation, based on housing value loss and contingent valuation surveys, which is sometimes pointed out to be insensitive to the health effects of noise and air pollution. No broad consensus exists among the experts concerning how this should be done [171].

7.13 Urban Morphology and Planning

The classical measures pertaining to the propagation path, at the receiver site or traffic management are primarily used, while only 10% measures are oriented toward land use planning [172]. Some of the architectural modifications like balcony design modification (6–10 dB(A) by ceiling absorption, parapet (angled), ceiling shield, ceiling (angled), and side wall absorption as pointed by Naish et al. (2012) [173]), inclined type eaves on facades in high rise buildings [174] can be very effective in traffic noise reduction in canyons. Figure 7.22 shows the balcony modifications of a facade in a high rise building studied by Lee P J et al. (2007) [175]. The combination of absorbing surface on the inclined ceiling and parapet showed a maximum noise reduction of 10 dB at 1 kHz in an apartment complex [175].

Thus, the planning policies should consider traffic management, urban morphology and building sound insulation in an integrated way for traffic noise control. Studies such as that by Sanchez (2016) demonstrated that building shape, street geometry, and the presence of street furniture strongly affect the noise exposure. Building shape can be responsible for variations of up to 7.0 dB(A) at pedestrians. The facade design can diminish the noise exposure at windows and the street geometry can enhance the positive effect of low barriers to 11.3 dB(A) along sidewalks. A careful design of building facades and street geometry can improve the sound climate for acoustic comfort of residents [176]. The joint multifractal spectrum analysis by Ariza-Villaverde et al. (2014) revealed a positive correlation between the total noise pollution and the street width to building height ratio for regular urban morphology [177]. Some studies correlate factors such as ground space index, floor space index, traffic volume, traffic speed, road area density, and the fraction of industrial area on noise levels and observed all of them to have significant impacts on the noise level. Ground space index was observed to have a negative correlation on road traffic noise, while other indicators had positive impact on traffic noise [178]. The sky view factor (SVF) describing the degree to which sky is visible for each given point, changing from zero, when the sky is totally obstructed, to one, when the sky is entirely visible is also correlated in one such study and is observed to be inversely proportional to noise [179].

Fig. 7.22 Balcony treatments for exterior noise reduction: (LHS) inclined ceiling, an absorbing surface, and a parapet; (RHS) inclined ceiling, parapet, and absorbing surfaces (Lee P J et al. 2017, [175])

A balcony equipped with reflectors reduced road traffic noise by 7–10 dB (A) compared to the ordinary balcony at incident angles of noise close to the angle for which the reflectors are designed. The efficiency is same or greater than that for a balcony with an absorbent ceiling. However, for larger incident angles in case of low-level floors, efficiency can be degraded [180]. The noise shielding efficiency of a balcony equipped with reflectors is greater than that for an absorbent balcony [181]. In one such study, the field measurements revealed that reflectors reduced the railway noise by 5 dB(A) approximately compared to an ordinary balcony [181]. In case of larger horizontal incident angles of noise, the efficiency is reduced [181]. Thus, an innovative design consideration in balcony design can achieve higher noise level reductions. For instance, a balcony designed with solid parapets and absorbent applied to ceiling can offer 2–3 dB(A) noise reductions. However, innovative development of an Enhanced Acoustic Balcony including extending the side parapets to full height side walls, inclusion of a laminated glass panel projecting out from the top of front parapet and adding sound absorptive panels on the side walls and ceiling in order to reduce the sound reflections can accomplish reductions up to 5 dB(A) or more [182]. Additionally, including a vertical panel above the front parapet and in front of balcony door, the noise reduction can be enhanced above 10 dB(A) on higher floors [182]. The implementation of baffle-type acoustic windows comprising of two layers and aligned in a way that the pane of inner window is behind the opened outdoor window for reducing noise had been reported [183]. Figure 7.23 shows the various acoustical treatments in balcony that can be helpful for significant noise reductions up to 10 dB(A) [173].

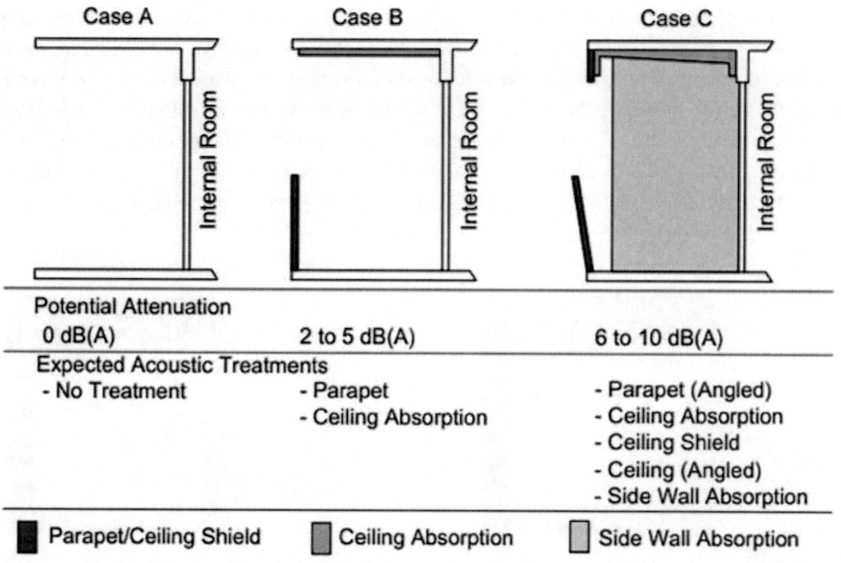

Fig. 7.23 Levels of balcony acoustic treatment and their generalized potential noise attenuation (Naish et al. 2012 [173])

The urban morphology influences the attenuation of traffic noise levels in residential areas. The various urban morphological parameters defined in Hao et al. (2014) and (2015) [184, 185] studies are as follows:

- Building Plan Area Fraction (BPAF): It is the ratio of the plan area of buildings to the total surface area of study region.
- Road Area Fraction (RAF): It is the ratio of plan area of roads to the total surface are of study region.
- Complete Aspect Ratio (CAR): It is the summed area of buildings and exposed ground divided by the total surface area of the study region.
- Height-to-Width Ratio (HWR): It is the ratio of building height to the horizontal distances between the two adjacent buildings.
- Building Frontal Area Index (BFAI): It is the ratio of total facade area parallel with the road direction divided by the total surface are of the study region.
- Horizontal Distance of First-row Building to Flight Path (HDFBFP): It is the mean of horizontal distances from the façade of first-row buildings to flight path.

The linear regression model correlating the acoustic features in an area based on urban morphological parameters for facade had been described as [184, 185]:

$$L_{10} = 57.50 + 32.75\ \mathrm{BPAF} + 38.47\ \mathrm{RAF}, R^2 = 0.70 \tag{7.29}$$

$$L_{\mathrm{avg}} = 78.71 + 28.85\ \mathrm{BPAF} - 24.29\ \mathrm{CAR} + 6.09\ \mathrm{HWR}, R^2 = 0.78 \tag{7.30}$$

Thus, from these studies [184, 185], the key aspects evident are as follows:

- Building Frontal Area Index significantly helps in attenuating outside noise.
- Street pattern, that is, HWR or canyon effect, has no significance on noise attenuation in urban texture.
- A site with higher building coverage has been observed to have more noisy indoor spaces due to traffic noise. The noise reduction occurs with an increase of facade areas along a road.
- Decrease of Building Plan Area Fraction and an increase of Complete Aspect Ratio can contribute to the noise attenuation on facades.
- In case of aircraft noise, BFAI and HDFBFP correlate with noise attenuation most, while the BPAF and HWR have no effect.

Thus, the traffic noise level attenuation by urban morphology and use of masking effects of natural sounds are two main aspects that can be considered for improved soundscape quality [185]. The analysis on correlations shows that the urban morphological indices generally have significant effects on noise levels, with different patterns in the cities. For instance, in one such study conducted for two cities: one of the UK and other of China revealed that the road coverage showed a significant positive correlation with most of the noise indices in the Greater Manchester samples in terms of ground as well as facade noise levels, while it had only a significant effect on facade noise levels in terms of L_{avg} and L_{50} in Wuhan city of China [186]. Another study for Singapore city observed the reflecting surface coverage as a positive factor and open space ratio as a negative factor, which suggested that the increase the ratio of open spaces and reduction in reflecting surfaces area or replacement with absorptive materials, for example, porous bricks can be effective in reducing the noise level on ground. Thus, for the creation of a pleasant acoustic environment, the consideration of the effects of urban morphology at a regional scale is vital [187]. The configuration of buildings also help to attenuate the traffic noise as the contiguous and connected forms of buildings along the sides of roads are preferable for traffic noise reduction. Also, high-density, aggregated distribution, and regular shape of vegetation is analyzed to be efficacious in traffic noise reduction [187]. Thus further studies need to be focused examining the relationship between building typology, urban morphology, and urban acoustic environment in different cities in Indian perspectives so that potential planning and design factors for mitigating the negative impact of traffic noise can be done prior development [188]. These investigations shall be thus very instrumental in development of *Smart Cities* in future in India with reduced noise pollution levels.

7.14 Noise Management: Soundscape Approach

Soundscape is defined in ISO 12913-1 (2014) standard as the acoustic environment as perceived or experienced and/or understood by a person or people, in context. Soundscape approach is used to ascertain the sounds in landscape from different sources at multiple temporal and spatial scales. It involves the design or management to control the acoustic environment of a place that results in improved human perception of the environment. The difference between this approach and traditional approach lies in the fact that the traditional approach considers the overall sound at a receptor by integrating all the sources, whereas soundscape approach stresses on differentiation between the sources, source identification, and human judgment on various sources. Also, the traditional approach considers minimizing the noise levels, whereas soundscape approach entails wanted sounds masking the unwanted one. Both these approaches have different measurement perspectives as the environmental noise management focuses on sounds of discomfort, while the soundscape approach focuses on sounds of preference [189]. Soundscape approach is relationship between the ear, human beings, sound environments, and society. For each sound, the sound pressure level, spectrum, temporal conditions, source location and movement, psychological, and social characteristics must be analyzed. Figure 7.24 shows the system for the soundscape description in urban open public spaces [190]. A recent study proposed 22 soundscape actions under three main themes: strategic localization of functions away from existing sources of noise considering spatial relationship, reduction of unwanted sounds, and introduction of wanted sounds as tool for landscape architects [191]. Chew and Wu (2016)

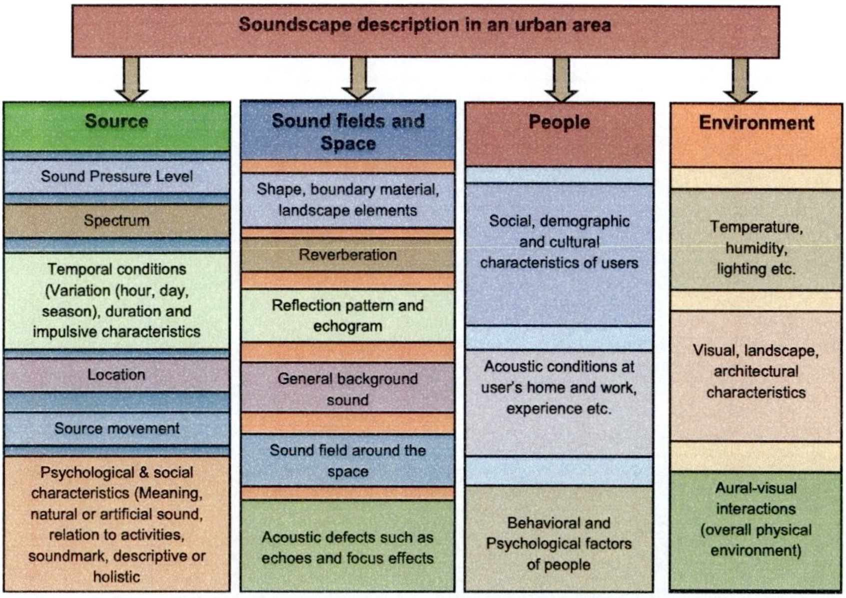

Fig. 7.24 Soundscape description in urban open public spaces. (Modified from Kang J (2019), [190])

presented an analysis of soundscape pattern and statistical description to understand the relationship between traffic noise and urban landscape so as to develop a framework for reducing the traffic noise impact in city of Taipei, Taiwan. The traffic noise from street is recorded and analyzed, and the soundscape of traffic noise was interpreted that leads to selection of artificial sound (e.g., changing the flow of water during different times of the day in fountains) for masking the traffic noise [192].

Thus, the paradigm shift from conventional environmental noise management approach that considers noise as "*resource*" rather than "*waste*" as highlighted by Kang (2019) [190] can be integrated with noise mapping methods in an actual urban redevelopment project. The preferred sounds like water features or bird song could be introduced to mask the traffic noise and thus reducing the traffic noise impact and making the sound environment more suitable. In this aspect, the landscape design modifications can play a vital role in enhancing the quality of soundscape. The integration of noise maps, sound maps, and soundscape maps offers a potential alternative procedure for town planners, policy-makers, and stake holders for future noise management perspectives in urban planning [193]. Figure 7.25 shows the various soundscape actions for noise control as outlined by Cerwén et al. (2017) [194]. There were 23 soundscape actions analyzed and divided into three main categories: localization of functions, reduction of unwanted sounds, and introduction of wanted sounds described [194, 195]. A recent study reported by Lugten et al. (2018) shows that soundscape interventions like moving water (water jets) and vegetation can contribute to a better perceptual quality of areas exposed to aircraft noise and can complement the other noise action plans to be executed [196]. Jeon

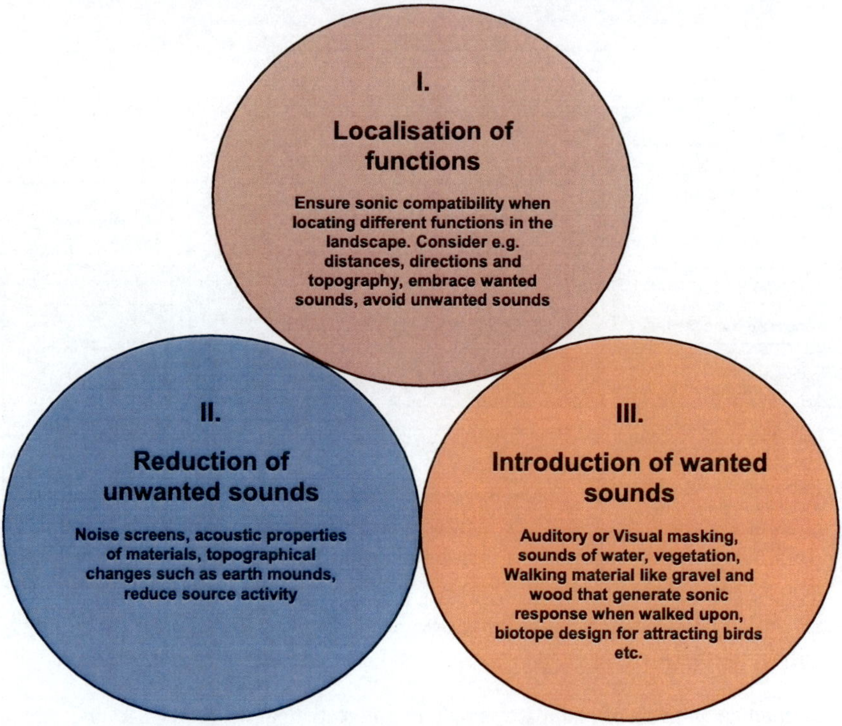

Fig. 7.25 Soundscape actions for noise control (Cerwén et al. 2017 [194])

et al. (2010) study recommended that water sounds ("Stream" and "Waves of lake") were the best sounds to be used for enhancing urban soundscape. The study proposed that the level of water sounds should be similar to or not less than 3 dB below the level of urban noises [197]. Thus, it is evident that soundscape design shall be crucial for planning and building of *Smart Cities* in future.

Ethnography is an approach to evaluate people's experience and perception of a sound environment. Soundscape of a place is context dependent and thus integration of soundscape into the placemaking will not only enhance the aural environment but will enhance people's positive soundscape experience. In addition to water features, some soundscape approaches that have been used for urban placemaking includes sound installations (recorded sound played by hidden loudspeaker) and sonic art, etc. [198].

7.15 Monetary Valuation of Noise Control

The social cost of road traffic noise in Europe has been estimated as €38 (30–46) billion per year, which is approx. 0.4% of the total GDP. For rail, the estimated social cost is about €2.4 (2.3–2.5) billion per year (0.02% of the GDP) [199]. The social cost of environmental noise in England had been estimated as GBP 7 to 10 billion per annum. The most severe health effect such as cardiovascular disease was estimated as GBP 2–3 billion per year [170]. The position paper on valuation of noise recommended a value of the perceived benefit of noise reduction of €25 per household/decibel/year [200]. The major elements of the external costs are: reduced property prices, reduced possibilities of land-use, increased medical costs and cost of lost productivity in workplace due to illness caused by effects of noise pollution, also called cost-of-illness (COI) [201]. It is sometimes difficult to assess the noise impact in terms of monetary valuation, but the saying "*Money speaks louder than words*" [202] is very important to understand and implement [13]. Monetization includes estimation of various factors such as Willingness-to-pay (WTP), Noise Depreciation Sensitivity Index (NDSI), house price, and threshold value for which research investigations are needed for calculating area-specific or country-specific values [13]. There are two methods for analyzing the noise impact in monetary terms [170]:

- Revealed approach (RP) or Indirect approach: It is based on the assumption that the price of goods and services are influenced by noise. So, the willingness to pay or to accept is evaluated. The Hedonic pricing method is utilized to compute the change in costs w.r.t change in noise levels.
- Stated Preferences approach or Direct approach: It consists of asking the individuals through a questionnaire about their willingness to pay or willingness to accept to live in a quieter environment. The contingent valuation method belongs to this category.

The costs in which values for noise could be presented are [203]:

- An economic value per dB per year measured by Noise Depreciation Sensitivity Index (NDSI) defined as average percentage change in the property prices per decibel.
- An economic value per year per person annoyed by noise expressed as (a) value per person highly annoyed and (b) value per person annoyed.

The Department of the Environment, Transport and the Regions, DETR in the UK (1999) concluded from a comprehensive review of 64 studies as [204]: 24–48 Euro per dB per household per year, 0.08–2.30% change in property price per dB; and 0.02–2.27% GDP. The health effects of noise are sometimes difficult to assess due to various confounding factors such as air pollution, diet intake, physical exercise, and other comorbidities. In one study in the USA, a 5 dB noise reduction scenario was estimated to reduce the prevalence of hypertension by 1.4% and coronary heart disease by 1.8%. The annual economic benefit was estimated as $3.9 billion [205]. Another study in Korea revealed that the people who were annoyed and extremely annoyed due to noise had a willingness to pay US $7.55 and US $8.83 annually per household for reducing the annoyance. The economic benefit of noise reduction policies to reduce noise by 1 dB(A) was estimated at US $2.91 million per year [206]. Thus, it is imperative that cost-benefit analysis should be undertaken prior planning for the noise action plans and most cost-effective options should be considered. Reducing the noise from vehicles is the most cost-effective measure and more than seven times less costly than any other measure

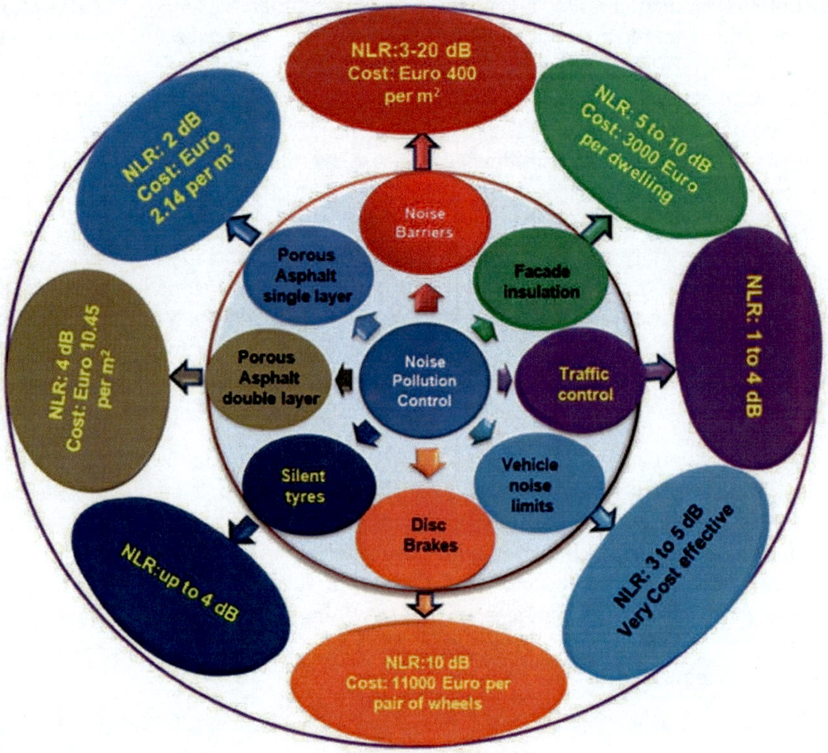

Fig. 7.26 Various noise control measures for noise abatement and control: noise level reduction (NLR) and associated costs (in Euro)

[207]. However, sometimes the multi-criteria analysis considering all aspects could be rather more reliable and helpful in decision making rather than cost-benefit analysis only. The implementation of different methods like cost-benefit analysis and cost-effectiveness may yield different conclusions. The Norwegian study focused on implementation of a national facade insulation program in Norway for the 2500 dwellings exposed to high road traffic noise levels recommended a mixed noise abatement policy employing the low noise asphalt and facade insulation. The cost-effectiveness showed that facade insulation was the least expensive noise control alternative, but the cost-benefit analysis revealed that benefit were less than 20% of the costs as the residents targeted dwellings are benefitted only [12]. Figure 7.26 summarizes the various noise control measures for noise control and the noise level reduction (NLR) and associated costs (in Euro) based on the costs mentioned in Euros in references (Nijland et al. 2003, [13], CEDR report 2013, [207]). The cost of reducing noise annoyance in 20-year perspectives had been estimated in CEDR report (2013) to vary from EUR 16 per person per year to EUR 4200 per year.

The average cost of 4 m high noise barrier is EUR 1600 per linear m or EUR 400 per m^2 and annual cost of maintenance is 77 EUR per linear m. In case of two-layer porous asphalt, the cost of investment is EUR 10.45 per m^2 and EUR 1.85 per m^2 for maintenance, while for the single layer porous road surface, the cost is estimated as EUR 2.14 per m^2 and EUR 0.25 per m^2 for maintenance [207]. The cost of replacing two windows is estimated as Euro 3000 per dwelling, while the average cost per apartment for facade insulation had been evaluated as EUR 28,125 in a Norwegian study [12]. Controlling the noise at source is evaluated to be the most cost-effective measure. Thus, the stricter noise limits for vehicles are in particular the most cost-effective measures to reduce noise annoyance. However, implementation of multiple measures shall be efficacious in bringing the levels below the legal limits and reduced noise annoyance [207]. Also, in recent times, noise mapping has been widely used exercise for noise management and control, and thus cost of noise mapping is an integral part of devising and implementing the noise action plans. The administrative costs had been calculated in EU to be low (€0.15 for noise maps and €0.03 for action plans per citizen, every 5 years). Also, the cost-benefit analysis showed that EU Directive (2002) had been efficacious with a favorable cost-benefit ratio of 1:29 wherein the noise action plans had been implemented [208].

7.15.1 Noise Exposure Score and Noise Costs

The Danish Environmental Protection Agency (EPA) valuation of noise in terms of noise exposure score (NES) is one of the excellent illustrative example of valuing the noise cost/benefits. Noise exposures score (NES) is accumulated noise load on all dwellings in an area calculated as the sum of weighted noise loads on all dwellings in an area. The computation is based on the noise levels outside the facade of dwelling that inhabitants are exposed while the windows are open. The NES is based on a dose-response relationship called the noise exposure unit (NEU) calculated as [170]:

$$\mathrm{NEU} = 0.01 \times 4.22\ 0.1^{(L_{\mathrm{den}}-\mathrm{C})} \tag{7.31}$$

where C is constant = 44 and L_{den} is the calculated noise exposure at facade. The calculation of total noise exposure for an area under study requires the multiplication of each dwelling by the corresponding NEU. For instance, the NEU from an exposure of 76 dB is 1 and 0.2 for an exposure of 65 dB. Also, 10 dwellings exposed to 76 dB will have NES of 10 and 50 dwellings with a noise level of 65 dB will have 10 NES (=50 × 0.2 NEU). The cost per one NEU had been calculated as 3070 per year in Denmark. Table 7.13 shows the costs of noise per NEU per year in Denmark (2015 values) [170]. Thus, the concept of NES correlated with costs facilitates the

Table 7.13 Costs of noise per NEU per year in Denmark (2015 values) [170]

Costs type	Costs per NEU per year (Euro per year)
Nuisance costs	1862
Health costs	1907
Total costs	3070

comparison of benefits of various noise action plans and prioritize them and additionally evaluate the socioeconomic costs and benefits associated with them.

7.16 Implications in Indian Perspectives

The studies pertaining to noise annoyance and health effects had been widely reported in European continent, but in Indian scenario, there had been a very few studies reported so far. Particularly in the Asian continent, there had been few studies reported so far on these aspects and economic considerations and benefits of noise abatement. Monetization and willingness to pay values for road, rail, and aircraft noise is difficult in Indian scenario. Apart from the economic considerations, other aspects such as structural integrity, maintenance, safety, and durability, compatibility with environment, and social security are the other issues which are equally important while considering the noise abatement measures, which are somehow not considered in cost-benefit or cost-utility or cost-effectiveness approach. For instance, in Indian scenario dwellings situated near to metro trains stations or near to the main roads are more in demand, although air pollution and noise pollution is higher. Installation of innovative noise barriers like photovoltaic noise barriers, noise barriers with TiO_2 coating for eliminating nitrogen oxides, Inox/corten steel noise barriers, coverage/tunnels, etc. should be possibly explored and installed especially among the sensitive receptors [209]. In many situations, MCDA (Multi Criteria Decision Making Analysis) may prove to be more suitable considering the various alternatives subject to various criteria's. Garg et al. (2015) [210] study on selection of optimal noise barrier using Fuzzy TOPSIS (*Technique for Order Preference by Similarity to an Ideal Solution*) approach in one such example. A recent study by Ruiz-Padillo et al. (2016) [29] proposed a multi-criteria decision methodology for selection of suitable alternatives against traffic noise. It may be interesting to explore that in Indian scenario, the MCDA approach may prove to be more flexible and implementable approach providing sound arguments from the standpoint of objectivity, rigor, and especially for the constitutional committees set up for recommending noise control action plans. Chapter 10 discusses the implementation and usefulness of this technique in details. The noise costs associated with health issues must also be quantified and translated in Indian currency similar to that developed in European perspectives. As various studies have revealed the noise control at the source to be most cost-effective, thus tightening the noise limits for vehicles must be periodically exercised in legal regulations after fixed time intervals so as to reduce ambient noise levels in cities. Also, the concept of soundscape planning can be implemented in areas prone to dense traffic and honking noise for reducing the noise impact and accomplishing better urban acoustic environments. The conventional building technology of using plastered bricks and concrete structures has to be modified with innovative sandwich light weight constructions and green building materials with enhanced sound insulation and absorption characteristics. for indoor and outdoor applications. The role of Indian building industry in collaboration with testing and research laboratories shall be pivotal in this regard.

CSIR-National Physical Laboratory, New Delhi being a premier research laboratory having reverberation chambers testing facility and sound impedance tube facility had been working on developing the light weight acoustical materials with enhanced acoustical characteristics and has been involved in technology development and transfer to some of the industries for developing better acoustical products for noise control. The National Building Codes (NBC) of India also have to be amended periodically to accommodate for the paradigm shift in the building construction activities, development of new/innovative construction materials and technologies and greater need for preservation of environment.

7.17 Conclusions and Recommendations

The chapter discusses the sound regulation requirements for strengthening the sound insulation of buildings and enhanced acoustic comfort. The use of drywall constructions in conjunction with the massive constructions is discussed for accomplishing the noise level reduction targets. Technological innovations on development of sandwich constructions, use of composite materials and green building materials of enhanced acoustical characteristics shall be helpful to meet the enhanced acoustic comfort criteria's. The design aspects for selection of noise barriers of optimum dimensions providing a high insertion loss is essential and as such selection and implementation of suitable noise barrier shall be helpful in accomplishing the desired noise level reductions especially for areas under the silence zones. More recently, the European Union is working on a new project on characterizing the acoustic performance of noise barriers named as SOPRANOISE (Securing and Optimizing the Performance of Road trAffic noise barriers with New methOds and In-Situ Evaluation, https://www.enbf.org/sopranoise). In Indian perspective, the testing/certification of in-situ performance evaluation of noise barriers can also be developed and the traffic spectrum used in single-number ratings recommended in EN 1793 standards may be developed exclusively for Indian scenario. "Noise Pollution Control is a never ending battle." It is imperative to use and implement passive noise control studies in Indian context to devise best sandwich constructions for controlling the dwellings from outside traffic, honking, metro trains, and aircraft noise. Also, the strict enforcement of ambient noise standards and enforcement of a noise abatement program with a noise abatement goal shall be very significant for controlling the noise pollution across the globe.

References

1. European Commission, Science for Environment Policy (2017) Future brief: noise abatement approaches. European Commission DG Environment by the science communication report, April 2017
2. Murphy E, King E (2014) Environmental noise pollution. Elsevier, Amsterdam, pp. 1–282, May 5

3. The SMILE Consortium (2003) Guidelines for road traffic noise abatement. Sustainable Mobility Initiatives for Local Environment, Valbonne
4. Guarinoni M, Ganzleben C, Murphy E, Jurkiewicz K (2012) Towards a comprehensive noise strategy. Directorate General for Internal Policies, Policy Department, Economic and Science Policy, Brussels
5. Münzel T, Gori T, Babisch W, Basner M (2014) Cardiovascular effects of environmental noise exposure. Eur Heart J 35(13):829–836
6. European Environment Agency report (2014) Noise in Europe. EEA report 10/2014. Copenhagen, Denmark
7. Central Pollution Control Board, India (2017) Delhi's ambient noise levels influenced by traffic flow – case studies. Control of Urban Pollution Series. CUPS/87/2017-18, September, 2017
8. World Economic Forum report (2017) 27 March 2017, https://www.weforum.org/agenda/2017/03/these-are-the-cities-withthe-worst-noise-pollution. Accessed on 28 Mar 2021
9. Garg N, Sharma O, Mohanan V, Maji S (2012) Passive noise control measures for traffic noise abatement in Delhi, India. J Sci Ind Res 71:226–234
10. Garg N, Sharma O, Maji S (2011) Noise impact assessment of mass rapid transit systems in Delhi city. Indian J Pure Appl Phys 49:257–262
11. Kloth M, Vancluysen K, Clement F, Lars Ellebjerg P (2008) Practitioner handbook for local noise action plans: recommendations from the SILENCE project. http://www.noiseineu.eu/fr/3527-a/homeindex/file?objectid=3161&objecttypeid=0. Accessed on 28 Mar 2021
12. Klæboe R, Veisten K, Amundsen AH, Akhtar J (2011) Selecting road-noise abatement measures: economic analysis of different policy objectives. Open Transp J 5:1–8
13. Nijland H, van Kempen EEMM, Van Wee GP, Jabben J (2003) Costs and benefits of noise abatement measures. Transp Policy 10(2):131–140
14. Gordon RT, Vining WD (1992) Active noise control: a review of the field. Am Ind Hyg Assoc J 53(11):721–725
15. Widrow B, Stearns SD (1985) Adaptive signal processing. Prentice-Hall, Englewood Cliffs
16. Morgan DR (1980) An analysis of multiple correlation cancellation loops with a filter in the auxiliary path. IEEE Trans Acoust Speech Signal Process ASSP-28:454–467
17. Kuo SM, Gan WS (2005) Digital signal processors: architectures, implementations, and applications. Prentice Hall, Upper Saddle River
18. Kuo SM, Kuo K, Gan WS (2010) Active noise control: open problems and challenges. The 2010 International conference on green circuits and systems, 21–23 June 2010, Shanghai, China
19. Guo J, Pan J (1998) Increasing the insertion loss of noise barriers using an active-control system. J Acoust Soc Am 104:3406–3416
20. Sas P, Bao C, Augusztinovicz F (1995) Active control of sound transmission through a double panel partition. J Sound Vib 180:609–625
21. Lam B, Shi C, Gan WS (2017) Active noise control systems for open windows: current updates and future perspectives. 24th international congress on sound and vibration, 23–27 July, 2017, London, pp. 1–7
22. Huang H, Qiu X, Kang J (2011) Active noise attenuation in ventilation windows. J Acoust Soc Am 130(1):176–188
23. Lam B, Elliott SJ, Cher J, Gan WS (2015) The physical limits of active noise control of open windows. WESPAC 2015, Singapore
24. Rubin M, Lampert CM (1983) Transparent silica aerogels for window insulation. Solar Energy Mater 7(4):393–400
25. Lane J (2013) Active control of noise through windows. Masters in engineering dissertation, University of Canterbury, Christchurch, New Zealand
26. Jakob A, Moser M (2003) Active control of double-glazed windows. Part I. Feedforward control. Appl Acoust 64:163–182
27. Ruckman C. Active noise control FAQ. http://chrisruckman.com/ancfaq.htm
28. Milosevic A, Schaufelberge U (2005) Active noise control. Diploma thesis. University of Applied Sciences Rapperswil HSR, December 14, 2005

29. Ruiz-Padillo A, Ruiz DP, Torija AJ, Ridao AR (2016) Selection of suitable alternatives to reduce the environmental impact of road traffic noise using a fuzzy multi-criteria decision model. Environ Impact Assess Rev 61:8–18
30. Ohio Department of Transportation (2006) Statewide Investigation of noise abatement alternatives, McCormick Taylor, Philadelphia, June 2006
31. Nielsen G, Solberg S (1988) Costs and effects of traffic noise abatement measures: basis for a national programme. Appl Acoust 25:149–168
32. Milford I, Aaseboe S J, Strommer K (2014) Value for money in road traffic noise abatement. Internoise 2014, Melbourne, Australia, 16 to 19 November, 2014
33. Morgan P A, Nelson P M and Steven H (2003) Integrated assessment of noise reduction measures in the road transport sector. TRL Ltd, Essen
34. Sandberg U, Kalman B (2004) The poroelastic road surface-Results of an experiment in Stockholm. Proceedings of forum acusticum 2004, Budapest, Ther Hung
35. Sandberg U (2003) The multi-coincidence peak around 1000 Hz in tyre/road noise spectra. Euronoise, Naples, p 2003
36. Ellebjerg L (2005) Effectiveness and benefits of traffic flow measures on noise control. European Commission DG Research report
37. Liao G, Sakhaeifar MS, Heitzman M, West R, Waller B, Wang S, Ding Y (2014) The effects of pavement surface characteristics on tire/pavement noise. Appl Acoust 76:14–23
38. Ögren M, Molnár P, Barregard L (2018) Road traffic noise abatement scenarios in Gothenburg 2015–2035. Environ Res 164:516–521
39. Sandberg U, Ejsmont JA (2002) Tyre/road noise reference book. Informex, Kisa
40. Pallas MA, Kennedy J, Walker I, Chatagnon R, Berengier M, Lelong J (2014) Noise emission of electric and hybrid electric vehicle. FOREVER deliverable D2.1. CEDR Transnational Road Research Programme, April, 2014
41. Sandberg U, Ainge M, Altekoester C, Glaeser K-P, Nelson P, Phillips S, Steven H, Treleven C, Watts G (2006) Tyre/road noise. Volume 2: appendices no. FEHRL, report SI2.408210. Forum of European National Highway Research Laboratories, FEHRL
42. Heutschi K, Bhlmann E, Oertli J (2016) Options for reducing noise from roads and railway lines. Transp Res Part A Policy Pract 94(Suppl. C):S308–S322
43. Vázquez V, Paje S (2016) Study of the road surface properties that control the acoustic performance of a rubberised asphalt mixture. Appl Acoust 102:33–39
44. Sirin O (2016) State-of-the-art review on sustainable design and construction of quieter pavements-part 2: factors affecting Tire-pavement noise and prediction models. Sustainability 8(7):692. https://doi.org/10.3390/su8070692
45. Ohiduzzaman MD, Sirin O, Kassem E, Rochat JL (2016) State-of-the-art review on sustainable design and construction of quieter pavements-part 1: traffic noise measurement and abatement techniques. Sustainability 8(8):742. https://doi.org/10.3390/su8080742
46. Freitas E, Mendonça C, Santos JA, Murteira C, Ferreira JP (2012) Traffic noise abatement: how different pavements, vehicle speeds and traffic densities affect annoyance levels. Transp Res Part D: Transp Environ 17(4):321–326
47. ASTM E413-87 (1999) Classification for rating sound insulation. American Society for Testing and Materials, West Conshohocken
48. Mitigating aircraft noise in new residential construction, builders guide, Metropolitan Council, Minnesota (2006). http://www.metrocouncil.org/planning/transportation/BuildersGuide.pdf. Accessed on 28 Mar 2021
49. Dunn IP (1989) Comparison of STC rating to dB(A) reduction for aircraft and traffic noise. Acoustics Australia 17(1):11
50. Quirt JD (1983) Sound transmission through windows II. Double and triple glazing. J Acoust Soc Am 74:534–542
51. Kim YT, Kim HC, Jung SS, Jho MJ, Suh SJ (2002) Dependence of coincidence frequency in double-glazed window on glass thickness and interpane cavity. Appl Acoust 63:927–936
52. Quirt JD (1982) Sound transmission through window I. Single and double glazing. J Acoust Soc Am 72:834–844

53. Garg N (2016) Investigations on modeling transportation noise and passive noise control measures for its abatement. Doctoral dissertation, Delhi Technological University, Delhi, February
54. Garg N, Sharma O, Maji S (2011) Experimental investigations on sound insulation through single, double and triple window glazing for traffic noise abatement. J Sci Ind Res 78:471–478
55. Garg N, Sharma S, Maji S (2012) Design consideration for enhancing sound insulation characteristics of window glazing for traffic noise abatement. J Buil Acoust 19(2):89–98
56. Garg N, Kumar A, Maji S (2013) Parametric sensitivity analysis of factors affecting sound insulation of double glazing using Taguchi method. Appl Acoust 74:1406–1413
57. Dry Wall Handbook, Saint Gobain Gyproc Ltd., http://www.gyproc.in/images/DrywallHandbook.pdf
58. Warnock ACC (1990) Sound transmission loss measurement through 190 mm and 140 mm blocks with added dry wall and through cavity block walls, NRC Canada. Internal Report No. 586
59. Quirt JD, Warnock ACC, Birta JA (1995) Sound transmission through gypsum board walls: sound transmission results. National Research Council, Internal report IRC-IR-693
60. Garg N, Sharma S, Maji S (2011) Design considerations of building elements for traffic and aircraft noise abatement. Indian J Pure Appl Phys 49:437–450
61. Bai G, Zhan P, Sui F, Yang J (2014) Research on sound insulation of multiple-layer structure with porous material and air-layer. Internoise, 16–19 November, Melbourne, Australia
62. Warnock ACC (1993) Sound transmission through slotted concrete blocks with attached gypsum board. J Acoust Soc Am 94:2713–2720
63. Warnock ACC (1985) Factors affecting sound transmission loss. Canadian Building Science Insight, CBD 239, National Research Council, Canada report. http://web.mit.edu/parmstr/Public/NRCan/CanBldgDigests/cbd239_e.html. Accessed on 28 Mar 2021
64. Guillen I, Uris A, Estelles H, Llinares J, Llopis A (2008) On the sound insulation of masonry wall facades. Build Environ 43:523–529
65. Amundsen AH, Klæboe R, Aasvang GM (2011) The Norwegian facade insulation study: the efficacy of facade sound insulation in reducing noise annoyance due to road traffic. J Acoust Soc Am 129(3):1381–1389
66. Garg N, Kumar A, Maji S (2014) Parametric sensitivity analysis of factors affecting the sound insulation of multi-layered building elements. Achiev Acoust 39(2):165–176
67. Bravo JM, Sinisterra J, Uris A, Linares J, Estelles H (2002) Influence of airlayers and damping layers between gypsum boards and sound transmission. Appl Acoust 63:1051–1059
68. Lin GF, Garrelick JM (1977) Sound transmission through periodically framed parallel plates. J Acoust Soc Am 61:1014
69. Bradley JS, Birta JA (2001) On the sound insulation of wood stud exterior walls. J Acoust Soc Am 110:3086–3096
70. Warnock ACC, Quirt JD (1997) Control of sound transmission through gypsum board walls. Construction Technology Update No. 1
71. Uris A, Llopis A, Llinares J (1999) Effect of the Rockwool bulk density on the airborne sound insulation of light weight double walls. Appl Acoust 58:327–331
72. Warnock ACC, Quirt JD (2021) Sound transmission through gypsum board walls. NRCC-39272. http://citeseerx.ist.psu.edu/viewdoc/download?doi=10.1.1.10.4668&rep=rep1&type=pdf. Accessed on 28 Mar 2021
73. Bradley JS, Birta JA (2000) Laboratory measurements of the sound insulation of building facade elements. IRC Internal Report, IRC IR-818
74. Bradley JS, Birta JA (2001) A simple model of the sound insulation of gypsum board on resilient supports. Noise Control Eng J 49:216–223
75. Nightingale T, Halliwell R, Quirt J (2001) Sound transmission through gypsum board walls – effect of shear membranes and framing details. Can Acoust 29(3):62–63
76. Quirt JD (1985) Sound transmission through building components. National Research Council, Canada report
77. Bradley JS, Gover BN, Halliwell RE, Nightingale TRT, Quirt JD, Warnock ACC (2004) Recent projects in building acoustics at NRC. Can Acoust 32(3):202–203

78. Davy JL, Fard M, Dong W, Loverde J (2019) Empirical corrections for predicting the sound insulation of double leaf cavity stud building elements with stiffer studs. J Acoust Soc Am 145(2):703–713
79. Warnock ACC (1998) Controlling sound transmission through concrete block walls. Construction Technology Update No. 13
80. Bradley JS, Gover BN (2011) Selecting walls for speech privacy. IRC report RR-314
81. Rasmussen B, Rindel JH (2010) Sound insulation between dwellings – descriptors applied in building regulations in Europe. Appl Acoust 71:171–180
82. Rasmussen B (2010) Sound insulation between dwellings-requirements in building regulations in Europe. Appl Acoust 71:373–385
83. National Building Code of India (2016) Bureau of Indian Standards, Delhi, volume 2
84. Ljunggren F, Simmons C, Öqvist R (2017) Correlation between sound insulation and occupants' perception – proposal of alternative single number rating of impact sound, part II. Appl Acoust 123:143–151
85. Machimbarrena M, Rasmussen B (2017) Acoustic regulations for housing and schools in Europe and South America – a pilot study in 2016. 22nd international congress on acoustics, ICA2016 – 487, proceedings of meetings on acoustics, 28, 015013
86. Simmons C (2004) Reliable building element sound insulation data for EN 12354 calculations facilitates analysis of Swedish dwelling houses. Inter-noise 2004, Prague, Czech Republic, August 22–25, 2004
87. Scholl W, Lang J, Wittstock V (2011) Rating of sound insulation at present and in future. The revision of ISO 717. Acta Acustica United Acust 97:686–698
88. Rindel JH (1998) Acoustic quality and sound insulation between dwellings. Conference in building acoustic-acoustic performance of medium-rise timber buildings, Dublin
89. Jagniatinskis A, Fiks B, Mickaitis M, Šukys R (2017) Features of sound classification scheme designated to label buildings in Lithuania. J Civ Eng Manag 23(3):409–420
90. Rindel JH (1998) The relationship between sound insulation and acoustic quality in dwellings. Inter-Noise 98, Christchurch, New Zealand
91. Rasmussen B (2019) Use of ISO 717 low-frequency sound insulation descriptors in acoustic regulations and recommendations for housing in Europe. Proceedings of DAGA 2019. Deutsche Gesellschaft für Akustik, DEGA, 45, 78-81. (Fortschritte der Akustik, Vol. 45), Berlin
92. Rasmussen B (2019) Sound insulation between dwellings – comparison of national requirements in Europe and interaction with acoustic classification schemes. In Ochmann M, Vorländer M, J Fels (eds) Proceedings of the ICA 2019 and EAA EuroRegio: 23rd international congress on acoustics, integrating 4th EAA EuroRegio 2019: September 2019, Aachen, Germany. Deutsche Gesellschaft für Akustik (DEGA e.V.), Berlin, pp. 5102–5109, Aachen, Germany, 9 September 2019
93. Rasmussen B, Machimbarrena M (2019) Developing an international acoustic classification scheme for dwellings – from chaos and challenges to compromises and consensus? In Proceedings of inter-noise 2019. Noise-Con Proceedings, 48th international congress and exposition on noise control engineering, Madrid, Spain, June 16–19, 2019
94. Fausti P, García TC, Ingelaere B, Machimbarrena M, Monteiro C, Santoni A, Secchi S, Smith S (2014) Chapter 11: Design and acoustic performance of building constructions for multi-storey housing: compendium. In: Rasmussen B, Machimbarrena M (eds) Building acoustics throughout Europe – volume 1: towards a common framework in building acoustics throughout Europe, COST Action TU 0901
95. Monteiro C, Machibarrena M, Tarrero AI, Smith RS (2017) Transition between existing and proposed harmonized airborne sound insulation descriptors; a statistical approach based on in-situ measurements. Appl Acoust 116:94–106
96. LoVerde JJ, Dong DW (2017) A dual-rating method for evaluating impact noise isolation of floor-ceiling assemblies. J Acoust Soc Am 141(1):428–440
97. Rasmussen B (2018) Building acoustic regulations in Europe – brief history and actual situation. In: Proceedings of the baltic-nordic acoustics meeting 2018, 2018 edition. Nordic

Acoustic Association. Joint Baltic-Nordic Acoustics Meeting (BNAM) Proceedings, Reykjavik, pp 1–16
98. Garg N, Kumar A, Maji S (2013) Significance and implications of airborne sound insulation criteria in building elements for traffic noise abatement. Appl Acoust 74(2013):1429–1435
99. Arenas JP (2008) Potential problems with environmental sound barriers when used in mitigating surface transportation noise. Sci Total Environ 405:173–179
100. Pigasse G, Kragh J (2011) Optimised noise barriers, a state-of-the-art report. Report 194 – 2011, 978-87-92094-77-3, Vejdirektoratet, December, 2011
101. Paige T (2015) Outdoor noise barriers: design and applications, kinetics noise control., http://www.enoisecontrol.com/wpcontent/uploads/2014/12/outdoor_noise_barrier_wall.pdf
102. U.S. DOT (2000) Federal Highway Administration, FHWA highway noise barrier design handbook. Final Report FHWA-EP-00-005
103. Watson D (2006) Evaluation of benefits and opportunities for innovative noise barrier designs. Final report 572, Arizona Department of Transportation
104. Grubeša S, Jambrošić K, Domitrović H (2012) Noise barriers with varying cross-section optimized by genetic algorithms. Appl Acoust 73:1129–1137
105. Parnell J, Samuels S (2005) Development and evaluation of roadside barriers to attenuate road traffic noise. Proceedings of acoustics, 9–11 November, 2005, Busselton, Western Australia, pp. 265–269
106. Morgan PA (2004) Review of Japanese noise barrier research. DWW-2004-081 RAP-04.000836, December, 2004
107. Ishizuka T, Fujiwara K (2004) Performance of noise barriers with various edge shapes and acoustical conditions. Appl Acoust 65:125–141
108. Daigle GA (1999) Technical assessment of the effectiveness of the noise walls. I-INCE Publication, 99-1, September, 1999, pp. 137–161
109. Oltean-Dumbrava C, Miah A (2016) Assessment and relative sustainability of common types of roadside noise barriers. J Clean Prod 135:919–931
110. Slutsky S, Bertoni HL (1988) Analysis and programs for assessment of absorptive and tilted parallel barriers. Technical report, Transportation Research Board
111. Hothersall DC, Crombie DH, Chandler-Wilde SN (1991) The performance of T-profile and associated noise barriers. Appl Acoust 32(4):269–287
112. Oldham DJ, Egan CA (2011) A parametric investigation of the performance of T-profiled highway noise barriers and the identification of a potential predictive approach. Appl Acoust 72(11):803–813
113. Monazzam MR, Lam YW (2005) Performance of profile single noise barriers covered with quadratic residue diffusers. Appl Acoust 66:709–730
114. Monazzam MR, Lam YW (2008) Performance of T-shape barriers with top surface covered with absorptive quadratic residue diffusers. Appl Acoust 69:93–109
115. Baulac M, Defrance J, Jean P (2008) Optimization with genetic algorithm of the acoustic performance of T-shaped noise barriers with a reactive top surface. Appl Acoust 69:332–342
116. ISO 10847 (1997) Acoustics: in-situ determination of insertion loss of outdoor noise barriers of all types
117. Eusébio Conceição VRE, Rosão C (2011) More than one insertion loss, throughout the day, for the same outdoor noise barrier. ICSV 18, Rio De Janeiro Brazil, 10, 14 July, 2011, pp 750–757
118. Kumar K, Parida M, Katiyar VK (2014) Optimized height of noise barrier for non-urban highway using artificial neural network. Int J Environ Sci Technol 11:719–730
119. Maekawa Z (2004) Shielding highway noise. Noise Control Eng J 52(3):86–92
120. Kurze UJ, Anderson GS (1971) Sound attenuation by barriers. Appl Acoust 4:35–53
121. Menounou P (2001) A correction to Maekawa's curve for the insertion loss behind barriers. J Acoust Soc Am 110:1828–1838
122. Bies DA, Hansen CH (2003) Engineering noise control: theory and practice, 3rd edn. Spon Press, London
123. ISO 9613-2 (1996) Acoustics-Attenuation of sound during propagation outdoors – part 2: general method of calculation

124. Jónsson G B (2007) A comparison of two numerical models for outdoor sound propagation Harmonoise and Nord 2000. MSc thesis, acoustic technology Ørsted DTU, July, 2007
125. Mun S, Cho Yoon-Ho (2009) Noise barrier optimization using a simulated annealing algorithm. Appl Acoust 70:1094–1098
126. Mohanan V, Sharma O (2001) Acoustical design of a noise barrier for an elevated rail corridor. J Acoust Soc India 28(1–4):95–100
127. Rominder R, Parida M, Jain SS, Khalid M (2013) Design of a noise barrier for Britania Chowk Flyover in Delhi, India. J Environ Res Dev 7:1644–1650
128. De Schepper E, Passel SV, Manca J, Thewys T (2012) Combining photovoltaics and sound barriers-a feasibility study. Renew Energy 46:297–303
129. Poe C, Plovnick A, Hodges T, Hastings A, Dresley S (2017) Highway renewable energy: photovoltaic noise barriers. FHWA-HEP-17-088, U.S. Department of Transportation, Washington D.C, August, 2017
130. Krynkin A, Umnova O, Chong AYB, Taherzadeh S, Attenborough K. Sonic crystal noise barriers, https://www.salford.ac.uk/research/sirc/research-groups/acoustics/materials-for-acoustics/sonic-crystal-noise-barriers. Accessed on 28 Mar 2021
131. Sanchez-Perez JV, Rubio C, Martinez-Sala R, Sanchez-Grandia R, Gomez V (2002) Acoustic barriers based on periodic arrays of scatterers. Appl Phys Lett 81(27):5240–5242
132. Sanchez-Perez JV, Caballero D, Martinez-Sala R, Sanchez-Dehesa CJ, Rubio F, Llinares MJ, Galvez F (1998) Sound attenuation by a two-dimensional array of rigid cylinders. Phys Rev Lett 80:5325–5328
133. Sanchez-Dehesa J, Garcia-Chocano VM, Torrent D, Cervera F, Cabrera S, Simon F (2011) Noise control by sonic crystal barriers made of recycled materials. J Acoust Soc Am 129: 1173–1183
134. Garcìa-Chocano VM, Cabrera S, Sanchez-Dehesa J (2012) Broadband sound absorption by lattices of microperforated cylindrical shell. Appl Phys Lett 101:184101
135. Santra SC, Chakrabarty D, Roy B (1998) Urban traffic noise abatement with vegetation barriers. J Acoust Soc India 26(1–4):1–10
136. Mohanan V, Sharma O, Singh M, Garg N (2009) NPL technical report, noise control measures for proposed Commonwealth Games Village near Noida Mor. NPL Technical Report No. AC. C.07(4)-01
137. Nilsson M, Bengtsson J, Klaeboe R (2014) Environmental methods for transport noise reduction. CRC Press, Hoboken
138. Kaplan R, Kaplan S (1989) The experience of nature: a psychological perspective. Cambridge University Press, New York
139. Renterghem TV (2018) Towards explaining the positive effect of vegetation on the perception of environmental noise. Urban For Urban Green 40:133–144
140. Tyagi V, Kumar K, Jain VK (2013) Road traffic noise attenuation by vegetation belts at some sites in the Tarai region of India. Archiv Acoust 38:389–395
141. Huddart L (1990) The use of vegetation for traffic noise screening. Department of transport TRRL report ISSN 0266-5247. Transport and Road Research Laboratory, Crowthorne
142. Renterghem TV, Bottledooren D, Verheyen K (2012) Road traffic noise shielding by vegetation belts of limited depth. J Sound Vib 331:2404–2425
143. Pathak V, Tripathi BD, Mishra VK (2008) Dynamics of traffic noise in a tropical city Varanasi and its abatement through vegetation. Environ Monit Assess 146:67–75
144. Huisman WHT, Attenborough K (1998) Reverberation and attenuation in a pine forest. J Acoust Soc Am 90:2664–2677
145. Aylor D (1972) Noise reduction by vegetation and ground. J Acoust Soc Am 51:197–205
146. Aylor D (1972) Sound propagation through vegetation in relation to leaf density, leaf width, and breadth of canopy. J Acoust Soc Am 51:411. https://doi.org/10.1121/1.1912852
147. Ow LF, Ghosh S (2017) Urban cities and road traffic noise: reduction through vegetation. Appl Acoust 120:15–20
148. Joynt JLR, Kang J (2010) The influence of preconceptions on perceived sound reduction by environmental noise barriers. Sci Total Environ 408:4368–4375

149. Herrington LP (1976) Effect of vegetation on the propagation of noise in the outdoors, USDA Forest Service General Technical Report. US Rocky Mountain Forest. Range Experimental Station, pp 229–233
150. Green Buffers for Screening and Noise Reduction (2008) Sustainable community forestry program. Georgia Forestry Commission, Georgia. https://www.urbanforestrysouth.org/resources/library/greenbuffers-for-screening-and-noise-reduction. Accessed on 28 Mar 2021
151. Fujiwara K, Hothersall DC, Kim CH (1998) Noise barriers with reactive surfaces. Appl Acoust 53(4):255–272
152. Pal AK, Kumar V, Saxena NC (2000) Noise attenuation by green belts. J Sound Vib 234(1): 149–165
153. Garg N (2018) Environmental noise pollution: measurement strategies, implications, effects and its control in urban perspectives. Workshop on noise pollution and its effects on human health, Ambedkar University, Kashmere Gate Campus, New Delhi, India, 26 October, 2018
154. Garg N, Maji S (2016) A retrospective view of noise pollution control policy in India: status, proposed revisions and control measures. Curr Sci 111(1):29–38
155. Münzel T, Schmidt FP, Steven MDS, Herzog J, Driber A, Sørensen M (2018) Environmental noise and the cardiovascular system. J Am Coll Cardiol 71:688–697
156. Chanaud RC (2014) Noise ordinances, tools for enactment, modification and enforcement of a community noise ordinance. https://www.noisefree.org/Noise-Ordinance-Manual.pdf
157. Road traffic signs. http://www.roadtrafficsigns.com/truck-sign/quite-please-no-honking-unless-for-danger-sign/sku-k-6179.aspx
158. Federal Highway Administration, FHWA (2017) Entering the quiet zone: noise compatible land use planning. https://www.fhwa.dot.gov/environment/noise
159. International Civil Aviation Organization, ICAO (2008) Guidance on the balanced approach to aircraft noise management. Doc 9829 AN/451
160. Vogiatzis K (2012) Airport environmental noise mapping and land use management as an environmental protection action policy tool. The case of the Larnaka International Airport (Cyprus). Sci Total Environ 424:162–173
161. Drollinger MT (1992) Land use planning approaches to mitigating general aviation aircraft noise. Transp Res Rec 1332:11–20
162. Peeters B, van Blokland G (2018) Decision and cost/benefit methods for noise abatement measures in Europe. M+P.BAFU.15.02.1. European Network of the Heads of Environment Protection Agencies (EPA Network), February 2018
163. "Kostenkentallen 2009", prepared by M+P for Rijkswaterstaat, The Netherlands, September 2009
164. Wirtschaftliche Tragbarkeit und Verhältnismässigkeit von Lärmschutzmassnahmen – optimierung der Interessenabwägung. BAFU report code UV-0609-D, Switzerland, 2006
165. Kuijpers A et al (2009) Beheersing railruwheid als geluidmaatregel, van proefproject naar implementatie. M+P report no. M+P.RAIL.06.15.11, The Netherlands, February 2009
166. Impact assessment study on rail noise abatement measures addressing the existing fleet (2007) European Commission: Directorate General Energy and Transport, REFERENCE TREN/A1/46-2005, December 10, 2007
167. Peeters B, van Blokland G, Schwanen W, Boegli H (2016) The role of monetization in decision methods for traffic noise abatement measures. Internoise, Hamburg
168. WHO-JRC (2011) Burden of disease from environmental noise-quantification of healthy life years last in Europe. European Center for Environment and Health, JRC EU, 2011
169. Peeters B, van Blokland G (2017) Best practice for cost/benefit based decisions on abatement of traffic noise. 12th ICBEN congress on noise as a public health problem, 18–22 June, Zurich, 2017
170. CEDR Technical Report (2017) State of the art in managing road traffic noise: cost benefit analysis and cost effectiveness analysis, June, 2017
171. Finegold LS (2011) Final report of the I-INCE technical study group on community noise: environmental noise impact assessment and mitigation (TSG 6). I-INCE Publication Number: 11-1, March, 2011

172. Lercher P (2017) Noise in cities: urban and transport planning determinants and health in cities. In: Nieuwenhuijsen M, Khreis H (eds) Integrating human health into urban and transport planning. Chapter, vol 22. Springer, Cham, pp 443–481
173. Naish DA, Tan ACC, Nur Demirbilek F (2012) Estimating health related costs and savings from balcony acoustic design for road traffic noise. Appl Acoust 73(5):497–507
174. Sakamoto S, Aoki A (2015) Numerical and experimental study on noise shielding effect of eaves/louvers attached on building facade. Build Environ 94(2):773–784
175. Lee PJ, Kim YH, Jeon JY, Song KD (2007) Effects of apartment building façade and balcony design on the reduction of exterior noise. Build Environ 42(10):3517–3528
176. Sanchez GME, Renterghem TV, Thomas P, Botteldooren D (2016) The effect of street canyon design on traffic noise exposure along roads. Build Environ 97:96–110
177. Ariza-Villaverde AB, Jiménez-Hornero FJ, De Ravé EG (2014) Influence of urban morphology on total noise pollution: multifractal description. Sci Total Environ 472:1–8
178. Ryu H, Park IK, Chun BS, Chang S Il (2017) Spatial statistical analysis of the effects of urban form indicators on road-traffic noise exposure of a city in South Korea. Appl Acoust 115:93–100
179. Silva LT, Fonseca F, Rodrigues D, Campos A (2018) Assessing the influence of urban geometry on noise propagation by using the sky view factor. J Environ Plan Manag 61(3): 1–18. https://doi.org/10.1080/09640568.2017.1319804
180. Ishizuka T, Fujiwara K (2012) Traffic noise reduction at balconies on a high-rise building. J Acoust Soc Am 131(3):2110–2117
181. Ishizuka T, Fujiwara L (2013) Full scale tests of reflective noise-reducing devices for balconies on high-rise buildings. J Acoust Soc Am 134(2):EL185–EL190
182. Lau K-K, Lee C-K, Yeung M (2017) Innovative means to protect residential developments from traffic noise-Hong Kong experience. World sustainable built environment conference 2017, Hong Kong, track 2: Practices and Policies for High-Performance Buildings, pp 585–590
183. Cheung KMC, Wong HYC, Hung WCT, Lau KK (2019) Development and application of specially designed windows and balconies for noise mitigation in Hong Kong. Proceedings Internoise, June 16–19, 2019, Madrid, Noise Control for a Better Environment
184. Hao Y, Kang J, Krijnders D, Wörtche H (2015) On the relationship between traffic noise resistance and urban morphology in low-density residential areas. Acta Acust United Acust 101(3):510–519
185. Hao Y (2014) Effects of urban morphology on urban sound environment from the perspective of masking effects. Doctoral thesis, School of Architecture, the University of Sheffield, April
186. Wang B, Kang J (2011) Effects of urban morphology on the traffic noise distribution through noise mapping: a comparative study between UK and China. Appl Acoust 72(8):556–568
187. Han X, Huang X, Liang H, Ma S, Gong J (2018) Analysis of the relationships between environmental noise and urban morphology. Environ Pollut 233:755–763
188. Zhang Ji, Lau SSY, Heng CK, Lau SK, Lai H (2017) A case study on the relationship between urban morphology and traffic noise distribution in high-density urban context. SimAUD 2017 May 22–24, Toronto, Canada, pp 281–287
189. Brown AL (2010) Soundscapes and environmental noise management. Noise Control Eng J 58(5):493–500
190. Kang J (2019) Noise management: soundscape approach. In: Encyclopedia of environmental health. Elsevier, Amsterdam, pp 683–694
191. Cerwén G, Kreutzfeldt J, Wingren C (2017) Soundscape actions: a tool for noise treatment based on three workshops in landscape architecture. Fron Archit Res 6(4):504–518
192. Chew YR, Wu BS (2016) A soundscape approach to analyze traffic noise in the city of Taipei, Taiwan. Comput Environ Urban Syst 59:78–85
193. Alletta F, Kang J (2015) Soundscape approach integrating noise mapping techniques: a case study in Brighton, UK. Noise Mapp 2:1–12
194. Cerwén G, Wingren C, Qviström M (2017) Evaluating soundscape intentions in landscape architecture: a study of competition entries for a new cemetery in Järva, Stockholm. J Environ Plan Manag 60(7):1253–1275

195. Cerwén G (2019) Listening to Japanese gardens: an autoethnographic study on the soundscape action design tool. Int J Environ Res Public Health 16:4648. https://doi.org/10.3390/ijerph16234648
196. Lugten M, Karacaoglu M, White K, Kang J, Steemers K (2018) Improving the soundscape quality of urban areas exposed to aircraft noise by adding moving water and vegetation. J Acoust Soc Am 144(5):2906–2917
197. Jeon JY, Lee PJ, You J, Kang J (2010) Perceptual assessment of quality of urban soundscapes with combined noise sources and water sounds. J Acoust Soc Am 127(3):1357–1366
198. Sheikh M, Mitchell A (2018) Design strategies for perceived acoustic comfort in urban environments – A literature review. Proceedings of ACOUSTICS 2018, 7–9 November 2018, Adelaide, Australia
199. den Boer LC, Schroten A (2007) Traffic noise reduction in Europe. Health effects, social costs and technical and policy options to reduce road and rail traffic noise. CE Delft, Delft
200. Position Paper of the Working Group on Health and Socio-Economic Aspects (2003) Valuation of noise. December, 2003. http://www.moa.gov.cy/moa/environment/environmentnew.nsf/69D658BCCDE2FDB5C2258041002D61F5/$file/Noise%20Monetisation.pdf. Accessed on 2 Apr 2021
201. European Commission Green Paper (1996) Future noise policy. Commission of the European Authorities, Brussels, 4 September 1996 COM (96) 540 final
202. Hoevenagel R (1994) The contingent valuation method: scope and validity. Thesis. Vrije, Universiteit, Amsterdam, April, pp 1–290
203. Navrud S (2002) The state-of-the-art on economic valuation of noise. Final report to European Commission DG Environment, April 14. https://ec.europa.eu/environment/archives/noise/pdf/020414noisereport.pdf. Accessed on 2 Apr 2021
204. Department of the Environment, Transport and the Regions, DETR (1999) Review of studies on external costs of noise. Prepared by Rhian Hawkins, Environment Protection Economics Division, DETR (now Department of Environment, Food and Rural Affairs, DEFRA). July 1999
205. Swinburn TK, Hammer MS, Neitzel RL (2015) Valuing quiet: an economic assessment of US environmental noise as a cardiovascular health hazard. Am J Prev Med 49(3):345–353
206. Kim K, Shin J, Oh M, Jung JK (2019) Economic value of traffic noise reduction depending on residents' annoyance level. Environ Sci Pollut Res 26:7243–7255
207. Conference of European Directors of Roads report (2013) Value for money in road traffic noise abatement. Final Report, v4, 16 May 2013, CEDR Project Group Road Noise: subgroup tyre/vehicle noise
208. ETC/ATNI (2019) Noise action plans. Managing exposure to noise in Europe, No ETC/ATNI Report No 8/2019. https://www.eionet.europa.eu/etcs/etc-atni/products/etc-atni-reports/etc-atni-report-8-2019-noise-actionplans-managing-exposure-to-noise-in-europe. Accessed on 3 Apr 2021
209. Conference of European Directors of Roads report (2017) State of the art in managing road traffic noise: noise barriers. CEDR Technical Report 2017-02. https://www.cedr.eu/download/Publications/2017/CEDR-TR2017-02-noise-barriers.pdf. Accessed on 28 Mar 2021
210. Garg N, Vishesh MS (2015) Fuzzy TOPSIS approach in selection of optimal noise barrier for traffic noise abatement. J Achiev Acoust 40:453–467

Chapter 8
Noise Mapping in Urban Environment

Abstract This chapter presents noise mapping procedures followed for conducting noise impact assessment studies. The noise mapping strategy, benefits, and concerned noise control action plans are discussed. This chapter critically reviews all the pros and cons and implications of noise mapping in urban environment. Noise mapping has become a vital tool for urban planning and devising noise control measures in the recent times. The concepts of soundscape mapping and planning as a potential tool for urban development and noise priority indices for noise action plans are also presented. It is notable that the traditional noise control methods and the soundscape approach are complementary methods rather than being a substitute or conflicting strategy.

8.1 Introduction

The societal cost of rail and traffic noise has been estimated to 40 billion Euros per year in the EU (0.4% of the GDP), with 90% of this cost being related to passenger cars and goods vehicles [1]. The European Noise Directive 2002/49/EC (END) has introduced two new instruments for urban planning and noise management: strategic noise maps and action plans. The declared aim of the Directive is "to define a common approach intended to avoid, prevent or reduce on a prioritized basis the harmful effects, including annoyance, due to exposure to environmental noise" [2]. One principle of the Directive is monitoring the environmental noise by requiring the competent authorities in Member States to draw up "strategic noise maps" for major roads, railways, airports, and agglomerations, using harmonized noise indicators L_{den} (day-evening-night equivalent level) and L_{night} (night equivalent level). These maps will be used to assess the number of people annoyed and sleep-disturbed respectively throughout Europe. A noise map is an instrument to inform the citizens about the number of people exposed to the different noise levels, with the aim to involve the general public into the development of action plans [3]. The noise maps are an information tool and a diagnosis of the sound environment for decision-making. These maps have to thus initiate the noise control actions in areas considered as critical according to this diagnosis and identify and protect "quiet" areas by

N. Garg, *Environmental Noise Control*,
https://doi.org/10.1007/978-3-030-87828-3_8

formulating strategies to keep these quiet [4]. The END distinguishes between two different kinds of quiet areas, “open country” (an area delimited by the competent authority undisturbed by noise from traffic, industry, or recreational activities) and “agglomerations” (an area delimited by the competent authority which is not exposed to greater than a certain set value of L_{den} by the Member State, from any noise source). Thus, the noise maps created are helpful for the management of environmental noise [5]. Public participation is required for noise action planning under the END in accordance with Directive 90/13/EEC on freedom of access of information to the environment. The concept of noise mapping is a paradigm shift from traditional methods, whereby the accuracy of simulation tools is often challenged by the complexity of modeled scenario particularly when broader perspectives are considered like the whole city [6]. In recent years, it has become an important and valuable tool for environmental noise assessment and control.

8.2 Noise Mapping: Smart Tool for Environmental Noise Control

Noise mapping is an important tool for action planning concerning action and abatement plans based on the analysis of noise maps for the area in question. The analysis may offer decision makers the necessary decision support tools in order to formulate and undertake the noise action planning. “Noise mapping” shall mean the presentation of data of an existing or predicted noise situation in terms of a noise descriptor, indicating breaches of any relevant limit value in force, the number of people affected in a certain area, or the number of dwellings exposed to certain values of a noise indicator in a certain area (European Directive 2002/49/EC, Article 3, letter q). “Strategic noise map” is a map designed for the global assessment of noise exposure in a given area due to different noise sources or for overall predictions for such an area (European Directive 2002/49/EC, Article 3, letter *r*). A strategic noise map describes:

- Noise scenario in terms of suitable indicators such as L_{den} and L_{night} as recommended in END and compliance of sites w.r.t legal limits recommended.
- Estimated number of dwellings exposed to noise levels.
- Estimated number of exposed people to noise levels.

Furthermore, strategic noise mapping (European Directive 2002/49/EC, Point 10) should be imposed in certain areas of interest as it can capture the data needed to provide a representation of the noise levels perceived within that area [6]. The END requires Member States to produce strategic noise maps for the main sources of environmental noise, that is, major roads, major railways, major airports, and agglomerations. Strategic noise map thus provides a representation of the noise levels perceived within this area, using the suitable noise indicators L_{den} and $L_{\mathrm{night.}}$ Thus, while the “noise mapping” is focused primarily on the presentation of noise

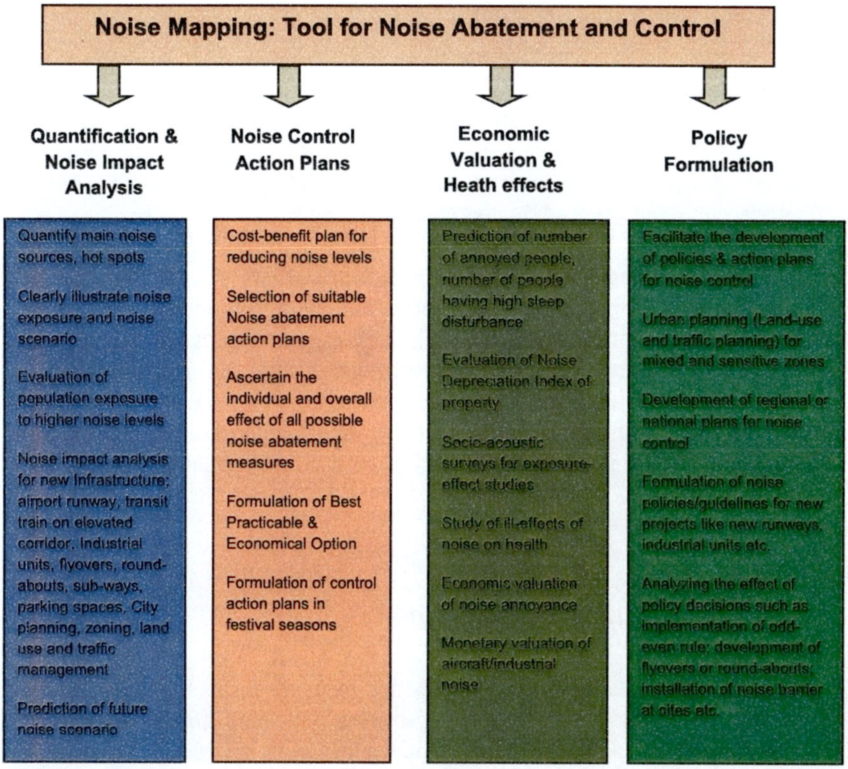

Fig. 8.1 Noise mapping: An indispensable tool for noise control in national and international perspectives

data, "strategic noise mapping" is more concerned with global assessment of noise exposure in a given area due to different noise sources under the terms of European Noise Directive [7]. Figure 8.1 describes the various aspects that make noise mapping an indispensable tool.

Strategic noise maps for agglomerations emphasize on the noise emitted by road, railway, airports, and industrial activity. It involves the following process [8]:

- Collecting raw data and preparing, storing, and analyzing the data.
- Computing the noise levels using validated computer models.
- Cumulating the noise levels (when there are different sources).
- Determining noise contours and effects.
- Presentation of noise impact.

8.3 Noise Mapping Strategy

Noise mapping strategy is not easy as it looks like. There are many input parameters required for developing an accurate and precise noise maps. The noise maps may be drawn based on actual measurements at various sites based on a certain sampling strategy or may be plotted using a validated noise model integrated with geographical interface systems (GIS). Usually, noise mapping is preferred to be done for all the sources for an area under study so as to present an overall noise scenario of the site. The noise mapping analysis focuses on three major issues as shown in Fig. 8.2: identification of noisy hot spots, identification of the most significant noise sources, and estimated number of people exposed in certain noise classes, for example, L_{den} and L_{night} and devise action plans for reducing noise levels at the noisy hot spots. The major considerations involved are measurement aspects related to microphone position and height, digital terrain model and input parameters required, and the sampling and spatial interpolation strategy to be used for developing accurate and precise noise maps.

8.3.1 Measurement Considerations

The ISO 1996-2 standard [9] recommends that for noise mapping the microphone height must be 4.0 ± 0.5 m in residential areas with multistorey buildings and 1.2 ± 0.1 m or 1.5 ± 0.1 m in residential areas with one floor buildings and recreational areas. Also, the END states that when calculations are carried out for developing strategic noise maps, the assessment point must be 4.0 ± 0.2 m in height above the ground level. Some studies such as that reported by Nicol and Wilson (2004) show that on average microphone situated at height 1.5 m height registered 0.2 dB more noise levels than microphone situated at height 4 m [10]. Walerian et al. [11] study also showed that measured noise levels increased with height from 1.4 to 4.2 m for a total of 0.5 dB, decreasing by 1.0 dB between 4.2 m and 5.6 m. Montes González et al. (2015) study also showed the differences of sound level between the mobile microphone located at heights of 1.2 m, 1.5 m, 2.5 m, and 6.0 m and the

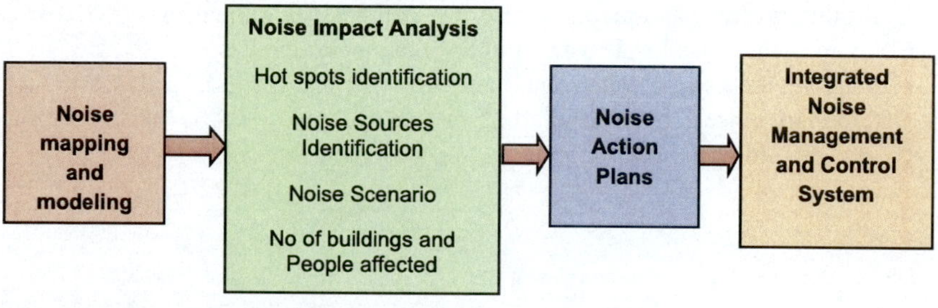

Fig. 8.2 Noise mapping strategy

reference microphone located at 4.0 m were 0.7 dB, −0.8 dB, −0.2 dB, and 0.4 dB, respectively. Thus, higher sound values were observed as the height increased [12]. ISO 1996-2 stipulates that the incident sound level for a fixed facade should be measured and corrected to exclude the corrections of reflected contribution at the facade. A correction of −3 dB (energy doubling) in noise levels for microphone position located between 0.5 and 2 m in front of reflecting facade and − 6 dB (pressure doubling) for microphone flush mounted on the reflecting surface is allowed. However, some published results show significant differences with values proposed by standards for these corrections [13, 14]. The European Commission Working Group Assessment of Exposure to Noise (WG-AEN) points out that for strategic noise mapping, it is generally impractical to measure at a sufficient number of positions for a sufficiently long period of time with sufficient degree of resolution. WG-AEN recommended that wherever possible strategic noise mapping should generally be carried out by computation, and noise measurements may be used to validate noise maps at selected sites [15].

8.3.2 Digital Terrain Model

The key first stage is to focus on the understanding of the area under review via Digital terrain model (geography, topology, urban and land use, traffic and population). The data set specifications required for noise computations involves [16]: 3D model environment (digital terrain model, building heights, embankments and cuttings, topography, bridges / underpasses, and barriers); road source (carriageway centerline, traffic flow, traffic speed, % heavy grade vehicles, road surface type, and road texture depth); rail source (rail centerline, traffic flow, train speed, and train type); industry source (location, process type, and noise emission level); and aircraft source (flight track, number of flights, aircraft type). The analysis environment will typically require a number of datasets, including several not required for the noise calculation process: information on residential population numbers; population distribution information; identification of buildings as dwellings or other noise sensitive premises, such as schools and hospitals; and location of premises with special noise insulation measures [16]. Google Earth with very vivid graphic capabilities allows the users to view and present geographic information [17]. Figure 8.3 shows the flow chart of noise mapping process. The digital terrain model to be used in noise mapping must be accurate and updated to the reference year. It must include all the major roads noise sources present in the area and the surfaces and objects influencing the noise propagation from traffic on these roads, such as building, noise barriers, walls, and bridges [18]. The building data should include the height and number of floors whereby each building should have population assigned. Large junctions as well as tunnels and viaducts may be considered as special elements. Noise barriers must be included in the noise model with their exact intrinsic and extrinsic parameters such as sound insulation and absorption. The average meteorological conditions can be also fed into the software in order to avoid

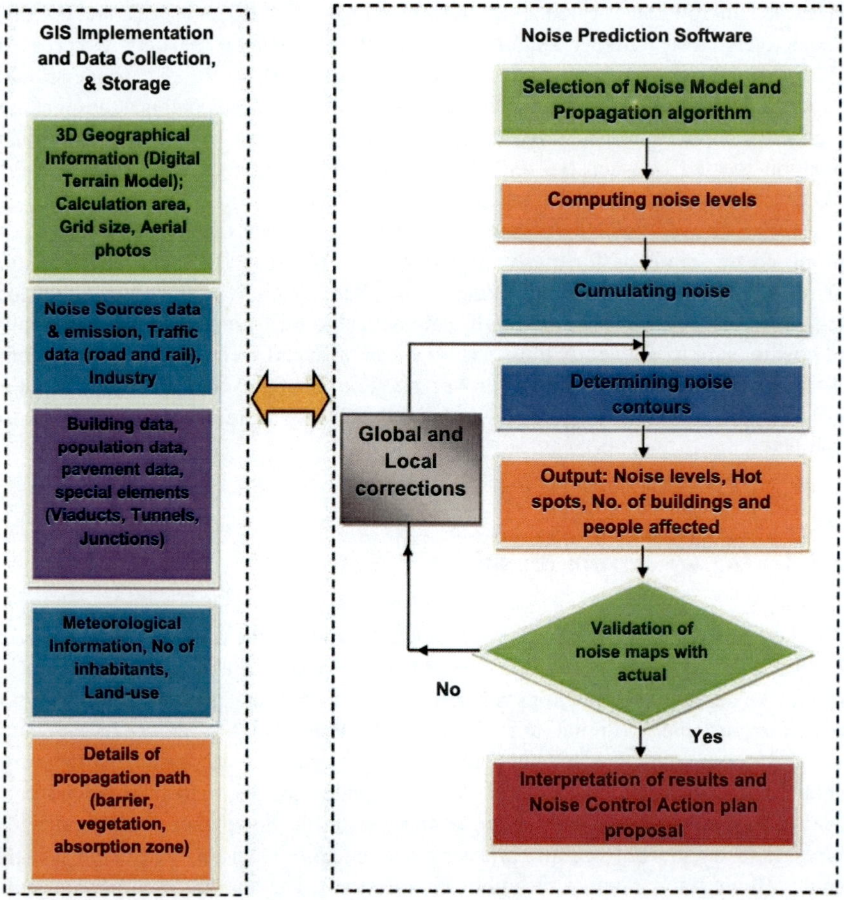

Fig. 8.3 Flow chart of noise mapping process

overestimation. The calculation grid size defines the detailed level of noise maps and is very crucial. Narrowing the grid size improves the accuracy of the maps, but at higher costs, in terms of calculation time, hardware, and human resources [18]. The number of people exposed to noise is calculated either by indirectly from noise contours, assigning to buildings the noise levels corresponding to the noise band in which buildings resides, or directly calculating sound pressure levels at receivers with a horizontal spacing ranging from 3 m to 10 m [18]. The resolution of the calculation grid and the number of receivers thus have an influence on the concluding map that results in a more accurate and precise isolines [19].

8.3.3 *Sampling and Spatial Interpolation Strategies*

A grid is superposed over a city map and the measurement points are located at the nodes of the square or at the nearest location when nodes are inaccessible. The grid method basically consists of superposing a certain grid over the map of the city such that the nodes are taken as sampling points. When a greater than 5 dB difference is measured between two adjacent points according to ISO 1996-2:1996, an additional sampling point should be measured between these two points. Table 8.1 shows the grid size selection and number of measurement points reported in recent noise mapping studies. The European Commission Working Group Assessment of Exposure to Noise (WG-AEN) Position paper recommends grid spacing should be more than 10 m in agglomerations. For aircraft noise contours, grid spacing up to 100 m may be acceptable [15]. However, many noise mapping studies [20–26] reported different grid size for developing noise maps as shown in Table 8.1. For instance, a recent study in the UK for soundscape mapping of inner city center of Sheffield of 3.6 km^2 area implemented a grid of 200 × 200 m with 90 measurement points [20]. However, Escobar et al's. (2012) study for the city of Cáceres, Spain, revealed that grid size coarser than 400 m leads to uncertainties of calculated mean values greater than 2 dB [24].

Grid size coarser than 283 m had been recommended as inadvisable for noise mapping of larger cities such as that of Cáceres (approx. 10 km^2) for which 215 measurement points were used for an analysis of 200 m grid [24]. Spatial interpolation techniques can be used to determine the noise levels with a greater resolution. The selection of number of grid points, type, and interpolation technique is very important for calculation of the noise levels. Interpolated noise levels are not the same as the calculated noise levels on exact positions [4]. However, some studies [27] proposed that urban noise survey locations may be categorized into four types, namely, the receptor-oriented sampling method; the second is source-oriented sampling method; the third is the randomly selected measurement locations using regular

Table 8.1 Grid size selection and number of measurement points in recent noise mapping studies [21–26]

Researcher	City	Area	Grid size selected	No of measurement points
Morillas et al. [21]	Extremadura, Spain	1.7 km^2	200 m	59
Sommerhoff et al. [22]	Valdivia, Spain	14.5 km^2	400 m	115
Martin et al. [23]	Valladolid, Spain	170 km^2	250 m	490
Escobar et al. [24]	Cáceres, Spain	10 km^2	200 m	215
Gozalo and Morillas [25]	Talca, Chile	29 km^2	800 m	52
Say et al. [26]	Seferihisar, Turkey	1.1 km^2	0.50 m	46

or nonregular grid systems; and in the last method, the study area is divided into zones according to population, receiver density, land use characteristics, and road functions such as importance of the roads.

Spatial interpolation methods such as inverse distance weighted (IDW) method, Kriging, radial basis function (RBF), spline, and natural neighbor are usually used in noise mapping software.

8.3.3.1 Inverse Distance Weighted (IDW) Method

IDW is the point-based interpolation method whereby the noise value at point *i* (N_0) is calculated as [28]:

$$N_0 = \frac{\sum_{i=1}^{n} N_i P_i}{\sum_{i=1}^{n} P_i} \tag{8.1}$$

where *n* represents the number of measurement points, N_i represents the noise value at position *i*, P_i represents the weight of the noise value at *i* position. P_i weights can be calculated with Eq. (8.2) as a function of distance between the reference point and interpolation point.

$$P_i = \frac{1}{d_i^k} \quad i = 1, 2, \ldots .n \tag{8.2}$$

where d_i is the horizontal distance between the interpolation point at (x_0,y_0) and the reference points at (x_i, y_i) and *k* is the power of the distance. In this method, the reference points around the interpolation point can be used to estimate noise value, whereby the reference points can be determined by a search circle whose center must be the point which will be interpolated [28].

8.3.3.2 Kriging Interpolation Method

The Kriging method used a weight pattern, where closer points have a greater weight. Kriging method is also called BLUE. The estimation error must be minimum in order to determine the weights. The general formula is as follows [28]:

$$N_0 = \sum_{i=1}^{n} W_i N_i \tag{8.3}$$

where N_0 represents the noise value for interpolation point at (x_0, y_0), and W_i represents the weights that correspond to each N_i at (x_i, y_i) used in the calculation of N_0. N_i represents the noise values of the reference points used in the calculation, and n shows the number of points used in the calculation. The weights directly affect the value of interpolation and must be unbiased. IDW method is more deterministic and should be used when there is an initial dense set of points as it can capture the local surface variation, whereas Kriging assumes a stationary and stochastic approach and provides the user with more options when controlling for the final outcome [20].

Harman et al. studies revealed that when all grid resolutions were taken into consideration, the Kriging method showed the smallest deviation. In order to generate a more reliable and sensitive noise map, the k value should be at least 3 in IDW method [28]. Can et al. studies [29] revealed that spatial interpolation techniques when tested fail in predicting accurately the noise level variations within streets as they don't offer a sufficient covering of the network and assume spatial variations which are not coherent with traffic dynamics or street configurations. Mobile measurements outperform the spatial interpolation techniques tested even when very short samples are used: taking 5 s of measurements every 5 m on the network is sufficient. For example, for a city of about 7.5 km^2 like Ghent, Belgium, nearly 200 fixed stations and 330 h of mobile measurements would be required to cover the whole city with same resolutions [29].

8.3.4 Self-Adaptive Grids

Smart interactive procedure to optimize the quality of map at low additional computational cost using self-adaptive grid has been also reported in recent times. The key to this algorithm is to properly select the locations for the new receivers as some locations can add more useful information to the map. Every new map can be created using all the receivers in previous grid and extra receivers, so it will be more accurate than previous one as the interpolator has made its prediction from an enhanced data input (self-adaptive grid). After using a noise model to make acoustic calculations in the positions defined by self-adaptive grid, spatial interpolation is performed to obtain enhanced map contour lines. Asensio et al. discussed about the self-adaptive grids that add new receivers to the sampling grid in those locations, where they are expected to be more useful, as a result of which the performance at the output of the interpolator is enhanced [30].

8.3.5 Calibration of Noise Maps

The calibration of noise maps is a vital task that involves comparison of measured and calculated results. There are two approaches [31]:

- *Global correction of noise levels*: Adjust the map and optimize the difference between the calculated and measured values.
- *Local correction of noise levels*: Measure close to sources to estimate the source levels by an iterative technique. This helps to determine the unknown source parameters and should be used as such. The technique is particularly useful when estimating emissions of several non-separable sources, for example, railways.

There are several indicators that could be used for assessment of results: mean squared error, difference of quantiles, uncertainty on the isolines, etc. [19]. Some of these techniques are known as "acceleration techniques" and are used to reduce calculation time while reasonable results are obtained [32]. The standard DIN 45687 [33] deals with quality requirements and test conditions for software products for the calculation of the sound propagation outdoors. Such techniques are to restrict the maximum search radius for sources around the receiver, to apply the time consuming projection method and the calculation of reflections only up to a maximum distance from source and/or receiver. In such cases, minimum 20 fixed receivers are distributed over the complete area, the planned acceleration configuration is applied, and the calculation is repeated [34].

8.4 Implications of Noise Mapping

Although noise maps are very powerful tools for communicating the results of assessment of environmental noise to the general public, yet some aspects such as grid size, use of interpolation techniques, counting the number of dwellings and residents, tunnels and viaducts, height of grid points, and other noise sources are very crucial issues to be focused for their accuracy and precision [4]. Many countries have developed their own traffic noise prediction models [35]. However, for the countries without their own method, the END recommends the use of the French method to calculate both the source and propagation model for road traffic [36]. Some of the noise models were evaluated (except NMPB-Routes-1996) in the framework of phase A of the CNOSSOS-EU process using six qualitative criteria, that is, *precision*, *accuracy*, *computational speed*, *flexibility*, *simplicity*, and *number of parameters* [37]. There are varied calculation optimization settings in the software, such as: source search radius, dynamic error margin, minimum section length, contour line utilization, reflection radius, grid interpolation, simplification of propagation, smoothing of sources, buildings, and contours [38].

There are various commercial software available for noise mapping such as CadnaA (Datakustik, Germany); Sound Plan (U.K); IMMI (Wölfel); Mithra-SIG (CSTB); NovaPoint (Vianova Systems); and Predictor-LIMA (B&K, Denmark). Figure 8.4 enlists the technical attributes considered in noise models used for noise mapping. It is also obvious that methodological differences and different calculation methods used results in different noise levels being predicted when applying different standards to an identical road [7]. Table 8.2 shows the various

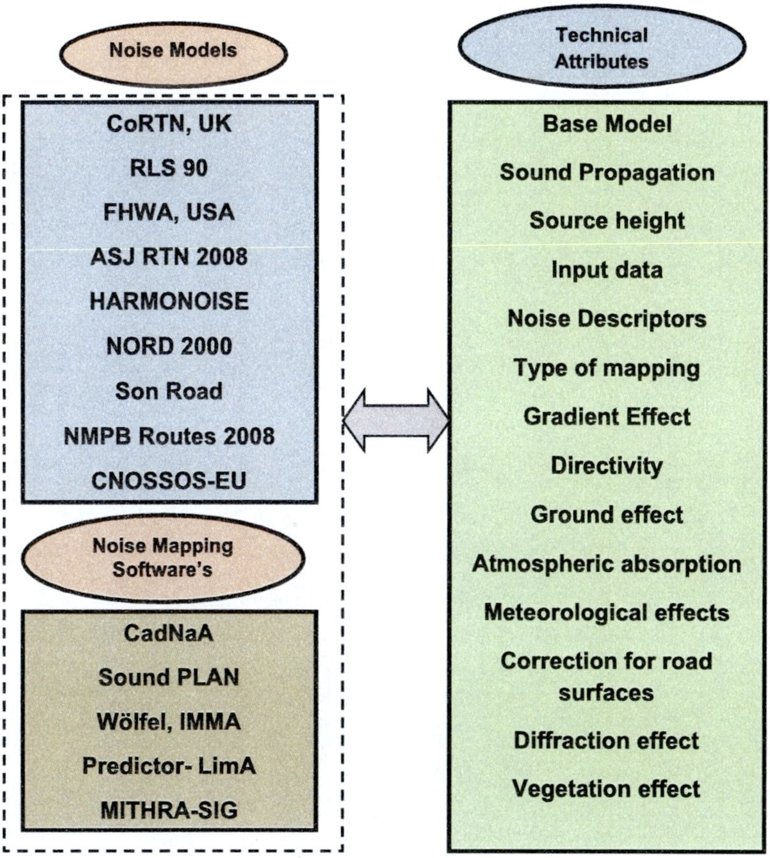

Fig. 8.4 Technical attributes considered in noise models used for noise mapping

calculation methods that are used among Member States in European Union for round three of strategic noise mapping [39]. Some studies summarized that differences of 6–10 dB(A) are observed when calculating different road traffic situation using Austrian, German, French, and Dutch methods [40, 41]. It has been evaluated that the differences in the outcome of noise calculations using different national methodologies exist up to 15 dB(A) [41]. The methodological differences in different calculation methods yield different noise levels when applied to an identical road [7]. Also, the use of software packages implementing the same standard with the same input data predicts different noise levels as shown in a study by Hepworth (2006), whereby the greatest mean difference of 2 dB(A) and individual difference of 11 dB(A) at a single point were observed for different commercial packages employing the CRTN standard [7, 43]. King and Rice (2009) recommended that in order to achieve true standardization in noise studies, competent authorities would be required not only to apply same calculation procedures but also to employ the

Table 8.2 Noise assessment methods and guidelines followed in the EU member states (MS) for the Round #3 of strategic noise mapping (2017) (Kumar et al. [39])

#	EU MS	Road noise	Rail noise	Industrial noise	Aircraft noise
1	Austria	RVS 04.02.11	ONR 305011	ÖNORM ISO 9613	ÖAL Guideline No 24–1
2	Belgium	NMPB-Routs-2008 (Wallonia) RMV(Flanders)	Reken-en Meetvoorschrift Railverkeerslawaai (RMR); RMR (Wallonia & Flanders)	ISO 9613-2(Wallonia & Flanders)	Integrated Noise MODEL (INM) version 7.0b
3	Bulgaria	NMPB-Routes-96	RMR	ISO 9613-2	ECAC.CEAC Doc 29-2nd Edition
4	Croatia	NMPB-Routes-96	RMR	ISO 9613-2	ECAC.CEAC Doc 29-3rd Edition
5	Cyprus	–	–	–	–
6	Czech Republic	NMPB-Routes-96	RMR	ISO 9613-2	–
7	Denmark	Nord2000	Nord2000	DanishEPA 5/1993	DANSIM
8	Estonia	NMPB-Routes-96	RMR	ISO 9613-2	–
9	Finland	CNOSSOS-EU	CNOSSOS-EU	CNOSSOS-EU	ECAC.CEAC Doc 29-3rd Edition
10	France	NMPB-Routes-2008	NMPB-Routes-2008	ISO 9613-2	ECAC.CEAC Doc 29-3rd Edition
11	Greece	NMPB-Routes-96	RMR	–	ECAC.CEAC Doc 29-2nd Edition
12	Germany	VBUS (Vorläufige Berechnungsmethode für den Umgebungslärm an Straßen)	VBUSch (Vorläufige Berechnungsmethode für den Umgebungslärm an Schienenwegen)	VBUI (Vorläufige Berechnungsmethode für den Umgebungslärm durch Industrie und Gewerbe)	VBUF (Vorläufige Berechnungsmethode für den Umgebungslärm an Flugplätzen)
13	Hungary	UT2.1–302	MSZ-07-2904:1990	Hungarian National computation method according to KvVM Decree 25/2004 (XII.20.) on the Required Form and Content of	Hungarian national computation method according to joint KHVM-KTM Decree18/1997. (X. 11.) on Detailed Technical

				Strategic Noise Maps Used for the Evaluation and Management of Environmental Noise and MSZ 15036:2002	Rules of Designation, Management and Termination of Noise-Protective Zones in the Vicinity of Airports
14	Ireland	UK CRTN 1988	RMR	–	ECAC.CEAC Doc 29 -3rd edition & Integrated noise model (INM) version 7.0d
15	Italy	NMPB-Routes-96	RMR	–	ECAC.CEAC Doc 29-3rd edition & Integrated noise model (INM) version 7.0d
16	Latvia	NMPB-Routes-96	RMR	ISO 9613-2	ECAC.CEAC Doc 29 -3rd edition
17	Lithuania	NMPB-Routes-96	RMR	ISO 9613-2	ECAC.CEAC Doc 29 -3rd edition
18	Luxembourg	NMPB-Routes-96	RMR	Courrier d'information de la Commission concernant la cartographie stratégique du bruit des sites d'activité' industrielle au sein de l'agglomération de la Ville de Luxembourg et environs	ECAC.CEAC Doc 29-3rd edition
19	Malta	NMPB-Routes-96	–	ISO 9613-2	ECAC.CEAC Doc 29 -3rd edition
20	Netherlands	RMV	RMR	Handleiding meten en rekenen Industrielawaai (HMRI)	ECAC.CEAC Doc 29 -3rd edition
21	Poland	NMPB-Routes-96	RMR	ISO 9613-2	ECAC.CEAC Doc 29 -3rd edition
22	Portugal	NMPB-Routes-96	RMR	ISO 9613-2	ECAC.CEAC Doc 29 -3rd edition Integrated Noise Model (INM) Version 7.0d, and AIR 1845

(continued)

Table 8.2 (continued)

#	EU MS	Road noise	Rail noise	Industrial noise	Aircraft noise
23	Romania	–	–	–	–
24	Slovakia	NMPB-Routes-96	Shall03	ISO 9613-2	ECAC.CEAC Doc 29-3rd Edition
25	Slovenia	NMPB-Routes-96	RMR	ISO 9613-2	ECAC.CEAC Doc 29-3rd Edition
26	Spain	NMPB-Routes-96	RMR	ISO 9613-2	ECAC.CEAC Doc 29-2nd Edition
27	Sweden	Nordic Pred. Method for Road Traffic Noise (RTN)	Nordic Pred. Method for Train Noise(NMT)	Environmental Noise From Industrial Plants .General Prediction Method	ECAC.CEAC Doc 29 -3rd edition & Integrated Noise Model (INM) Version 7.0d
28	UK	CRTN (Calculation of Road Traffic Noise)	CRN (Calculation of Railway Noise)	–	Integrated Noise Model (INM) Version 7.0d

Note: "–" means that data is not available

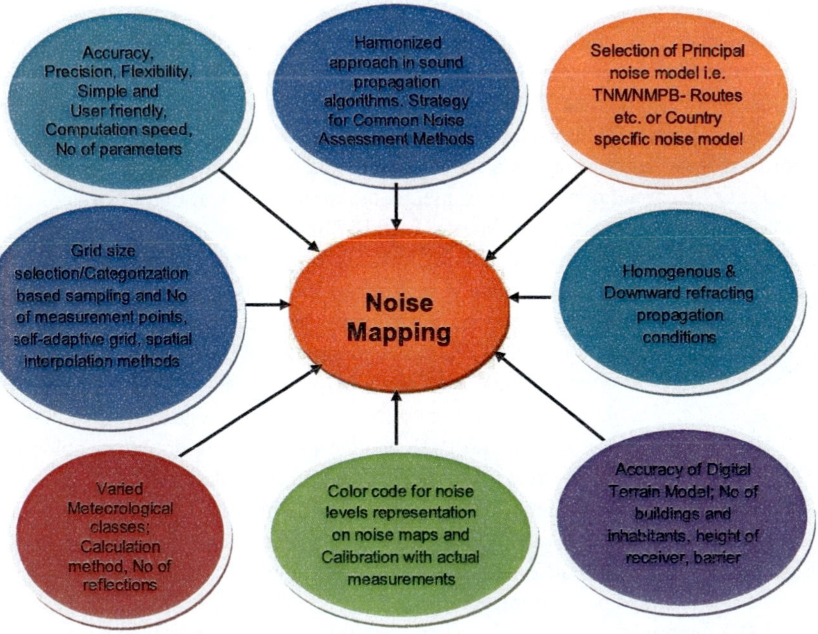

Fig. 8.5 Implications of noise mapping for noise pollution control

same calculation software [42]. The concept of Fresnel zones is applied to account for the extension of the reflecting surface element. The recent models include meteorological effects that can be seen if the normalized sound pressure level from road surface is plotted versus distance source-receiver [42, 44]. A number of 25 classes have been recommended that cause a determination of long-term average sound levels with an error of at most 2 dB at 1000 m range. Comparison of noise maps calculated with different meteorological conditions shows only negligible differences [44]. The European Commission Working Group Assessment of Exposure to Noise (WG-AEN) Position paper recommends that within a dense urban setting due to closeness of building, the varying width of roads, etc., meteorological conditions when compared to other variables don't have dominant effect on sound pressure levels in most situations, and thus can be ignored. The exceptions are for large open areas, aviation noise and elevated sources/receiver. The utilization of separate long-term averages (for humidity, temperature, wind velocity/direction) for different periods of a 24-hour period will be essential, that is, for day/evening/night. The effect of turbulence can be ignored for low frequencies and distances up to a few hundred meters in free-field conditions [15]. Figure 8.5 shows the implications of noise mapping for noise level predictions.

The recent models have recommended the segregation of vehicle noise into propulsion and rolling noise. The separation of emission in engine and tire noise allows to relate the increase of emission with slope only to engine noise and increase in speed to tire noise. The road surface corrections are also varied in different models [45]. The correction term is formulated in terms of running speed of vehicle, running mode including acceleration and deceleration and the number of years after pavement is laid in some models like ASJ RTN 2008. It is quite obvious that the source model may be different for all these models; however, there should be a harmonized approach for sound propagation modeling especially. The consideration of the new CNOSSOS-EU methodology for implementation in the next round of strategic noise mapping in 2022 can be a right step in this direction [46]. Also, the absence of standardized color scheme makes it difficult to compare the maps from different EU states. This issue may be problematic as different methods affect the visual impact of noise mapping results when they are presented to general public [7]. Table 8.3 shows the color schemes to be used in END noise mapping for European major roads and recommended in Conference of European Directors of Roads held on August 2013. Diverging color schemes use a light/neutral color to represent average values and contrasting dark color for low to high values [18, 47]. A common color chart followed shall make noise maps comparable and more readable.

The main components determining the uncertainty of noise map are uncertainty in input data, propagation, model uncertainty, and uncertainty of evaluation data. Figure 8.6 shows the cause-and-effect analysis diagram for factors affecting the uncertainty in noise predictions.

Model uncertainties are those issues related to how well the documented calculation standard represents the real-world situation, and what uncertainties it introduces due to the simplifications made in order to present a solution which is relatively simple to implement. The evaluation data that is used to confirm the accuracy of the calculated noise level also has some uncertainty. Corrections to the directly measured noise levels are necessary in order to compare them with the calculated, long-term noise levels [48]. The type of noise model and sound propagation algorithm used in noise mapping also inculcates uncertainty in reporting the noise levels. The application of sampling and interpolation techniques also contribute to the uncertainty of results [19]. There are many other elements whose relative accuracy and precision constitute similar sources of error or uncertainty. For instance, a common approach to assign uniform height to the buildings shall obviously induce an error. Also, the type of vehicle traveling on the road, their flow characteristics, directivity pattern, etc., multiple reflections from building facades, meteorological effects, varied height of source and receiver, vehicular horn noise in some countries, and different sections of road such as tunnels/bridges/viaducts may induce an uncertainty in noise level predictions. Thus, it is indispensable to reduce the uncertainties in noise mapping for urban noise management and control. ISO 1996-2 standard estimates a minimum uncertainty of 2 dB for the measured noise levels that accounts instrumentation, operating conditions (repeatability), weather and terrain conditions, and residual sound. The evolution of GIS technologies makes it possible to share the results with citizens and decision-makers distributed over the Internet, through Spatial Data Infrastructure (SDI) platforms. A recent study by Bocher et al. (2019) proposed the prediction method implemented in an open source GIS ecosystem written in JAVA language compliant for agglomeration of Nantes, France [49].

Table 8.3 Color proposal recommended in Conference of European Directors of Roads held on August 2013 to be used for END strategic noise mapping [18, 47]

Noise band	Name	Color
35-40	Moderate sea green	
40-44	Greyish green	
45-49	Light greyish chartreuse green	
50-54	Pale yellow	
55-59	Light brilliant amber	
60-64	Brilliant tangelo	
65-69	Light brilliant red	
70-74	Moderate amaranth	
75-79	Dark rose	
80 or more	Deep blue violet	

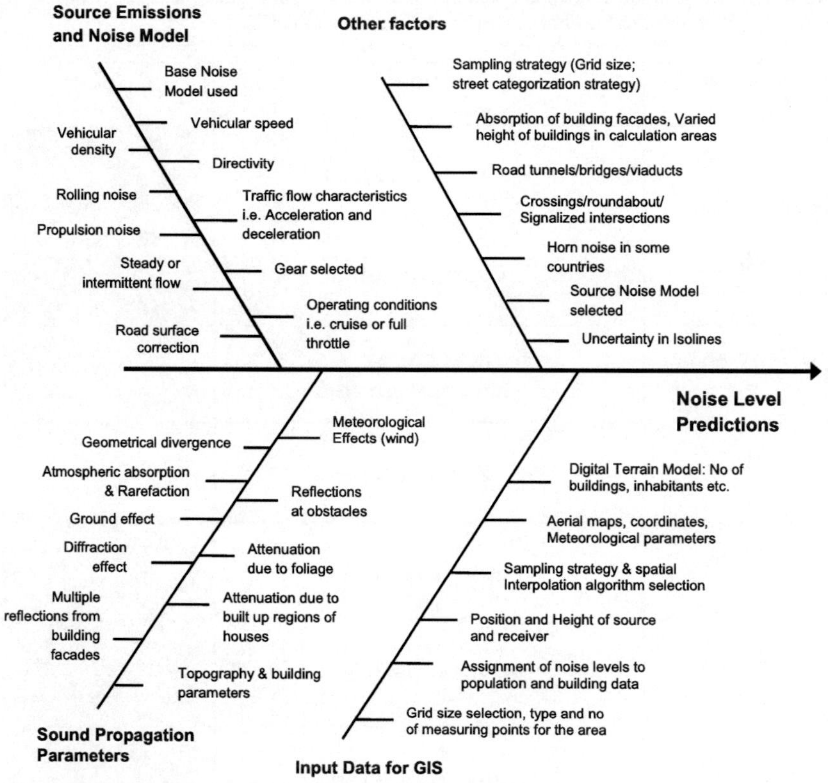

Fig. 8.6 Cause-and-effect analysis diagram for factors affecting the uncertainty in noise predictions using noise model integrated with GIS interface

8.5 Dynamic Noise Mapping

The principles of dynamic noise mapping uses the technique of reverse engineering whereby the measurements are used to update the calculated noise maps, primarily intended to improve local noise maps and supplement noise monitoring results. The basic steps of dynamic noise mapping are as follows [50]:

1. Assume nominal emission values for the unknown or variable sources.
2. Calculate the noise level at the receiver (i.e., measurement) positions.
3. Add the difference in the measured and calculated sound pressure levels to the source emission power level.
4. Perform noise contour calculations based on the new source emission.

Combining measurements and calculations can provide more flexible problem-solving, enabling automated calibration of noise maps and producing fast and cheap real noise maps that represent both long-term and short-term levels [51]. One such study based on this methodology had been reported for the city of Madrid, Spain [51]. A recent study shows a "real-time" noise mapping project, named DYNAMAP, dynamic noise mapping system that performs the update of noise maps by means of a combination of dedicated monitoring sensors, software tools, and a GIS platform, implemented along the ring road encircling the city of Rome, Italy [52].

8.6 Noise Control Action Plans

In the European Directive concerning noise action plans, it is a requirement that competent authorities draw up action plans for the major roads, railways, and agglomerations within their remit and that these plans are reviewed every five years once adopted and on an ongoing basis by accounting for major new developments [53]. The END thus advocates the adoption of Noise Action Plans, on the basis of noise mapping results, for managing the environmental noise and its effects, including noise reduction if necessary and preservation of quiet areas. The acoustical cost-benefit criteria can be distinguished in three groups [8]: best guess method, cluster definition, and interactive criteria method. The best guess method involves the selection of an optimal criteria and checking whether it fulfills the criterion or not. The second method defines the clusters of houses based on the density of houses. Noise measures will be taken to reduce the noise level for the clusters, and no measures will be taken for houses outside the clusters. The iterative criteria compare the benefits and costs for each step iteratively, and effect of each step is compared to the costs. Murphy and King (2011) investigations on the usefulness of three mitigation measures for "acoustical planning"—traffic reductions, speed reductions, and erection of acoustical barriers—revealed that reducing travel demand leads to most significant reduction in population exposure for L_{den} and L_{night}, while considerable reductions can also be achieved by reducing the travel speeds. The study had also highlighted the potential reduction that can be achieved by using acoustical noise barriers in areas that may have unusually high levels of noise exposure and are not visually sensitive [53]. Dintrans and Préndez (2013) presented a study of road traffic noise abatement strategy suggesting three measures: traffic flow reduction, maximum speed limit of 40 km/h for heavy vehicles, and current road surface improvement/replacement for city of Santiago, Chile. Scenarios that considered speed limits for heavy vehicles were the most effective reducing by 30% the number of facades in conflict [54]. One such study was reported by Coelho and Alarcão for a school located close to a high traffic road where noise levels were higher exceeding 60 dB(A) on building facades and even in the playground. The noise action plan comprised new sound-absorbing porous road pavements and the installation of acoustical barriers on the facing sides of the roads [55]. It plays an important role in Environmental Impact Assessment studies as it not only shows the present noise levels but also the increased levels due to new project such as construction sites with due consideration to sound propagation in complex

Fig. 8.7 Noise control measures for ambient noise level control in urban areas

topographic conditions of surrounding buildings [56]. Figure 8.7 shows the noise control measures for controlling the ambient noise levels in cities. It may be noted that the cost-benefit analysis is prior required for implementing the optimal action plan. Recently, the European NADIA (Noise Abatement Demonstrative and Innovative Actions and information to the public) project reported a new indicator named CBI (Cost-Benefit Indicator) that can be utilized to identify the most technically and economically efficient noise reduction measure in each area where an action is needed. The sound-insulating windows were evaluated to be the highest ranking with highest CBI score [57]. A noise reduction policy with prominence on source measures is usually the most cost-effective. Also, it must be ensured at regional and local levels to understand the noise scenario from the very beginning as a part of the spatial planning process [41]. Noise compatible land-use planning is one of the vital community planning methods for ambient noise control in which the land surrounding an airport or a highway is planned and managed by encouraging the location of less noise-sensitive land uses next to highways and thus *using adjacent land for activities, services, or businesses that are not disrupted by noise. It can be a good substitute to noise barriers provided it is adopted at the beginning stage before the development occurs. Thus, an* effective noise compatible land use planning can eliminate the requirement of noise barriers especially for the highways and can be used to attractively design open space next to a road or highway [58]. Public consultation and dissemination of noise mapping results in general is a very important issue in noise action planning as advocated by END [7]. The public participation shall facilitate in enhancing the awareness of citizens and potential stakeholders by focused communication and dissemination activities.

8.7 Sampling Strategies

The sampling methods most commonly used by different researchers (grid, legislative road types, and categorization methods) may be used judiciously considering the economic and resources (both infrastructural and manpower) constraints. To obtain a noise map, different methods or strategies may be considered. These methods differ largely from each other in methodological aspects associated with the selection of sampling points. However, even when a computerized method is used, "*in situ*" measurements are necessary for calibration or validation [13, 59]. The grid method is widely used as it guarantees the statistical principle of equal probability and, moreover, provides a uniform coverage of the area under study. The categorization method is based on the concept of street functionality, that is, the functionality of the streets of the city as a communication path between different parts of the city and between the city and other urban areas and as such the other variables such as the flow of vehicles, the type of traffic, the average speed, and urban variables may have a clear relationship with functionality [25, 60]. It is based on the generally accepted assumption that road traffic is the most important source of noise in cities, and for most streets it can be considered the main cause of the spatial and temporal variability of that noise [25, 61, 62].

Table 8.4 presents the *SWOT* (strengths, weaknesses, opportunities, and threats) of the two sampling strategies based on the review of the various studies conducted on this aspect in European region in recent years [25, 63–66]. Morillas et al. [63] presented the use of categorization method and categorized the streets of Cáceres, Spain, in six types with total number of 60 sampling points (rather than 215 points (168 squares) for grid method). It was observed that compared with the grid procedure, the categorization method required considerably fewer measurements and showed a better predictive capacity [63]. Apart from the sampling strategy, the input database: digital maps in the form of GIS files containing street centerlines and building boundaries, as well as coordinate system information, building height (including number of floors and inhabitants), traffic data, traffic volume, average speed, and vehicle size during the three periods daytime, evening, availability of screens, average meteorological data, population information, terrain characteristics, etc., is essentially required for developing accurate noise maps [67].

Table 8.4 *SWOT* analysis (strengths, weaknesses, opportunities, and threats) of the two sampling methods based on the review of the various studied (Gozalo and Morillas [25]; Morillas et al. [63]; Gozalo et al. [64]; Gozalo et al. [65] and Gozalo et al. [66])

Methods	Strengths	Weaknesses	Opportunities	Threats
Grid method	Initial selection of sampling points is easier as a grid is to be overlaid on the map of town/ city. It is accepted in International Standard, ISO 1996-2.	The grid squares may have a high variability of sound levels that lead to differences in sound levels registered at adjacent points of more than 5 dB.	Uniform coverage of area under study; the number of sampling points were almost the same for small towns using both the methods in one such study.	For larger agglomerations, this method is a big challenge.
Categorization method	Noise levels are associated with each mapped street rather than with an area. This approach requires considerably fewer measurements and showed a better predictive capacity. While only 60 sampling points were needed for categorization method, for the grid method 215 sampling points were required for city of Cáceres, Spain (10 km^2 approx. 95,000 inhabitants). 21 measurements points were considered for Madrid, Spain (605.8 km^2, 3,255,944 inhabitants) using this approach.	Total length of each category must be measured and random selection of points must be made taking into account the elimination of equivalent points.	This method can be suitable for resolving some of the problems of grid method such as effect of future reorganization of traffic flow as in that case streets would be reassigned to a different category which otherwise in grid method necessitates new measurements It seems to be a more suitable predictor for new measurements particularly for noisiest streets of town/ city. It is easy to estimate number of citizens affected by given noise levels.	More streets are required to confirm that the categorization method sufficiently estimates the long-term indicators recommended in European Directive. Efficacy of this method in Asian continent needs to be explored.

8.8 Smartphone Applications in Noise Mapping

The present paradigm of noise monitoring uses specialized sound level meters using sensitive microphones that involve huge costs as well as infrastructural and manpower requirements. A prodigious growth of mobile phone technology offers a unique opportunity of noise monitoring through community participation. However, the key issues of environmental noise mapping using smartphones are reliability of the sensors and noise monitored data. Kumar et al. [68] utilized a system that involved a client application on smartphones which recorded noise, processed the information and communicated to a server, and shared the information as visual noise levels on Google Maps. The combined use of multiple smartphones endowed with several types of sensors gives the possibility to monitor a certain area with fine spatial and temporal granularity, a procedure typically known as crowdsensing. Experimental results showed that, if the smartphone application is well tuned, it is possible to measure noise levels with an accuracy degree comparable to professional devices for the entire dynamic range typically supported by microphones embedded in smartphones, that is, 35–95 dB. Overall, some studies such as Zamora et al. [69] demonstrated that it is possible to accurately assess noise levels by taking relatively short samples (from 1 to 3 s) while introducing a minimal estimation error. Experimental results showed that both the sampling rate and the selected buffer size can have a significant impact on the accuracy of noise level estimations, being that estimation errors can vary from 1 to 12% in the best cases [69]. Kanjo presented the design, implementation, evaluation, and user experiences of the NoiseSpy application, a sound sensing system that turns the mobile phone into a low-cost data logger for monitoring environmental noise and also allowing the users to explore a city area while collaboratively visualizing noise levels in real time. The software combines the sound levels with GPS data in order to generate a map of sound levels [70]. Kardous and Shaw examined 192 sound measurement *apps* on the iOS and Android platforms and found only four iOS *apps* that had the means of their differences with a type 1 sound level meter to be within ±2 dB over a 65–95 dB SPL test range [71, 72]. Figure 8.8 shows box plots of the differences between the reference system and app measurements by app and by nominal sound level for both the iMM-6 and i436 microphones. Visual inspection of the graphs suggests that the medians of the differences for the iMM-6 microphones were slightly higher than those for the i436 microphones. The study suggested that using external calibrated microphones greatly improves the overall accuracy and precision of smartphone sound measurements and removes much of the variability and limitations associated with the built-in smartphone microphones [72].

However, a major weakness remains the micro-electro-mechanical-system (MEMS) built-in microphone used in smartphones. MEMS microphones continue to have certain limitations because of their miniature size and circuit board placement, which affects their dynamic range and signal-to-noise ratio response [73] and also inability to perform periodic or pre-measurement calibration [72]. A Type 2 SLM is typically appropriate for use in meeting occupational noise monitoring

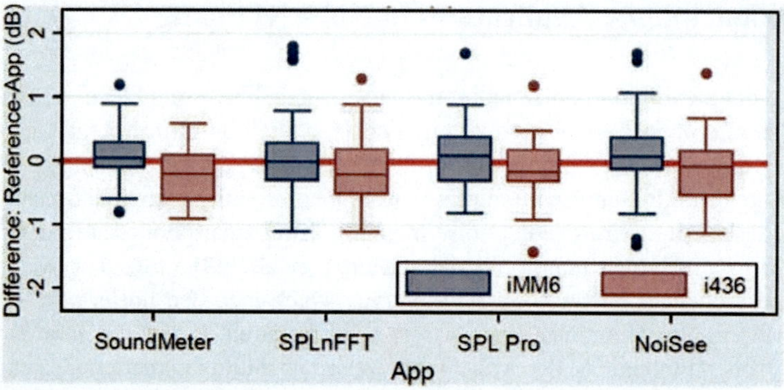

Fig. 8.8 Box plots of the differences between the reference and app measurements for both iMM-6 and i436 microphones by app (Kardous and Shaw [72])

requirements set by the Occupational Safety and Health Administration (OSHA,) and ideally smartphone-based apps would meet the Type 2 criteria if they are to be useful in estimating noise hazard. Nast et al. [74] assessed the accuracy of five SLM applications by comparing smartphone-based readings with measurements made using a Type 1 SLM. The accuracy of sound level measurements made using five mobile phone applications or "apps" on an Apple iPhone 4S, one of the most widely used mobile phones was assessed. It was reported that most of the apps erred by reporting higher sound levels, one application measured levels within 5 dB of a calibrated SLM across all frequencies tested. King and Murphy focused on research undertaken to assess the possibility for smartphone-based noise mapping data to be integrated into the traditional strategic noise mapping process. The study compared the maps generated using the traditional approach with those generated using smartphone-based measurement data as illustrated in Fig. 8.9 [75]. The advantage of the smartphone approach based on its potential to eliminate the need of exhaustive input data into the source calculation model for noise prediction was highlighted. The magnitude of differences between the smartphone and traditional sound level meter was observed to be in range of 0.1 to 15.1 dB (A). The average difference was measured to be 4.4 dB(A) for 26 locations outside the acceptable degree of error of ±2 dB(A), which suggested a relatively high degree of volatility associated with smartphone in field measurements when compared with traditional sound level meters. Also, the study reported that smartphone app or its microphone was geared toward over measuring the environmental noise levels in the field [76]. The crowdsourcing approach proves to be the best alternative for collecting noise data with the help of an android-based noise applications. More recently, a study by Lokhande et al., 2021 reported the use of an indigenously developed "Noise Tracker" app that records one noise value per second or 300 measurements over 300 seconds. The study acquired the real-time noise data across the city through crowd sourcing technique and analyzed the changes in sound levels during the lockdown and post lockdown in Bhandara city in Maharashtra state, India [77]. Because different mobile applications produce different outputs, user-generated noise measurements cannot replace professional surveys, but they can provide additional data for strategic noise mapping and monitoring

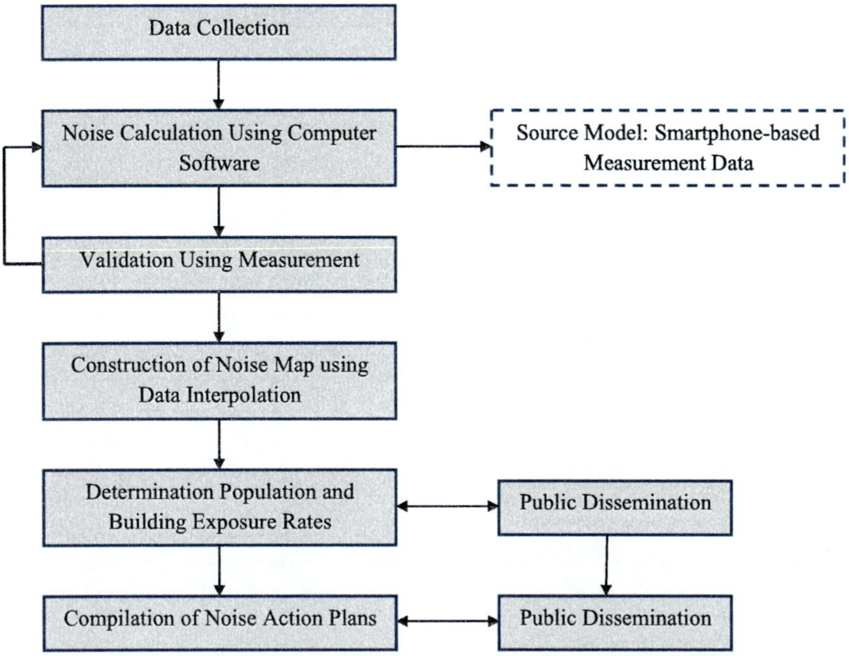

Fig. 8.9 Traditional strategic noise mapping process (gray fill) and the potential for smartphone-based integration (non-gray fill) (King and Murphy [75])

[78]. However, the accuracy and precision associated with the reported sound levels, traceability to the national standards of sound pressure and recording of long time hourly sound levels data like the conventional sound level analyzers stand as the biggest challenges in suitability and applicability of smart phone applications in noise monitoring.

These studies however demonstrated the power of crowd sourcing through smartphones in monitoring environmental noise levels. It reduces the cost and thus the smartphones can successfully be used as a pocket-sized noise level monitor [78]. It is envisaged that in the near future the advancements made in app design and external microphones availability and the gap between professional instruments and smartphone-based apps shall rapidly narrow down, and smartphones can become a promising substitute for noise pollution monitoring and mapping. A recent study by Picaut et al. [79] reported the use of crowdsourcing approach for environmental noise assessment using smartphones. This interesting study developed an open science framework (freely available on http://noise-planet.org) for smartphone applications and a spatial data infrastructure developed specifically for collecting the physical data (noise indicators, GPS positions, etc.) and perceptual data (pleasantness). Thus, smartphone-based noise mapping has potential for the future especially for the developing countries like India for noise mapping of larger parts of the cities without much infrastructural and economic investment.

8.9 Noise Mapping and Action Plans: Cost-Benefit Analysis

The costs associated with noise mapping and noise action plans for abatement and control of noise pollution is one of the biggest challenges and hurdles faced by many countries especially the developing countries. Generating noise maps of the cities involves both the infrastructural, economic, and scientific constraints especially for developing countries like India. The infrastructural constraints pertain to the establishment of noise monitoring terminals and manpower resources for monitoring and assessment, while the economic constraints refer to the complete financial expenses borne on monitoring, developing noise maps and planning and execution of noise action plans. The scientific constraints are limited only for some of the countries that do not have an authorized or validated model for noise modeling and predictions and also have a limited experience for devising and implementing the noise abatement action plans. The European Union has estimated an average cost of noise mapping as 0.18 € per capita and action planning costs of 0.06 € per capita. Different Member States have estimated varying cost per capita ranging from 0.05 € per capita for the UK to 0.56 € per capita for Slovakia. Similarly, the action planning costs range from 0.01 € per capita for the UK to 0.29 € per capita for Germany. The cost-benefit ratio for the individual measures is also varied for the Member States. The spread in cost-benefit ratios is large, ranging from a situation in which costs appear to exceed the benefits for noise barriers in Munich (1:0.3) to a ratio of 1:14,335 for speed enforcement in Augsburg. Overall, measures to reduce the speed of road traffic and to reduce the numbers of heavy road vehicles appear to be the most cost-effective [80]. The implementation of the END analyzed that although the directive has not yet achieved its full prospective, yet calculations show a favorable cost-benefit ratio of 1:29, that is, the benefits have outweighed the costs (EC [81]). The administrative costs are low (EUR 0.15 for noise maps and EUR 0.03 for action plans per citizen, every 5 years). The cost-benefit analysis estimated a cost-benefit ratio of 1:29 wherein action plans have been implemented [81]. It may be noted that in order to undertake a cost-benefit analysis for noise control measures, at least two different options or scenarios are required for comparison. To demonstrate that the benefits offered by the noise control measures are effective, each scenario should be considered in terms of a design target. In addition, a "do-minimum" assessment is required if there are planned activities which may affect the noise levels. Apart from the economic aspects, other assessments, approvals, and consents such as planning permission, transport assessment, sustainability assessment, infrastructural costs, user costs and benefits, indirect socioeconomic costs, externalities and environment assessments such as ecology, landscape, and townscape, and visual effects and community effects should be considered for selecting the noise action plans [82]. In such cases, the multi-criteria decision-making (*MCDM*) techniques discussed elaborately in Chap. 10 is very helpful in finding the optimal strategy and prioritizing the noise action plans.

8.10 Noise Mapping in Indian Scenario

The European Union has adopted harmonized noise indicators all along the Member States suggesting L_{den} as an appropriate metric to assess annoyance and L_{night} as metric to assess sleep disturbance. Strategic noise maps using these harmonized noise metrics were used throughout Europe to assess the number of people exposed to different noise levels. In Indian scenario, the noise maps may be developed using the day equivalent level, L_{day}, and night equivalent level, L_{night}, in accordance with ambient noise standards. Noise mapping is at very primitive stages in Indian scenario. Hon'ble National Green Tribunal (NGT) vide order dated 15 March 2019 had issued numerous directions to Central Pollution Control Board (CPCB) to prepare noise maps of the cities, identify the hotspots, categorize the cities with specified hotspots, and propose remedial action plans for prevention of noise pollution.

However, in recent times many studies have been reported in Indian perspectives focused on noise mapping and noise abatement plans for reducing the ambient noise levels. A recent study reported by Das et al. [83] explored the principal sources of noise, spatial noise mapping, and diurnal noise cycle in the residential and heavy traffic area. It was observed that traffic (58.7%) and construction works (21.2%) were the major sources of noise. Around 30% of the area to the total study area was emerged as very high to high vulnerable noise pollution, and the areas were mostly alongside the major roads in general and national highway in particular. Spatial pattern of noise over the town in different times (Fig. 8.10a–f) clearly exhibited that commercial core of the town and alongside the major roads were prone to high noise level > 65 dB(A), which was much higher than the ambient noise standards. Banerjee et al. [84] conducted a study to compute the temporal and spatial distribution of road traffic noise pollution in an urban environment by monitoring and mapping for Asansol city of West Bengal, India. The study revealed that noise levels at all the locations exceeded the ambient noise standards.

Recently, the Directorate General of Mines Safety (DGMS), India, has made noise mapping compulsory in the mining region. The continuous exposure to noise levels leads to irreversible damage to the ears and may cause noise-induced hearing loss, which is the most common disease in mining occupation [85]. Lokhande et al. [86] conducted a noise mapping study in nearby areas of mines in northern Keonjhar district of Odisha in India. The study showed that noise levels observed in residential and sensitive zones exceeded the limits set by Central Pollution Control Board (CPCB), India. The mitigation measure was proposed and demonstrated using a contour map showing a significant reduction in the noise levels. The noise contour map of survey area was regenerated based on traffic flow relocation plan by nullifying the noise levels contribution of heavy vehicles on existing highways and haulage roads that showed a significant decrease in noise levels up to 5.3 dB (A). Also, a recent study reported by Lokhande et al. [87] focused to measure and assess the effects of the existing noise levels of major operational mines in the Keonjhar, Sundergadh, and Mayurbhanj districts of Odisha, India. It was observed

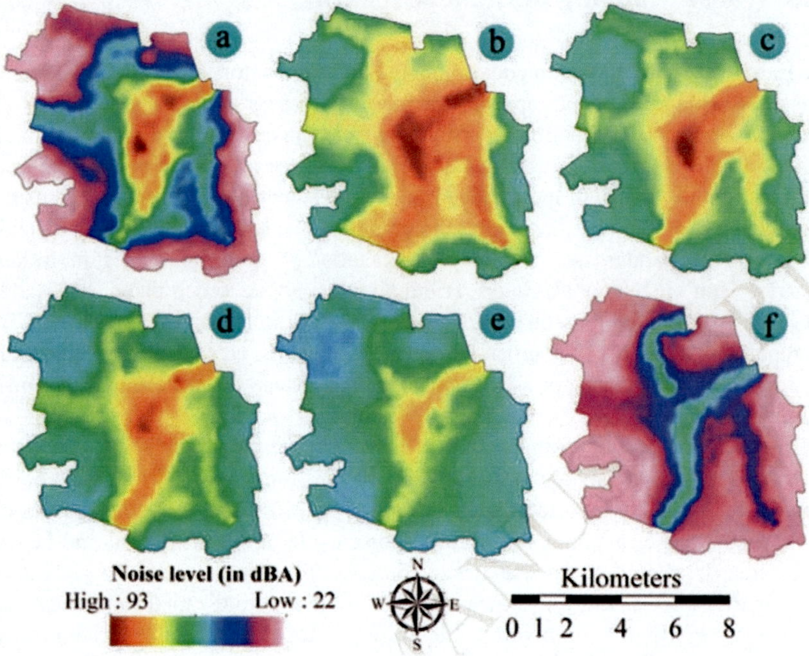

Fig. 8.10 (**a–f**). Spatial pattern of noise level in English Bazar Municipality, West Bengal, India, and its surroundings in different period of times, for example, (**a**) 6–8 a.m. (**b**) 10 a.m.–12 noon (**c**) 2 p.m.–4 p.m. (**d**) 6 p.m.–8 p.m. (**e**) 10 p.m.–12 midnight (**f**) 12 midnight–2 a.m. (Das et al. [76])

that noise induced by heavy earth moving machinery (HEMM) such as rock breakers, jackhammers, dumpers, and excavators, blasting noise in the mining terrain, as well as associated transportation noise became a major source of annoyance to the habitants living in proximity to the mines. The noise produced by mechanized mining operations was observed between 74.3 and 115.2 dB(A). Thus, not only the quantification of noise levels and noise impact analysis can be undertaken using noise mapping, but also it helps in the development of policy framework for controlling noise pollution in the metropolitan cities. In the Indian scenario, a variety of software and modeling tools had been reported for noise mapping, such as:

- Noise level-based digital elevation model (DEM) technique in ERDAS Imagine (v.9.2) environment is used by Das et al. [83].
- Banerjee et al. [84] reported the use of Multispec (Purdue Research Foundation, USA), a multispectral image data analysis system along with GRAM++ (C.S.R.E., IIT-Bombay, India) GIS software. The base map of the study area was loaded, delineated, and digitized in AutoCADMap-2000i platform in various layers. The survey of India (SOI) satellite imagery obtained from National Remote Sensing Agency, India, was used.

- B&K Predictor-LimA (7810) was used by Lokhande et al. [86, 87] to predict the noise levels in the various types of occupational areas for the major operational mines in the Keonjhar, Sundergadh, and Mayurbhanj districts of Odisha, India.
- Vijay et al. [88] reported the GIS-based noise simulation module developed for point sources (N-Dhwani) with attenuation due to barrier, vegetation, and atmosphere. Quantum GIS (Q-GIS) and "Python," an open source software and programming language, were used.

8.10.1 What Strategy Should Be Followed for Noise Mapping of Indian Cities?

Noise mapping is an indispensable tool for quantification and impact evaluation of ambient noise levels and facilitates the selection of noise action plans. Ultimately, the European Noise Directives (END) that establishes clearly a set of minimum requirements listed in Table 8.5 for noise action planning should be considered.

However, developing the noise maps in Indian scenario is really a great challenge for big metropolitan cities like Delhi, and thus the sampling methods most commonly used by different researchers (grid, legislative road types, and categorization methods) may be used judiciously. Validity of traffic noise models (TNMs) used, harmonized representation, appropriate sampling strategy, calibration with actual measurements, etc. are the major hurdles to be tackled for developing precise maps with lowest levels of uncertainty. Figure 8.11 shows the major technical considerations associated with noise mapping for control of noise pollution in developing

Table 8.5 Minimum required elements for actions plans and potential noise mitigation measures (Annex V of the END (EU, 2002), Taken from Murphy and King [7])

Minimum Required Elements for Action Plans	Potential Mitigation Measures
• Description of agglomeration, major roads, major railways or airports • The authority responsible • The legal context (national legislative compliance) • Summary of noise mapping results • Evaluation of human exposure • Actions to be taken in next five years • Estimates of the reduction of the people affected • A long-term strategy • Financial information (cost effectiveness assessments etc) • Provision for evaluation of action plan implementation and results	• Traffic planning • Land-use planning • Technical measures at noise sources • Selection of quieter sources • Reduction of sound transmission • Regulatory or economic measure or incentives

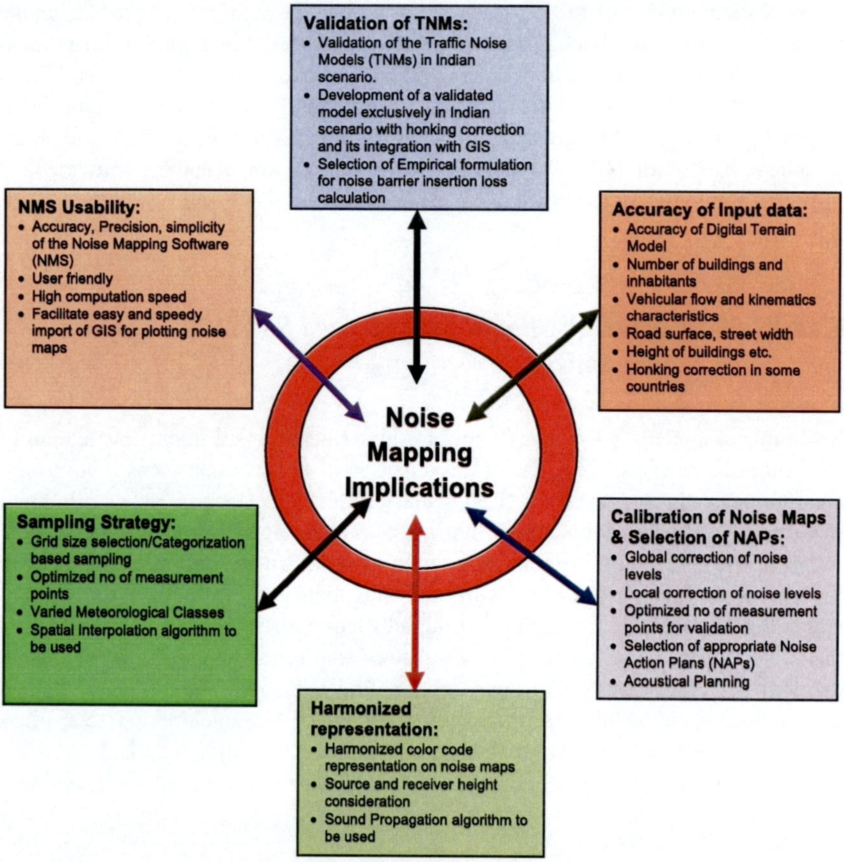

Fig. 8.11 Major technical considerations associated with noise mapping for control of noise pollution in developing countries like India (Garg et al. [89])

countries like India [89]. The aspects pertaining to developing of noise maps for cities in Indian scenario are recommended as follows:

- Noise mapping should be made mandatory for new project establishments especially new runways, elevated transit train corridors, metropolitan cities, and smart cities. Noise maps should be developed in terms of the day equivalent level, L_{day}, and night equivalent level, L_{night},. as per the ambient noise standards.
- Mapping area and sampling strategy: There is not a set process for this, but relevant considerations may include [90]:

  (a) Strategic importance: Areas of high strategic importance may warrant mapping.

  (b) Availability of data: Areas with high-quality geospatial data available may be better to group together.

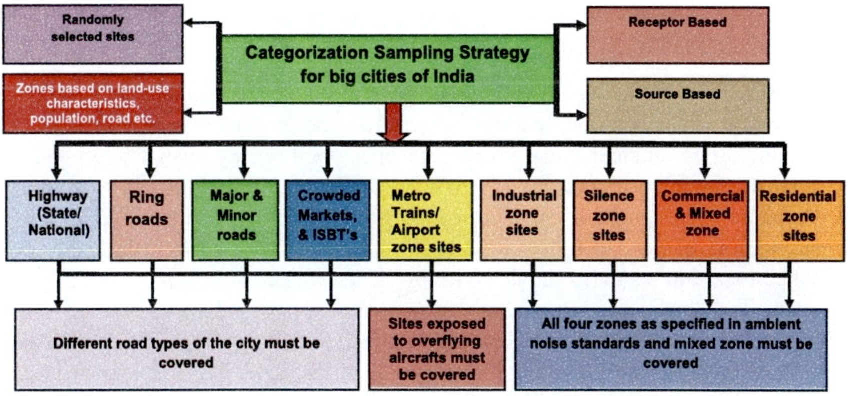

Fig. 8.12 Categorization sampling strategy recommended for big metropolitan cities

(c) Source of data: Minimizing the number of data sources may reduce the amount of preprocessing (e.g., avoid crossing territorial authority boundaries).
(d) Type of area: Urban areas defined by statistics or national census will include the areas of highest population density where strategic noise mapping might provide the most benefit.
(e) Grid-based sampling of area into equal subunits.
(f) Area segmentation based on both source and sensitive receptors especially the silence zone and residential zone areas.
(g) Division of area based on road size/traffic flow/highway road/residential and other zones, etc.

The sampling strategy adopted can be either of the above approaches depending upon the city, site-specific conditions, and economic, political, and infrastructural constraints. Categorization sampling may be preferred as compared to the grid-based sampling. As such the sampling strategy such as dividing the roads (national highway, state highway, ring road, major and minor roads), railway, airport, and land-use pattern as per ambient noise standards (industrial, residential, commercial and silence zones), areas with high traffic density or population density, and areas with several noise sources may be considered for the big metropolitan cities of India.

A recent study reported for Nagpur city of India [91, 92] categorized 700 noise monitoring locations for noise mapping as: national highway (137), state highway (68), ring road (100), major road (188), minor road (88), industrial (54), commercial (24), and residential zones (41). Figure 8.12 recommends an illustrative example of categorization sampling strategy for big metropolitan cities of India. An optimal strategy covering all the various road types including highways; major, minor, and ring roads; crowded markets; and interstate bus terminals (ISBTs); metro trains;

airport zone sites; and all the four zone sites (silence/industrial/residential/commercial) including the mixed zone sites should be employed.

- A combined approach involving the use of calibrated noise monitoring terminals traceable to the national standards for noise monitoring of various sites and prediction of ambient noise levels for some points using spatial interpolation algorithms may be exercised.
- A harmonized approach for noise mapping (like color codes and range of sound pressure level), use of traffic noise models that are validated in Indian scenario, and integration with GIS should be followed. The heterogeneity in noise mapping methods makes it difficult to obtain comparable results as faced by EU in the first round of noise mapping [7]. Thus, harmonization of mapping methods, calculation methods, and noise indices is must. Fig. 8.13 shows the technical attributes and suggested strategy for noise mapping of cities in India (Garg et al. [89]). A coherent national strategy for assessment and management of environmental noise must be adopted [93].
- Noise maps should also assess the number of people highly annoyed in the area under consideration. Also, studies should be focused on assessing the effect of implementing noise abatement measures on noise exposure and annoyance along the major roads.

Thus, a comprehensive approach utilizing the noise mapping of various parts of cities especially the noisy hot spots and inventory of noise control measures shall be undertaken in next few years for reducing the ambient noise levels. Figure 8.14 shows the flow chart of a noise mapping and control strategy for reducing the ambient noise levels in Indian cities.

Thus, after understanding and analyzing the noise scenario, a balanced approach utilizing the inventory of noise action plans (*NAP*s) can be beneficial in realization and accomplishment of noise abatement goals. Also, the concept of "acoustical planning" as highlighted in European noise directives for controlling noise by planned measures such as land-use planning and traffic management should be considered. The potential applications of smartphones can be explored especially for metropolitan cities of India for noise mapping of the larger parts of Indian cities. Vaibhav et al. reported the development of a smartphone app developed indigenously by Council of Scientific and Industrial Research-National Physical Laboratory, New Delhi, India for environmental noise monitoring as shown in Fig. 8.15 [94]. The developed app has to however undergo a rigorous validation using different smart phones and establishing a mechanism of its traceability to the national standards of sound pressure for ensuring its reliability for reporting the ambient noise level measurements. Council of Scientific and Industrial Research-National Environmental Engineering Research Institute (CSIR - NEERI), Nagpur, India recently also developed an android based smartphone app called as "Noise Tracker" [77] for noise monitoring available freely to the users [77]. The use of smartphones in conjunction with GIS interface, data mining, Cloud, and Internet of Things (IOT) shall provide an indispensable, economical, and effective solution for environmental noise monitoring, mapping, management and control especially for developing countries like India.

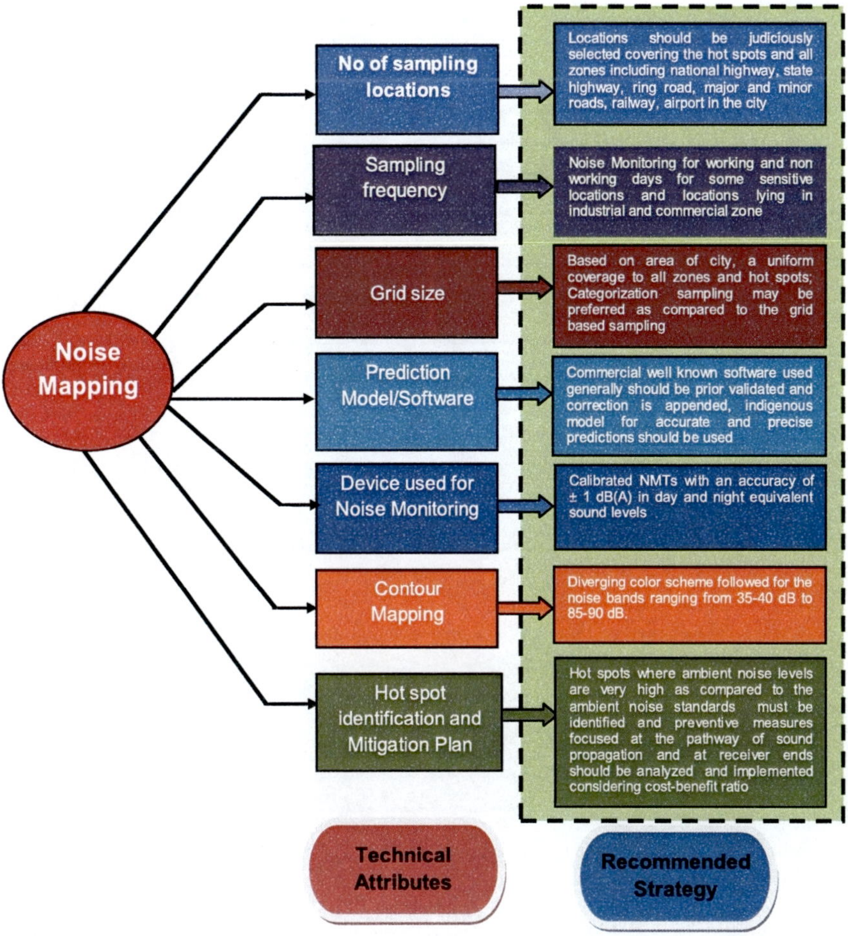

Fig. 8.13 Technical attributes and suggested strategy for noise mapping of cities in India (Garg et al. [89])

8.11 Soundscape Mapping: Potential Tool for Urban Development

Soundscape management focuses on the sounds of preference and reducing the unwanted sounds and thus necessitates the segregation of the various sound sources. The noise action plans in conjunction with soundscape action plans offer an expanded view for a sustainable sound environment [95]. Noise maps can be supplemented with psychoacoustic maps, sound maps, and soundscape maps for preservation, care, and protection of good-quality soundscapes contributing to more better and livable cities [96]. The joint holistic approach of noise mapping with

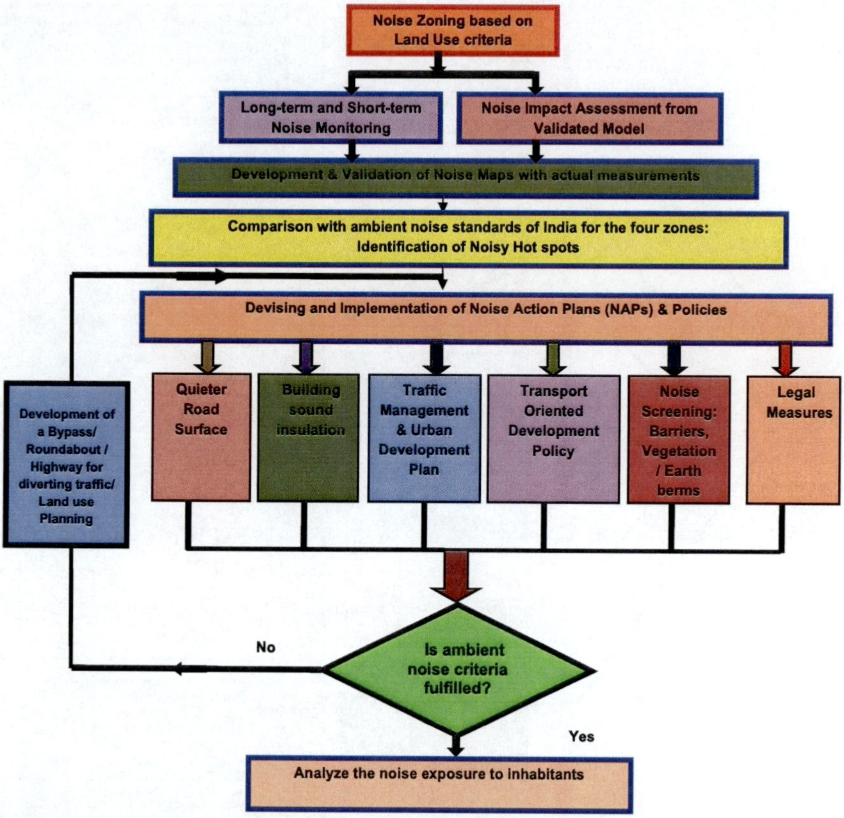

Fig. 8.14 Flow chart of a noise mapping and control strategy for reducing the ambient noise levels in Indian cities

soundscape approach has been endorsed by the European Environmental Agency in the Good Practice Guide on Quiet Areas [97]. The basic idea behind this approach is to correlate and integrate the quantitative data (e.g., from measurements, simulations, urban forms, topography, and noise model) with qualitative data (e.g., from type of sources, interviews, survey reports based on perception) by creating quantitative and qualitative maps. This methodology requires a multidisciplinary approach inviting the viewpoints of not only the experts, but also the city planners, authorities, and citizens together. The qualitative maps indicate the various qualities of the space and the kind of sounds that may be heard beyond traffic noise, and complement the usual strategic noise maps and noise action plan maps for developing a sustainable sound environment. Vogiatzis and Remy discussed the cross-analysis of six distinct overlapping layers of noise maps: strategic noise map depicting traffic and industrial noise in the study area; urban typology map depicting the sound environment as closed/semi-closed/open; spatio-acoustic typology map describing the built-up environment that might influence the sound propagation such

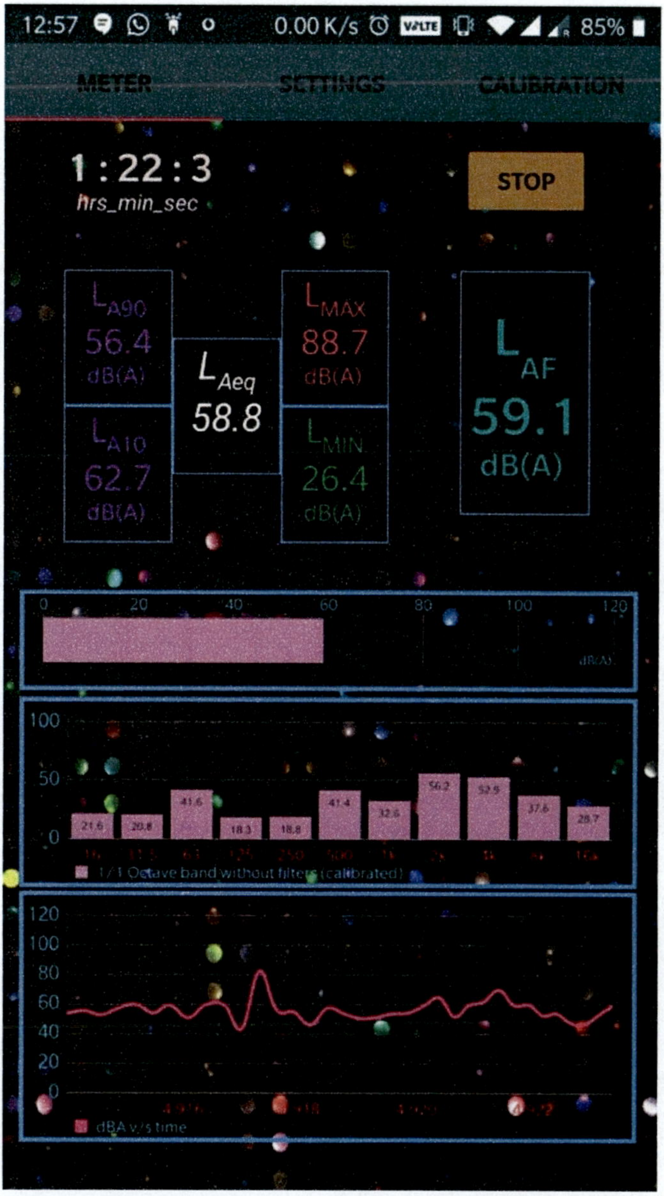

Fig. 8.15 Smartphone app developed for environmental noise measurements [94]

as acoustic fence, sound territories, and sound rhythms; predominant usage map describing the information on activities in the built environment and the public spaces and usage of public space and buildings; sound signal and sound markers map; and sound identity map based on the interviews. These maps in conjunction with noise maps and perceptual attributes are instrumental for the city planners in depicting the general geography of the area for implementation of noise action plan and soundscape design [98]. A recent study [20] of Margaritis and Kang reported the soundscape mapping by analyzing the spatial variability of the sounds sources in three categories: technological (manmade sounds), anthropic (human speech, singing, footsteps, music from shops church bells, laughter, etc.), and natural. This study described a mapping model for soundscape planning and implementation and evaluation of its effectiveness in two cities of the UK. Thus, soundscape mapping has been perceived as the process of visualizing the prominent aspects: sound sources, psychoacoustic parameters, and perceptual attributes relevant to soundscape quality [20]. It is important to understand and analyze that although the traditional control approach and soundscape approach differ significantly, yet both are complementary methods and not mutually exclusive [99].

8.12 2 ½ D and 3D Noise Mapping

Majority of the noise mapping studies reported across the globe are based on 2D approach. 2D noise maps are deficient to differentiate in noise impacts at various heights. However, some studies had been also reported that presented a 3D approach in noise mapping that offers noteworthy insight in the urban areas. A precise noise impact assessment is feasible particularly when different floors of a building close to the noise source are considered. In some situations, wherein the noise levels are invariant with height due to absence of obstacles, a 2D representation can be adequate [100]. However, a 3D approach can offer fundamental improvements in conjunction with better visualization in the urban areas. A 3D noise map can be developed by integration of a 3D city model with a noise surface representing noise levels. Photorealistic 3D noise model methodology had been utilized for visualizing the efficacy of mitigation measures and comparison of the resulting noise environment with and without the mitigation measures as shown in Fig. 8.16 [101]. The latest technological advancements in computer graphics, hardware, software, and interactive techniques particularly the VRGIS (Virtual Reality Geographic Information System) platform enables to simulate the three-dimensional noise maps of cities [102]. The 2D noise models present the noise levels at a particular height, while the 2 ½ D model presents the noise levels of different height in 3D space. 3D noise models show the noise effect in form of solid 3D grids depicting the volumetric information about the noise. In a 3D noise model, the noise levels are shown at different height of road surfaces as well as on the face of the building. 2 ½ D model can be built by projecting the 2D surface on height model (z axis) from the limitation of calculating the efficacy of noise mitigation measure such as noise barrier effect and building sound insulation. The European Commission Working Group Assessment of Exposure to Noise (WG-AEN) recommended to develop additional noise maps using different assessment heights, and 4 m may sometime lead to inaccuracies in the noise exposure assessment. The 2 ½ D noise maps are constructed using 2D

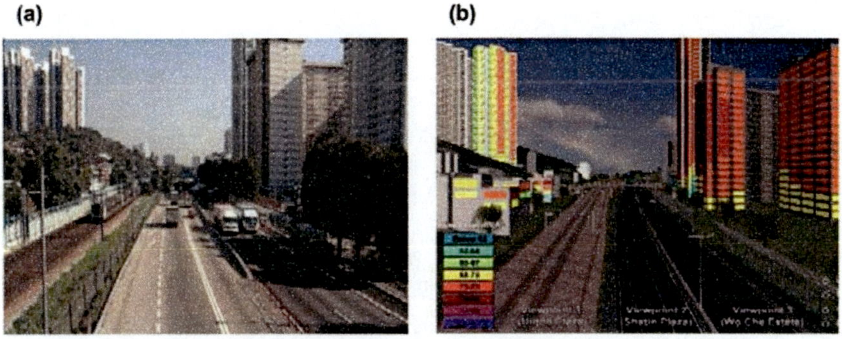

Fig. 8.16 Noise mapping of a road in Hong Kong (**a**) Photo. (**b**) Photorealistic 3D noise model with proposed noise barriers (Law C. –w. et al., 2011, [101])

interpolation techniques available in GIS, while 3D noise maps with 3D interpolation techniques. Kurakula (2007) study for 2 ½ D and 3D noise models show that triangulated irregular network (TIN) is the most suitable interpolation method among inversed distance weighted (IDW), natural neighborhood (NN), and Kriging interpolation algorithms for building 2 ½ D noise maps. Accurate 3D models of higher accuracy can be developed using laser altimetry [103].

8.13 Long-Term Noise Action Planning

An effective noise policy for sustainable environment and integrating noise vulnerability assessment into the urban planning shall be effective for noise pollution prevention and control. The policy measures may differ at local and national levels in terms of cost efficacy and suitability. For instance, spatial planning as preventive noise control measure may be considered as an optimal approach at local levels but has different implications on national levels. The formulation and enforcement of a harmonized national policy at both regional and national level would be the best approach. The Working Group on Health and Socio-economic Aspects recommends a value of the perceived benefit of noise reduction of 25 euro/household/dB/year in Europe [104]. Eventually, such a quantification of perceived benefits of noise control shall be instrumental for justifying the annual expenditure incurred on noise control action plans. Noise action plans (NAPs) help to structure and prioritize the noise abatement measures based on the assessment of noise situation and involve the basic principles of [105]:

- Participatory approach involving the public and all stakeholders.
- Cooperative approach for implementation and execution of measures.
- Open approach linking to the ongoing programs such as clean air and land-use planning.
- Measurable approach for quantification of noise level reductions.

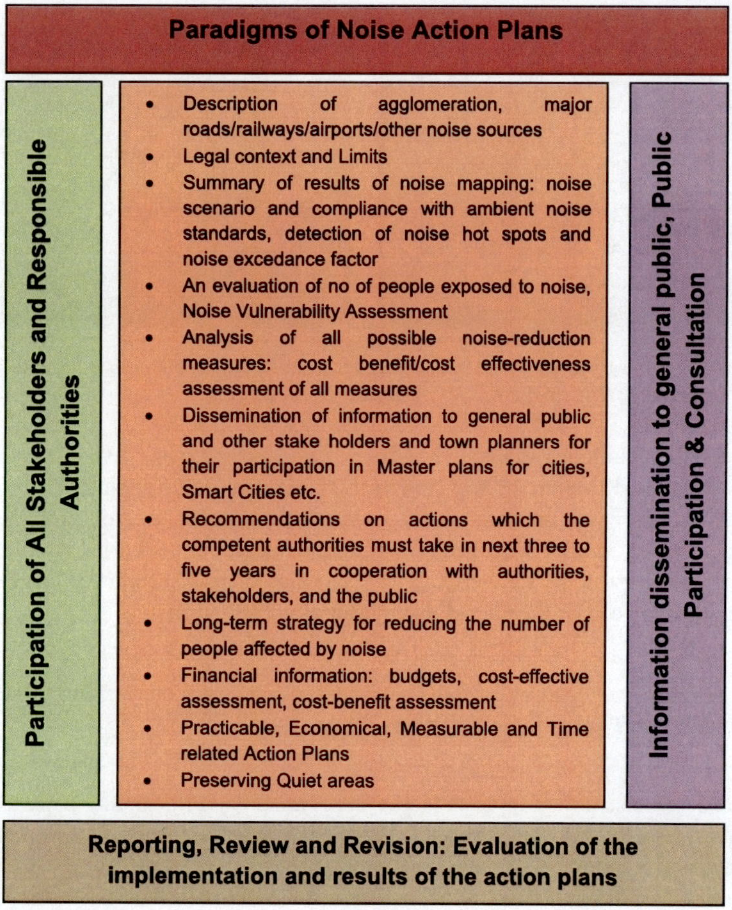

Fig. 8.17 Major paradigms of noise mapping and noise action plans for controlling the noise pollution in cities

Figure 8.17 shows the major paradigms of noise mapping and noise action plans for controlling the noise pollution in cities. The paradigms are similar to the minimum requirements that action plan must fulfill in European Noise Directives [105]. The action plan should include estimations in terms of the diminution of the number of people affected by noise and financial information of all the control measures. Cost-benefit analysis of various noise control measures, recommendations of long-term strategy and action plans that must be taken in next three-to-five years in cooperation with the relevant authorities and all stake holders, and dissemination of the information to general public such as that done by CPCB (http://cpcbnoise.com) must be the prime considerations of the noise mapping studies. The plan for noise mapping and noise action plans should thus consider the various aspects [106]:

- Noise action plans and noise level reductions to be accomplished and reduction in annoyance level or population exposed to noise levels above the legal limits.
- Cost-benefit and cost-effectiveness of the NAPs, prioritization, implementation methodology, and schedule.
- Comprehensive analysis and reporting of the costs incurred and benefits, administrative and social implications of executing NAPs, and any other legal matters pertaining.
- Description of responsibilities and roles of various participating agencies and departments and monitoring and reviewing the progress.
- Description of public consultation and dissemination of information to public.

The European Union reports revealed a diversification in the action planning approaches followed by the Member States including the noise control measures followed. However, the administrative costs were observed to be low (€0.15 for noise maps and €0.03 for action plans per citizen, every 5 years) and favorable cost-benefit ration of 1:29 was assessed with the implementation of END 2002/49/EC [106]. An action plan should define actions in a "smart" way (specific, measurable, achievable, realistic, time related). So, it should not merely be a policy statement but should define the targets to be achieved and must be accountable, that is, must compare the expected and real effects and inculcate the policy revisions if required [107]. The harmonization of action plans for the development of integrated strategic action plans at regional and national levels like that in H.U.S.H. (Harmonization of Urban noise reduction Strategies for Homogeneous action plans) project co-funded by Life+2008 program shall be a cost-effective approach for the accomplishment of noise abatement goals [108].

8.14 Quiet Areas

The Environmental Noise Directive (END) stresses the need to preserve environmental noise quality where it is good, that is, for the Quiet areas. Article 3 of the END defines a quiet area in an agglomeration as an area, delimited by a competent authority, for instance, which is not exposed to a value of L_{den} or another appropriate noise indicator greater than a certain value fixed by the Member State, from any noise source. A quiet area in open country shall mean an area, delimited by a competent authority, that is, undisturbed by noise from traffic, industry, or recreational activities. Different legal limits are recommended in different countries like that of $L_{den} < 55$ dB and $L_{night} < 45$ dB in Portuguese legislations. For some Member States, $L_{den} < 50$ dB is also defined, while for some states like Czech Republic, $L_{day} < 40$ dB and $L_{night} < 40$ dB is also recommended. However, the definition is not explicit of the properties and characteristics of quiet areas, and criteria are left to the competent authority. Figure 8.18 summarizes the various such criteria for defining the potential quiet urban areas [109, 110].

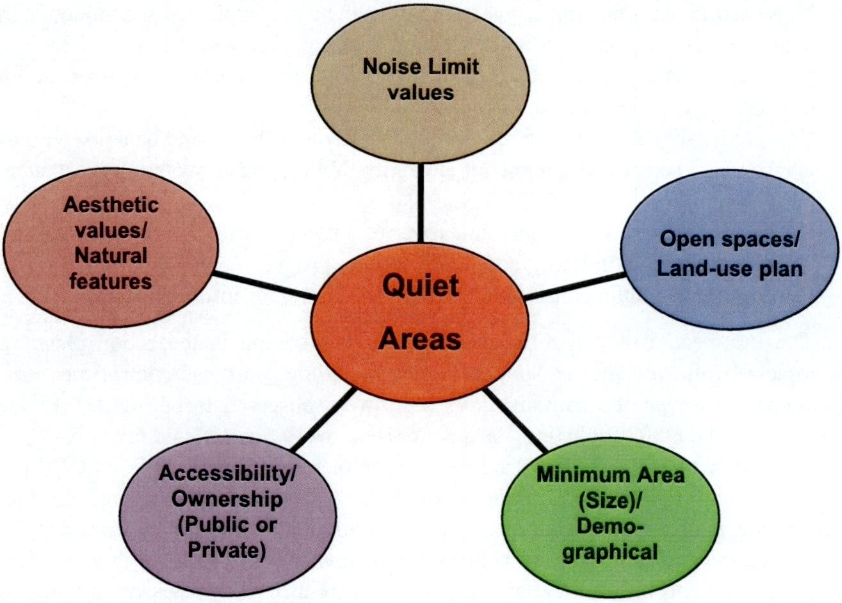

Fig. 8.18 Criteria's for defining the potential quiet urban areas

The delineation of quite areas is a national obligation for EU Member States. Apart from the legal noise limits, the perception indicators such as pleasant nature, nice color, and natural features like vegetation, small calm spaces (< 1 Ha), low-density residential areas, ownership in terms of space freely available, land-use plan like natural parks, and protected landscapes are some of the criteria adopted for defining the quiet areas. The spatial pattern of quiet areas in EU depends on the environment, topographic, and demographic profile of the Member States. A wider description of quiet areas suitable in Indian perspectives also can be as: *areas undisturbed by noise from traffic, industry, aircraft or recreational activities or any other noise sources* [111].

8.15 Noise Priority Indices for Noise Action Plans

The identification of most critical areas that require the execution and implementation of NAPs is the prime consideration of noise mapping exercise followed. There had been various noise scoring or priority indices developed in recent times that can be used to prioritize the mitigation measures based on subjective response and ambient noise levels. Table 8.6 summarizes the various noise priority indices based on a comparative analysis described by D'Alessandro and Schiavoni (2015) [112] used in some countries in recent times [112–126]. Some of these assign a

Table 8.6 Noise Priority Indices used in Europe [112–126]

Name of the index	Country/city using the index	References	Empirical formula
Building Prioritisation Score (BPS)	Scotland	Scottish Government (2009)	$BPS = L + 10\ \log(N_A)$ L = noise level (L_{den} or L_{night}) at the considered building generated by the considered source $N_A = \frac{N \times P_A \times A}{100}$ N = number of address points within the building P_A = population per address A = percentage of people annoyed depending on the source
Multi Annoyance Building Prioritisation Score (MABPS)	Italy	Licitra et al. (2011)	$MABPS = L_{tot} + 10\ \log(N_A)$ L_{tot} = estimated value of the noise level in the most exposed façade due to all the impacting sources in terms of L_{den}
Gden	–	Jabben et al. (2010)	$\text{Gden, norm} = 10\log\left(\frac{1}{N_{tot}}\sum_i n_i^{L_{den,i}/10}\right)$ N_{tot} = population of the studied area n_i = number of dwellers exposed to the i-th class of exposure $L_{den,\ i}$ = representative noise level of the i-th class
Qcity noise scoring proposal	–	Petz et al. (2007)	$QC = \begin{cases} \sum_i n_i 10^{0.15*(L_{den,i}-50)} \text{for} L_{den,i} \leq 65\ \text{dB(A)} \\ \sum_i n_i 10^{0.30*(L_{den,i}-57.5)} \text{for} L_{den,i} \geq 65\ \text{dB(A)} \end{cases}$ n_i = number of people exposed to the noise level at the relevant facade
Dutch Population Annoyance Index (PAI)	–	de Ruiter (2009)	$PAI = n_i \times 0.0323 \times (L_{centralvalue,\ den} - 42)^2$ n_i = number of residents in each building or dwelling $L_{centralvalue,\ den}$ = central value of the noise class assigned to each dwelling or building
Norwegian Støyplageindeks (SPI)	Norway	Gjestland et al. (2003)	$SPI = n_i\ (1.58\ L_{den} - 62.5)$ n_i = number of residents in each building or dwelling
Prioritization matrices of the noise action plans of Dublin and Kilkenny	Dublin and Kilkenny City	Dublin Local Authorities (2013)	Prioritization of the areas to be acoustically mitigated is made using a decision matrix that considers the noise

(continued)

Table 8.6 (continued)

Name of the index	Country/city using the index	References	Empirical formula
		Kilkenny City Council (2013)	exposure, land use, and impacting sources. Kilkenny city council uses L_{den} indicator instead of L_{day}. Major emphasis is focused on quiet areas and worst noise polluted areas
Number of people annoyed and highly annoyed	–	European Environmental Agency (2010)	Annoyance is defined as an “emotional state connected to feeling of discomfort, anger, depression and helplessness” that should be evaluated by means of ISO 15666 questionnaires (ISO 15666, 2013)
House Depreciation Index	–	European Environmental Agency (2010)	House depreciation in terms of lost €/dB(A)
Italian Priority Index	Italy	Italian Government (2000)	$IP_i = k_i \times n_i \times \text{MAX}(L_{diurno} - L_{lim,\,diurno}, L_{notturno} - L_{lim,\,notturno})$ k_i = correction coefficient k_i = correction coefficient $k_i = 4$ for hospitals $k_i = 3$ for schools and kndergardens $k_i = 1$ for other residential buildings n_i = number of people affected by noise; L_{diurno} and $L_{notturno}$ = average values of the A-weighted noise level estimated at the i-th building façade in the Italian diurno (06–22) and notturno (22–06) time period; $L_{lim,\,diurno}$ and $L_{lim,\,notturno}$ = Italian noise limits for the i-th building in the Italian diurno (06–22) and notturno (22–06) time period.

higher weightage to the number of citizens affected by noise exposure, other to the noise levels, and some to the urban characteristics such as quiet areas to be preserved and land-use. For instance, the Dublin and Kilkenny (2013) prioritization matrix prioritize not only the rehabilitation of noise areas but also the preservation of the quieter ones [112, 121, 122]. The benefit of such an evaluation is the single value description of the noise climate of a city or an area. An important consideration on the comparison of estimates of noise exposure across the EU Member States revealed contrasting results that made it cumbersome for comparing the results. The five methods used for estimating the noise exposure reported are END, VBEB, NEAREST, ALE, and CLOSEST. All these methods have the prime objective of estimation of the percentage of people exposed to a range of noise levels but differ in calculations and assumptions. The END and CLOSEST method provides a single value for all inhabitants in a building, while the VBEB, NEAREST, and ALE methods provide a range of noise exposure for all inhabitants in a building [127]. It is evident that erroneous estimations of the noise exposure may lead to ambiguous allotment of economical resources in the noise action plans [128]. The END method advocates the assumption of noise exposure to be the maximum value of the incident sound levels calculated at the outer building facade. In VBEB method [129], the inhabitants are equally distributed over the entire receiver's ring located at facades. In NEAREST method [130], the incident sound level is assumed to be equal to that at the nearest grid point (reduced by 3 dB). In case of the average level exceedance (ALE) method, inhabitants in a building are equally distributed over the receivers exposed to incident sound levels greater than the arithmetic mean value of sound levels calculated over the receiver's ring [131]. The CLOSEST method [128] assumes all inhabitants in a building are exposed to the sound level measured outside the building at a geographical grid point closest to the building façade. The sound level is reduced by 3 dB to account for the reflections from the building facades. The study by Licitra et al. [128] on comparison of these methods showed that END method overestimates the level of noise exposure, while the VBEB method calculates the lowest noise exposure. This study recommended the use of the CLOSEST approach which is a simple method considering the noise levels at the grid point closest to the building and thus serves as transparent and useful method for informing the public about noise exposure. Figure 8.19 summarizes all the five methods to estimate the population noise exposure [127–131].

8.16 Tranquillity Rating Prediction Tool (TRAPT)

Tranquillity implies a quiet as well as a natural environment, helps in relieving stress and improving feelings of well-being, and is thus free from emotional or mental disturbance [132]. Tranquillity rating prediction tool is thus very helpful for the designers and town planners to optimize the levels of tranquillity in valuable spaces and provide adequate relief from stress induced by living and working in city environment; this supports in the design of "restorative environments"

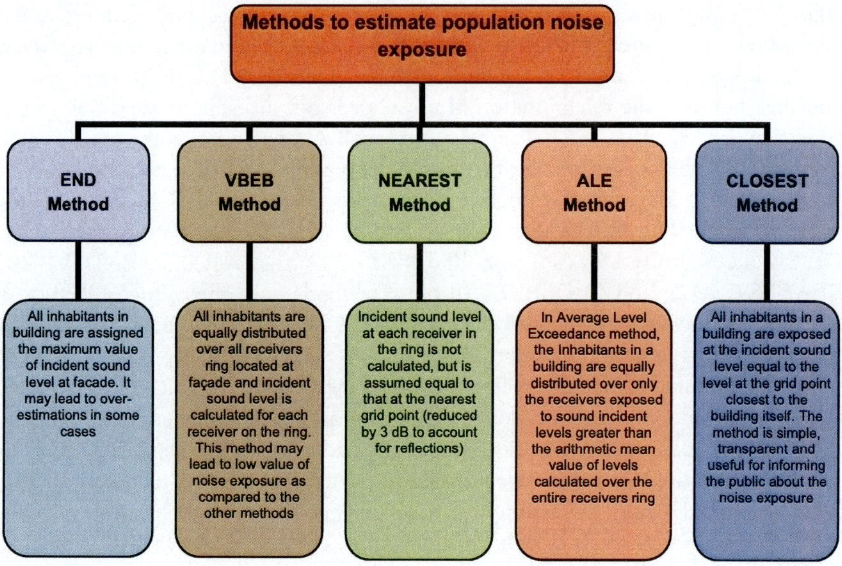

Fig. 8.19 Methods to estimate population noise exposure [127–131]

[132, 133]. The tranquillity rating (*TR*) is calculated in terms of L_{Aeq}, the equivalent continuous sound pressure levels of anthropogenic noise and *NCF*, and the percentage of natural and contextual features in the scene excluding the sky. The moderating factors (*MF*) account the presence of litter and water sounds. The empirical formulation for calculation of *TR* is [134]:

$$TR = 10.55 - 0.146L_{Aeq} + 0.041NCF + MF \tag{8.4}$$

TR value less than 5 points is unacceptable; in range 5.0–5.9 is just acceptable; in range 6.0–6.9 is fairly good; in range 7.0–7.9 is good, and $TR \geq 8.0$ is considered as excellent. Various factors affect the tranquillity rating such as litter can decrease the *TR* values significantly, while water sounds due to natural sources enhance the tranquillity rating [133, 135]. Improvements had been noticed even when the water sounds were 7 dB(A) below the level of background traffic noise [135]. The moderating factors account a minor adjustment designed to consider the actual environmental conditions at the time of assessment and influence the calculated *TR* by ±1 scale point [136]. The quiet areas and tranquillity are significant as they promise health benefits and enhance the economic value of surrounding agglomerations. The audio-visual interaction in soundscape such as visibility of sound sources, aesthetic values of the landscape, and harmoniousness between the sound and visual environment strongly affects the soundscape. For instance, water sound is one of the most effective masking sounds for reducing annoyance in urban squares [136].

8.17 Conclusions

The chapter reviews the various aspects of noise mapping process and presents an overall view in advantages and limitations associated with developing noise maps of the cities. Noise maps are powerful tools for dissemination of information pertaining to noise levels at various sites for general public and serve as best simulative tool for devising noise control measures and raising awareness among the community. It provides a baseline data to the town planners, pollution control bodies, and researchers for planning and execution of new projects. Thus, it is suggested that noise maps should be developed for every big cities across the globe for decision-making process concerning controlling the noise pollution. Developing noise maps for large agglomerations is a challenge. In Indian scenario, it is imperative to develop a harmonized strategy for developing the noise maps for metropolitan cities for understanding the noise scenario and devising suitable steps for controlling the noise pollution. Undoubtedly, this is an expensive approach with large infrastructural requirements; but considering the effects of noise pollution on human health, the economic and infrastructural issues are really shadowed. Also, in spite of all the challenges faced in developing noise maps for big cities, the concept and strategy of noise mapping has proven to be a promising technique for urban planning.

References

1. European Federation for Transport & Environment (2008) Can you hear us? Why is it finally time for the EU to tackle the problem of noise from road and rail traffic. http://www.epha.org/IMG/pdf/2008-02_traffic_noise_can_you_hear_us.pdf. Accessed on 17.05.2020
2. Directive 2002/49/EC of the European Parliament and of the Council of 25th June 2002, relating to the assessment and management of environmental noise, Official Journal of the European Communities, June, 2002
3. Working Group on the Assessment of Environmental Noise (WG-AEN) (2008) Position paper on Presenting Noise Mapping Information to the Public
4. van Leeuwen HJA, van Banda SEH (2015) Noise mapping – State of the art – is it just as simple as it looks? Euro Noise 2015, 31 May – 3 June, Maastricht, 433–438
5. Fiedler PEK, Zannin PHT (2015) Noise mapping as a tool for urban planning. J Sci Ind Res 74: 114–116
6. Borelli D, Repetto S, Schenone C (2014) Noise mapping of the flyover highway in Genoa: comparison of different methods. Noise Mapping 1:59–73
7. Murphy E, King EA (2010) Strategic environmental noise mapping: methodological issues concerning the implementation of the EU environmental noise directive and their policy implications. Environ Int 36(3):290–298
8. de Kluijver H, Stoter J (2003) Noise mapping and GIS: Optimizing quality and efficiency of noise effect studies. Comput Environ Urban Syst 27:815–102
9. ISO 1996-2 (2007) Acoustics-Description, measurement and assessment of environmental noise-Part 2: Determination of environmental noise levels. International Organization for Standardization, Geneva, Switzerland
10. Nicol F, Wilson M (2004) The effect of street dimensions and traffic density on the noise level and natural ventilation potential in urban canyons. Energ Buildings 36:423–434

11. Walerian E, Janczur R, Czechowicz M, Smyrnova Y (2011) Validation of a complex urban noise model close to a road. Appl Acoust 72(10):790–802
12. González DM, Morillas JMB, Gozalo GR (2015) The influence of microphone location on the results of urban noise measurements. Appl Acoust 90:64–73
13. Morillas JMB, González DM, Gozalo GR (2016) A review of the measurement procedure of the ISO 1996 standard. Relationship with the European noise directive. Sci Total Environ 565: 595–606
14. Memoli G, Paviotti M, Kephalopoulos S, Licitra G (2008) Testing the acoustical corrections for reflections on a facade. Appl Acoust 69:479–495
15. European Commission Working Group Assessment of Exposure to Noise (WG-AEN)-Position Paper (2006) Final draft-good practice guide for strategic noise mapping and the production of associated data on noise exposure-version 2–13th January, 2006
16. Guidance Note for Strategic Noise Mapping (2011) For the environmental noise regulations 2006. Environmental Protection Agency, August, 2011
17. Kaliski K (2007) Community and Regional Noise Mapping in the United States. Sound and Vibration:14–17
18. Conference of European Directors of Roads, CEDR (2013) Best practice in strategic noise mapping, August, 2013
19. Asensio C, Ruiz M, Pav6n I, Ausejo M, Recuero M (2011) Uncertainty in noise maps isolines: the effect of sampling grid. Acta Acust United Acust 97:237–242
20. Margaritis E, Kang J (2017) Soundscape mapping in environmental noise management and urban planning: case studies in two UK cities. Noise Mapping 4:87–103
21. Barrigon Morillas JM, Gomez Escobar V, Mendez Sierra JA, Vílchez Gomez R (2002) Study of noise in a small Spanish Town. Int J Acoust Vib 7:231–237
22. Sommerhoff J, Recuero M, Sùarez E (2004) Community noise survey of the city of Valdivia, Chile. Appl Acoust 65:643–656
23. Martín MA, Tarrero A, Gonźale J, Machimbarrena M (2006) Exposure-effect relationships between road traffic noise annoyance and noise cost valuations in Valladolid, Spain. Appl Acoust 67:945–958
24. Escobar VG, Morillas JMB, Gozalo GR, Gómez RV, del Río JC, Sierra JAM (2012) Analysis of the grid sampling method for noise mapping. Arch Acoust 37(4):499–514
25. Gozalo GR, Morillas JMB (2016) Analysis of sampling methodologies for noise pollution assessment and the impact on the population. Int J Environ Res Public Health 13(5):490
26. Say N, Ökten SÖ, Aysu A, Yalçinkaya NM (2017) Seasonal change and spatial distribution of traffic noise in Seferihisar. Turkey Arch Acoust 42(4):631–642
27. Doygun H, Gurun DK (2007) Analyzing and mapping spatial and temporal dynamics of urban traffic noise pollution: a case study in Kahramanmaras, Turkey. Environ Monit Assess 142: 65–72
28. Harman BI, Koseoglu H, Yigit CO (2016) Performance evaluation of IDW, Kriging and multiquardic interpolation methods in producing noise mapping: a case study at the city of Isparta, Turkey. Appl Acoust 112:147–157
29. Can A, Dekoninck L, Botteldooren D (2014) Measurement network for urban noise assessment: comparison of mobile measurements and spatial interpolation approaches. Appl Acoust 83:32–39
30. Asensio C, Recuero M, Ruiz M, Ausejo M, Pav6n I (2011) Self-adaptive grids for noise mapping refinement. Appl Acoust 72:599–610
31. Stapelfeldt H, van Banda EH, Manvell D Quality assurance for the calculation of large noise maps. https://www.environmental-expert.com/Files/1442/articles/6145/bn0145.pdf. Accessed online 15.03.2018
32. Notario A (2016) Time and cost saving software techniques and their application in large-scale noise mapping projects. In: Proceedings of 22nd international Congress on acoustics, Buenos Aires, 5–9th September 2016

33. DIN 45687: Acoustics – software products for the calculation of the sound propagation outdoors – quality requirements and test conditions, 2006–05
34. Probst W, Rabe I, Techniques to accelerate noise mapping calculations for larger areas and cities. In: 15th International Congress on sound and vibration, 6–10th July 2008, Daejeon, Korea
35. Steele C (2001) A critical review of some traffic noise prediction models. Appl Acoust 62(3): 271–287
36. Arana M, San Martín R, Nagore I, Pérez D, San Martín ML (2008) Noise map from two computer programs. Analysis of differences. Acústica 2008, 20–22 de Outubro, Coimbra, Portugal, Universidade de Coimbra
37. Kephalopoulos S, Paviotti M, Anfosso-Lédée I, Maercke DV, Shilton S, Jones N (2014) Advances in the development of common noise assessment methods in Europe: the CNOSSOS-EU framework for strategic environmental noise mapping. Sci Total Environ 482-483:400–410
38. Manvell D, van Banda EH (2011) Good practice in the use of noise mapping software. Appl Acoust 72(8):527–533
39. Kumar K, Ledoux H, Schmidt R, Verheij T, Stoter J (2020) A harmonized data model for noise simulation in the EU. ISPRS Int J Geo-Inf 9:121. https://doi.org/10.3390/ijgi9020121
40. Van Den Berg M, Gerretsen E (1996) Comparison of road traffic noise calculation models. J Build Acoust 3:13–24
41. Nijland HA, Van Wee GP (2005) Traffic noise in Europe: a comparison of calculation methods, noise indices and noise standards for road and rail road traffic in Europe. Transp Rev 25:591–612
42. King EA, Murphy E (2009) The development of a practical framework for strategic noise mapping. Appl Acoust 70:1116–1127
43. Hepworth P (2006) Accuracy implications of computerized noise predictions for environmental noise mapping. In: Proceedings of 35th International congress on noise control engineering, Hawaii (Inter-Noise 2006)
44. Probst W (2010) New techniques in noise prediction. In: Proceedings of 20th international Congress on acoustics, ICA, 23–27 August, 2010, Sydney
45. Descornet G, Goubert L (2006) Noise classification of road pavements, Task 1: technical background information. Draft report, European Commission Directorate-General Environment, June 2006
46. Murphy E, Faulkner J P, Douglas O (2020) Current state-of-the-art and new directions in strategic environmental noise mapping. Current pollution reports. https://doi.org/10.1007/s40726-020-00141-9.
47. Alberts W, Alférez JR (2012) The use of colors in END noise mapping for major roads. Euronoise, Prague, 10–13th June 2012
48. Shilton S (2017) Uncertainty in noise modeling. 24th International Congress on sound and vibration, 23–27th July, 2017, London
49. Bocher E, Guillaume G, Picaut J, Petit G, Fortin N (2019) Noise modelling: An open source GIS based tool to produce environmental noise maps. ISPRS Int J Geo-Inf 8:130. https://doi.org/10.3390/ijgi8030130.
50. Stapelfeldt H, Manvell D (2011) Using dynamic noise mapping for pro-active environment noise management. internoise, September, 4–7, 2011, Osaka, Japan
51. Manvell D, Ballarin M L, Stapelfeldt H, Sanz R (2004) SADMAM-Combining measurements and calculations to map noise in Madrid. In: Proceedings of the 33rd inter-noise, Prague, Czech Republic, August 22–25, 2004
52. Benocci R, Bellucci P, Peruzzi L, Bisceglie A, Angelini F, Confalonieri C, Zambon G (2019) Dynamic noise mapping in the suburban area of Rome (Italy). Environments 6:79. https://doi.org/10.3390/environments6070079
53. Murphy E, King EA (2011) Scenario analysis and noise action planning: modelling the impact of mitigation measures on population exposure. Appl Acoust 72:487–494

54. Dintrans A, Préndez M (2013) A method of assessing measures to reduce road traffic noise: a case study in Santiago, Chile. Appl Acoust 74:1486–1491
55. Coehlo JLB, Alarcão D (2005) Noise mapping and noise action plans in large areas. Forum Acusticum 2005, Budapest, Hungary, 1039–1044
56. Lee S-W, Chang SI, Park Y-M (2008) Utilizing noise mapping for environmental impact assessment in a downtown redevelopment area of Seoul, Korea. Appl Acoust 69:704–714
57. Schiavoni S, D'Alessandro F, Conte A (2015) The contribution of LIFE+NADIA project on the implementation of the European Directive on Environmental Noise. Noise Mapping 2:13–30
58. Federal Highway Administration, FHWA (2017) Entering the quiet zone: noise compatible land use planning. https://www.fhwa.dot.gov/environment/noise
59. WG-AEN (Working Group Assessment of Exposure to Noise) (2007) Good practice guide for strategic noise mapping and the production associated data on noise exposure. Version 2. European Commission, Brussels, Belgium (Position paper 2)
60. Gozalo GR, Morillas JMB, Carmona JT, González DM, Moraga PA, Escobar VG, Vílchez-Gómez R, Sierra JAM, Prieto-Gajardo C (2016) Study on the relation between urban planning and noise level. Appl Acoust 111:143–147
61. Carmona del Río FJ, Escobar VG, Carmona JT, Vílchez-Gómez R, Sierra JAM, Gozalo GR, Morillas JMB (2011) A street categorization method to study urban noise: The Valladolid (Spain) study. Environ Eng Sci 28:811–817
62. Morillas JMB, Ortiz-Caraballo C, Gajardo CP (2015) The temporal structure of pollution levels in developed cities. Sci Total Environ 517:31–37
63. Morillas JMB, Escobar VG, Carmona JT, Sierra JAM, Vílchez-Gómez R, Carmona del Río FJ (2011) Analysis of the prediction capacity of a categorization method for urban noise assessment. Appl Acoust 72:760–771
64. Gozalo GR, Morillas JMB, Gajardo CP (2015) Urban noise functional stratification for estimating average annual sound level. J Acoust Soc Am 137:3198–3208
65. Gozalo GR, Morillas JMB, Escobar VG (2012) Analysis of noise exposure in two small towns. Acta Acust United Acust 98:884–893
66. Gozalo GR, Morillas JMB, Escobar VG, Vílchez-Gómez R, Sierra JAM, Carmona del Río FJ, Gajardo CP (2013) Study of the categorisation method using long-term measurements. Arch Acoust 38:397–405
67. Zytoon MA (2016) Opportunities for environmental noise mapping in Saudi Arabia: A case of traffic noise annoyance in an urban area in Jeddah City. Int J Environ Res Public Health 13:496
68. Kumar R, Mukherjee A, Singh VP (2017) Traffic noise mapping of Indian roads through smartphone user community participation. Environ Monit Assess 189:17
69. Zamora W, Calafate CT, Cano JC, Manzoni P (2017) Accurate ambient noise assessment using smartphone. Sensors 17:E917
70. Kanjo E (2010) NoiseSPY: a real-time mobile phone platform for urban noise monitoring and mapping. Mobile Netw Appl 15(4):562–574
71. Kardous CA, Shaw PB (2014) Evaluation of smartphone sound measurement applications. J Acoust Soc Am 135:EL186–EL192
72. Kardous CA, Shaw PB (2016) Evaluation of smartphone sound measurement applications (apps) using external microphones – a follow-up Study. J Acoust Soc Am 140(4):EL327
73. Robinson D, Tingay J (2014) Comparative study of the performance of smartphone-based sound level meter apps, with and without the application of a ½″ IEC-61094-4 working standard microphone, to IEC-61672 standard metering equipment in the detection of various problematic workplace noise environments, Inter-noise 2014, Melbourne, Australia, 16th to 19th November, 2014, 1–10
74. Nast DR, Speer WS, Le Prell CG (2014) Sound level measurements using smartphone "apps": Useful or inaccurate? Noise Health 16:251–256
75. King EA, Murphy E (2014) Environmental noise pollution: noise mapping, public health and policy. Elsevier, Amsterdam

76. King EA, Murphy E (2016) Smartphone-based noise mapping: Integrating sound level meter app data into the strategic noise mapping process. Sci Total Environ 562:852–859
77. Lokhande S K, Chopkar P F, Jain M C, Hirani A (2021) Environmental noise assessment of Bhandara City in the adversity of COVID-19 pandemic: A crowdsourcing approach. Noise Mapp. 8:249-259
78. Pődör A, Révész A (2014) Noise map: professional versus crowdsourced data. In: Huerta J, Schade S, Granell C (eds) Proceedings of the 17th AGILE International conference on geographic information science, Castellón, Spain
79. Picaut J, Fortin N, Bocher E, Petit G, Aumond P, Guillaume G (2019) An open-science crowdsourcing approach for producing community noise maps using smartphones. Build Environ 148:20–33
80. Evaluation of Directive 2002/49/EC Relating to the Assessment and Management of Environmental Noise (2016) Final report, Directorate-General for Environment Knowledge, Risks and Urban Environment. Luxembourg: Publications Office of the European Union
81. European Commission (2017) Report from the Commission to the European Parliament and the Council on the implementation of the Environmental Noise Directive in accordance with Article 11 of Directive 2002/49/EC (COM(2017) 151 final)
82. Defra NANR 201 (2011) Environmental noise valuation-the costs and benefits of remediation measures. Final report, August, 2011
83. Das P, Talukdar S, Sk Z, Das S, Pal S (2019) Noise mapping and assessing vulnerability in meso level urban environment of Eastern India. Sustain Cities Soc 46:101416
84. Banerjee D, Chakraborty SK, Bhattacharyya GA (2009) Appraisal and mapping the spatial-temporal distribution of urban road traffic noise. Int J Environ Sci Technol 6:325–335
85. Bauer ER, Kohler JL (2000) Cross-sectional survey of noise exposure in mining industry. In: Proceedings of 31st annual Institute of Mining, Safety and Research, Blacksburg, Virginia, USA, pp 17–31
86. Lokhande SK, Dhawale SA, Pathak SS, Gautam R, Jain MC, Bodhe GL (2017) Appraisal of noise level dissemination surrounding Mining and industrial areas of Keonjhar, Odisha: a comprehensive approach using noise mapping. Arch Acoust 42(3):423–432
87. Lokhande SK, Jain MC, Dhawale SA, Gautam R, Bodhe GL (2018) Realizing modeling and mapping tools to study the upsurge of noise pollution as a result of open-cast mining and transportation activities. Noise Health 20:60–67
88. Vijay R, Lingote D, Sharma A, Gupta R (2018) Development of GIS-based noise simulation module for point sources: N-Dhwani. In: Geospatial infrastructure, applications and technologies: India case studies. Springer, Singapore, pp 329–338
89. Garg N, Chauhan BS, Singh M (2021) Normative framework of noise mapping in India: Strategies, Implications and Challenges ahead. Acoustic Australia. 49, 23-41 https://doi.org/10.1007/s40857-020-00214-1
90. Guide to State Highway Noise Mapping (2013) New Zealand Transport Agency. November, 2013, Version 1.0
91. Vijay R, Upadhaya V, Dey J, Sakhre S, Khandelwal V, Gupta V, Singh T, Kalawapudi K, Kumar R (2018) Cycle based noise monitoring of urban city: an innovative approach. In: Proceedings of WESPAC 2018 Conference at CSIR-National Physical Laboratory, New Delhi, India, 11th to 15th November, 2018
92. Laxmi V, Dey J, Kalawapudi K, Vijay R, Kumar R (2019) An innovative approach of urban noise monitoring using cycle in Nagpur. India Environ Sci Pollut Res 26(36):36812–36819
93. King EA, Murphy E, Rice HJ (2011) Implementation of the EU environmental noise directive: lessons from the first phase of strategic noise mapping and action planning in Ireland. J Environ Manag 92(3):756–764
94. Vaibhav K, Saini P, Garg N (2020) Development of smartphone app for environmental noise measurements. In: Proceedings of International Conference on Industrial and Manufacturing Systems (CIMS-2020), Dr. B R Ambedkar National Institute of Technology, Jalandhar, Punjab, India, 26th to 28th June 2020, CIMS-358

95. Vogiatzis K, Remy N (2019) Environmental noise mapping as a smart urban tool development.In: Smart urban development. https://doi.org/10.5772/intechopen.88449
96. Lippold M, Lawrence B (2019) Soundscape planning as a tool for urban planning. In: Proceedings of 23rd International Congress on Acoustics, Aachen, Germany, 9th to 13th September, 2019, pp 7035–7042
97. European Environment Agency (2014) Good practice guide on quiet areas. EEA report No. 4/2014
98. Vogiatzis K, Remy N (2017) Soundscape design guidelines through noise mapping methodologies: an application to medium urban agglomerations. Noise Mapping 4:1–19
99. Welsh Department report (2018) Noise and soundscape action plan, 2018–23. https://gov.wales. Accessed on 02.04.2020
100. Stoter J, de Kluijver H, Kurakula VK (2008) 3D noise mapping in urban areas. Int J Geogr Inf Sci 22(8):907–924
101. Law CW, Lee CK, ASW L, Yeung MKL, Lam KC (2011) Advancement of three-dimensional noise mapping in Hong Kong. Appl Acoust 72:534–543
102. Bin Z, Wencheng H, Linhui P (2014) A framework for improving urban noise map. Comput Model New Technol 18(12C):222–226
103. Kurakula VK (2007) A GIS-based approach for 3D noise modeling using 3D city models, MSc thesis. International Institute for Geo-information Science and earth Observation, The Netherlands
104. Position Paper of the Working Group on Health and Socio-economic aspects, Valuation of noise, December, 2003
105. Practitioner Handbook for local noise action plans, Recommendations from the SILENCE project, The SILENCE Handbook, https://www.polisnetwork.eu/wp-content/uploads/2019/06/silence_handbook_local-noise-action-plans-2.pdf. Accessed on 04.12.2020
106. ETC/ATNI (2019) Noise action plans. Managing exposure to noise in Europe. No ETC/ATNI report no 8/2019, https://www.eionet.europa.eu/etcs/etc-atni/products/etc-atni-reports/etc-atni-report-8-2019-noise-actionplans-managing-exposure-to-noise-in-europe. Accessed 25 December 2020
107. Luzzi S (2019) Noise reduction and control in urban planning: the strategic action plan of Florence. Lecture – St. Petersburg 22nd March 2011
108. Borchi F, Carfagni M, Governi L, Curcuruto S, Silvaggio R, Bellomini R, Luzzi S, Licitra G, Palazzuoli D, Melloni A (2016) LIFE+2008 HUSH project results: a new methodology and a new platform for implementing an integrated and harmonized noise action plan and proposals for updating Italian legislation and environmental noise directive. Noise Mapping 3:71–85
109. European Environment Agency (2004) Good practice guide on quiet areas. EEA technical report no 4/2014, Denmark
110. Blanes N, Sáinz de la Maza M, Fons-Esteve J, Peris E (2019) Potential quiet areas in Europe inside urban areas. In: Proceedings Internoise 2019, Noise Control for a better environment, June 16–19, Madrid
111. Votsi NEP, Kallimanis AS, Pantis JD (2017) The distribution and importance of Quiet Areas in the EU. Appl Acoust 127:207–214
112. D'Alessandro F, Schiavoni S (2015) A review and comparative analysis of European priority indices for noise action plans. Sci Total Environ 518-519:290–301
113. Scottish Government (2009) Consultation on draft technical guidance for noise management areas, under strategic noise action plans
114. Licitra G, Gallo P, Rossi E, Brambilla G (2011) A novel method to determine multiexposure priority indices tested for Pisa action plan. Appl Acoust 72:505–510
115. Jabben J, Verheijen E, Schreurs E (2010) Group noise exposure level Gden/Gnight; applications to airport noise. Report RIVM (in Dutch)
116. Licitra G, Ascari E (2014) Gden: an indicator for European noise maps comparison and to support action plans. Sci Total Environ 482–483:411–419

117. Petz M, Stenman A, Malm P, Desanghere G (2007) Deliverable D1.2 of Quiet City Transport (FP6–516420)
118. de Ruiter EPJ (2009) A tool for environmental noise control in urban planning: the Population Annoyance Index. In: Proceeding of NAG/DAGA conference, Rotterdam
119. Gjestland T, Olsen H, Randeberg RT (2003) Noise in Norwegian cities and rural areas. In: Proceedings of EuroNoise conference, Naples
120. European Environmental Agency (2010) Good practice guide on noise exposure and potential health effects. Technical report
121. Dublin Local Authorities (2013) Dublin agglomeration-environmental noise action plan December 2013–November 2018
122. Kilkenny City Council (2013) Noise action plan 2014–2018
123. ISO 15666 (2013) Acoustics – assessment of noise annoyance by means of social and socio-acoustic surveys
124. Italian Government (2000) Decreto Ministeriale Criteri per la predisposizione, da parte delle società e degli enti gestori dei servizi pubblici di trasporto o delle relative infrastrutture, dei piani degli interventi di contenimento e abbattimento del rumore. Gazzetta Ufficiale della Repubblica Italiana 285 (in Italian, 29/11)
125. D'Alessandro F, Asdrubali F, Schiavoni S (2014) The noise abatement plan of an Italian road network: a comparison between standard and innovative methodologies. Open Transp J 8:26–38
126. Ascari E, Chiari C, Gallo P, Licitra G, Palazzuoli D (2010) Comparison of methods to identify hot spots in Pisa Municipality. In: 37th AIA National conference; 2010. (in Italian)
127. Science for Environment Policy (2012) European Commission DG Environment News Alert Service, edited by SCU, The University of the West of England, Bristol. https://ec.europa.eu/environment/integration/research/newsalert/pdf/303na5_en.pdf. Accessed on 21.12.2020
128. Licitra G, Ascari E, Brambilla G (2012) Comparative analysis of methods to estimate urban noise exposure of inhabitants. Acta Acust United Acust 98:659–666
129. Vorläufige Berechnungsmethode zur Ermittlung der Belastetenzahlen durch Umgebungslärm. VBEB, Federal German Gazette, 20/04/2007
130. Arana M, Martin RS, Nagore I, Pérez D (2009) Using noise mapping to evaluate the percentage of people affected by noise. Acta Acust United Acust 95:550–554
131. Brambilla G, Poggi A (2007) Indicazioni operative per la costruzione dell' indicatore "popolazione esposta al rumore" in riferimento alla direttiva 2002/49/CE. I quaderni di ARPA Emilia Romagna "Inquinamento acustico", Bologna. ISBN 88-87854-20-3, 65-71
132. Watts G, Pheasant R, Horoshenkov K (2010) Tranquil spaces in a metropolitan area. In: Proceedings of 20th international congress on acoustics, ICA 2010, 23rd–27th August 2010, Sydney, Australia
133. Watts G, Marafa L (2017) Validation of the Transquillity Rating Prediction Tool (TRAPT): comparative studies in UK and Hong Kong. Noise Mapping 4:67–74
134. Watts GR, Miah A, Pheasant RJ (2013) Transquillity and soundscapes in urban green spaces-predicted and actual assessments from a questionnaire survey. Environ Plann B Plann Des 40(1):170–182
135. Watts GR, Pheasant RJ, Horoshenkov KV, Ragonesi L (2009) Measurement and subjective assessment of water generated sounds. Acta Acust United Acust 95:1032–1039
136. D'Alessandro F, Evangelisti L, Guattari C, Grazieschi G, Orsini F (2018) Influence of visual aspects and other features on the soundscape assessment of a university external area. Build Acoust 1–19. https://doi.org/10.1177/1351010X18778759

Chapter 9
Effects of Noise on Health

Abstract This chapter presents the effects of noise on human health. A number of studies on the adverse health effects, such as annoyance, sleep disturbance, cardiovascular diseases, tinnitus, cognitive impairment, etc., have linked the exposure to noise levels. These health effects are the most investigated non-auditory health endpoints of noise exposure. Sleep disturbance, cardiovascular diseases, and annoyance mostly related to road traffic noise comprise the main burden of disease from environmental noise. The chapter reviews the various health effects associated with environmental noise and the empirical formulations proposed in recent years for correlating annoyance and sleep disturbances with noise exposure. Also, the concept of disability-adjusted life years (DALYs) and case study for estimation of DALYs are presented.

9.1 Introduction

Noise is pervasive in urban environment and the availability of quiet places is decreasing globally. Several studies have been conducted so far that showed that noise levels below the hearing damaging criterion cause annoyance, sleep disturbance, cognitive impairment, physiological stress reactions, and other health disorders. High blood pressure and ischemic heart disease (including myocardial infarction) have a high prevalence in industrialized countries and are major causes of death [1]. Road traffic noise contributes substantially to the burden of disease in the European Union (EU). The World Health Organization (WHO) "Night Noise Guidelines for Europe" (2009) confirmed that nighttime noise levels (L_{night}) above 55 decibels (dB) are "increasingly dangerous for public health" and cause adverse health effects [2]. Noise levels above 65 dB during daytime have been shown in some studies to be associated with a substantial increase in risk of cardiovascular disorders, including high blood pressure, ischemic heart diseases, and stroke [3, 4]. The World Health Organization (WHO Regional Office for Europe) has estimated the burden of disease from environmental noise for the WHO Euro-A region. The results are published in the most recent report "Burden of Disease from Environmental Noise" [5]. According to the report, sleep disturbance and annoyance

N. Garg, *Environmental Noise Control*,
https://doi.org/10.1007/978-3-030-87828-3_9

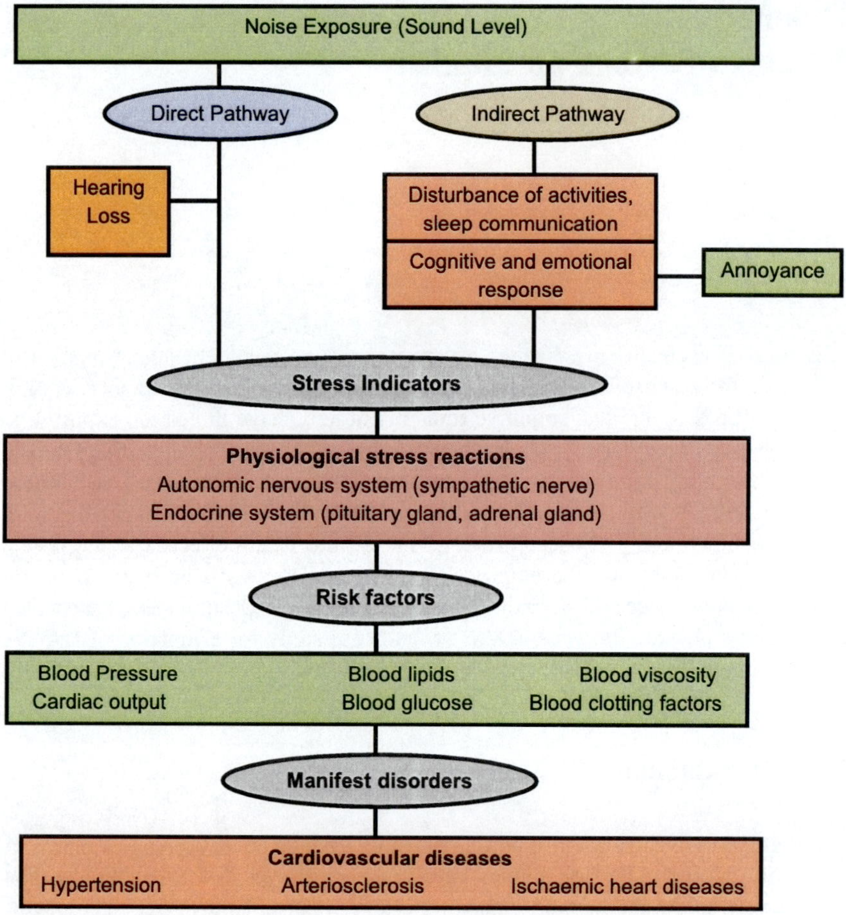

Fig. 9.1 Noise effects reaction scheme (Babisch, 2002 [6])

related to road traffic noise constitute most of the burden of environmental noise. Large-scale epidemiological studies have been carried out for a long time [6]. Figure 9.1 shows the various consequences of noise exposure particularly pertaining to health effects.

The cardiovascular risk is one of the prime considerations in analyzing the effect of noise on human health due to its high prevalence in the community across the world. However, the magnitude of effect, dose-response relationship, biological plausibility, and consistency of findings among the studies are the issues of epidemiological reasoning. Thus, it is imperative that noise mitigation measures, made at the source, the propagation path, or on the recipient's side, should be addressed and effectively implemented so as to control the increasing environmental noise pollution in urban cities.

9.2 Noise Annoyance

Environmental noise is a major health problem recognized across the globe. Transportation noise has been considered among the most significant causes of community annoyance in a large number of community attitudinal surveys. Research studies on annoyance from rail traffic noise in residential areas had shown that rail traffic is less disturbing then road traffic noise. The American National Standards Institute [7] and the US National Research Council recommended day-night average sound level (L_{dn}) for the assessment of environmental noise as do most federal agencies and administrations. The spatial modeling of environmental impact assessment is an indispensable tool for acousticians to quantify the impact and correlate with annoyance caused to residential community. The annoyance induced due to traffic noise is not only governed by acoustic factors but also with personal and situational factors and the relationship to the noise source. Figure 9.2 shows a simplified model of annoyance caused by road traffic noise similar to that proposed by Morihara [8]. The attitudinal survey for determining the relation between road traffic noise and annoyance has been done in various studies in various parts of India in the recent past (Chakraborty et al. 2002 [9], Gorai et al. 2006 [10], Agarwal et al. 2009 [11] and Banerjee et al. 2009 [12]). However, not much work had been reported on annoyance characterization and correlating dosage-response relationships in Indian scenario. The relationship between the day-night average sound level, L_{dn}, and percentage of highly annoyed people (% HA) was initiated by Schultz based on the results of 11 of the reviewed surveys. Schultz (1978) recommended the empirical formulation based on social survey concerning community annoyance attributed to the various transportation noise sources as [13]:

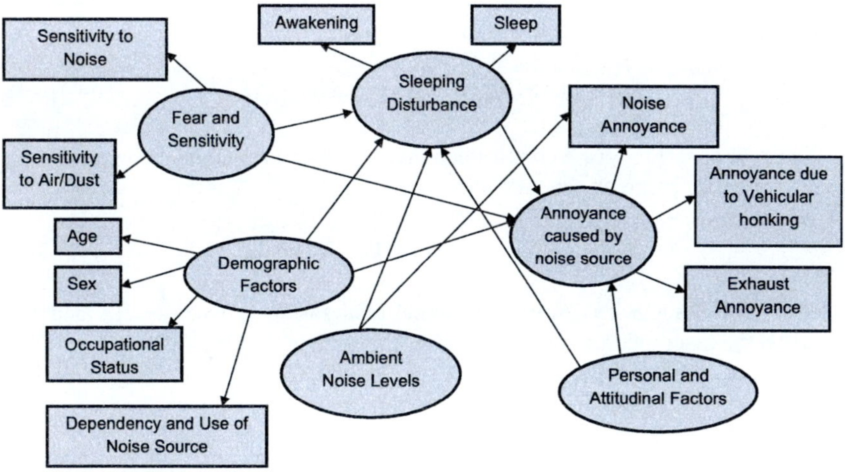

Fig. 9.2. Annoyance model caused by road traffic noise in Indian scenario

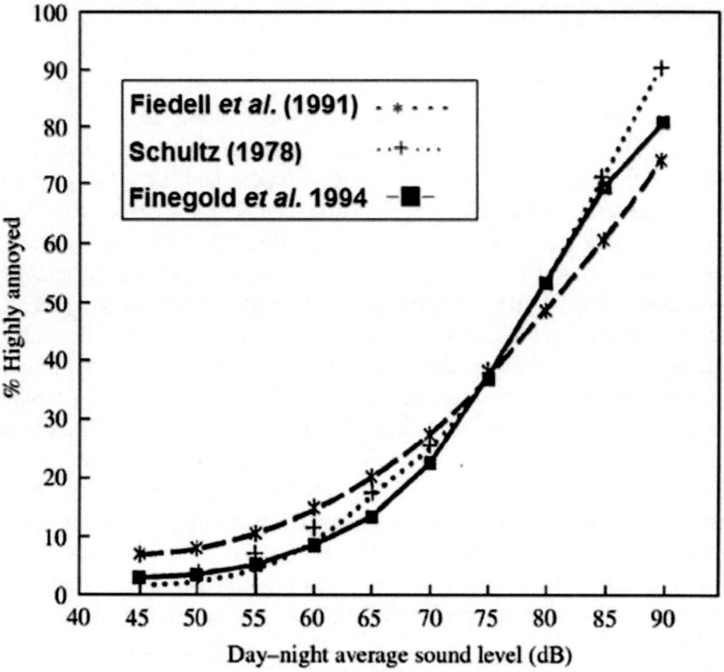

Fig. 9.3 % HA versus L_{dn} curve proposed by Schultz, Fidell, and Finegold (Ouis [18])

$$\%HA = 0.8553\,L_{dn}\text{-}0.0401\,L_{dn}{}^2 + 0.00047\,L_{dn}{}^3 \tag{9.1}$$

The Schultz curve has been updated by Finegold and Fidell et al. In their updated analysis, Fidell et al. 1991 reported a correlation of % HA with L_{dn} using the following least squares quadratic fitting function as [14, 15]:

$$\%HA = 78.9181\text{-}3.2645\,L_{dn} + 0.0360\,L_{dn}{}^2 \tag{9.2}$$

Finegold et al. [15, 16] recommended the following logistic fit as the prediction curve of choice, based on the final 400 data points:

$$\%HA = 100/1 + \exp\,[(11.13\text{-}0.14\,L_{dn})] \tag{9.3}$$

Ouis (2002) suggested another empirical relationship between % HA and L_{dn} based on the assumption that annoyance level is zero at 42 dB as [17]:

$$\%HA = 0.24 \times (L_{dn}\text{-}42) + 0.0277 \times (L_{dn}\text{-}42)^2 \tag{9.4}$$

Figure 9.3 shows the curves representing the percentage of people highly annoyed by noise as a function of day-night average sound level [18].

Table 9.1. % A and % HA approximations for day-night average sound level and day-evening-night average sound level (Miedema and Oudshoorn [19])

Source	Day-night average sound level, L_{dn}	Day-evening-night average sound level, L_{den}
% Annoyance (% A)		
Aircraft	$1.460 \times 10^{-5}(L_{dn}-37)^3 + 1.511 \times 10^{-2}(L_{dn}-37)^2 + 1.346 \times (L_{dn}-37)$	$8.588 \times 10^{-6}(L_{den}-37)^3 + 1.777 \times 10^{-2}(L_{den}-37)^2 + 1.221 \times (L_{den}-37)$
Road traffic	$1.732 \times 10^{-4}(L_{dn}-37)^3 + 2.079 \times 10^{-2}(L_{dn}-37)^2 + 0.566 \times (L_{dn}-37)$	$1.795 \times 10^{-4}(L_{den}-37)^3 + 2.110 \times 10^{-2}(L_{den}-37)^2 + 0.5353 \times (L_{den}-37)$
Railways	$4.552 \times 10^{-4}(L_{dn}-37)^3 + 9.400 \times 10^{-3}(L_{dn}-37)^2 + 0.212 \times (L_{dn}-37)$	$4.538 \times 10^{-4}(L_{den}-37)^3 + 9.482 \times 10^{-3}(L_{den}-37)^2 + 0.2129 \times (L_{den}-37)$
% Highly Annoyed (% HA)		
Aircraft	$-1.395 \times 10^{-4}(L_{dn}-42)^3 + 4.081 \times 10^{-2}(L_{dn}-42)^2 + 0.342 \times (L_{dn}-42)$	$-9.199\times10^{-5}(L_{den}-42)^3 + 3.932\times10^{-2}(L_{den}-42)^2 + 0.2939\times (L_{den}-42)$
Road traffic	$9.994 \times 10^{-4}(L_{dn}-42)^3 - 1.523 \times 10^{-2}(L_{dn}-42)^2 + 0.538 \times (L_{dn}-42)$	$9.868 \times10^{-4}(L_{den}-42)^3 - 1.436\times10^{-2}(L_{den}-42)^2 + 0.5118\times (L_{den}-42)$
Railways	$7.158 \times 10^{-4}(L_{dn}-42)^3 - 7.774 \times 10^{-3}(L_{dn}-42)^2 + 0.163 \times (L_{dn}-42)$	$7.239 \times10^{-4}(L_{den}-42)^3 - 7.851\times10^{-3}(L_{den}-42)^2 + 0.1695\times (L_{den}-42)$

Miedema et al. developed exposure response equation for L_{dn} and percentage "highly annoyed" [19]. The European Commission Position Paper (2002) proposed changing from the historical use of L_{dn} for predicting community responses to the use of day-evening-night sound level, L_{den} [20]. The use of L_{dn} or L_{den} is supported by Miedema et al. (2000) that investigated which noise metrics best predicts the annoyance from aircraft noise [21]. Miedema and Oudshoorn (2001) proposed some empirical formulations to correlate L_{dn} and L_{den} as [19]:

$$\text{Aircraft}: L_{den} = L_{dn} + 0.6 \tag{9.5}$$

$$\text{Road traffic}: L_{den} = L_{dn} + 0.2 \tag{9.6}$$

$$\text{Railway}: L_{den} = L_{dn} \tag{9.7}$$

The study recommended that value of differences L_{den} - L_{dn} outside the range of 0 to 1.5 dB will be rare. Table 9.1 summarizes these approximations for day-night average sound level and day-evening-night average sound level [19].

The change in % HA for change in L_{den} is thus calculated as:

$$\begin{aligned}\%\text{HA} = 9.868 \times 10^{-4}(L_{den}\text{-}42)^3 - 1.436 \times 10^{-2}(L_{den}\text{-}42)^2 \\ + 0.5118\ (L_{den}\text{-}42) \text{ for road traffic}\end{aligned} \tag{9.8}$$

$$\frac{d(\%HA)}{d(L_{den})} = 2.9604 \times 10^{-3}(L_{den}\text{-}42)^2 - 0.0287\ (L_{den}\text{-}42) + 0.5118 \tag{9.9}$$

$$d(\%HA) = d(L_{den}) \times \left(2.9604 \times 10^{-3}(L_{den}\text{-}42)^2 - 0.0287\,(L_{den}\text{-}42) + 0.5118\right) \quad (9.10)$$

Thus, for change of $\Delta\ L_{den}$ by 3 dB(A) for a site having 70 dB(A) L_{den}, the Δ % HA changes by 6.1%. However, change of $\Delta\ L_{den}$ by 3 dB(A) for an site having 50 dB(A) L_{den}, the Δ % HA changes by 1.4%.

9.2.1 Low Frequency Noise Annoyance

The assessment of road traffic noise is reported in terms of A-weighted sound pressure level in most of the national standards. However, a study reported that at equal dB(A) levels, the noise dominated by the low-frequency component was perceived as 4–7 dB louder and 5–8 dB more annoying than the other noise [22]. Vos [23] and Kjellberg et al. [24] suggested the indicators based on C- and A-weighted levels for annoyance of impulse sounds and occupational noise. Studies showed that A-weighted sound pressure level is a useful indicator of short-term loudness and annoyance of road-traffic noise without prominent low-frequency content [25, 26]. Low-frequency content had been observed to influence the perceived loudness and annoyance of road-traffic sound. Sounds with similar A-weighted sound pressure levels having greater low-frequency content ($L_{C\text{-}A}$) in general were louder and more annoying [26]. Nilsson (2007) reported that the change in perceived loudness or annoyance associated with a 1 dB change in difference between C- and A-weighted sound pressure levels ($L_{C\text{-}A}$) corresponded to approximately a 0.4 dB change in A-weighted sound pressure level [26].

9.2.2 Noise Annoyance Models in Indian Scenario

In Indian scenario, only few studies had been reported on correlating the noise annoyance with the noise exposure levels particularly in the metropolitan cities of India. A health survey reported by Agarwal and Swami (2011) for Jaipur city revealed that about 52% of population was suffering from frequent irritation, 46% respondent felt hypertension, and 48.6% observed loss of sleep due to noise pollution [27]. Mean dissatisfaction score (MDS) had been proposed in order to define the noise annoyance quantitatively. Vehicular traffic had been observed to be the main source of noise pollution and annoyance [28]. Banerjee (2012) presented a comprehensive review of various research conducted in the past two decades on traffic noise impacts in India [29]. The study revealed that road traffic noise is a cause for annoyance to a variety of degree among the respondents. A recent study conducted by Garg et al. (2012) revealed that > 50% people are disturbed from road traffic and horn noise and 17% are annoyed due to aircraft noise [30]. However, in Indian scenario, there is a need of comprehensive studies to be conducted for defining the exposure-effect relationship in different parts of the country. The following issues may be considered while developing these models in Indian perspectives:

- The onset of annoyance of 42 dB as taken in European scenario may not be appropriate in Indian scenario. However, the European models represent the maximum values of% annoyed and% highly annoyed due to the higher noise sensitivity of Indian community as compared to the European community.
- A sound level of 50 dB may be taken as the onset of noise annoyance as it is recommended daytime noise level limit for areas lying under silence zone.

9.2.3 Effect of Demographic Factors on Noise Annoyance

The effect of demographic factors such as sex, age, education level, occupational status, and size of household and attitudinal variables such as the noise sensitivity and fear of the noise source had been a prime consideration focused in many studies so far. Miedema and Vos (1999) study showed that fear and noise sensitivity have a large impact on annoyance, while the demographic factors are much less important [31]. A recent study reported by Mc Sweeney (2016) showed that annoyance response due to vibration from the construction of a light rapid transit (LRT) system was strongly influenced by two attitudinal variables, concern about property damage and sensitivity to vibration. The study recommended the consideration of attitudinal factors as most cost-effective reduction of the impact of construction vibration and annoyance [32]. Noise sensitivity is associated with an elevated susceptibility to non-auditory effects [33] and is singular personal attribute which demonstrably is the most consistent predictor of noise annoyance [34]. The extent of annoyance may also depend on the socio-demographic factors, including gender, age, education, and income levels [35]. Figure 9.4 shows the cause-and-effect analysis diagram for factors affecting the noise annoyance based on the review of previous studies conducted on noise annoyance. The main conclusions derived from Miedema and Vos (1999) study is helpful in understanding the effect of demographic and attitudinal factors on noise annoyance as follows [31]:

- Demographic factors are less important than fear and noise sensitivity. The difference in annoyance between the noise sensitive and highly sensitive persons is equivalent to a L_{dn} difference of 11 dB.
- Age is an important factor rather than gender for annoyance reported due to noise.
- Higher annoyance is reported in case of higher education and the occupational status; dwelling is owned instead of rented, if a person doesn't depend upon the noise source and if the use of transportation noise source is nominal.

However, there had been only a few studies that have considered socio-demographic factors as potential modifiers of the association of interest between

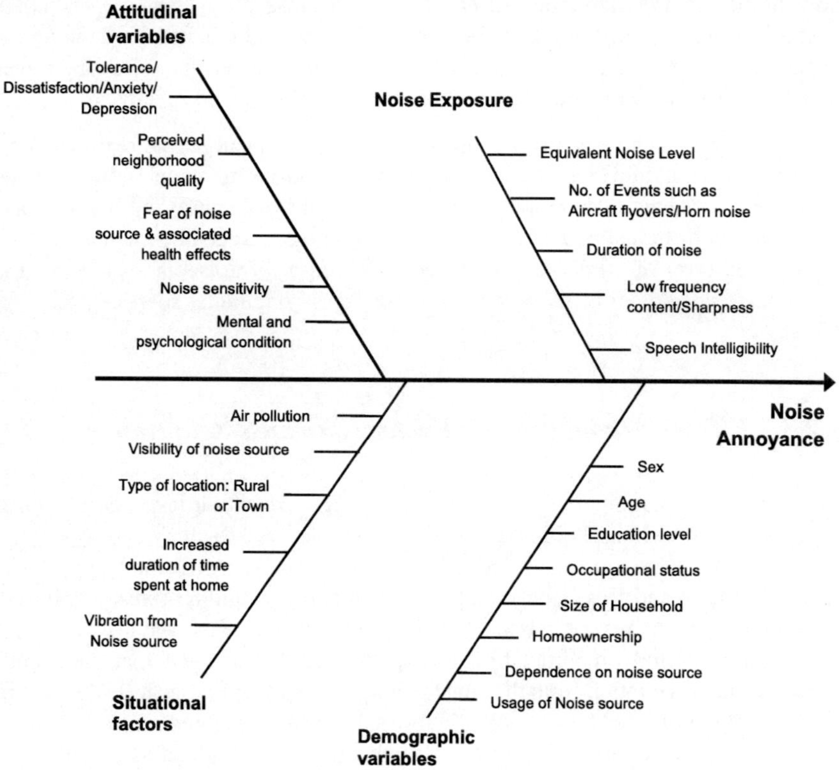

Fig. 9.4 Cause-and-effect analysis diagram for factors affecting the noise annoyance

the transportation noise and annoyance [36–38]. Gradually, ISO/TS 15666:2003 had been published that provides specifications for socio-acoustic surveys and social surveys which include questions on effects of noise [39].

9.2.4 Noise Depreciation Index

Noise Depreciation Index (NDI) is used to predict the annoyance costs related to noise. NDI is calculated as the percent increase in the loss of property values due to a unit increase in noise exposure. These are generally determined using hedonic price (HP) techniques that utilize the trade-off between varying property prices and corresponding ambient noise levels in the real estate market [40]. Schipper et al. (1998) studied noise impact of aircraft using a meta-analysis of 30 previously completed hedonic price studies and estimated the mean NDI to be 0.83 [41]. Levesque (1994) estimated the noise impacts of aircrafts by segregating the noise effects into loudness and event frequency and estimated NDI of about 1.3% [42]. There had been many studies focused on one of the three methods in estimating the noise impact of aircraft: hedonic price method, meta-analysis, or contingent valuation method (based on the willingness to pay), whereby the hedonic price method is the most commonly utilized [43]. Nelson (2004) concluded that the NDI estimates in the USA lies between 0.5% and 0.6%, and in Canada it is around 0.9% [44]. The social costs associated with noise exposure may be categorized as [45, 46]:

- Resource costs in the form of medical and health care that includes costs financed by taxes and direct payments by the individual.
- Opportunity costs in the form of loss of production; the sum of resource and opportunity costs is called "Cost of illness.
- Disutility in the form of other negative influences resulting from noise exposure, estimated by willingness to pay (WTP).

The hedonic price model considers that the property price, P_i, is dependent on four main characteristics classified as: (1) structural characteristics S_i of the house such as size and number of rooms; (2) socio-economic neighborhood N_i; (3) accessibility characteristics A_i such as the nearness to employment and market; (4) specific environmental amenities E_i such as noise and air pollution. The non-explained part of the price is represented by ε_i, a normally distributed error term [47]:

$$P_i = f(S_i, N_i, A_i, E_i) + \varepsilon_i \tag{9.11}$$

The functional forms such as linear, log-log, and semi-log can be applied for the development of model. Lavandier et al. (2016) [47] proposed a new descriptor: real estate tolerance level (RETL) that corresponds to the sound exposure level whereby a property price is 50% depreciated compared to the price of the same property which would be situated in an area whose L_{den} is below 50 dB(A). The methodological approach and unit used to evaluate the economic value of noise annoyance is varied in different countries. In some cases, it is measured in terms of the Noise Depreciation Sensitivity Index (NDSI), calculated as the average percentage change in property prices per decibel. It is also expressed as an economic value per year per person (or household) annoyed by noise as: (a) value per person "highly annoyed" (HA), and (b) value per person "annoyed," independent of the level of annoyance [48]. NDSI for road traffic noise had been reported to vary from 0.08% to 2.22% and from 0.29% to 2.30% for aircraft noise [49].

9.2.5 *Community Tolerance Level*

The immense variability of annoyance prevalence in communities with similar levels of transportation noise exposure sometimes complicates the interpretation of exposure-response relationships for assessment and prediction of transportation noise impact. Fidell et al. (2011) demonstrated that the variability in annoyance prevalence rates is well accounted by a single parameter: Community tolerance level (CTL) represented symbolically as L_{ct} [50]. Community tolerance level, L_{ct}, is defined in ISO 1996-1:2016 [51] as the day-night average sound level at which 50% of the people in a particular community are predicted to be highly annoyed by noise exposure. The L_{ct} parameter is typically used to account for differences among the sources and/or communities when predicting the percentage of individuals highly annoyed by noise exposure [52]. CTL reflects all non-L_{dn} related factors in simple decibel form that influence annoyance judgments for a particular noise annoyance survey data. Equation 9.15 expresses the percent highly annoyed directly as a function of L_{dn} and CTL [53]:

$$\%\text{HA} = 100\,\text{EXP}\left(-\left(1/\left(10^{((L_{dn}-L_{ct}+5.306)/10)}\right)^{0.3}\right)\right) \tag{9.12}$$

Practically, the empirical formulation is developed that minimizes the root mean square error between the equation (9.15) and % HA values observed from socio-acoustic survey of various sites [54].

The exposure-response relationship in each dataset can be estimated using equation 9.16 [52]:

$$\%\text{HA} = e^{-A/m} \tag{9.13}$$

Parameter m is estimated using equation 9.14 as [50, 52]:

$$m = \left(10^{\left(\frac{L_{dn}}{10}\right)}\right)^{0.3} \tag{9.14}$$

The community-specific constant A is a scalar variable whose value is that which minimizes the root mean square error between the predicted and observed annoyance. The L_{ct} value is calculated from A [50, 52]:

$$L_{ct} = 5.31 + 33.33\ \log_{10} A \tag{9.15}$$

The average CTL value for the 43 aircraft noise studies included in the original analysis by Fidell et al. was 73.3 dB. The dose-response curve described by the CTL 73.3 dB function is very similar to the so-called Miedema curve recommended by the EU for assessing noise annoyance. If the CTL value for a certain community is higher than 73.3 dB, this community can be considered more tolerant to noise than the average. CTL values lower than 73.3 dB means less tolerance to noise than the average [55]. With respect to road traffic, aircraft are 5 dB more annoying; conventional trains accompanied by appreciable vibration are about 2–3 dB more annoying; and conventional trains unaccompanied by appreciable vibration are about 9–10 dB less annoying. Table 9.2 shows the predicted annoyance prevalence rates for three levels of noise exposure and three degrees of community tolerance for noise

Table 9.2 Predicted annoyance prevalence rates for three levels of noise exposure and three degrees of community tolerance for noise exposure by aircraft noise

L_{dn}	% HA for $L_{ct} = 73.3$ dB	% HA for $L_{ct} = 66.3$ dB	% HA for $L_{ct} = 80.3$ dB
55 dB	8.6%	1.9%	22.0%
60 dB	17.6%	6.0%	34.3%
65 dB	29.3%	13.6%	46.9%

Source: Reproduced from Fidell et al., 2011, J of Acoustical Society of America 130 (2):791–806 [50] with the permission of the Acoustical Society of America

exposure by aircraft noise (Fidell et al., 2011 [50]). The collection of data by selecting the respondents that are representative for the community of interest is very important, and it is strongly recommended to measure the degree of annoyance for each respondent [56]. A recent study reported by Gjestland et al. (2015) [57] for analyzing the exposure-response relationship based on community tolerance level showed that Vietnam population tolerated 5–10 dB elevated noise levels from road traffic so as to furnish a response similar to the dose response curve recommended in the European Noise Directives [20].

9.3 Noise-Induced Hearing Loss

Noise-induced hearing loss (NIHL) is a public health issue that has become a serious concern in recent years. Hearing loss is one of the most common neurological disorders affecting 10% of the general population and more than 40% of those over 60 years of age [58, 59]. Noise exposure is one of the major causes of acquired senorineural hearing loss in adults [60]. Approximately 5% of the population worldwide suffers from industrial, military, or recreational noise-induced hearing loss (NIHL) at great economic cost and detriment to the quality of life of affected individuals [61]. Noise-induced hearing loss (NIHL) usually affects the higher frequencies (3k, 4k, or 6 kHz) and then spreads to the lower frequencies (0.5k, 1k, or 2 kHz) [62]. It is the most common form of acquired hearing loss after age-related loss (presbycusis) and denotes the permanent hearing loss that develops gradually after prolonged exposure to higher noise levels. Impairment of hearing at high frequencies will initially cause a loss of clarity in perceived speech and then interfere with daily activities as hearing loss progresses [63]. Hearing loss-related symptoms, such as trouble in normal and telephone conversation, turning up the radio/television volume, and tinnitus, usually occur in the early stages of NIHL [64]. The main auditory effects include [65]: acoustic trauma, tinnitus (ringing or buzzing in the ear), temporary hearing loss/TTS, and permanent hearing loss/PTS. Exposure to sound above a level of approximately 85 dB may cause a temporary hearing loss or dullness of hearing known as temporary threshold shift (TTS). A repeated or sustained noise exposure leads to degenerative changes and the TTS becomes permanent threshold shift (PTS). The excessive noise may cause permanent memory

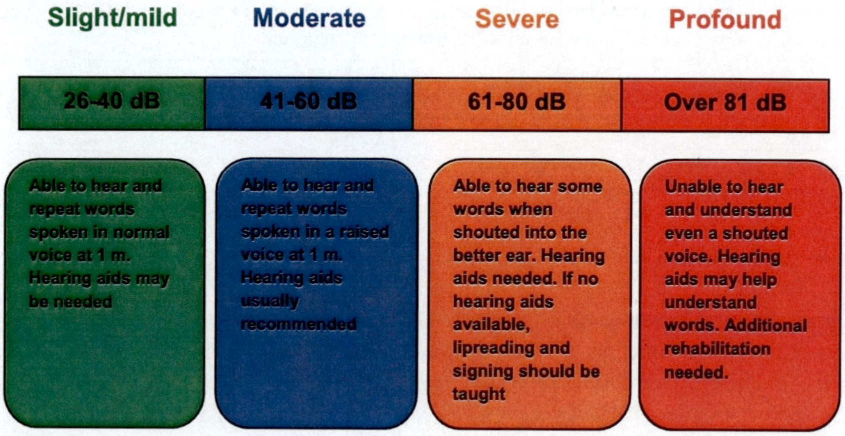

Fig. 9.5 Grades of hearing impairment described by WHO [69]

loss or psychiatric disorder [65]. Two publications in *Indian Journal of Otology* stated that use of mobile phone may lead to sensorineural deafness [66] and if continuously used for more than 30 min, may lead to 10 dB transient hearing loss [67]. There are two types of hearing impairment defined according to where the problem occurs [68]:

- Conductive hearing impairment, which is a problem in the outer or middle ear, which is often medically or surgically treatable; for example, childhood middle ear infection.
- Sensorineural hearing impairment is usually with the inner ear, and occasionally with the auditory nerve going from there to the brain. It is a permanent problem caused due to aging, exposure to high noise levels, etc. and can be solved with a hearing aid.

Hearing impairment may result from a long-lasting buzzing or ringing known as tinnitus and may remain a constant irritation for the listener [68]. The hearing threshold level, using audiometry, is to be taken as the better ear average for four frequencies 0.5, 1, 2, and 4 kHz. There is hardly any or very slight hearing problems observed for 25 dB hearing loss, and as such one is able to hear whispers. The WHO grades of hearing impairment (Fig. 9.5) pertain to the remediation after the acquisition of hearing loss [69]. Prolonged exposure to chronic aircraft noise had been shown to be associated with cognitive impairment in children, but there is also an indication that children may recover from the noise-induced cognitive deficit when the noise exposure stops. Tinnitus can affect quality of life in several ways, including sleep disturbance, depression, or the inability to sustain attention [70]. Some studies such as one reported by Fligor and Cox (2004) [71] enunciated that reasonable guidelines should include a recommendation to limit headphone of personal music players for its use to 1 hour or less per day if using supra-aural style headphones at a gain control setting of 60% of maximum. The Occupational Safety and Health

Administration's (OSHA's) noise standard (29 CFR 1910.95) stipulates employers to have a hearing conservation program in place if workers are exposed to a time-weighted average (TWA) noise level of 85 decibels (dBA) or higher over an 8-hour work shift. OSHA's permissible exposure limit for noise exposure is 90 dB(A) for an 8-hour time-weighted average, and the standard uses a 5 dB(A) exchange rate, which means that when the noise level is increased by 5 dB(A), the amount of time a person can be exposed is cut in half. NIHL has the following characteristics [72–75]:

- It is always sensorineural, mainly affecting the cochlear hair cells in the inner ear.
- Its first sign is a "notching" of the audiogram at the high frequencies of 3000, 4000, or 6000 Hz with recovery at 8000 Hz [73].
- The risk of NIHL is felt to be low at exposures below 85 dB (8-hour time-weighted average) but increases significantly as exposures rise above this level [74].
- It will not progress once noise exposure is stopped.
- Continuous noise is more damaging than intermittent noise.

9.4 Cognitive Performance

Noise pollution is an aspect of the urban environment that had been shown to impair cognitive performance in nonclinical populations [76]. WHO recommended that to avoid hearing impairment, impulse noise exposure should never exceed a peak sound level of 140 dB peak in adults, while daily 1 hour exposure of listening to music through headphones should not exceed 85 dB [77]. Noise that is stressful to one individual may be tolerable to another [78]. Individuals that are highly noise sensitive have been found to have reduced work ability and attention when compared to those who report low noise sensitivity [79]. The cognitive performance of a human being such as attention, short- and long-term memory, and decision-making may be affected to a varying degree by different types of noise stressors [78, 80]. For example, recall and recognition are more impaired by meaningful irrelevant speech compared to traffic noise [80]. In addition, the same type of noise stressor may impair one cognitive domain, while having no effect on another [78]. Some studies have shown that noise has a distracting effect on short-term memory for visual-verbal and visuo-spatial items [81, 82]. A fewer number of studies indicated that intentional memory can be adversely affected by chronic noise exposure (Evans et al., 1995 [83]; Haines, Stansfeld, Job, et al., 2001 [84]; Hygge et al., 2002 [85]; Meis et al., 1998 [86]), although Haines et al. (2001) study did not replicate this effect [87]. Haines et al. (2001) reported that the chronic noise exposure is associated with raised noise annoyance in children. The cognitive study results indicated that chronic aircraft noise exposure does not always lead to generalized cognitive effects, but rather more selective cognitive impairments on difficult cognitive tests in children. There had been numerous studies that had shown that environmental noise exposure has a negative effect on children's learning outcomes and cognitive performance and children with chronic aircraft, road traffic, or rail noise exposure at school have poorer reading ability, memory, and performance on national standardized tests than do children who are not exposed to noise at school [88, 89]. The cross-sectional RANCH (Road-traffic and Aircraft Noise exposure and children's Cognition and Health) study of 2844 children aged 9–10 years attending 89 schools around Heathrow (London, UK), Schiphol (Amsterdam, the Netherlands), and Madrid-Barajas (Spain) airports showed a linear exposure-effect relation between aircraft noise exposure at school and a child's reading comprehension and recognition memory after adjusting for a range of socio-economic factors. A 5 dB L_{eq} increase in aircraft noise exposure was associated with a 2 month delay in reading in children in the UK and 1 month delay in those in the Netherlands [90–92]. Some studies suggested that chronic aircraft noise has a deleterious effect on memory, sustained attention, reading comprehension, and reading ability [93]. A study reported by Clark et al. (2013) [94] revealed that aircraft noise exposure at primary school was associated with a significant increase in noise annoyance and with a nonsignificant decrease in reading comprehension. However, some studies such as the one reported recently by Dreger et al. (2015) [95] revealed that exposure to noise at children's home is associated with mental health problems such as emotional symptoms, conduct problems, and hyperactivity. Thus, conclusions from the various cross-sectional studies are clearly not definitive. However, these inferences should be viewed as suggestive warranting follow-up as suggested by Lercher et al. [89].

9.5 Sleep Disturbances

One of the most serious effects of community noise is sleep disturbance [96]. Night-time noise can negatively affect people's sleep. The relationship between environmental noise and different aspects of sleep, and long-term health effects is a complex one [97]. The repeated disturbances over a whole night may cause cumulative effects as [98]: reduction in sleep duration, slow-wave sleep (SWS) loss pertaining to reduced time spent in deep sleep, and sleep fragmentation due to regular disturbances. Figure 9.6 shows the flow chart on the effects of traffic noise on sleep [99].

The restorative power of sleep may be reduced by repeatedly occurring activations called sleep fragmentation. The amounts of wake and decrease of slow wave sleep (SWS) and REM (rapid eye movement) sleep is associated with number and acoustic properties of the noise events [99]. The standard for measurement of sleep is polysomnography. The Night Noise Guidelines (*NNG*) of WHO [2] concluded that children with disturbed sleep present cognitive dysfunction and behavioral disturbances, abnormal growth hormone release, increase in diastolic BP, and an increased risk of accidents and use of sleeping pills. In case of night noise, 30 dB $L_{night,outside}$ is equivalent to the no observed effect level (NOEL), whereas 40 dB $L_{night,outside}$ is equivalent to the lowest observed adverse effect level (*LOAEL*). Above 55 dB, the cardiovascular effects become the major public health concern. An interim target

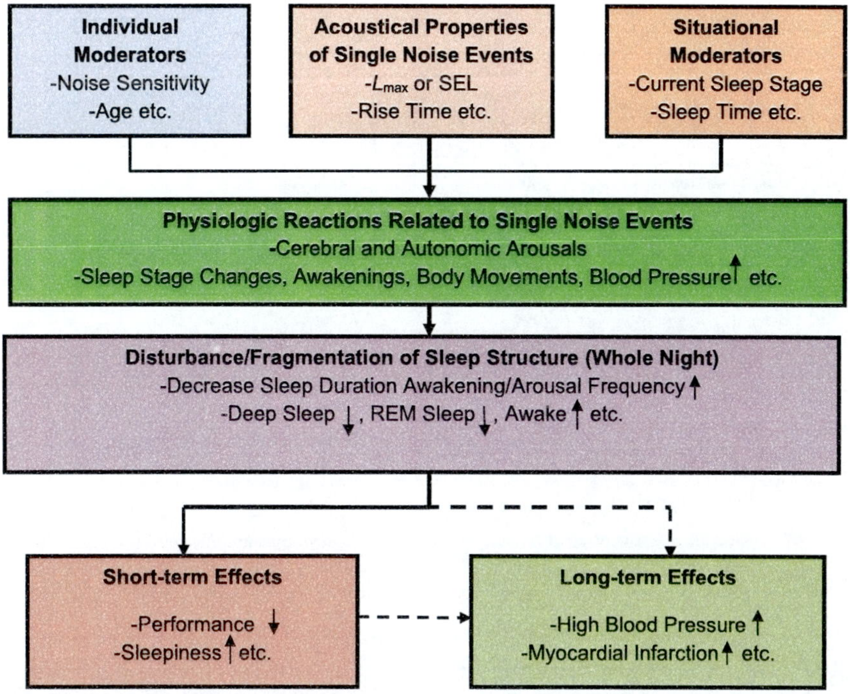

Fig. 9.6 Flow chart on the effects of traffic noise on sleep. L_{max} = maximum sound pressure level, SEL = single event level (Basner et al. 2010 [99])

(IT) of 55 dB $L_{night,outside}$ has been suggested in the situations where the achievement of *NNG* is not feasible in the short run for various reasons. The Community Guidelines 2000 from the WHO recommends 50/55 $L_{Aeq,16\ hrs}$ as health-based threshold [77]. The CALM network with a vision of avoiding harmful effects of noise exposure and preserving quiet areas considered L_{den}/L_{night} values of 50/40 dB as an optimum target [100]. A socio-acoustic survey in Hong Kong among 10,077 respondents [101] observed an exposure-response curve similar to both the Miedema and Vos [102] and Hong et al. [103] curves. The results indicated a similarity in reported sleep disturbance due to road traffic noise between European and Asian countries. Miedema and Vos (2010) study showed that at the same average nighttime noise exposure level, aircraft noise is associated with more self-reported sleep disturbance than road traffic, and road traffic is associated with more sleep disturbance than railways. The association of noise-induced sleep disturbance with age is observed to have an inverse U-shape with the strongest reaction found between 50 and 56 years of age [102]. However, another study reported by Basner et al. (2011) [104] based on polysomnography measures of sleep disturbance analyzed that road traffic noise was associated with greater disturbance than aircraft noise. Only a limited number of previous studies examined the effects of noise on children's sleep [105–108], which had primarily used self-reports of sleep disturbance,

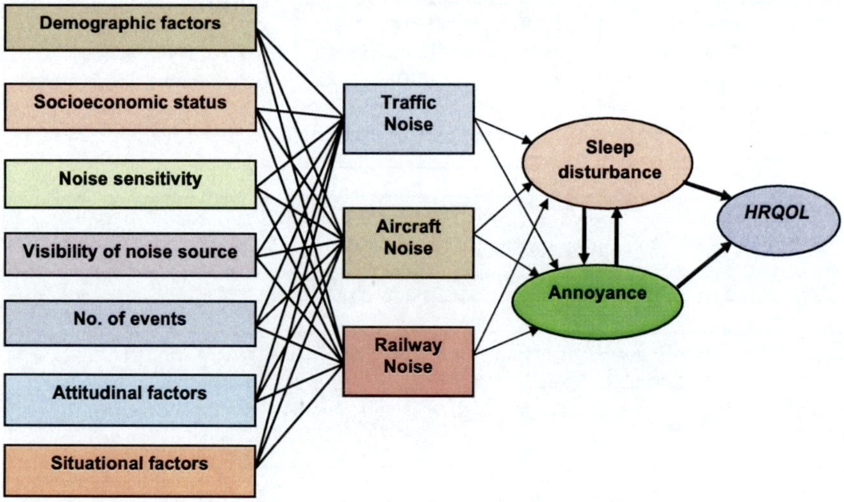

Fig. 9.7 Association between road traffic, aircraft, and railway noise and *HRQOL* (Health Related Quality of Life)

and further studies using subjective and objective assessments are required [109]. The recent study reported by Foraster et al. (2016) [110] emphasized that noise annoyance might lead to behavioral patterns not captured by noise levels, which could reduce the physical activity either directly or through impaired sleep and constitute a noise pathway toward cardiometabolic diseases. A recent study in Western India (in Asansol industrial city) supported an association between higher nocturnal road traffic noise and self-reported sleep disturbance in comparison to the areas with quieter nights. Further sleep disturbance among women was significantly associated with higher night noise exposure [111]. Poor sleep is associated with a number of cardiometabolic, psychiatric, and social negative outcomes both in adults and children. Nocturnal environmental noise clearly affects sleep architecture as well as subjective sleep quality [112]. Poor sleep may act as a mediator between nocturnal noise pollution and increased risk of cardiovascular morbidity, through impaired endocrine and metabolic functions [113]. The various factors that affect noise-disturbed sleep must also be identified for correlating with annoyance and potential health effects. Figure 9.7 shows the factors affecting noise annoyance, sleep, and finally the Health Related Quality of Life (*HRQOL*) based on the review of some recent studies [114, 115].

Fong et al. (2018) conducted a multivariable analysis that showed that a 1 dB increase in noise tolerance was associated with a 0.1-unit increase in the degree of feeling refreshed after sleep on a 0–10 scale after adjusting for other factors such as age, education, stress, anxiety, and depression. A 1 dB(A) increase in most comfortable level (MCL) of hearing was associated with feeling 1% more refreshed after sleep [116]. The German Aerospace Center (DLR) investigated nighttime noise effects on sleep due to the introduction of a night curfew (11:00 p.m.–5:00 a.m.)

in the *NORAH* (Noise-Related Annoyance, Cognition and Health) study at Frankfurt Airport and previously conducted STRAIN (Study on Human Specific Response to Aircraft Noise) sleep study at Cologne–Bonn Airport in 2001. A considerable increase in the annoyance probability with increasing number of overflights was observed reaching approximately 80% at 80 fly-overs per night [117]. A recent study on Wind Turbine Noise Effects on Sleep (WiTNES) revealed that the presence of beats and amplitude modulation were the important constituents of wind turbine noise contributing to sleep disruption [118]. Vibration and noise from railway freight traffic was also observed to impact more on sleep frequent awakenings and increased wakefulness [119]. Thus, advancement in knowledge and understanding on metabolic, vascular, and neuro-physiological effects of disturbed sleep could pave way for future research on long-term effects of nighttime noise on health aspects such as obesity, cardiovascular disease, and neurodegenerative disorders [109, 120–122]. The publications of WHO, "Night Noise Guidelines for Europe" and "Burden of disease from environmental noise," both stressed on the importance of nocturnal noise exposure for health and well-being. The summary of the findings from the past three year's research by Humme et al. (2012) on the effects of environmental noise on sleep suggested that nocturnal noise exposure may be more relevant for the genesis of cardiovascular disease than daytime noise exposure. However, studies signifying a causal pathway that directly link noise and sleep disturbance with cardiovascular disease and other health issues are still missing [123]. Thus, further longitudinal studies exploring the correlation between noise-induced sleep disturbances and its other health outcomes such as with cardiovascular effects are required for establishing definite relationships.

9.5.1 Noise Metrics and Sleep Disturbance Measures

The EU Directive (END) specified L_{night} as indicator of sleep disturbance [124]. The L_{night} limit values for road traffic noise in new residential areas recommended in some countries range from 40 dB for Latvia and Netherlands to 62 dB for France [2]. The effects of noise on sleep and awakenings are related to individual event noise descriptors such as maximal noise level ($L_{\max}$) or sound exposure level [125]. The relationships for self-reported sleep disturbance were based on 12,000 individual observations of exposure–response combinations, from 12 field studies [125, 126]. The curves were based on L_{night} (outside, most exposed facade) levels ranging from 45–65 dB(A). Sleep disturbance is measured by means of a questionnaire with questions regarding sleep disturbance with all set of response categories translated into a scale from 0 to 100. When the cut-off is 72 on scale 0 to 100, then it is percentage "highly sleep disturbed" persons (%HSD); with a cut-off at 50, it is sleep disturbed (% SD); and with a cut-off at 28, it is percentage little sleep disturbed (% LSD). The formulae of these polynomial approximations are as follows (SD = sleep disturbance; H = high; L = low) [2]:

$$\%HSD = 20.8 - 1.05\,L_{night} + 0.01486\left(L_{night}\right)^2 \text{for road traffic} \tag{9.16}$$

$$\%HSD = 18.147 - 0.956\,L_{night} + 0.01482\left(L_{night}\right)^2 \text{for aircraft} \tag{9.17}$$

$$\%HSD = 11.3 - 0.55\,L_{night} + 0.00759\left(L_{night}\right)^2 \text{for railways} \tag{9.18}$$

The *NNG* provides this relation between reported awakenings and noise level as [2]:

$$\begin{aligned}&\text{Percentage of noise-induced awakenings}\\&= -0.564 + 1.909 \times 10^{-4}\left(SEL_{inside}\right)^2\end{aligned} \tag{9.19}$$

where SEL_{inside} is the sound exposure level of an aircraft noise event in the bedroom. This relation is confined to commercial aircraft noise over the intervals $54<SEL<90$ ($37<L_{Amax}<82$) and the number of events per night $1< N< 10$.

9.6 Cardiovascular Effects

Exposure to noise may result in stress reaction with activation of the sympathetic and endocrine system that causes changes in blood pressure (BP), heart rate, and release of stress hormones [127–130]. Noise exposure has been reported to increase the systolic and diastolic blood pressure, affect the heart rate, and lead to the release of stress hormones (including catecholamines and glucocorticoids) [131]. Basner et al. (2014) [92] reported an increase in risk of between 7% and 17% per 10 dB increase in equivalent noise level L_{Aeq} based on review from previous studies. Münzel et al. (2018) [132] review highlighted the process underlying noise-induced vascular damage. The nighttime noise augments the stress hormones levels and vascular oxidative stress thus causing endothelial dysfunction and arterial hypertension [132]. Noise affects the hypothalamus-pituitary-adrenal axis leading to augmented levels of cortisol [133]. The most important meta-analysis reported by Vienneau et al. (2015) [134] observed a 6% significant increase in risk for every 10 dB (A) increase in traffic noise (L_{den} in dB(A)) initially as low as 50 dB(A). Some cohort studies observed noise pollution to be significantly associated with coronary heart disease [135] after the adjustment of other confounding factors such as air pollution. Some of the studies find the association between road traffic noise and cardiovascular disease to be relatively independent of air pollution [136, 137]. Cardiovascular effects of noise rank second in terms of disability adjusted life years (*DALYs*) after annoyance. Lercher et al. (2011) [138] findings based on review of many studies supported the relevant moderation by age, gender, and family history and suggested a strong need for consideration of nonlinearity in the exposure-response analysis. High blood pressure (BP) is a major risk factor for cardiovascular disease and, therefore, even small increment in BP from road traffic noise may have high impact on public health [139]. Sleep disturbances also contribute to cardiovascular risk [140, 141], and it is, therefore, believed that night noise exposure is more detrimental than daytime exposure [2]. There is no evidence available on the relationship between noise and stroke [5]. Chronic noise exposure is associated with adverse pathophysiological effects and may contribute to the progression of cardiovascular disease [142]. Hypertension has been valued using the odds ratios presented by WHO (2011). Willich et al. (2006) [143] reported a marginal increased risk of myocardial infarction associated with annoyance by environmental noise in women (adjusted odds ratio 1.47, 95% confidence interval 0.95–2.25, $P = 0.081$) but not in men, and not associated with annoyance by work noise. The "Good Practice Guide" published by European Environmental Agency (EEA) stipulates the threshold for hypertension as that 50 dB (L_{den}) outdoors and that for ischemic heart disease as 60 dB (L_{den}) [144]. Banerjee (2014) meta-analysis based on studies conducted during last three decades observed that the pooled meta-analysis results do not reveal a statistically significant association between exposure to traffic noise and ischemic heart disease, although the relationship is positive. The risk estimate (95% CI) have increased over the three decades from 0.97 (0.37–1.67) in 1980s to 1.28 (1.05–1.62) in 2010s, which supports the facts that the risk of cardiovascular disease due to exposure to the transportation noise has increased. Aircraft noise is positively and significantly associated to prevalence of IHD cases, compared with road traffic as the pooled subgroup effect estimate yielded a positive association [145]. Cardiovascular diseases are more frequent in middle-aged male subjects [146]. Babisch et al. (2005) [147] reported a hospital-based study in the city of Berlin and observed no higher MI risk among women. However, one of the confounding factors pertaining to the hormonal/menopausal status was not considered, as it dilutes the effect [148, 149]. Sørensen et al. (2012) [150] reported that residential exposure to road traffic noise was associated with a 12% higher risk for incident myocardial infarction (MI) per 10 dB higher noise exposure. van Kempen et al. (2017) [151] reported a systematic and comprehensive review in terms of relative risk (RR) per 10 dB L_{den} calculated from odd ratio per noise unit as explained in National Institute for Public Health and the Environment, The Netherlands report in 2017 [151]. An RR of 1.05 (95% CI: 1.02–1.08) per 10 dB L_{den} for the association between road traffic noise and the prevalence of hypertension and RR of 1.09 (95% CI: 1.04–1.15) per 10 dB L_{den} for the association between aircraft noise and the prevalence of ischemic heart disease (IHD) were reported. A significant increase in waist circumference of 3.46 (95% CI: 2.13–4.77) cm during 8 to 10 years of follow-up was associated with an increase in air traffic noise of 10 dB L_{den} [152].

The recent study named as Hypertension and Exposure to Noise near Airports (HYENA) funded by European Commission was carried out in six European countries with an objective of analyzing the relationship between aircraft or road traffic noise near airports and the risk of hypertension. For nighttime aircraft noise (L_{night}), a 10 dB(A) increase in exposure was associated with an odds ratio of 1.14

(95% CI: 1.01–1.29). For road traffic noise ($L_{eq,24h}$), a 10 dB(A) increase in exposure was associated with an odds ratio of 1.10 (95% confidence interval: 1.00–1.20). The exposure-response relationships for road traffic noise were stronger for men with an odds ratio of 1.54 (95% CI: 0.99–2.40). The study results indicated that excess risks of hypertension related to long-term noise exposure, primarily for nighttime aircraft noise and daily average road traffic noise [153, 154]. The recent study reported by Evrard et al. (2017) [155] confirmed the findings of the HYENA study suggesting the effect of aircraft noise on the risk of hypertension was stronger in men than in women. Babisch (2006 [156]) reported that below 60 dB(A) for the road traffic noise level during the day (L_{day}: 6–22 hr), no notifiable increase in myocardial infarction (MI) risk could be detected. For noise levels greater than 60 dB(A), the MI risk increases continuously, with relative risks (odds ratios) ranging from 1.1 to 1.5 (in reference to ≤60 dB(A)). Most of the epidemiological studies but not all (Floud [157] and de Kluizenaar [158]) have reported associations between chronic exposure to environmental noise and health outcomes. The empirical formulation recommended in "Burden of Disease from Environmental Noise Report" (2011) [5] associated the road traffic noise and incidence of myocardial infarction as:

$$OR = 1.63 - 6.13 \cdot 10^{-4} \cdot L^2{}_{day,16h} + 7.36 \cdot 10^{-6} \cdot L^3{}_{day,16h} \tag{9.20}$$

where $L_{day,16h}$ is 16 hours ambient day level and OR is the odds ratio used to compare the relative odds of the occurrence of the outcome of disease for a given noise exposure. An OR>1 represents exposure associated with higher odds of outcome, while an $OR = 1$ represents that exposure does not affect odds of outcome [159]. A cohort study reported by Gan et al. (2012) [160] showed a 10-dB(A) elevation in residential noise levels was associated with a 9% increase in the risk of death from coronary heart disease (CHD). The findings suggested that both community noise and traffic-related fine particulate air pollution indicated by black carbon concentration may be partly responsible for the observed associations between road traffic exposure and adverse cardiovascular outcomes [160]. Some studies had reported that the effects of noise on cardiovascular disease are strongest among men [147, 153, 161], whereas others had indicated no sex disturbances [162, 163]. Also, with regard to effect modification by age, the results are inconclusive.

Figure 9.8 summarizes the potential confounding factors apart from noise source for risk of coronary heart disease and cardiometabolic diseases based on review of previous studies. Pyko et al. (2017) [164] reported an association between road traffic noise and increase in waist circumference of 0.03 cm/year (95% CI: 0.02–0.05) per 10 dB L_{den}. For aircraft noise, the waist circumference increase was estimated as 0.14 cm/year (95% CI: 0.12–0.16) per 10 dB L_{den}. No clear association was observed between railway noise and waist circumference increase. Christensen et al. (2015) study [165] revealed that baseline body mass index (BMI) and waist circumference were significant effect modifiers for the association between road traffic noise and waist circumference, with gain in waist circumference only among

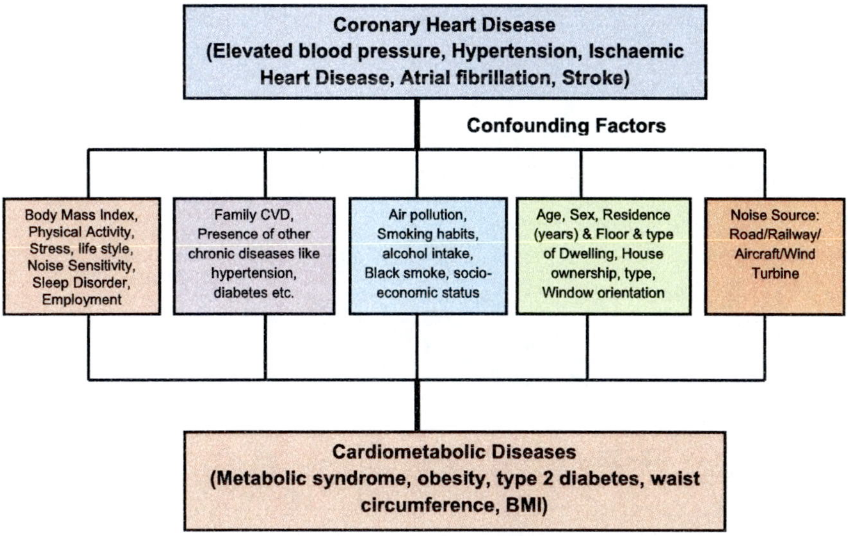

Fig. 9.8 Potential confounding factors apart from noise source for risk of coronary heart disease and cardiometabolic diseases

the obese (BMI≥30) participants (1.20 mm/year (95% CI: 0.68; 1.72)) and participants with a large waist circumference (0.83 mm/year (95% CI: 0.42; 1.23)). van Kempen et al. (2018) [152] reviewed eight studies investigating the impact of noise on children's blood pressure and concluded that the quality of evidence supporting an association between traffic noise and blood pressure in children as "very low" indicating that any estimate of effect is very uncertain.

Exposure to road traffic noise exposure has been associated with cardiometabolic diseases, for example, elevated blood pressure and glucose dysregulation. Long-term exposure to road traffic noise was associated with increased incidence of hypertension and diabetes mellitus in Toronto, Canada. The results showed that every 10-dBA increase of long-term exposure to road traffic noise was associated with an 8% increased risk of incident diabetes mellitus and a 2% increased risk of incident hypertension among individuals, aged 35–100 years, who resided in Toronto [166]. Münzel et al. (2017) [167] reviewed the association between noise exposure and health effects such as ischemic heart disease, including fatal myocardial infarction and stroke, and concluded that the odd's ratios describing these relationships are in the range of 1.07–1.17 per 10 dB increase. Figure 9.9 shows the exposure-response relationships of the associations between transportation noise and cardiovascular health outcomes as described by Babisch (Cardiovascular Effects of Noise on Man-Wolfgang Babisch, https://acoustics.org/cardiovascular-effects-of-noise-on-man-wolfgang-babisch) and Münzel et al. 2017 [167]).

Münzel et al. (2018) [168] pathophysiological insight into the mechanism of noise-induced health effects stipulated that noise exposure leads to oxidative stress, vascular dysfunction, autonomic imbalance, and metabolic abnormalities, thus

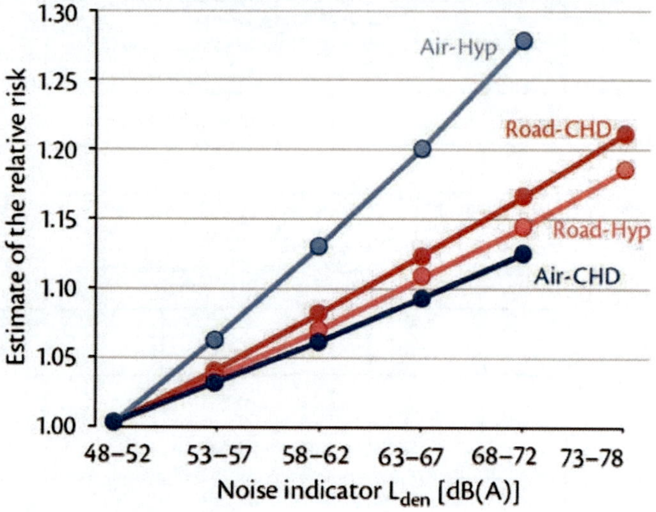

Fig. 9.9 Exposure–response relationships of the associations between transportation noise and cardiovascular health outcomes. Air, aircraft noise; CHD, coronary heart disease; Hyp, hypertension; Road, road traffic noise (https://acoustics.org/cardiovascular-effects-of-noise-on-man-wolfgang-babisch, Münzel et al. 2017 [167])

increasing the risk of cardiometabolic diseases such as arterial hypertension, coronary artery disease, diabetes mellitus type 2, and stroke. Some of the recent studies utilize the GRADE methodology (Grading of Recommendations Assessment, Development and Evaluation) that ranks the quality of evidence as high, moderate, low, or very low to deduce the body of evidence for each noise source and outcome. A recent systematic review study of UK Department for the Environment, Food and Rural Affairs (Defra) shows a low-quality evidence for detrimental effect of traffic noise on medication use and interview measures of depression and anxiety and on cancer outcomes [169]. Stronger associations with hypertension had been observed for the indoor noise as compared to the outdoor noise [170]. Thus, the epidemiological studies suggested a higher risk of cardiovascular diseases, in subjects chronically exposed to high levels of road traffic or aircraft noise. Research work focused on noise exposures relatively independent of air pollution such as aircraft noise and railway noise is pivotal in ascertaining about traffic noise as an independent predictor of cardiovascular risk [167]. The common set of covariates considered in the descriptive and analytical studies had been: age, sex (males only) social class, body mass index, smoking, family history of ischemic heart disease, physical activity during leisure time, and prevalence of preexisting diseases, school education, employment status, shift work, smoking, etc. [5]. Figure 9.10 summarizes the effects of noise on health and well-being with sufficient evidence and thresholds in terms of acoustic indicators as highlighted in European Environmental Agency (EEA) technical report (No. 11/2010 [144]). The threshold levels in terms of descriptors L_{den} and L_{night} are defined as the outside noise exposure levels.

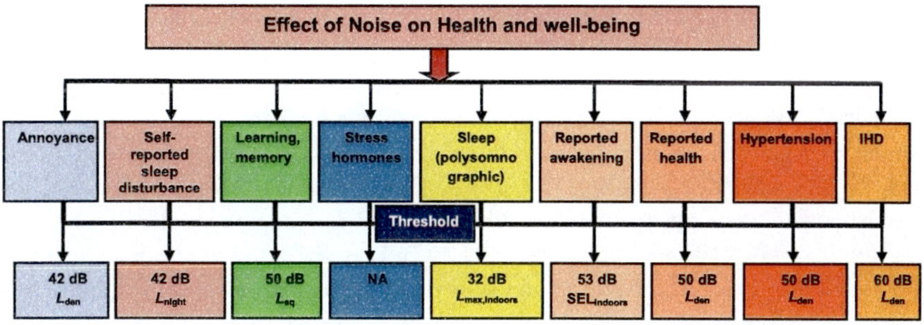

Fig. 9.10 Effects of noise on health and well-being (European Environmental Agency (EEA) Technical report No. 11/2010 [144])

9.6.1 Studies in Indian Scenario

Cardiovascular disease, especially coronary heart disease (CHD), is a type of plague in India. Noise pollution is increasing in Indian cities with alarming increase in vehicular population. However, a limited number of studies have been reported in Indian scenario for ascertaining the health effects of noise pollution. Banerjee et al. (2014) [171] study suggested the epidemiological evidence that exposure to road traffic noise of $L_{den} > 65$ dB(A) may be associated with occurrence of CHD in adult subjects. The subjects in ground or first floor apartments were observed to be at more risk for coronary heart disease than those residing on higher floors. Singh and Davar (2004) also reported the impact of noise pollution on human health leading to stress, high blood pressure, etc. and a general reduction in quality of life [172]. However, due to paucity of data pertaining to health effects of noise pollution, more studies need to be done for understanding the adverse effects of noise on health of exposed residents, especially on sleep disturbance, hypertension, abnormal heart beat, and stress, and also to develop exposure-effect relationships for conducting environmental impact assessment (EIA) studies in reference to outdoor noise exposure level.

9.7 Hospital Noise

Medical equipment, for example, infant incubators, anesthesia units, and alarms, are noise sources that produce high noise levels. Infant incubators are used mainly for premature infants and for newborns to regulate the body temperature. In some instances, it may be required that they stay in the incubators for a prolonged period of time. While operating, if the incubator produces a large amount of sound pressure level, then it may cause hearing loss. The acceptance criteria defined in IEC 60601-2-19:2009 standard (Particular Requirements for the Basic Safety and Essential Performance of Infant Incubators) require that sound pressure level in the infant incubator should not exceed 60 dB(A). A recent study in Thailand [173] for ascertaining the sound pressure level of 26 infant incubators used in three hospitals showed that sound pressure level was within range of 40 dB(A) to 60 dB(A). The majority of incubators (65.4%) generated sound pressure level from 45 dB(A) to 50 dB(A). In one such study, measurements near the patients showed average L_{Aeq} values of 53–58 dB. Ninety percent of the time, the L_{AFMax} levels exceeded 50 dB and $L_{C,Peak}$ exceeded 70 dB [174]. The WHO guidelines recommended that for wardrooms in hospitals, the indoor values are 30 dB L_{Aeq} together with 40 dB L_{Amax} during the night (with instrument set at "Fast"). Hospital noise could be an increasing threat to patient rehabilitation and staff performance. Annoyance, irritation, and headaches have been reported by many staff exposed to the noisy workplace environment. A recent study reported by Khaiwal et al. (2016) [175] conducted in a 1800 bedded tertiary medical institute in India showed that about three-fourths (74%) reported irritation with loud noise, whereas 40% of respondents reported headache due to noise. Less than one-third of respondents (29%) reported loss of sleep and 8% reported hypertension due to noise. One such study reported by MacLeod et al. (2007) that involved the treatment of hospital unit with sound-absorbing panels of 2 inch thick fiberglass wrapped in an antibacterial fabric installed in ceiling and corridor walls reduced the noise by 5 dB(A) [176].

9.8 Environmental Noise Guidelines for Europe

The WHO report on burden of disease from environmental noise estimated annoyance to be the second major health effect of environmental noise after sleep disturbance [5]. The WHO Regional Office for Europe has recently developed some guidelines with an objective to provide recommendations for shielding human health from noise exposure due to different sources: transportation (road traffic, railway, and aircraft) noise, wind turbine noise, and leisure noise. For average noise exposure, the Guidelines Development Group (GDG) strongly recommended reducing noise levels produced by road traffic below 53 dB L_{den}, that for railway as 54 dB L_{den}, and for aircraft noise as 45 dB L_{den} as shown in Table 9.3. The GDG recommended reducing noise levels produced by wind turbines below 45 dB L_{den} and that from all leisure noise sources combined to 70 dB $L_{Aeq,24h}$ as higher leisure noise levels are related with undesirable health effects [177].

Four guiding principles have been recommended: the first principle is to reduce noise exposure to the population; the second principle is to promote interventions to reduce noise exposure; the third principle is to synchronize various approaches to

Table 9.3 WHO environmental noise guidelines for reducing the noise below the recommended levels (WHO 2018 [177])

Noise indicator	Road	Rail	Air
L_{den}	53	54	45
L_{night}	45	44	40

Table 9.4 Exposure response relationships between exposure to transportation noise sources and annoyance and highly sleep disturbed and probability of wake [177–179]

Model	Source	References
% Highly annoyed (HA)		
% HA = $78.9270 - 3.1162 \times L_{den} + 0.0342 \times L_{den}^2$	For traffic noise	WHO report on Environmental Noise Guidelines for the European Region (2018) [177] and Guski et al. (2017) [178]
% HA = $38.1596 - 2.05538 \times L_{den} + 0.0285 \times L_{den}^2$	For railway noise	
% HA = $-50.9693 + 1.0168 \times L_{den} + 0.0072 \times L_{den}^2$	For aircraft noise	
% Highly sleep disturbed (HSD)		
% HSD = $19.4312- 0.9336\ L_{night} + 0.0126\ L_{night}^2$	For traffic noise	
% HSD = $67.5406\text{-}3.185\ L_{night} + 0.0391\ L_{night}^2$	For railway noise	
% HSD = $16.79\text{-}0.9293 \times L_{night} + 0.0198 \times L_{night}^2$	For aircraft noise	
Probability of wake		
Prob. of Wake =$-3.3188\text{-}0.0478\ L_{AS,max} + 0.0037\ (L_{AS,max})^2$	For traffic noise	Basner and McGuire (2018) [179]
Prob. of Wake = $-1.7768\text{-}0.0529\ L_{AS,max} + 0.0033\ (L_{AS,max})^2$	For railway noise	
Prob. of Wake = $-3.0198\text{-}0.0449\ L_{AS,max} + 0.0034\ (L_{AS,max})^2$	For aircraft noise	

control noise sources and other environmental health risks; and the fourth principle is to inform and involve communities (such as environmental and public health experts with policy-makers and citizens) that may be affected by a change in noise exposure. The GDG recommended using the term "environmental noise" instead that account for all noise sources except occupational noise source in workplaces. Table 9.4 shows the association between exposure to transportation noise sources and annoyance (% HA and % HSD) based on Guski et al. (2017) analysis [178].

9.9 Noise-Induced Pathophysiological Effects: Summary

The Nobel Prize laureate Robert Koch said in 1910 that "One day man will have to fight noise as fiercely as plague and cholera" [180]. It is evident that epidemiological studies have shown that noise from aircraft, road traffic, and rail traffic represents a significant risk factor for the development of cardiovascular disease [181]. The key conclusions derived from a review study clearly show the noise-induced pathophysiological effects [181]:

- Epidemiological studies showed that traffic noise raises the risk of cardiovascular diseases such as hypertension, coronary heart disease, myocardial infarction, heart failure, and stroke.
- Sustained noise causes stress reactions that bring about activation of the autonomic and endocrine system, resulting in increased cardiovascular risk
- At the pathophysiological level, particularly traffic noise at night causes oxidative stress and inflammatory processes in the brain, which in turn impact negatively on the vascular system and result in endothelial dysfunction and hypertension.

A recent study [152] of WHO in which the investigators analyzed longitudinal studies observed that road traffic noise, starting at 50 dB(A) and per increase of 10 dB(A), raised the incidence of coronary heart disease by 8% [1.01; 1.15]. Analysis of a Danish cohort study showed that road traffic noise, starting at 42 dB (A) and per increase of 10 dB(A) L_{den}, increases the incidence of myocardial infarction by 12% (incidence rate ratio [IRR] [1.02; 1.22]) independent of air NO_X concentration, smoking behavior, education, and style of nutrition [150]. A similar analysis in this study [150] pertaining to stroke for noise levels starting at 55 dB (A) and per increase of 10 dB(A) L_{den}, road traffic noise increases the risk of stroke by 14% (IRR [1,03; 1,25]). Long-term exposure to aircraft noise, particularly during the night, is associated with incident hypertension and possibly, also, cardiovascular effects. The *OR* for hypertension per 10 dB increase in L_{night} aircraft noise exposure was 2.63 (95% CI 1.21 to 5.71) observed in a cohort study in Athens, Greece [182]. Recent epidemiological evidences indicate that environmental noise exposure may not only be associated with cardiovascular but also with cardiometabolic outcomes. A prospective longitudinal German cohort study [183] from 3,396 participants of age 45–75 years investigated the effect of outdoor and indoor residential road traffic noise on incident type 2 diabetes mellitus (T2DM). Long-term exposure to indoor and outdoor road traffic noise may increase the risk of developing T2DM, independent of air pollution exposure. A 10-dB(A) increase in outdoor road traffic noise (L_{den}) was associated with an RR of 1.09 (95% CI: 0.96–1.24) for T2DM in the fully adjusted model. Potential biological pathways of traffic noise–related health effects include noise-induced stress, activating the autonomic nervous system causing noise annoyance [183]. Some studies have also shown that both short-term and long-term exposure to environmental traffic noise were associated with an increase in body mass index (BMI) and/or waist circumference (WC) [184–186]. Long-term aircraft noise exposure may be linked to metabolic outcomes, in particular increased waist circumference. An increase in 5 dB(A) had been associated with a greater waist circumference of 1.5 cm as aircraft noise stimulates the release of cortisol which is known to contribute central obesity [186]. With regard to diabetes, a large cohort study including 4000 diabetes cases, found a 10 dB higher level of long-term exposure to road traffic noise to be associated with an 11% higher risk of diabetes after adjusting for a number of confounders including air pollution [187]. Similarly, a twofold higher risk of type 2 diabetes (252 cases) was demonstrated for people exposed to intense road traffic in a cohort study [188]. Also, the epidemiological research on low-frequency noise (LFN) and associated health effects is limited. Some associations were observed between exposure to LFN and annoyance, sleep-related problems, concentration difficulties and headache in the adult population living in the vicinity of a range of LFN sources. However, evidence, especially in relation to chronic medical conditions, was very limited [189].

9.10 Potential Health Effects and Sustainable Action Planning

The review of various studies strengthens the concept that transportation noise contributes to the development of cardiovascular risk, hypertension, stroke, sleep disorders, and even heart failure. On the basis of results of cohort and case control studies, a high-quality evidence of association of traffic noise to increased incidence of IHD is observed. Air pollution is one of the important confounding factors apart from other factors such as socioeconomic status, life style, family history of cardiovascular diseases, age, sex, and endogenous risk factors like blood cholesterol, glucose levels, and cortisol. However, recent studies were consistent in suggesting that both air pollution and noise are likely independent risk factors of cardiovascular disease [190]. Both daytime and nighttime aircraft noise above 55 dB (A) significantly increased the risk for stroke hospitalization, suggesting that nighttime noise may be especially hazardous [191]. However, the generalization of results, that is, development of exposure-effect relationship, may be dealt continuously keeping in mind the limitations and all the confounding factors, such as gender, age, body mass index, education level, physical activity, smoking habits, alcohol consumption, life style, stress, employment, family history of disease, and presence of other chronic diseases. [170]. Further studies are required particularly from the Asian population as well as to answer the questions: magnitude and time course of noise pollution on blood pressure and metabolic risk; duration of effect/time course of reversal; and effect on lifestyle, for example, diet, stress. 60 dB (A) L_{day} can be seen as *NOAEL* (no observed adverse effect level) for relationship between road traffic noise and myocardial infarction and for noise levels higher than 60 dB(A), the risk of MI increases continuously [2, 6]. 40 dB $L_{\mathrm{night,outside}}$ is equivalent to the *LOAEL* (lowest observed adverse effect level) for night and above 55 dB the cardiovascular effects become the major public concern, which are likely to be less dependent on the nature of noise. An interim target of 55 dB $L_{\mathrm{night,outside}}$ is recommended in the situations where the achievement of 40 dB $L_{\mathrm{night,outside}}$ is not feasible in the structures for various reasons [2]. Nighttime aircraft noise markedly impairs endothelial function in patients with or at risk for cardiovascular disease [192]. Also, from a cardiovascular point of view, there is no rationale known for weighing factors such as +5 dB (A) and +10 dB (A) for evening and night periods. It should be better approach to consider day and night exposures separately with respect to its effects particularly for noise sources other than road traffic noise. Studies should also try to distinguish between the exposure of living room (during daytime) and exposure of the bedroom (during nighttime) [5].

The action plans should be sustainable in nature and the economic, social, and environmental impact of the plans should be prior considered before their execution. Hedonic pricing data is generally evaluated from studies of the real estate markets. The best estimate is that house prices lose 0.5% of their value per decibel over 50–55 L_{den}. The range of research results is between 0.2% and 1.5%, with a tendency for higher values for aircraft noise [144]. The willingness to pay (WTP) is the prime factor for social cost from noise exposure [193]. In Indian perspectives, the situational factors for the property like nearness to the main roads, metro stations, etc. are more decisive rather than the air pollution or noise pollution concerns. However, with the growing awareness toward the potential health effects of the air and noise pollution, the concept of *noise free community* is increasing gradually and gaining more importance as compared to other situational factors.

9.11 Estimation of DALYs: A Case Study

The burden of disease is expressed in deaths and disability-adjusted life-year (DALY) calculated through the addition of years of life lost (YLL) and years lost due to disability (YLD). The YLD calculation depends upon disability weights (DWs) that are socially derived values based on how the majority of people perceive living with a disease or condition for a one year period, where 0 is optimal health and 1 is equivalent to death. DALY = YLL + YLD. The years of life lost (YLL) is calculated as number of deaths (N) multiplied by the standard life expectancy at the age at which death occurs (L), YLL= N×L. YLD is the number of "years lived with disability" estimated as YLD = I×DW×D, where I is the number of incident cases multiplied by a disability weight (DW) and an average duration D of disability in years. DALY combines years lived with disability (YLD) and years of life lost due to premature mortality (YLL). One DALY can be interpreted as one lost year of healthy life [194–196]. The following case study of calculation of DALYs from myocardial infarction due to road traffic noise in a city in India shall present a broader aspect of calculation of DALYs. Table 9.5 shows an illustrative case study of road traffic noise

Table 9.5 Statistical data of road traffic noise exposure and number of exposed citizen and cases of myocardial infarction (MI)

S. No	Traffic noise (dB(A))	Number of exposed citizens
1	< 60.0	3740
2	60-64.9	1200
3	65.0-69.9	1000
4	70.0-74.9	196
5	>75.0	89

Absolute number of newly diagnosed cases of MI = 4083
Absolute number of deaths from MI = 800

Table 9.6 Estimated road traffic noise exposure, population relative risk, and attributable fraction for myocardial infarction (MI)

S. No	Traffic noise (dB (A))	Number of exposed citizens	Percentage of exposed citizens	Population relative risk of MI	Attributable fraction (%)
1	< 60.0	3740	60.1	1.000	0.00
2	60–64.9	1200	19.3	1.032	3.13
3	65.0–69.9	1000	16.1	1.101	9.14
4	70.0–74.9	196	3.1	1.213	17.54
5	>75.0	89	1.4	1.374	27.23
Total		6225	100		

exposure and number of exposed citizen and cases of myocardial infarction (MI) in a city in northern part of India.

Step 1: Calculation of Population Relative Risk and Population-Attributable Fraction (%):

The relative risk (*RR*) due to myocardial infarction (MI) is calculated in terms of odds ratios (*OR*) from equation (9.21) as:

$$OR = 1.63 - 6.13 \cdot 10^{-4} \cdot L^2{}_{\text{day,16h}} + 7.36 \cdot 10^{-6} \cdot L^3{}_{\text{day,16h}} \tag{9.21}$$

The *OR* may be calculated from the middle sound level in a band. The attributable risk percentage (*AR* %) and population-attributable fraction (PAF) is calculated from the relative risk as:

$$\text{AR\%} = [(RR\text{-}1)/RR]\ 100\ [\%] \tag{9.22}$$

$$\text{PAF} = \left[\frac{\sum(P_i RR_i) - 1}{\sum(P_i RR_i)}\right] \tag{9.23}$$

where P_i is the proportion of population at exposure level *i* and RR_i is the relative risk at exposure level *i*. Table 9.6 thus shows the estimated road traffic noise exposure, population relative risk, and attributable fraction for myocardial infarction using all these relations.

Thus, the PAF for MI due to road traffic noise is calculated as:

$$\text{PAF} = \frac{[(1.000 \times 0.601 + 1.032 \times 0.193 + 1.101 \times 0.161 + 1.213 \times 0.031 + 1.374 \times 0.014)] - 1}{[(1.000 \times 0.601 + 1.032 \times 0.193 + 1.101 \times 0.161 + 1.213 \times 0.031 + 1.374 \times 0.014)]}$$

$$= \mathbf{0.0331}$$

The resulting contribution of road traffic noise to the occurrence of MI for the population of the city is 0.0331 or 3.31%.

Step 2: Calculation of YLL (Years of Life Lost)

The calculation of years of life lost is obtained by multiplying the number of deaths from MI in each age category with standard life expectancy at age of death. According to the WHO data, life expectancy in India for male is 67.4 and 70.3 for female and 68.8 as total life expectancy. Thus, life expectancy of a 60-year-old male dying will cause a loss of 7.4 years full life potentially lived in optimal health. YLL can be calculated as $1 \times 7.4 = 7.4$. In this way, the YLL was calculated for men and women separately within the different age groups as 3600. The YLL due to road traffic noise was calculated by multiplying total YLL with PAF as $3600 \times 0.0331 = 119.16$. The average number of years of life lost due to MI is thus calculated as: $3600/800 = 4.5$.

Step 3: Calculation of YLD (Years Lived with Disability)

The disability weight, D = 0.405, has been assumed in accordance with WHO guidelines [5] and one year of disability for each nonfatal case of MI for estimation of YLD as follows:

- Absolute number of newly diagnosed cases of MI = 4083
- Absolute number of deaths from MI = 800
- Absolute number of nonfatal cases of MI = (4083-800) = 3283
- YLD attributable to road traffic noise $(I \times DW \times D) \times PAF = (3283 \times 0.405 \times 1) \times 0.0331 = 44.01$

Step 4: Calculation of DALY

$$\text{DALY} = \text{YLL} + \text{YLD} = 119.16 + 44.01 = 163.17$$

The contribution of road traffic noise to the occurrence of MI for the population of an urban district of India is estimated as 3.31%. The total DALYs for MI due to road traffic noise result in 163 years. This can be calculated in terms of per million inhabitants also. A similar analysis is presented for the estimation of DALYs for highly sleep disturbed people due to road traffic noise in the city as shown in Table 9.6. Figure 9.11 shows the flow chart of calculation of the burden of disease due to environmental noise. The distribution of the population in the five-decibel classes as shown in Table 9.7, exposure response function, and disability weights correlating the potential heath effect with the noise levels are must for such an analysis. Table 9.8 summarizes the various exposure response functions and disability weights (DW) for different source of noise for calculation of DALYs as recommended in various studies. However, it may be noted that while implementing the concept of DALY, the fundamental uncertainties affecting each calculation step should be borne in mind [197].

Noise from road traffic is major health issue as 8,900 of the estimated 10,000 premature deaths per year (89%) are attributable to road traffic noise exposure. Also, it contributes to a total of about 6,700 premature deaths per year due to coronary heart disease and about 3,300 premature deaths due to stroke [198]. The DALYs lost

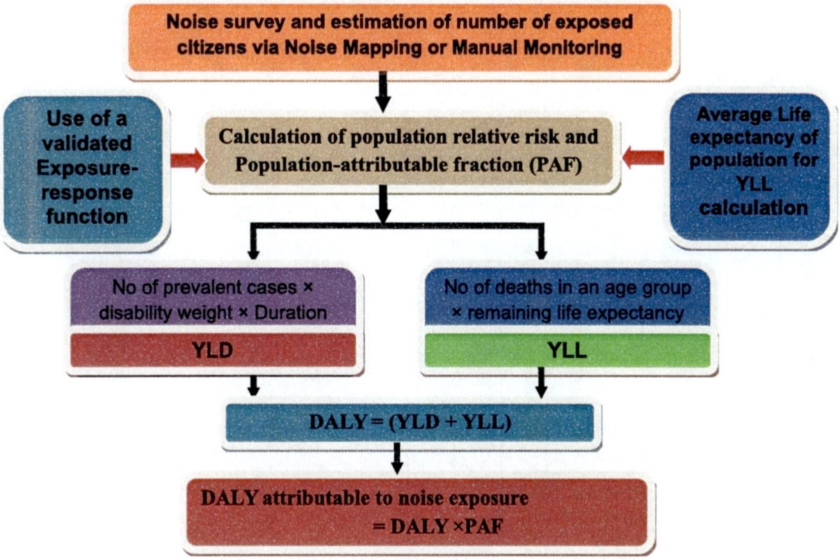

Fig. 9.11 Flow chart of calculation of the burden of disease due to environmental noise

Table 9.7 Estimated DALYs for highly sleep disturbed due to road traffic noise

S. No	L_{night} in (dB (A))	Percentage of exposed citizens	Percentage of highly sleep disturbed	No. of case per million	DALYs lost		
					DW=0.04	DW=0.07	DW=0.10
1	<45	42	–	–	–	–	–
2	45–49	19	4.5	8550	342	598.5	855
3	50–54	23	6.6	15180	607.2	1062.6	1518
4	55–59	9	9.6	8640	345.6	604.8	864
5	60–64	5	13.2	6600	264	462	660
6	65–69	2	17.6	3520	140.8	246.4	352
Total				**42490**	**1699.6**	**2974.3**	**4249**

from environmental noise in the western European countries are 61,000 years for ischemic heart disease (IHD), 45,000 years for cognitive impairment of children, 903,000 years for sleep disturbance, 22,000 years for tinnitus, and 654,000 years for annoyance. When taken together, the range of burden would be 1.0–1.6 million DALYs [5]. The major contributors (92.4%) are observed to be annoyance and sleep disturbance. A similar analysis was presented in a study by Eriksson et al. (2017) [199] whereby the annoyance and sleep disturbance contribute 83.6% to DALYs

Table 9.8 Exposure response functions and disability weights (DW) for different noise sources for calculation of DALYs

Health Outcomes		Road Traffic noise	Aircraft noise	Railway noise	DW and Source
Annoyance		$\%HA = 78.9270 - 3.1162\, L_{den} + 0.0342\, L_{den}^2$	$\%HA = -50.9693 + 1.10168\, L_{den} + 0.072\, L_{den}^2$	$\%HA = 38.1596 - 2.05538\, L_{den} + 0.0285\, L_{den}^2$	0.02 (0.01-0.12), [177]
		$\%\ HA = 9.868 \times 10^{-4}(L_{den} - 42)^3 - 1.436\times10^{-2}(L_{den} - 42)^2 + 0.5118\,(L_{den} - 42)$	$\%\ HA = -9.199\times10^{-5}(L_{den} - 42)^3 + 3.932\times10^{-2}(L_{den} - 42)^2 + 0.2939\,(L_{den} - 42)$	$\%\ HA = 7.239 \times10^{-4}(L_{den} - 42)^3 - 7.851\times10^{-3}(L_{den} - 42)^2 + 0.1695\,(L_{den} - 42)$	[5]
Sleep Disturbance		$19.4312 - 0.9336\, L_{night} + 0.0126\, L_{night}^2$	$16.7885 - 0.9293\, L_{night} + 0.0198\, L_{night}^2$	$67.5406 - 3.185\, L_{night} + 0.0391\, L_{night}^2$	0.07 (0.04-01.0), [177]
		$20.8 - 1.05L_{night} + 0.01486\, L_{night}^2$	$18.147 - 0.956L_{night} + 0.01482\, L_{night}^2$	$11.3 - 0.55L_{night} + 0.00759\, L_{night}^2$	[5]
Myocardial Infarction		$OR = 1.63 - 6.13\cdot10^{-4}\cdot L^2_{day,\,16h} + 7.36\cdot10^{-6}\cdot L^3_{day,\,16h}$; $L_{day,16h} = L_{den} - 2.5$ dB; $L_{den} = L_{night} + 7.5$ dB			0.23, [197]
IHD morbidity		1.24 (95% CI: 1.04-1.42)*	1.07 (95% CI: 0.94-1.23)*	1.24 (95% CI: 0.82-1.68)*	0.114, [196]
IHD incidence		$RR = \exp(\ln(1.08)/10 \times (L_{den} - 53))$ if $L_{den} \geq 53$ dB, and RR = 1 if $L_{den} < 53$ dB	--	--	[152]
Hypertension	Prevalence	1.05 (95% CI: 1.02–1.08)*	1.05 (95% CI: 0.95–1.17)*	1.05 (95 % CI: 0.88–1.26)*	0.117, [177]
	Incidence	0.97 (95 % CI:0.90–1.05)*	1.00 (95% CI: 0.77–1.30)*	0.96 (95 % CI:0.88–1.04)*	
Stroke	Incidence	1.14 (95% CI: 1.03 – 1.25)*	1.05 (95 % CI:0.96 – 1.15)*	--	Acute stroke=0.92, chronic effects of stroke =0.171, [151]
	Mortality	0.87 (95% CI: 0.71–1.06)*	0.99 (95 % CI:0.94–1.04)*	--	
CHD (Incidence)		1.12 (95 % CI: 0.85 – 1.48)	1.09 (95 % CI:1.04 – 1.15)*	--	--
Cognitive effect		--	1.38 (95%CI: 1.09 – 1.75)*	--	0.006, [151]

*per 10 dB linear increase, CI-Confidence interval, IHD-Ischaemic Heart Disease, CHD-Coronary Heart Disease.

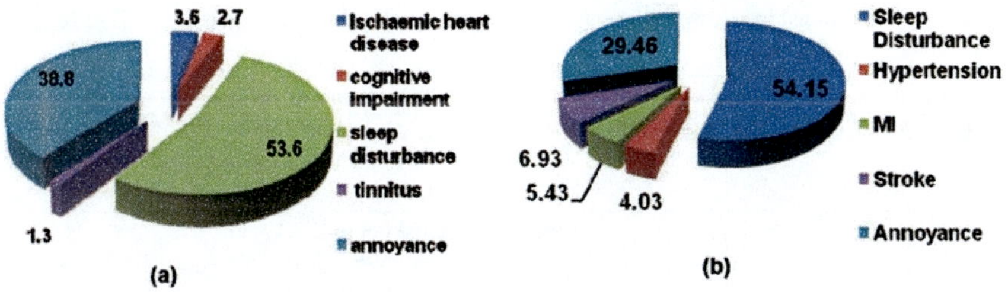

Fig. 9.12 (**a**) Percentage contribution of various health outcomes to the DALYs calculated in WHO study [5], (**b**) Percentage contribution of various health outcomes to the DALYs in Sweden for road and railway traffic noise combined (Eriksson et al. 2017 [199])

attributed to road and railway traffic noise combined. Figure 9.12 shows the percentage contribution of various health outcomes to the DALYs calculated in WHO study and that conducted in Sweden [199]. However, the recent German study by Tobollik et al. (2019) [196] accounts IHD morbidity and mortality to be the major contributor (69.5%), rather than annoyance and sleep (30.5%). Also, a recent study on the evaluation of burden of disease using the Urban and Transport Planning Health Impact Assessment (UTOPHIA) tool for physical activity (PA), air pollution, noise, heat, and access to green spaces in Barcelona, Spain, showed that 36% of burden of disease (i.e., 18,951 DALYs) was due to traffic noise with sleep disturbance and annoyance contributing largely [200]. Table 9.9 enlists the relationships between various noise sources and health effects as summarized in a European Environment Agency recent report [201–204].

9.12 Noise Pollution Health Effects Evaluation and Reduction (NOPHEER)

The socioeconomic impact of road and rail traffic noise in the European Union had been evaluated as €40 billion, which represents around 0.4% of the EU's total GDP [205]. A recent study valued the loss of healthy life due to noise exposure in the UK of €1.34 billion (£1.09 billion). Exposure to environmental noise levels above the recommended levels resulted in 1169 cases of dementia, 788 strokes, and 542 heart attacks every year in the UK alone [206]. It is evident that noise exposure leads to mental stress and cognitive decline that may manifest psychological disorders via stress hormone release and induce potent vasoconstrictors, to vascular dysfunction and damage thus initiating cardiometabolic risk factors [207]. A recent study of European Commission estimated the societal costs for the combined noise and air pollution as nearly 1 trillion Euros, while the costs for alcohol and smoking are considerably less (50–120 and 540 billion Euro) [208]. A very recent study in Taiwan [209] revealed the hazard ratio for metabolic syndrome (including abdominal obesity, hypertension, hyperglycemia, hypertriglyceridemia, and low high-

Table 9.9 Relationships between various noise sources and health effects [91, 102, 152, 178, 179, 201–204]

Health effect	Population	Source	Relationship
High annoyance	Adults	Road	Guski et al. (2017) $(78.927 - 3.1162 \times L_{den} + 0.0342 \times L_{den}^2)/100$
		Rail	Guski et al. (2017) $(38.1596 - 2.05538 \times L_{den} + 0.0285 \times L_{den}^2)/100$
		Air	Guski et al. (2017) $(-50.9693 + 1.0168 \times L_{den} + 0.0072 \times L_{den}^2)/100$
		Industry	Miedema and Vos (2004) 1-*normal* $((72 - (-126.52 + (L_{den}) \times (2.49)))/ \mathrm{sqrt}(2054.43))$
High sleep disturbance	Adults	Road	Basner and McGuire (2018) $(19.4312 - 0.9336 \times L_{night} + 0.0126 \times L_{night}^2)/100$
		Rail	Basner and McGuire (2018) $(67.5406 - 3.1852 \times L_{night} + 0.0391 \times L_{night}^2)/100$
		Air	Basner and McGuire (2018) $(16.7885 - 0.9293 \times L_{night} + 0.0198 \times L_{night}^2)/100$
		Industry	Miedema and Vos (2007) 1-*normal* $((72 - (-90.70 + (L_{night}) \times (1.80)))/ \mathrm{sqrt}(1789 + 272))$
Reading comprehension	Children	Air	Clark et al. (2006) and van Kempen (2008) $1/(1 + \exp(- (\ln(0.1/0.9) + (\ln(1.38)/10 \times (L_{den} - 50))))$ if $L_{den} \geq 50$ dB and 0.1 if $L_{den} < 50$ dB
Ischemic heart disease incidence	Adults	Road, rail, air, industry	van Kempen et al. (2018) relative risk (RR) derived from road noise $RR = \exp(\ln(1.08)/10 \times (L_{den} - 53))$ if $L_{den} \geq 53$ dB, and RR = 1 if $L_{den} < 53$ dB
Premature mortality due to Ischemic heart disease	Adults	Road, rail, air, industry	van Kempen et al. (2018) relative risk (RR) derived from road noise $RR = \exp(\ln(1.05)/10 \times (L_{den} - 53)$ If $L_{den} \geq 53$ dB, and RR = 1 if $L_{den} < 53$ dB

density lipoprotein (HDL) cholesterol) was 1.13 (95% CI: 1.04–1.22) for medium noise exposure and 1.24 (95% CI: 1.13–1.36) for high noise exposure. Occupational noise exposure may be associated with an increased incidence of hyperglycemia, with the highest risk observed at 31.5 Hz [210]. Noise impacts – positive or negative – can be measured and compared by using health-related quality of life (HRQOL). The inhabitants of the Greater Paris Metropolis, Paris, lose on average 8 months of healthy life due to transport noise, which reached 18 months for some municipalities [211]. People living in quiet locations had a better quality of life.

Preserving quiet areas improves health [212]. Night noise levels may have more of an impact on cardiovascular health than daytime noise as the sleep is disturbed [213]. The non-auditory effects induced due to noise including annoyance, sleep disturbance, heart disease, and cognitive impairment are all the major concerns in national and international scenario. Air and noise pollution are both the major environmental risks whereby the primary source is the road traffic in urban environment [214]. Higher levels of NO_2, $PM_{2.5}$ (particulate matter <2.5 μm in diameter), and noise are related to an increased risk of developing diabetes mellitus type 2 [215]. Both road traffic noise and NO_2 were associated with a higher risk of myocardial infarction, wherein the major risk is associated due to traffic noise [216]. Also, in case of noise-induced hearing loss (NIHL), 10% of the population worldwide suffer from hearing loss and about half of these cases can be attributed to auditory damage caused by intense noise exposure [217]. The WHO 2018 guidelines recommends that levels below 55 dB L_{den} and 50 dB L_{night} are likely to cause health problems and thus national and local noise action plans should target these goals. Reducing noise at the source (51.2%) for agglomerations is the most extensively reported measure in END action plans, regulating routes and re-routing aircraft flight paths to a significant extent (20%). While changes in infrastructure, such as a bypass or re-routing (8%), land use and urban planning, and education and communication measures (15%), are used to a small extent [202]. Thus, a program similar to that of European consortium on Noise Pollution Health Effects Reduction (NOPHER) is required in Indian perspectives also for protection against noise pollution in next decades [218]. These studies in conjunction with studies focused on air pollution should be conducted in parallel for identifying the combined effects, as well as potential confounding factors [137]. A proposed solution for this problem is to investigate noise sources which are less correlated with air pollution, such as railway or aircraft noise [167]. The Noise Policy Statement for England (NPSE) inspires long-term vision of a national noise policy on: “Promote good health and quality of life through the effective noise management within the context of Government policy on sustainable development” with an objective of avoiding and reducing the adverse impacts and improvement of health and quality of life [219]. In Indian perspectives, the evaluation of health aspects associated with noise exposure is rarely reported, and thus the program must include analysis and may be called as Noise Pollution Health Effects Evaluation and Reduction (NOPHEER). Prevention-oriented health systems are vital for meeting the challenges and inculcate the need of strong noise policy with an objective of preventable health effects [220].

9.13 Correlation of Noise Levels with Saliva Cortisol

Exposure to the aircraft noise increases morning saliva cortisol levels in women which could be of relevance for noise-related cardiovascular effects. The difference in influence of noise exposure on saliva cortisol in men and women was however difficult to explain. Also, it had been observed that the effect is connected directly to the noise exposure and relatively independent of the subjective annoyance experience. The women who were employed had a higher cortisol level than did the retired women [221]. Baudin et al. (2019) study showed that the exposure to aircraft noise, at night in particular, could make the hypothalamic pituitary-adrenal (HPA) axis less flexible, especially in women, resulting in higher average values for the evening cortisol concentration, and therefore a flattening of the difference between morning and evening levels. However, further biological research is needed to deepen knowledge of the pathway between noise exposure and disturbed hormonal regulation, and specially the difference in effects between genders [222]. Many studies such as the HYENA (HYpertension and Exposure to Noise near Airports) study and DEBATS (Discussion on the health effects of aircraft noise) projects have different observations. The HYENA study analysis found increases in morning saliva concentrations with aircraft noise exposure in women only [221], while DEBATS found higher evening but not morning cortisol in both men and women [223]. Thus, aircraft noise exposure resulting in HPA dysregulation and a flattened cortisol rhythm causes the cardiovascular diseases [224]. Saliva cortisol levels appeared particularly high in one such study [224] among those who were highly annoyed and exposed to road traffic noise levels ≥ 55 dB L_{den}. Road traffic noise exposure was not associated with saliva cortisol; however, annoyance to noise tended to increase the levels [224]. Industrial noise exposure with levels higher than 80 dB(A) has a significant effect on salivary cortisol elevation [225]. Some studies suggested that exposure to low-frequency noise caused alterations in cortisol levels among subjects categorized as high-sensitive to noise in general [226]. It is important to consider that salivary cortisol responses to the diverse acute challenge paradigms show large intra- and inter-individual variability. Thus, an extensive analysis of HPA axis stress responses is essential to understand the mechanisms related to stress-related disorders [227]. Further studies are required to correlate the community noise exposure and cortisol levels for developing exposure-response relationships and understand the implications of gender differences.

9.14 Noise Pollution in COVID-19 Pandemic and Cardiovascular Risk

COVID-19 pandemic in year 2020 and lockdowns imposed had an unintended effect of pronounced decrement in air and noise pollution in some cities and a substantial decrease of annoyance and stress related to sound pollution and, therefore, benefit to cardiovascular disease [228]. Reduction in seismic noise up to 50% was observed, and it was observed as one of the quietest recorded periods on earth [229]. In Italy, the lockdown considerably reduced ambient noise in the 1–10 Hz frequency range [230], and a reduction of life-threatening diagnoses such as 26% for stroke, 40% for heart failure, 22% in trauma, and 45% for acute coronary syndrome had been recorded as compared to the same period in the previous year [231]. Apart from reduced noise pollution, this reduction might be attributed to other factors such as fear of getting infected by COVID-19 at the hospital or to lower regulatory access to hospitals in order to avoid crowdedness, lifestyle changes, increased physical activity, reduced work stress, and increased incidence of CVD among patients infected from COVID-19 [232]. Further studies need to be focused in future based on large population trials to understand the effect of noise on cardiovascular risks. Also, the indoor nocturnal noise had been found in one of the study to be associated with body mass index (BMI) and blood pressure in females, but only blood pressure in males [233]. It is thus very essential to control the nocturnal noise as nighttime aircraft noise can trigger acute cardiovascular mortality [234].

9.15 Conclusions and Outlook

The chapter presents various studies that had been carried out to correlate the noise exposure with the health effects. The negative impact of environmental noise on human health and well-being and is a growing concern among both the general public and policy-makers across the globe. Recently, the Guidelines Development Group (GDG) of World Health Organization (WHO) provided recommendations for protecting human health from exposure to environmental noise originating from various sources: transportation (road traffic, railway and aircraft) noise, wind turbine noise, and leisure noise. However, still there is high uncertainty involved in the exposure–response relationships between the exposure to environmental noise and the proportion of people with a validated measure of health outcome, when adjusted for the confounders. The GDG set self-reported sleep disturbance as a primary health outcome. For IHD and hypertension, relative risk (RR) is considered; for annoyance and sleep disturbance, absolute risks of% HA and % HSD were considered; and for reading and oral comprehension, an average delay of reading age was defined. Thus, longitudinal studies assessing the exposure to environmental noise levels and health outcomes objectively should be initiated, taking into account the most relevant confounders. Thus, proper health risk assessments utilizing the decision support instruments such as health impact and health risk assessments for quantification and potential benefits associated with measures for environmental noise are essentially required in every country in order to understand and analyze the health outcomes and initiating more and more studies in order to evaluate the effectiveness of interventions to reduce noise exposure and their health impact.

References

1. Traffic Noise Reduction in Europe (2007) CE Delft, The Netherlands, August 2007.
2. World Health Organization (2009) Night noise guidelines for Europe. WHO Regional Office for Europe, Copenhagen, p 2009
3. Babisch W (2006) Transportation noise and cardiovascular risk: updated review and synthesis of epidemiological studies indicate that the evidence has increased. Noise Health 8(30):1–29

4. Sörensen M, Hvidberg M, Andersen ZJ, Norsborg RB, Lillelund KG, Jakobsen J, Tjönneland A, Overvad K, Raaschou-Nielsen O (2011) Road traffic noise and stroke: a prospective cohort study. Eur Heart J 32:737–744
5. WHO-JRC (2011) Burden of disease from environmental noise-Quantification of healthy life years last in Europe. European Center for Environment and Health JRC EU, 2011
6. Babisch W (2002) The noise/stress concept, risk assessment and research needs. Noise Health 4(16):1–11
7. American National Standards Institute (ANSI) (1996) Quantities and Procedures for Description and Measurement of Environmental Sound – Part 4: Noise Assessment and Prediction of Long-term Community Response. Standards Secretariat, Acoustical Society of America, New York, 1996
8. Morihara T, Sato T, Yano T (2004) Comparison of community responses to railway and road traffic noises using structural equation model. Internoise, Czech Republic, 22nd to 25th August, 2004.
9. Chakraborty D, Santra SC, Mukherjee AL, Roy BM, Das P (2002) Road traffic noise in Calcutta metropolis, India. Indian J Environ Health 3:173–180
10. Gorai AK, Pal AK (2006) Noise impact assessment of an iron ore mining complex. Noise Control Engg Journal 54:352–359
11. Agarwal S, Swami BL (2009) Road traffic noise annoyance in Jaipur city. Int J Eng Stud 1:39–46
12. Banerjee D, Chakraborty SK, Bhattacharya S, Gangopadhyay A (2009) Attitudinal response towards road traffic noise in the industrial town of Asansol. India Environ Monit Assessment 151:37–44
13. Schultz TJ (1978) Synthesis of social surveys on noise annoyance. J Acoustical Soc Am 64(2): 377–405
14. Fidell S, Barber DS, Schultz TJ (1991) Updating a dosage-effect relationship for prevalence of annoyance due to general transportation noise. J Acoustical Soc Am 89(1):221–233
15. Finegold L S, Finegold M S (2002) Development of exposure-response relationships between transportation noise and community annoyance. Japan Net Symposium on Annoyance, Stress and Health effects of environmental noise, Nov-Dec. 2002.
16. Finegold LS, Harris CS, von Gierke HE (1994) Community annoyance and sleep disturbance: Updated criteria for assessing the impacts of general transportation noise in people. Noise Control Eng J 42(1):25–30
17. Ouis D (2002) Annoyance caused by exposure to road traffic noise: an update. Noise Health 4: 69–79
18. Ouis D (2001) Annoyance from road traffic noise: a review. J Environ Psychol 21:101–120
19. Miedema HME, Oudshoorn CGM (2001) Annoyance from transportation noise: relationships with exposure metrics DNL and DENL and their confidence intervals. Environ Health Perspect 109(4):409–416
20. European Commission (2002) Position paper on Dose-response relationships between noise and annoyance – EU's future policy, WG2-dose/effect 2002.
21. Miedema HME, Vos H, de Jong RG (2000) Community reaction to aircraft noise: time-of-day penalty and trade off between level of overflights. J of Acoustical Society of America 107(6): 3245–3253
22. Kjellberg A, Goldstein M, Gamberale F (1984) An assessment of dB(A) for predicting loudness and annoyance of noise containing low frequency noise. J Low Freq Noise Vib Act Cont 3:10–16
23. Vos J (2001) On the annoyance caused by impulse sounds produced by small, medium-large and large firearms. J Acoust Soc Am 109:244–253
24. Kjellberg A, Tesarz M, Holmberg K, Landström U (1997) Evaluation of frequency-weighted sound level measurements for prediction of low frequency noise annoyance. Environ Int 23: 519–527

25. Kuwano S, Namba S, Miura H (1989) Advantages and disadvantages of A-weighted sound pressure level in relations to subjective impression of environmental noises. Noise Control Eng J 33:107–115
26. Nilsson ME (2007) A-weighted sound pressure level as an indicator of short-term loudness or annoyance of road traffic sound. J Sound Vib 302:197–207
27. Agarwal S, Swami BL (2011) Road traffic noise, annoyance and community health survey – a case study for an Indian city. Noise Health 13(53):272–276
28. Agarwal S, Swami BL (2009) Noise annoyance under interrupted traffic flow condition for Jaipur city. Int J Appl Sci & Eng 7(2):159–168
29. Banerjee D (2012) Research on road traffic noise and human health in India: Review of literature from 1991 to current. Noise Health 14(58):113–118
30. Garg N, Sharma O, Mohanan V, Maji S (2012) Passive noise control measures for traffic noise abatement in Delhi, India. J Sci Ind Res 71:226–234
31. Miedema HME, Vos H (1999) Demographic and attitudinal factors that modify annoyance from transportation noise. J of Acoustical Society of America 105(6):3336–3344
32. Mc Sweeney DW, Woodcock J, Waddington D, Peris E, Koziel Z, Moorhouse A, Redel-Macías MD (2016) Effect of attitudinal, situational and demographic factors on annoyance due to environmental vibration and noise from construction of a Light Rapid Transit System. Int J Environ Res Public Health 13:1237–1264
33. Park J, Chung S, Lee J, Sung JH, Cho SW, Sim CS (2017) Noise sensitivity rather than noise level predicts the non-auditory effects of noise in community samples: a population based survey. BMC Public Health 17:315
34. van Kamp I, Rob JF, Hatfield J, Haines M, Stellato RK, Stansfield SA (2004) The role of noise sensitivity in the noise-response relation: a comparison of three international Airports studies. J of Acoustical Society of America 116:3471–3479
35. Thomas JR, Jones DM (1982) Individual differences in noise annoyance and the uncomfortable loudness level. J Sound Vib 82:289–304
36. Méline J, Hulst AV, Thomas F, Karusisi N, Chaix B (2013) Transportation noise and annoyance related to road traffic in the French RECORD study. Int J Health Geogr 12:44
37. Whitfield A (2003) Assessment of noise annoyance in three district communities living in close proximity to a UK regional airport. Int J Environ Health Res 13:361–372
38. Fyhri A, Klaeboe R (2006) Direct, indirect influences of income on road traffic noise annoyance. J Environ Psychol 26:27–37
39. ISO/TS 15666:2003. Acoustics-Assessment of Noise annoyance by means of social and socio-acoustic surveys, Geneva, Switzerland.
40. Wadud Z (2013) Using meta-regression to determine Noise Depreciation Indices for Asian airports. Asian Geogr 30(2):127–141
41. Schipper Y, Nijkamp P, Rietveld P (1998) Why do aircraft noise value estimates differ? A meta-analysis. J Air Transp Manag 4:117–124
42. Levesque TJ (1994) Modelling the effects of airport noise on residential housing markets: a case study of Winnipeg International Airport. J Transport Economic & Policy 28:199–210
43. Lazic A, Golaszewski R (2006) A Technical note on Aircraft noise and its cost to society, GRA, Incorporated, Economic Counsel to the Transportation Industry, April, 2006.
44. Nelson JP (2004) Meta-analysis of Airport noise and Hedonic property values: problems and prospects. J Transport Economic & Policy 38(1):1–28
45. Anderson H, Jonsson L, Ögren M (2013) Benefit measures for noise abatement: calculations for road and rail traffic noise. Eu Transp Res Rev 5:135–148
46. Andersson H, Ögren M (2007) Noise charges in rail infrastructure: a pricing schedule based on the marginal cost principle. Transp Policy 14(3):204–213
47. Lavandier C, Sedoarisoa N, Desponds D, Dalmas L (2016) A new indicator to measure the noise impact around airports: the Real Estate Tolerance Level (RETL)-Case study around Charles de Gaulle Airport. Appl Acoust 110:207–217

48. Navrud S (2002) The state-of-the-art on economic valuation of noise. Final report to European Commission DG Environment, April 14th, 2002.
49. Bateman I J, Day B, Lake I, Lovett A A (2000). The effect of road traffic noise on residential property values: A literature review and Hedonic pricing study. Report to the Scottish office, Development Department, Edinburg School of Environmental Sciences, University of East Anglia, Norwich, March, 2000.
50. Fidell S, Mestre V, Schomer P, Berry B, Gjestland T, Vallet M, Reid T (2011) A first-principles model for estimating the prevalence of annoyance with aircraft noise exposure. J of Acoustical Society of America 130(2):791–806
51. ISO 1996-1:2016, Acoustics-Description, measurement and assessment of environmental noise-part 1: Basic quantities and assessment procedures, 2016.
52. Morinaga M, Yokoshima S, Morihara T, Kawai K, Yano T (2017) Community tolerance level for transportation noises derived from the Socio-Acoustic Survey Data Archive. SASDA, 12th ICBEN Congress on Noise as a Public Health Problem, 18 to 22nd June, 2017, Zurich.
53. Schomer P, Mestre V, Fidell S, Berry B, Gjestland T, Vallet M, Reid T (2012) Role of community tolerance level (CTL) in predicting the prevalence of the annoyance of road and rail noise. J Acoust Soc Am 131(4):2772–2786
54. I-INCE Publication No.:2015-1, Supplemental Metrics for Day/Night average sound level and Day/Evening/Night average sound level. Final report of the I-INCE Technical Study Group on Metrics for Environmental Noise Assessment and Control (TSG 9), April, 2015.
55. Gjestland T, Gelderblom F, Granøien I L N (2016) Noise surveys at five Norwegian airports. Internoise 2016, Hamburg, 2571–2577.
56. Fields M, Jong RG d, Gjestland T, Flindell IH, Job RFS, Kurra S, Lercher P, Valet M, Yano T, Guski R, Felscher-Suhr U, Schumer R (2001) Standardized general-purpose noise reaction questions for community noise surveys: research and a recommendation. Community response to noise team of ICBEN (International Commission on the Biological Effects of Noise). J Sound Vib 242:641–679
57. Gjestland T, Nguyen TL, Yano T (2015) Community response to noise in Vietnam: Exposure response relationships based on the community tolerance level. J Acoust Soc Am 137(5): 2596–2601
58. Tucci D, Merson MH, Wislon BS (2010) A summary of the literature on global hearing impairment: current status and principles for action. Otol Neurotol 31(1):31–41
59. Agrawal Y, Platz EA, Niparko JK (2008) Prevalence of hearing loss and differences by demographic characteristics among US adults: data from the National Health and Nutrition Examination Survey, 1999–2004. Ach Intern Med 168:1522–1530
60. Nelson DT, Nelson RY, Concha-Barrientos M, Fingerhut M (2005) The global burden of occupational noise-induced hearing loss. Am J Ind Med 48(6):446–458
61. Oishi N, Schacht J (2011) Emerging treatments for noise-induced hearing loss. Expert Opin Emerg Drugs 16:235–245
62. National Institute of Health (1990) Consensus Conference. Noise and Hearing loss. JAMA, 263, June, 1990, 3185–3190.
63. Nandi SS, Dhatrak SV (2008) Occupational noise-induced hearing loss in India. Indian J Occup Environ Med 12(2):53–56
64. Dobie RA (1995) Prevention of noise induced hearing loss. Arch Otolaryngol Head Neck Surg 121:385–391
65. Taneja MK (2014) Noise-induced hearing loss. Indian Journal of Otology 20:151–154
66. Sahoo GC, Sebastian H (2011) Prevalence of sensorineural deafness in habitual mobile phone users. Indian Journal of Otology 17:97–100
67. Ramya CS, Karthiyanee K, Vinutha S (2011) Effect of mobile phone usage on hearing threshold: a pilot study. Indian Journal of Otology 17:159–161
68. Scientific Committee on Emerging and Newly Identified Health Risks (SCENIHR) (2008) Potential health risks of exposure to noise from personal music players and mobile phones including a music playing function, European Commission report, 2008.

69. World Health Organization Grades of hearing impairment, http://www.who.int/pbd/deafness/hearing_impairment_grades/en
70. Davis A, El Refaie A (2000) The epidemiology of tinnitus. In: Tyler R (ed) Tinnitus handbook. Cengage Learning, Hampshire, pp 1–23
71. Fligor BJ, Cox C (2004) Output levels of commercially available portable compact disc players and the potential risk to hearing. Ear Hear 25:513–527
72. Kirchner DB, Evenson CE, Dobie RA, Rabinowitz P, Crawford J, Kopke R, Hudson TW (2012) Occupational noise-induced hearing loss: ACOEM task force on occupational hearing loss. J Occup Environ Med 54(1):106–108
73. McBride DI, Williams S (2001) Audiometric notch as a sign of noise induced hearing loss. Occup Environ Med 58:46–51
74. American National Standards Institute (1996) Determination of Occupational Noise Exposure and Estimation of Noise-Induced Hearing Impairment, ANSI S3.44–1996. New York, NY: Acoustical Society of America; 1996.
75. Coles RR, Lutman ME, Buffin JT (2000) Guidelines on the diagnosis of noise-induced hearing loss for medicolegal purposes. Clin Otolaryngol Allied Sci 25:264–273
76. Szalma JL, Hancock PA (2011) Noise effects on human performance: a meta-analytic synthesis. Psychol Bull 137(4):682–707
77. Berglund D, Lindvall T, Schwela D (1999) Guidelines for Community noise. World Health Organization, Geneva, Switzerland
78. Wright B, Peters E, Ettinger U, Kuipers E, Kumari V (2014) Understanding noise stress-induced cognitive impairment in healthy adults and its implications for schizophrenia. Noise Health 16(70):166–176
79. Jelínková Z, Picek M, Hyncica V (1988) Psychophysiological factors determining responses to noise load. Act Nerv Super (Praha) 30(2):146–147
80. Boman E, Enmarker I, Hygge S (2005) Strength of noise effects on memory as a function of noise source and age. Noise Health 7(27):11–26
81. Salamé P, Baddeley A (1982) Disruption of short-term memory by unattended speech: implications for the structure of working memory. J Verb Learn Verb Behav 21(2):150–164
82. Tremblay S, Macken WJ, Jones DM (2011) The impact of broadband noise on serial memory: changes in band-pass frequency increase disruption. Memory 9:323–331
83. Evans GW, Hygge S, Bullinger M (1995) Chronic noise and psychological stress. Psychol Sci 6:333–338
84. Haines MM, Stansfeld SA, Job RFS, Berglund B, Head J (2001) Chronic aircraft noise exposure, stress responses, mental health, and cognitive performance in school children. Psychol Med 31(2):265–277
85. Hygge S, Evans GW, Bullinger M (2002) A prospective study of some effects of aircraft noise on cognitive performance in school children. Psychol Sci 13:469–474
86. Meis M, Hygge S, Evans G W, Bullinger M (1998) Dissociative effects of traffic noise on implicit and explicit memory: Results from field and laboratory studies. In N.L. Carter and R. F. S. Job (Eds.), Noise as a public health problem-Noise Effects 98 Conference, 7th International Congress on Noise as a Public Health Problem 22–26, November 1998, 389–394.
87. Haines MM, Stansfeld SA, Brentnall S, Head J, Berry B, Jiggins M, Hygge S (2001) The West London Schools Study: the effects of chronic aircraft noise exposure on child health. Psychol Med 31(8):1385–1396
88. Evans GW (2006) Child development and the physical environment. Annu Rev Psychol 57: 423–451
89. Lercher P, Evans GW, Meis M (2003) Ambient noise and cognitive process among primary school children. Environ Behav 35:725–735
90. Stansfeld SA, Beglund B, Clark C, Lopez-Barrio I, Fischer P, Ohrström E, Haines MM, Head J, Hygge S, van Kamp I, Berry BF. RANCH study team(2005) Aircraft and road traffic

noise and children's cognition and health: a cross national study. Lancet 365(9475): 1942–1949
91. Clark C, Martin R, van Kempen E, Alfred T, Head J, Davies HW, Haines MM, Lopez Barrio I, Matheson M, Stansfeld SA (2006) Exposure-effect relations between aircraft and road traffic noise exposure at school and reading comprehension: the RANCH project. Am J Epidemiol 163(1):27–37
92. Basner M, Babisch W, Davis A, Brink M, Clark C, Janssen S, Stansfield S (2014) Auditory and non-auditory effects of noise on health. Lancet 383:1325–1332
93. Jones K (2010) Aircraft noise and Children's learning, Environmental Research and Consultancy Department, Directorate of Airspace Policy, Civil Aviation Authority, CAA House, 45–59 Kingsway, London, February 2010.
94. Clark C, Head J, Stansfield SA (2013) Longitudinal effects of aircraft noise exposure on children's health and cognition: a six-year follow-up of the UK RANCH cohort. J Environ Psychol 35:1–9
95. Dreger S, Meyer N, Fromme H, Bolte G (2015) Environmental noise and incident mental health problems: a prospective cohort study among school children in Germany. Environ Res 143:49–54
96. Griefahn B (2002) Sleep disturbances related to environmental noise. Noise Health 4:57–60
97. van Kamp I, Waye KP, Gunnarsson AD (2015) The effects of noise disturbed sleep in children on cognitive development and long term health. J Child Adolesc Behav 3:179
98. Porter ND, Kershaw AD, Ollerhead JB (2000) Adverse effects of night-time aircraft noise. Environ Res Consult Depart R&D Report 9964:2000
99. Basner M, Müller U, Griefahn B (2010) Practical guidance for risk assessment of traffic noise effects on sleep. Appl Acoust 71(6):518–522
100. CALM (2007) Research for a quieter Europe in 2020. An updated strategy paper of the CALM II Network. Brussels: European Commission Research Directorate-General, September, 2007.
101. Brown AL, Lam KC, van Kamp I (2015) Quantification of the exposure and effects of road traffic noise in a dense Asian city: a comparison with western cities. Environ Health 14:22
102. Miedema HME, Vos H (2007) Associations between self-reported sleep disturbance and environmental noise based on re-analyses of pooled data from 24 studies. Behav Sleep Med 5(1):1–20
103. Hong J, Kim J, Lim C, Kim K, Lee S (2010) The effects of long-term exposure to railway and road traffic noise on subjective sleep disturbance. J of Acoustical Society of America 128: 2829–2835
104. Basner M, Müller U, Elmenhorst EM (2011) Single and combined effects of air, road, and rail traffic noise on sleep and recuperation. Sleep 34:11–23
105. Tiesler CMT, Birk M, Thiering E, Kohlböck G, Koletzko S, Bauer CP, Berdel D, von Berg A, Babisch W, Heinrich J, GINIplus and LISAplus Study Groups (2013) Exposure to road traffic noise and children's behavioural problems and sleep disturbance: results from the GINIplus and LISAplus studies. Environ Res 123:1–8
106. Ising H, Ising M (2002) Chronic cortisol increases in the first half of the night caused by road traffic noise. Noise Health 4:13–21
107. Lercher P, Eisenmann A, Dekonick L, Botteldooren D (2013) The relation between disturbed sleep in children and traffic noise exposure in alpine valleys. Proceedings of the 42nd International Congress and Exposition on Noise Control Engineering InterNoise, Innsbruck, Austria, 2013.
108. Öhrström E, Hadzibajramovic E, Holmes M, Svensson H (2006) Effects of road traffic noise on sleep: studies on children and adults. J Environ Psychol 26:116–126
109. Guire S M, Aasvang G M (2017) Review of research on the effects of noise on sleep over the last 3 years. 12th ICBEN Congress on Noise as a Public Health Problem, 18th to 22nd June, 2017.
110. Foraster M, Eze IC, Vienneau D, Brink M, Cajochen C, Caviezel S, Héritier H, Schaffner E, Schindler C, Wanner M, Wunderli JM, Röösli M, Hensch NP (2016) Long-term transportation

noise annoyance is associated with subsequent lower levels of physical activity. Environ Int 91:341–349
111. Banerjee D (2013) Road traffic noise and self-reported sleep disturbance: results from a cross-sectional study in Western India. Noise Vib Worldw 44(2):10–17
112. Halperin D (2014) Environmental noise and sleep disturbances: a threat to health? Sleep Sci 7: 209–212
113. Spiegel K, Leproult R, Van Cauter E (1999) Impact of sleep debt on metabolic and endocrine function. Lancet 354(9188):1435–1439
114. Héritier H, Viennenu D, Frei P, Eze IC, Brind M, Hensch NP, Röösli M (2014) The association between and traffic noise exposure annoyance and Health related Quality of life (HRQOL). Int J Environ Res Public Health 11:12652–12667
115. Evandt J, Oftedal B, Krog NH, Nafstad P, Schwarc PE, Aasvang M (2017) A population based study on night time road traffic noise and Insomnia. Sleep 40(2):1–10
116. Fong DYT, Wong JYH, Huang L (2018) Effect of noise tolerance on non-restorative sleep: a population-based study in Hong Kong. BMJ Open 8:e020518
117. Quehl J, Müller U, Mendolia F (2017) Short-term annoyance from nocturnal aircraft noise exposure: results of the NORAH and STRAIN sleep studies. Int Arch Occup Environ Health 90(8):765–778
118. Smith M G, Ögren M, Thorsson P, Pedersen E, Wage K P (2016) Physiological effects of wind turbine noise on sleep. Proceedings of 22nd International Congress on Acoustics, ICA 2016, Buenos Aires, 2016.
119. Smith MG, Croy I, Hammar D, Waye KP (2016) Vibration from freight trains fragments sleep: a polysomnographic study. Sci Rep 6:24717
120. Broussard JL, Cauter EV (2016) Disturbances of sleep and circadian rhythms: Novel risk factors for obesity. Curr Opin Endocrinol Diabetes Obes 23(5):353–359
121. Spira AP, Chen-Edinoboro LP, Wu MN, Yaffe K (2014) Impact of sleep on the risk of cognitive decline and dementia. Curr Opin Psychiatry 27(6):478–483
122. Bubu OM, Brannick M, Mortimer J, Umasabor-Bubu O, Sebastião YV, Wen Y (2017) Sleep, Cognitive impairment and Alzheimer's disease: a systematic review and meta-analysis. Sleep 40(1)
123. Hume KL, Brink M, Basner M (2012) Effects of environmental noise on sleep. Noise Health 14:297–302
124. Directive 2002/49/EC of the European Parliament and of the Council of 25th June 2002, relating to the assessment and management of environmental noise, Official Journal of the European Communities, June, 2002.
125. Miedema H M E, Passchier-Vermeer W, Vos H (2003) Elements for a Position Paper on night-time transportation noise and sleep disturbance. TNO Inro report 2002-59, January, 2003.
126. Miedema H M E (2004) Self reported sleep disturbance caused by aircraft noise, TNO Inro, Delft, 2004-15, April, 2004.
127. Ekstedt M, Akerstedt T, Soderstrom M (2004) Microarousals during sleep are associated with increased levels of lipids, cortisol, and blood pressure. Psychosom Med 66(6):925–931
128. Ising H, Kruppa B (2004) Health effects caused by noise: evidence in the literature from the past 25 years. Noise Health 6:5–13
129. Lusk SL, Gillespie B, Hagerty BM, Ziemba RA (2004) Acute effects of noise on blood pressure and heart rate. Arch Environ Health 59(8):392–399
130. Sørensen M, Hvidberg M, Hoffmann B, Andersen ZL, Nordsborg RB, Lillelund KG, Jakobsen J, Tjønneland A, Kim O, Nielsen OR (2011) Exposure to road traffic and railway noise and associations with blood pressure and self reported hypertension: a cohort study. Environ Health 10:92
131. Babisch W (2011) Cardiovascular effects of noise. In: Nriagu JO (ed) Encyclopedia of environmental health. Elsevier, Burlington, pp 532–542
132. Münzel T, Schmidt FP, Steven MDS, Herzog J, Driber A, Sørensen M (2018) Environmental noise and the cardiovascular system. J Am Coll Cardiol 71(6):688–697

133. Wagner J, Cik M, Marth E, Satner BI, Gallasch E, Lackner A, Raggam RB (2010) Feasibility of testing three salivary stress biomarkers in relation to naturalistic traffic noise exposure. Int J Hyg Environ Health 213:153–155
134. Vienneau D, Schindler C, Perez L, Probst-Hensch N, Roosti M (2015) The relationship between transportation noise exposure and ischemic heart disease: a metaanalysis. Environ Res 138:372–380
135. Roswall N, Raaschou-Nielson O, Ketzel M, Gammelmark A, Overvad K, Olsen A, Sørensen M (2017) Long-term residential and traffic noise and NO_2 exposure in relation to risk of incident myocardial infraction: a Danish cohort study. Environ Res 156:80–86
136. Stansfield SA (2015) Noise effects on health in the context of air pollution exposure. Int J Environ Res Public Health 12:12735–12760
137. Tètreault LF, Perron S, Smarglassi A (2013) Cardiovascular health, traffic related air pollution and noise: Are associations mutually confounded? A systematic review. Int J Public Health 58: 649–666
138. Lercher P, Botteldooren D, Widmann U, Uhrner U, Kammeringer E (2011) Cardiovascular effects of environmental noise: research in Austria. Noise Health 13(52):234–250
139. Lewington S, Clarke R, Qizilbash N, Peto R, Collins R (2002) Age-specific relevance of usual blood pressure to vascular mortality: a meta-analysis of individual data for one million adults in 61 prospective studies. Lancet 360:1903–1913
140. Meier-Ewert HK, Ridker PM, Rifai N, Regan MM, Price NJ, Dinges DF, Mullington JM (2004) Effect of sleep loss on C-reactive protein, an inflammatory marker of cardiovascular risk. J Am Coll Cardiol 43:678–683
141. Portela PC, Fumado JC, Garcia HQ, Borrego FR (2009) Sleep-disordered breathing and acute stroke. Cerebrovasc Dis 27:104–110
142. Babisch W (2000) Traffic noise and cardiovascular disease. Epidemiological review and synthesis. Noise Health 8:9–32
143. Willich SN, Wegscheider K, Stallmann M, Keil T (2006) Noise burden and the risk of myocardial infarction. Eur Heart J 27:276–282
144. European Environment Agency (2010) Good practice guide on noise exposure and potential health effects. European Environment Agency, Copenhagen, p 2010
145. Banerjee D (2014) Association between transportation noise and cardiovascular disease: a meta-analysis of cross-sectional studies among adult populations from 1980 to 2010. Indian J Public Health 58(2):84–91
146. Yusuf S, Reddy S, Ounpuu S, Anand S (2001) Global burden of cardiovascular diseases, Part I: General considerations, the epidemiologic transition, risk factors and impact of urbanization. Circulation 104:2746–2753
147. Babisch W, Beule B, Schust M, Kersten N, Ising H (2005) Traffic noise and risk of myocardial infarction. Epidemiology 16:33–40
148. Babisch W (2008) Road traffic noise and cardiovascular risk. Noise Health 10(38):27–33
149. Farley TM, Meirik O, Chang CL, Poulter NR (1998) Combined oral contraceptives, smoking and cardiovascular risk. J Epidemiol Community Health 52:775–785
150. Sørensen M, Andersen ZJ, Nordsborg RB, Jensen SS, Lillelund KG, Beelen R, Schmidt EB, Tjønneland A, Overvad K, Nielsen OR (2012) Road traffic noise and incident myocardial infarction: a prospective cohort study. PLoS One 7(6):e39283
151. van Kempen EEMM, Casas M, Pershagen G, Foraster M (2017) Cardiovascular and metabolic effects of environmental noise, Systematic evidence review in the framework of the development of the WHO environmental noise guidelines for the European Region, RIVM Report 2017-0078. National Institute for Public Health and the Environment, The Netherlands
152. van Kempen E, Casas M, Pershagen G, Foraster M (2018) WHO environmental noise guidelines for the European region: a systematic review on environmental noise and cardiovascular and metabolic effects: a summary. Int J Environ Res Public Health 15(2):379
153. Jarup L, Dudley M-L, Babisch W, Houthuijs D, Swart W, Pershagen G, Bluhm G, Katsouyanni K, Velonakis M, Cadum E, Vigna-Taglianti F (2005) Hypertension and exposure

to noise near airports (HYENA): study design and noise exposure assessment. Environ Health Perspect 113:1473–1478
154. Babisch W, Houthuijs D, Pershagen G, Katsouyanni K, Velonakis M, Cadum E, Jarup L – for the HYENA Consortium (2008) Hypertension and exposure to noise near airports – results of the HYENA study, Non-auditory: 9th International Congress on Noise as a Public Health Problem (ICBEN) 2008, Foxwoods, p 8.
155. Evrard AS, Lefèvre M, Champelovier P, Lambert J, Laumon B (2017) Does aircraft noise exposure increase the risk of hypertension in the population living near airports in France? Occup Environ Med 74(2):123–129
156. Babisch W (2006) Transportation noise and cardiovascular risk, review and synthesis of epidemiological studies. Dose-effect Curve and Risk Estimation. Federal Environmental Agency, Berlin, Germany. January, 2006
157. Floud S, Blangiardo M, Clark C, de Hoogh K, Babisch W, Houthuijs D, Swart W, Pershagen G, Katsouyanni K, Velonakis M, Vigna-Taglianti F, Cadum E, Hansell AL (2013) Exposure to aircraft and road traffic noise and associations with heart disease and stroke in six European countries: a cross-sectional study. Environ Health 12:89
158. de Kluizenaar Y, van Lenthe FJ, Visschedijk AJ, Zandveld PY, Miedema HM, Mackenbach JP (2013) Road traffic noise, air pollution components and cardiovascular events. Noise Health 15:388–397
159. Szumilas M (2010) Explaining odds ratios. J Can Acad Child Adolesc Psychiatry 19(3): 227–229
160. Gan WQ, Davies HW, Koehoorn M, Brauer M (2012) Association of long-term exposure to community noise and traffic-related air pollution with coronary heart disease mortality. *Am J Epidemiol* 175(9):898–906
161. Huss A, Spoerri A, Egger M, Roosli M (2010) Aircraft noise, air pollution and mortality from myocardial infarction. Epidemiology 21:829–836
162. Selander J, Nilson ME, Bluhm G, Rosenlund M, Lindqvist M, Nise G, Pershagen G (2009) Long-term exposure to road traffic noise and myocardial infarction. Epidemiology 20:272–279
163. Beelen R, Hoek G, Houthuijs D, van den Brandt PA, Goldbohm RA, Fischer P, Schouten LJ, Armstrong B, Brunekreef B (2009) The joint association of air pollution and noise from road traffic with cardiovascular mortality in a cohort study. Occup Environ Med 66:243–250
164. Pyko A, Eriksson C, Wallas A, Ögren M, Pershagen G (2017) Long-term exposure to traffic noise and development of obesity. 12th ICBEN Congress on Noise as a Public Health Problem, 18–22nd June, 2017, 1–7.
165. Christensen JS, Nielsen OR, Tjønneland A, Nordsborg RB, Jensen SS, Sørensen TIA, Sørensen M (2015) Long-term exposure to residential traffic noise and changes in body weight and waist circumference: a cohort study. Environ Res 143(Part A):154–161
166. Shin S, Bai L, Oiamo TH, Burnett RT, Weichenthal S, Jerrett M, Kwong JC, Goldberg MS, Copes R, Kopp A, Chen H (2020) Association between road traffic noise and incidence of diabetes mellitus and hypertension in Toronto, Canada: a population-based cohort study. J Am Heart Assoc 17 9(6):e013021
167. Münzel T, Sørensen M, Gori T, Schmidt FP, Rao X, Brook J, Chen LC, Brook RD, Rajagopalan S (2017) Environmental stressors and cardio-metabolic disease: part I-epidemiologic evidence supporting a role for noise and air pollution and effects of mitigation strategies. Eur Heart J 38(8):550–556
168. Münzel T, Sørensen M, Schmidt F, Schmidt E, Steven S, Kröller-Schön S, Daiber A (2018) The adverse effects of environmental noise exposure on oxidative stress and cardiovascular risk. Antioxid Redox Signal 28(9):873–908
169. Clark C, Clare C, Notley H (2020) Evidence for environmental noise effects on health for the United Kingdom policy context: a systematic review of the effects of environmental noise on mental health, wellbeing, quality of life, cancer, dementia, birth, reproductive outcomes, and cognition. Int J Environ Res Public Health 17:393. https://doi.org/10.3390/ijerph17020393

170. Foraster M, Künzli N, Aguilera I, Rivera M, Agis D, Vila J, Bouso L, Deltell A, Marrugat J, Ramos R, Sunyer J, Elosua R, Basagaña X (2014) High blood pressure and long-term exposure to indoor noise and air pollution from road traffic. Environ Health Perspect 122: 1193–2000
171. Banerjee D, Das PP, Foujdar A (2014) Association between road traffic noise and prevalence of coronary heart disease. Environ Monit Assess 186(5):2885–2893
172. Singh N, Davar SC (2004) Noise pollution-sources, effects and control. J Human Ecol 16:181–187
173. Plangsangmas V, Leeudomwong S, Kongthaworn P (2012) Sound pressure level in an infant incubator. MAPAN-J Metrol Soc India 27(4):199–203
174. Ryherd EE, Waye KP, Ljungkvist L (2008) Characterizing noise and perceived work environment in an neurological intensive care unit. J Acoustical Soc Am 123:747–756
175. Khaiwal R, Singh T, Tripathy JP, Mor S, Munjal S, Patro B, Panda N (2016) Assessment of noise pollution in and around a sensitive zone in North Indian and its non-auditory impacts. Sci Total Environ 566-567:981–987
176. MacLeod M, Dunn J, Busch-Vishnaic IJ, West JE, Reedy A (2007) Quieting Weinberg 5C: a case study in hospital noise control. J of Acoustical Society of America 121(6):3501–3508
177. Environmental Noise Guidelines for the European Region (2018) World Health Organization. Regional Office for Europe, Denmark, p 2018
178. Guski R, Schreckenberg D, Schuemer R (2017) WHO environmental noise guidelines for the European region: a systematic review on environmental noise and annoyance. Int J Environ Res Public Health 14:1539
179. Basner M, McGuire S (2018) WHO environmental noise guidelines for the European region: a systematic review on environmental noise and effects on sleep. Int J Environ Res Public Health 15:519. https://doi.org/10.3390/ijerph15030519
180. Bull HP: Was sagt die medizinische Wissenschaft zur Geräuschbe- lästigung? Ausgabe vom 11. January 1963. www.zeit.de/1963/02/wassagt-die-medizinische-wissenschaft-zur-geraeuschbelaestigung (Accessed on July, 2019).
181. Hahad O, Kröller-Schön S, Daiber A, Münzel T (2019) The cardiovascular effects of noise. Dtsch Arztebl Int 116:245–250
182. Dimakopoulou K, Koutentakis K, Papageorgiou I, Kasdagli MI, Haralabidis AS, Sourtzi P, Samoli E, Houthuijs D, Swart W, Hansell AL, Katsouyanni K (2017) Is aircraft noise exposure associated with cardiovascular disease and hypertension? Results from a cohort study in Athens, Greece. Occup Environ Med 74:830–837
183. Ohlwein S, Hennig F, Lucht S, Matthiessen C, Pundt N, Moebus S, Jöckel KH, Hoffmann B (2019) Indoor and outdoor road traffic noise and incident diabetes mellitus: Results from a longitudinal German cohort study. Environ Epidemiol 3:e037
184. Oftedal B, Krog NH, Pyko A, Oftedal B, Krog NH, Pyko A, Eriksson C, Graff-Iversen S, Haugen M, Schwarze P, Pershagen G, Aasvang GM (2015) Road traffic noise and markers of obesity – a population-based study. Environ Res 138:144–153
185. Pyko A, Eriksson C, Oftedal B, Hilding A, Östenson CG, Krog NH, Julin B, Aasvang GM, Pershagen G (2015) Exposure to traffic noise and markers of obesity. Occup Environ Med 72(8):594–601
186. Eriksson C, Hilding A, Pyko A, Bluhm G, Pershagen G, Östenson CG (2014) Long-term aircraft noise exposure and body mass index, waist circumference, and type 2 diabetes: a prospective study. Environ Health Perspect 122(7):687–694
187. Sorensen M, Andersen ZJ, Nordsborg RB, Becker T, Tjønneland A, Overvad K, Raaschou-Nielsen O (2013) Long-term exposure to road traffic noise and incident diabetes: a cohort study. Environ Health Perspect 121:217–222
188. Heidemann C, Niemann H, Paprott R, Du Y, Rathmann W, Scheidt-Nave C (2014) Residential traffic and incidence of type 2 diabetes: the German Health Interview and Examination Surveys. Diabet Med 31:1269–1276

189. Baliatsas C, van Kamp I, van Poll R, Yzermans J (2016) Health effects from low-frequency noise and infrasound in the general population: is it time to listen? A systematic review of observational studies. Sci Total Environ 557–558:163–169
190. Davies H, van Kamp I (2011) Noise and cardiovascular diseases: a review of the literature 2008–2011. Noise Health 14:287–291
191. Hansell AL, Blangiardo M, Fortunato L, Floud S, Hoogh K, Fecht D, Ghosh RE, Laszlo HE, Pearson C, Beale L, Beevers S, Gulliver J, Best N, Richardson S, Elliott P (2013) Aircraft noise and cardiovascular disease near Heathrow airport in London: small area study. BMJ 347: f5432
192. Schmidt F, Kolle K, Kreuder K, Schnorbus B, Wild P, Hechtner M, Binder H, Gori T, Münzel T (2015) Nighttime aircraft noise impairs endothelial function and increases blood pressure in patients with or at high risk for coronary artery disease. Clin Res Cardiol 104(1):23–30
193. Andersson H, Jonsson L, Ögren M (2013) Benefit measures for noise abatement: calculations for road and rail traffic noise. Eu Transp Res Rev 5(3):135–148
194. Paunović K, Belojević G (2014) Burden of myocardial infarction attributable to road-traffic noise: a pilot study in Belgrade. Noise Health 16(73):374–379
195. Devleesschauwer B, Havelaar AH, Maertens de Noordhout C, Haagsma JA, Praet N, Dorny P, Duchateau L, Torgerson PR, Van Oyen H, Speybroeck N (2014) Calculating disability-adjusted life years to quantify burden of disease. Int J Public Health 59(3):565–569
196. Tobollik M, Hintzsche M, Wothge J, Myck T, Plass D (2019) Burden of disease due to traffic noise in Germany. Int J Environ Res Public Health 16:2304. https://doi.org/10.3390/ijerph16132304
197. Hellmuth T, Classen T, Kim R, Kephalopoulos S (2012) Methodological guidance for estimating the burden of disease from environmental noise. European Commission Joint Research Centre, Institute for Health and Consumer Protection.
198. European Environment Agency (2014) Noise in Europe 2014. EEA Report No 10/2014.
199. Eriksson C, Bodin T, Selander J (2017) Burden of disease from road traffic and railway noise – a quantification of healthy life years lost in Sweden. Scand J Work Environ Health 43(6): 519–525
200. Mueller N, Rojas-Rueda D, Basagaña X, Cirach M, Cole-Hunter T, Dadvand P, Donaire-Gonzalez D, Foraster M, Gascon M, Martinez D, Tonne C, Triguero-Mas M, Valentín A, Nieuwenhuijsen M (2017) Health impacts related to urban and transport planning: a burden of disease assessment. Environ Int 107:243–257
201. Houthuijs D, Swart W, van Kempen E (2019) Implications of environmental noise on health and wellbeing in Europe. European Topic Centre on Air Pollution and Climate Change Mitigation, The Netherlands, European Environment Agency, Eionet Report – ETC/ACM 2018/10
202. European Environment Agency (2020) Environmental noise in Europe-2020. EEA report No 22/2019, Publications Office of the European Union, Luxembourg.
203. Miedema HME, Vos H (2004) Noise annoyance from stationary sources: Relationships with exposure metric day-evening-night level (DENL) and their confidence intervals. J Acoust Soc Am 116(1):334–343
204. van Kempen EEMM (2008) Transportation noise exposure and children's health and cognition. Thesis Utrecht University. Utrecht, the Netherlands., http://dspace.library.uu.nl/handle/1874/25891 (Accessed 14 January 2021).
205. Noise in Europe., http://www.noiseineu.eu/en/14-socioeconomic_impact/subpage/view/page/57 (Accessed on 9 Jan 2021).
206. Harding A-H, Frost GA, Tan E, Tsuchiya A, Mason HM (2013) The cost of hypertension-related ill-health attributable to environmental noise. Noise Health 15(67):437–445
207. Münzel T, Miller MR, Sørensen M, Lelieveld J, Daiber A, Rajagopalan S (2020) Reduction of environmental pollutants for prevention of cardiovascular disease: it's time to act. Eur Heart J 41:3989–3997

208. European Commission report (2016) Links between noise and air pollution and socioeconomic status. Science for Environment Policy. https://ec.europa.eu/environment/integration/research/newsalert/pdf/air_noise_pollution_socioeconomic_status_links_IR13_en.pdf (Accessed on 17 Dec 2020).
209. Huang T, Chan TC, Huang YJ, Pan WC (2020) The Association between Noise Exposure and Metabolic Syndrome: A Longitudinal Cohort Study in Taiwan. Int J Environ Res Public Health 17:4236. https://doi.org/10.3390/ijerph17124236
210. Chang TY, Yu TY, Liu CS, Young LH, Bao BY (2020) Occupational noise exposure and its association with incident hyperglycaemia: a retrospective cohort study. Sci Rep 10:8584. https://doi.org/10.1038/s41598-020-65646-1
211. Carlos R, Fanny M, Pierre J, Maria S, Nicolas R (2019) Health impact of noise in Greater Paris Metropolis: assessment of healthy life years lost. Proceedings Internoise, 16th to 19th June, 2019, Madrid.
212. Shepherd D, Welch D, Dirks KN, McBride D (2013) Do quiet areas afford greater health related quality of life than noisy areas? Int J Environ Res Public Health 10(4):1284–1303
213. Münzel T, Gori T, Babisch W, Basner M (2014) Cardiovascular effects of environmental noise exposure. Eur Heart J 35(13):829–836
214. Allen R W, Adar S D (2011) Are both air pollution and noise driving adverse cardiovascular health effects from motor vehicles. Environ Res 111 (1):184-185.
215. Dendup T, Feng X, Clingan S, Astell-Burt T (2018) Environmental risk factors for developing type 2 Diabetes Mellitus: a systematic review. Int J Environ Res Public Health 15:78. https://doi.org/10.3390/ijerph15010078
216. Roswall N, Raaschou-Nielsen O, Ketzel M, Gammelmark A, Overvad K, Olsen A, Sørensen M (2017) Long-term residential road traffic noise and NO2 exposure in relation to risk of incident myocardial infarction – a Danish cohort study. Environ Res 156:80–86
217. Daniel E (2007) Noise and hearing loss: a review. J Sch Health 77(5):225–231
218. Prasher D (2000) Noise Pollution Health Effects Reduction (NOPHER): an European Commission Concerted Action Workplan. Noise Health 2:79–84
219. Noise Policy Statement for England (NPSE), Department for Environment, Food and Rural Affairs (DEFRA), UK, 2010.
220. European Commission Science for Environment Policy (2015) Thematic issue: noise impacts on health 47, 1–15.
221. Selander J, Bluhm G, Theorell T, Pershagen G, Babisch W, Seiffert I, Houthuijs D, Breugelmans O, Taglianti FV, Antoniotti MC, Velonakis E, Davou E, Dudley ML, Järup L, for the HYENA Consortium (2009) Saliva cortisol and exposure to aircraft noise in six European countries. Environ Health Perspect 117(11):1713–1716
222. Baudin C, Lefèvre M, Selander J, Babisch W, Cadum E, Carlier MC, Champelovier P, Dimakopoulou K, Huithuijs D, Lambert J, Laumon B, Pershagen G, Theorell T, Velonaki V, Hansell A, Evrard AS (2019) Saliva cortisol in relation to aircraft noise exposure: pooled-analysis results from seven European countries. Environ Health 18:102
223. Lefèvre M, Carlier M-C, Champelovier P, Lambert J, Laumon B, Evrard A-S (2017) Effects of aircraft noise exposure on saliva cortisol near airports in France. Occup Environ Med 74(8): 612–618
224. Wallas A, Eriksson C, Gruzieva O, Lind T, Pyko A, Sjöström M, Ögren M, Pershagen G (2018) Road traffic noise and determinants of saliva cortisol levels among adolescents. Int J Hyg Environ Health 221(2):276–282
225. Fouladi Dehaghi B, Nassiri P, Monazam MR, Ebrahimi Ghavam Abadi L, Farahani ES, Hassanzadeh G, Hoseini M, Hesampor M (2014) Noise-induced stress assessment by salivary cortisol measurement. Jundishapur J Health Sci 6(3):e21719. https://doi.org/10.5812/jjhs.21719
226. Waye KP, Bengtsson J, Rylander R, Hucklebridge F, Evans P, Clow A (2002) Low frequency noise enhances cortisol among noise sensitive subjects during work performance. *Life Sci* 70(7):745–758

227. Kudielka BM, Hellhammer DH, Wüst S (2009) Why do we respond so differently? Reviewing determinants of human salivary cortisol responses to challenge. Psychoneuroendocrinology 34(1):2–18
228. Dutheil F, Baker JS, Navel V (2020) COVID-19 and cardiovascular risk: Flying toward a silent world? J Clin Hypertens 22:1945
229. Lecocq T, Hicks SP, Noten KV, van Wijk K, Koelemeijer P, De PRSM, Massin F, Hillers G, Anthony RE, Apoloner MT, Solórzano MA, Assink JD, Büyükakpınar P, Cannata A, Cannavo F, Carrasco S, Caudron C, Chaves EJ, Cornwell DG, Craig D, den Ouden OFC, Diaz J, Donner S, Evangelidis CP, Evers L, Fauville B, Fernandez GA, Giannopoulos D, Gibbons SJ, Girona T, Grecu B, Grunberg M, Hetényi G, Horleston A, Inza A, Irving JCE, Jamalreyhani M, Kafka A, Koymans MR, Labedz CR, Larose E, Lindsey NJ, McKinnon M, Megies T, Miller MS, Minarik W, Moresi L, Márquez-Ramírez VH, Möllhoff M, Nesbitt IM, Niyogi S, Ojeda J, Oth A, Proud S, Pulli J, Retailleau L, Rintamäki AE, Satriano C, Savage MK, Shani-Kadmiel S, Sleeman R, Sokos E, Stammler K, Stott AE, Subedi S, Sørensen MB, Taira T, Tapia M, Turhan F, van der Pluijm B, Vanstone M, Vergne J, Vuorinen TAT, Warren T, Wassermann J, Xiao H (2020) Global quieting of high-frequency seismic noise due to COVID-19 pandemic lockdown measures. Science 369:1338–1343
230. Poli P, Boaga J, Molinari I, Cascone V, Boschi L (2020) The 2020 coronavirus lockdown and seismic monitoring of anthropic activities in Northern Italy. Sci Rep 10(1):9404
231. Giamello JD, Abram S, Bernardi S, Lauria G (2020) The emergency department in the COVID-19 era. Who are we missing ? Eur J Emerg Med 27(4):305–306
232. Antza C, Stabouli S (2020) Reduction in environmental noise during COVID-19 pandemic and cardiovascular disease: a mystery for further investigation. J Clin Hypertens 22:1947–1948
233. Li S, Fong DYT, Wong JYH, Pherson BM, Lau EYY, Huang L, Man IPMS (2021) Indoor nocturnal noise is associated with body mass index and blood pressure: a cross-sectional study. BMC Public Health 21:815
234. Saucy A, Schäffer B, Tangermann L, Vienneau D, Wunderli JM, Röösli M (2021) Does nighttime aircraft noise trigger mortality? A case-crossover study on 24 886 cardiovascular deaths. Eur Heart J 42(8):835–843

Chapter 10
MCDM Approach in Noise Control

Abstract The chapter presents a novel application of one of multiple-criteria decision-making (MCDM) approach known as Fuzzy TOPSIS (Technique for Order Preference by Similarity to an Ideal Solution) in selection of optimal road traffic noise abatement strategy. This approach can be useful in the determination of the ranking order of the various alternatives for road traffic noise abatement with respect to the various criteria considered, especially when there is a varied opinion of the experts for finding out the Best Practicable and Economical Option (BPEO). It is proposed that application of this approach can be very helpful in selection and application of regional road traffic noise management strategies, ranking the noise action plans (NAPs), selection of the optimal noise barriers, specific source noise control, and for the selection of optimal aircraft noise abatement and control strategies. The approach finds immense applications for expert committees in situations when the experts opinion vary for the selection and implementation of BPEO for the noise management and control.

10.1 Introduction

Traffic noise has accentuated to very high levels in recent years owing to the rapid increase in vehicular density. With rapid urbanization and economic advancements, vehicular density has considerably increased causing serious noise and air pollution in metropolitan cities of the world. It is thus imperative to monitor the ambient noise levels for devising methods for combating the radiated noise and also planning of guidelines to be implemented for the new projects. A calibrated noise model with highest accuracy is thus indispensable for town planners and architects for conducting the environmental impact assessment (*EIA*) studies in respect of noise. There has been a great emphasis on noise and its effects in developed nations, and each nation has now come up with a validated road traffic noise model [1–3]. The European Directive 2002/49/EC lays emphasis on environmental noise and related problems and obligates the member states to provide an access to the information on noise pollution [4]. The increase in vehicular population has created an urgency of devising strategic noise control action plans and noise abatement goal for managing

N. Garg, *Environmental Noise Control*,
https://doi.org/10.1007/978-3-030-87828-3_10

noise issues and effects, including noise abatement particularly for the places near the vicinity of roads. Some studies reveal that significant reductions can be achieved by implementing traffic management noise mitigation measures in urban areas. Reducing the travel demand leads to the most significant reduction in population exposure to L_{den} and L_{night}, while considerable reductions can also be achieved by reducing travel speeds [5].

10.2 Traffic Noise Abatement Measures

The various strategies for road traffic noise abatement are summarized in Chapter 7. The strict implementation of a noise control program utilizing an inventory of noise control methods is a must to combat the high ambient noise levels. Efficient traffic management and intelligent transport system [6] is one of the vital reforms in this regard. Restrictions in use of horns especially near sensitive spaces like hospitals, schools, etc., reducing speed limits, timing traffic lights to achieve smooth traffic flow, smooth driving avoiding frequent acceleration and deceleration, and prohibiting heavy vehicles in certain roads and streets can be very fruitful in achieving the desired objectives. Reducing the speed of the vehicle by 10 km/h, a noise reduction by 2–3 dB(A) is achieved [7]. Land use planning and management is also another factor minimizing the traffic noise impact. Placing as much distance and noncompatible activities like parking bays, open areas, etc. between source and receiver is also helpful. The noise control at the receiver is another important aspect to focus. The shape, orientation, and location of the building and the arrangement of the internal spaces should be chosen to reduce potential noise problems. Reduction of noise by the building itself should only be considered as a last resort, that is, when insufficient reduction can be made at the source or between the source and the building [8]. Source noise control, that is, propulsion and rolling noise radiated by vehicles, is very essential in achieving the targets. An optimal combination of tire and road surface is instrumental in reducing road noise [9, 10]. Enhancing the sound insulative characteristics of building elements, namely, walls, windows, roof, ceilings, and exterior facade, will considerably reduce the outside noise and protect the residents from hazards of noise pollution [11, 12]. The prescriptive and verification criteria for sound insulation of building elements can be very instrumental in achieving the acoustic comfort in buildings [13, 14]. Noise barriers are also one of the most effective methods of mitigating roadway, railway, and industrial noise sources especially when there is not much control over source noise. Noise barriers derive their performance by blocking the line of sight, thus creating a sound shadow [15, 16]. An opinion survey [17] conducted for investigating the need of noise barriers has explicitly presented a clear public perception with a majority of respondents (80%) demanding for erecting noise barrier along national highways and willingness of 66% of respondents for incorporating the concept of noise barrier in future design or modification of existing boundary walls. Vegetation can be used in urban areas for transportation noise control as relatively inexpensive and aesthetically pleasing noise barriers. The visual shielding of a noise source has considerable psychological advantage as well. A minimum of 2–5 m wide dense shrubs followed by medium height trees are effective in offering 2–5 dB(A) noise reduction from traffic noise in residential areas [18, 19]. Earth mounds also serve as an optimal solution attributed to less cost, easiness to maintain, and unlimited life span. The development of a specific road traffic noise management strategy or noise action plan selection is a necessary part of pro-active approach to environmental noise management [20]. Dintrans and Préndez presented a study of road traffic noise abatement strategy suggesting three measures: traffic flow reduction, maximum speed limit of 40 km/h for heavy vehicles, and current road surface improvement/ replacement for city of Santiago, Chile. Scenarios that considered speed limits for heavy vehicles were the most effective reducing by 30% the number of facades in conflict [21]. It may be noted that there can be some constraints also in implementation of the noise abatement strategies. Practical feasibility of the proposed plan, maintenance requirements, structural integrity and installation perspectives for noise barriers, administrative and legal viability, compatibility with environment, safety and durability, economic considerations, and cost-to-benefit ratio, that is, the costs incurred on a noise abatement strategy versus the expect noise abatement in dB (A) are the various criteria to be considered for selection of best practicable and economical option (*BPEO*). It is thus imperative for the town planners to consider all these criteria before finally selecting the road map for noise abatement so as to bring the ambient noise levels well within the ambient standards [22–24] or noise abatement criteria [25] as recommended by Federal Highway Administration (FHWA), US Department of Transportation.

There is no simple solution to such a problem wherein among the several alternatives one has to find the optimal strategy considering the various benefit criteria. *TOPSIS* (Technique for Order Preference by Similarity to Ideal Solution) developed by Hwang and Yoon [26] is one such probable solution to this problem. This chapter illustrates a case study of application of *TOPSIS* approach in the selection of optimal regional road traffic noise management strategy (*RTNMS* described in [20]) for road traffic noise abatement using the fuzzy database of the linguistic variable assigned by three experienced decision-makers. Such an approach has been presented recently in the selection of optimal noise barriers for traffic noise abatement [27]. It is envisaged that application of this approach can be instrumental in solving this problem with minimum numerical complexities and in minimum time and shall be indispensable for deciding optimal road traffic noise abatement strategy [28–30]. The methodology is also flexible enough to accommodate the varied opinion of any number of decision-makers in assigning the linguistic variables to the criteria versus alternatives available for a particular problem so as to arrive at an optimal solution.

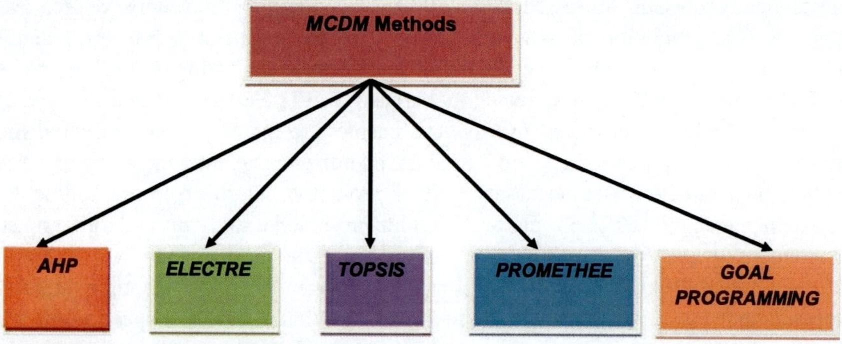

Fig. 10.1 Various *MCDM* methods for decision-making

10.3 *MCDM* Approach

Multiple-criteria decision-making (*MCDM*) is one of the fastest growing areas used often for solving various decision-making problems. The primary concern for decision aid is choosing the most preferred alternative to decision-makers, ranking the alternatives in order of importance for selection problems as screening the alternatives for the final decision [31]. The multiple criteria method is used for identifying optimal solution from a finite number of alternatives. *MCDM* is a function of alternatives (available options), criteria (performance parameters), and their comparison with an objective of achieving an optimal or the best solution. There are various techniques such as Weighted Sum method (*WSM*); Analytical Hierarchical Process (*AHP*); *ELECTRE* (ELimination EtChoix Traduisant la REalité) meaning Elimination and Choice Expressing the REality); Technique for Order Preference by Similarity to an Ideal Solution (*TOPSIS*); Analytic Network Process (*ANP*); Grey analysis; Preference Ranking Organization METHod for Enrichment of Evaluations (*PROMETHEE*), and Goal programming. The merits and demerits of all these techniques had been explicitly discussed by Aruldoss (2013) [32]. Kim et al. highlighted the advantages of *TOPSIS* as a major *MCDM* technique over other techniques like *AHP*, *ELECTRE*, etc. [33]. Figure 10.1 shows the various *MCDM* methodologies used in research.

10.3.1 TOPSIS *Method*

Technique for Order Preference by Similarity to an Ideal Solution (*TOPSIS*) is one of the classical approaches developed by Hwang and Yoon [26] and is a rational, logical, and easily understandable method. The basic methodology behind *TOPSIS* approach is that the chosen alternative should have the shortest distance from the positive ideal solution and farthest distance from the negative solution. Chen and Hwang [34] transformed this method with the crisp data to method with fuzzy data [34]. Chen [35] extended the *TOPSIS* for group decision-making in fuzzy environment [35]. The importance weights of various criteria and ratings of alternatives with respect to criteria are considered as linguistic variables. These variables are transformed into triangular fuzzy numbers. Final values of fuzzy numbers that contribute fuzzy decision matrix and fuzzy weight vector are calculated as a means of values assessed by the members of decision group. Normalization of decision matrix is performed and fuzzy positive ideal solution (*FPIS*) and fuzzy negative ideal solution (*FNIS*) are determined. The distances of each alternative to *FPIS* and *FNIS* are calculated and corresponding closeness coefficient of each alternative to these solutions is evaluated. Thus, the alternatives are ranked based on the closeness coefficient [36]. The lower value of the closeness coefficient indicates that an alternative is close to *FPIS* and farthest from *FNIS* simultaneously [37]. So, rank alternatives according to the closeness coefficient in the ascending order. The procedure of application is simple and easy to understand and as such it finds application in solving decision-making problems in various fields. *TOPSIS* has been widely used as described by Shih et al. [38] in areas of transportation [39], product design [40], manufacturing [41], quality control [42], etc. The application of *TOPSIS* in the field of noise control engineering has been rarely reported. Garg et al. (2015) [27] utilized this innovative technique in selection of optimal noise barrier for road traffic noise abatement. The merits of the integrated fuzzy *TOPSIS* is that not only the subjective and objective criteria are simultaneously considered in the real life, but also it releases the limitations of crisp values and facilitates its implementation as a computer based decision support system for ranking the alternatives in fuzzy environment [43]. It is imperative to integrate fuzzy data in conjunction with linguistic variables for solution of such problems wherein the best alternative is to be selected with respect to various criteria.

10.3.2 Fuzzy Versus Real-Time Data for MCDM

In many situations, the real-time data for various parameters such as compatibility with environment, safety, durability, installation, corrosion resistance, and structural integrity cannot be precisely measured or described in a single unit due to the subjective perception of individuals and as such in these situations; use of Fuzzy approach can serve the purpose. Fuzzy numbers can be assigned to these parameters based on a particular problem as described in Tables 10.1 and 10.2 for the solution of

Table 10.1 Linguistic variables for the importance weight of each criterion [37]

S. no.	Weight	Linguistic variables
1	Very Low (VL)	(0; 0; 0.1)
2	Low (L)	(0; 0.1; 0.3)
3	Medium Low (ML)	(0.1; 0.3; 0.5)
4	Medium (M)	(0.3; 0.5; 0.7)
5	Medium High (MH)	(0.5; 0.7; 0.9)
6	High (H)	(0.7; 0.9; 1.0)
7	Very High (VH)	(0.9; 1.0; 1.0)

Table 10.2 Linguistic variables for the ratings [37]

S. no.	Ratings	Linguistic variables
1	Very Poor (VP)	(0; 0; 1)
2	Poor (P)	(0; 1; 3)
3	Medium Poor (MP)	(1; 3; 5)
4	Fair (F)	(3; 5; 7)
5	Medium Good (MG)	(5; 7; 9)
6	Good (G)	(7; 9; 10)
7	Very Good (VG)	(9; 10; 10)

Table 10.3 The importance weight of the criteria

Criteria for noise abatement		Decision-makers D1	D2	D3
C1	Practical Feasibility	VH	VH	VH
C2	Compatibility with Environment	VH	H	MH
C3	Maintenance	M	ML	M
C4	Safety & Durability	MH	M	MH
C5	Structural Integrity & Installation Perspectives	M	MH	MH
C6	Economic Considerations	MH	H	VH
C7	Administrative & Legal Viability	H	MH	H
C8	Expected Noise Abatement in dB(A)	VH	H	H

a particular problem. The benefit of this methodology is that varied opinions of different experts can be used to find out a single optimized solution using *MCDM* approach.

10.3.3 Application of Fuzzy TOPSIS *Approach*

The application of fuzzy *TOPSIS* method proposed by Mahdavi et al. [37] is described in detail in Garg et al. (2015) [27] pertaining to its applications in the selection of optimal noise barriers as follows:

"Consider the possible alternatives and criteria for the given problem for which optimal solution is to be evolved [37].

Step 1: Assign the linguistic variables for the weight of the criteria. The decision makers use the linguistic weighting variables to assess the importance of the criteria for the given problem as shown in Table 10.3.

Step 2: Assign the linguistic ratings for alternatives with respect to criteria. The decision makers use the linguistic rating variables to evaluate the rating of alternatives with respect to each criterion for the given problem as shown in Table 10.4.

Step 3: Convert the linguistic ratings (x_{ij}, $i = 1, 2 \ldots m; j = 1, 2 \ldots n$) for alternatives with respect to criteria and the appropriate linguistic variables (w_j, $j = 1, 2 \ldots n$)

Table 10.4 The ratings of traffic noise action plans (alternatives) by decision-makers under all criteria

Alternative versus criteria's	Practical feasibility and ease of implementation (C1)			Environmental Compatibility (C2)			Maintenance (C3)			Safety and durability (C4)			Structural integrity, installation perspectives (C5)			Economic considerations (C6)	Administrative and legal viability (C7)			Expected Noise Abatement in dB(A) (C8)
	D1	D2	D3	D1	D2	D3	D1	D2	D3	D1	D2	D3	D1	D2	D3	D1, D2, D3	D1	D2	D3	D1, D2, D3
Improvement in road surface (A1)	G	VG	VG	G	VG	VG	MP	F	P	G	MG	G	F	P	MP	P	MG	G	MG	F
Low height barrier (A2)	F	MG	G	G	MG	G	F	P	MP	F	MG	MG	P	MP	P	VP	P	MP	P	VG
Green belt (A3)	VG	VG	G	VG	VG	VG	G	G	MG	VG	VG	VG	G	MG	MG	MG	G	MG	G	G
Sound insulation of facades (A4)	VP	MP	P	F	MG	MG	P	MP	P	G	G	MG	P	P	VP	VP	P	P	VP	VG
Traffic management (A5)	MG	G	F	G	G	G	G	G	VG	VG	G	G	G	G	VG	MG	F	MG	MG	G
Prohibiting horn noise (A6)	VG	VG	VG	VG	VG	VG	VG	VG	G	G	G	G	VG	VG	VG	VG	G	VG	VG	G
Enhancing sound insulation of window glazing (A7)	G	VG	G	G	G	VG	MG	G	MG	G	MG	G	G	VG	G	MP	G	MG	G	G

for the weight of the criteria into triangular fuzzy numbers using Tables 10.1 and 10.2. The fuzzy decision matrix, D is described by equation (10.1) and weight vector, W is defined by equation (10.2).

$$D = \begin{matrix} A_1 \\ A_2 \\ A_3 \\ \cdot \\ \cdot \\ A_m \end{matrix} \begin{bmatrix} x_{11} & x_{12} & x_{13} & \cdot & \cdot & x_{1n} \\ x_{21} & x_{22} & x_{23} & \cdot & \cdot & x_{2n} \\ x_{31} & x_{32} & x_{33} & \cdot & \cdot & x_{3n} \\ \cdot & \cdot & \cdot & \cdot & \cdot & \cdot \\ \cdot & \cdot & \cdot & \cdot & \cdot & \cdot \\ x_{m1} & x_{m2} & x_{m3} & \cdot & \cdot & x_{mn} \end{bmatrix} \tag{10.1}$$

$$W = [W_1 \quad W_2 \quad W_3 \quad \ldots \quad W_n] \tag{10.2}$$

where (x_{ij}; $i = 1; 2; \ldots ;m; j = 1; 2; \ldots ; n$) and ($w_j$; $j = 1; 2; \ldots ; n$) are linguistic triangular fuzzy numbers such that $x_{ij} =(a_{ij}; b_{ij}; c_{ij})$ and $w_j = (w_{j1} ;w_{j2}; w_{j3})$. Note that w_j represents the weight of the jth criteria, C_j and x_{ij} is the performance rating of the ith alternative, A_i, with respect to the jth criteria, C_j, evaluated by k evaluators. The present study considers the method of average value to integrate the fuzzy performance score x_{ij} for k evaluators concerning the same evaluation criteria, i.e.

$$x_{ij} = \left(x_{ij}^{-1} + x_{ij}^{-2} + \ldots\ldots.. + x_{ij}^{-k}\right)/k \tag{10.3}$$

where x_{ij}^{-k} is the rating of alternative A_i with respect to criterion C_j evaluated by kth evaluators, and $x_{ij}^{-k} = (a_{ij}^{-k}, b_{ij}^{-k}, c_{ij}^{-k})$. Linguistic evaluations (shown in Tables 10.3 and 10.4) are converted into triangular fuzzy numbers to ascertain the fuzzy decision matrix and the fuzzy weight of each criterion given by Eq. (10.3) is determined.

Step 4: Construct the normalized fuzzy decision matrix. The normalized fuzzy decision matrix denoted by R is shown as [37]:

$$R = \left[r_{ij}\right]_{m\times n} \tag{10.4}$$

If (x_{ij}; $i = 1; 2; \ldots ;m; j = 1; 2; \ldots ; n$) are triangular fuzzy numbers, then the normalization process can be performed by:

$$r_{ij} = \left(a_{ij}/c_j^*, b_{ij}/c_j^*, c_{ij}/c_j^*\right); i = 1, 2, \ldots.., m; j \in B \tag{10.5}$$

where B is the set of benefit criteria.

$$c_j^* = \max_i c_{ij}; j \in B \tag{10.6}$$

Step 5: Construct the weighted normalized fuzzy decision matrix. The weighted normalized value V is calculated by:

$$V = \begin{bmatrix} v_{11} & v_{12} & v_{13} & \cdot & \cdot & \cdot & v_{1n} \\ v_{21} & v_{22} & v_{23} & \cdot & \cdot & \cdot & v_{2n} \\ v_{31} & v_{32} & v_{33} & \cdot & \cdot & \cdot & v_{3n} \\ \cdot & \cdot & \cdot & \cdot & & & \cdot \\ \cdot & \cdot & \cdot & & \cdot & & \cdot \\ \cdot & \cdot & \cdot & & & \cdot & \cdot \\ v_{m1} & v_{m2} & v_{m3} & \cdot & \cdot & \cdot & v_{mn} \end{bmatrix} \tag{10.7}$$

$$= \begin{bmatrix} w_1 r_{11} & w_2 r_{12} & \cdot & \cdot & \cdot & w_j r_{1j} & \cdot & \cdot & \cdot & w_n r_{1n} \\ w_1 r_{21} & w_2 r_{22} & \cdot & \cdot & \cdot & w_j r_{2j} & \cdot & \cdot & \cdot & w_n r_{2n} \\ \cdot & \cdot & \cdot & & & \cdot & \cdot & & & \cdot \\ \cdot & \cdot & & \cdot & & \cdot & & \cdot & & \cdot \\ \cdot & \cdot & & & \cdot & \cdot & & & \cdot & \cdot \\ w_1 r_{i1} & w_2 r_{i2} & \cdot & \cdot & \cdot & w_j r_{ij} & \cdot & \cdot & \cdot & w_n r_{in} \\ \cdot & \cdot & \cdot & & & \cdot & \cdot & & & \cdot \\ \cdot & \cdot & & \cdot & & \cdot & & \cdot & & \cdot \\ \cdot & \cdot & & & \cdot & \cdot & & & \cdot & \cdot \\ w_1 r_{m1} & w_2 r_{m2} & \cdot & \cdot & \cdot & w_j r_{mj} & \cdot & \cdot & \cdot & w_n r_{mn} \end{bmatrix} \tag{10.8}$$

Step 6: Identify the set of positive ideal (A^*) and negative ideal (A^-) solutions. For the order of selecting the fuzzy positive ideal solution (*FPIS*, A^*) and the fuzzy negative ideal solution (*FNIS*, A^-), is thus utilized as [37]:

$$\begin{aligned} FPIS = A^* &= (v_1{}^*, v_2{}^*, v_3{}^*, \ldots, v_n{}^*) \\ &= \left\{ \left(\max_i v_{ij} | i = 1, 2, \ldots, m \right), j = 1, 2, \ldots, n \right\} \end{aligned} \tag{10.9}$$

$$\begin{aligned} FNIS = A^- &= (v_1{}^-, v_2{}^-, v_3{}^-, \ldots, v_n{}^-) \\ &= \left\{ \left(\min_i v_{ij} | i = 1, 2, \ldots, m \right), j = 1, 2, \ldots, n \right\} \end{aligned} \tag{10.10}$$

The situations of *FPIS* and *FNIS*, and the corresponding triangular fuzzy numbers of those situations are determined.

Step 7: The similarity degree of each alternative from A^* and A^- as S^* and S^-, can be calculated by using following equation [37, 44]:

$$S(a, b) = \frac{1}{1 + D_{2,\frac{1}{2}}(a, b)} \tag{10.11}$$

where $D_{p,\,q}$ is the distance, indexed by parameters $1 < p < \infty$ and $0 < q < 1$, between two fuzzy numbers a and b which is a non-negative function as follows:

For triangular fuzzy numbers $a = (a_1, a_2, a_3)$ and $b = (b_1, b_2, b_3)$ the above distance with $p = 2$ and $q = 1/2$ is calculated as:

$$D_{2,\frac{1}{2}}(a,b) = \sqrt{\frac{1}{6}\left[\sum_{i=1}^{3}(b_i - a_i)^2 + (b_2 - a_2)^2 + \sum_{i\in\{1,2\}}(b_i - a_i)(b_{i+1} - a_{i+1})\right]} \tag{10.12}$$

Step 8: Calculate the closeness coefficient calculated as [35, 37]:

$$CC_{\rm i} = \frac{S_i^-}{S_i^- + S_i^*} \tag{10.13}$$

where $0 \le CC_{\rm i} \le 1$.

Step 9: Choose the alternative with minimum $CC_{\rm i}$ and rank them in ascending order. Fig. 10.2 summarizes the procedure of application of Fuzzy *TOPSIS* methodology [27, 37] in a flow chart".

10.3.4 Case Study of Road Traffic Noise Abatement

The application of the fuzzy *TOPSIS* approach is thus utilized for selection of optimal road traffic noise abatement strategy. The hierarchical structure of problem is shown in Fig. 10.3. The various benefit criteria considered for the problem are summarized as:

- Practical Feasibility (C1)
- Compatibility with Environment (C2)
- Maintenance (C3)
- Safety and Durability (C4)
- Structural Integrity and Installation Perspectives (C5)
- Economic Considerations (C6)
- Administrative and Legal viability (C7)
- Expected Noise Abatement in dB(A) (C8)

The various **alternatives** considered for road traffic noise abatement are:

- Improvement in Road Surface (A1)
- Installation of Low Height Barrier (A2)
- Installation of Green Belt (A3)

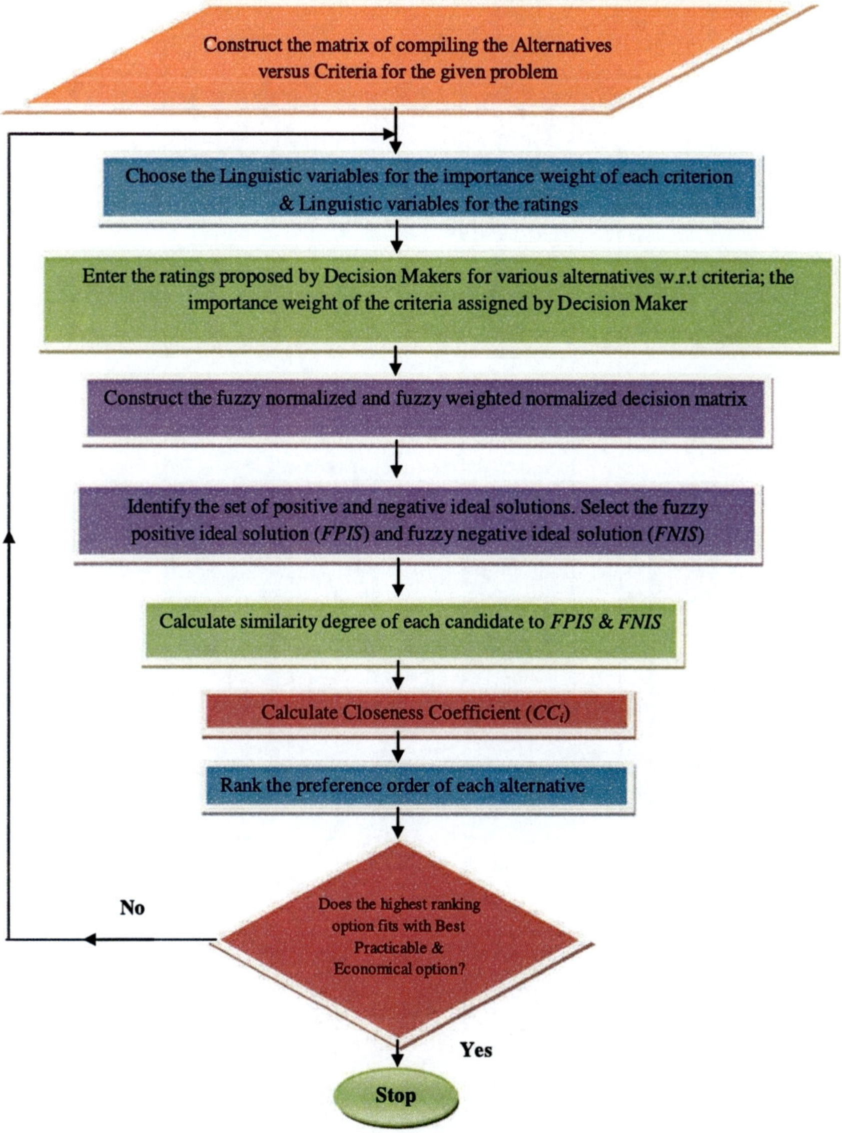

Fig. 10.2 Flow chart of the Fuzzy *TOPSIS* approach (Garg et al. [27])

- Enhancing Sound Insulation of Facades (A4)
- Traffic Management-Speed Limit for Heavy Vehicles (A5)
- Prohibiting or Banning Horn Noise (A6)
- Enhancement in Sound Insulation of Window Glazing (A7)

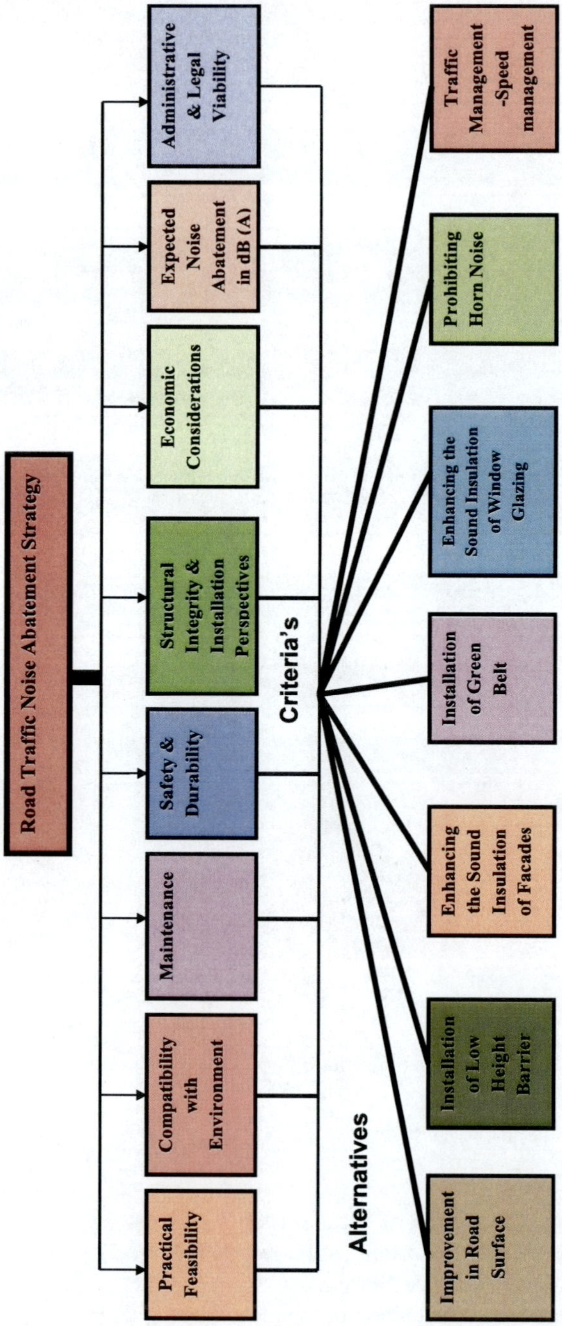

Fig. 10.3 Various alternatives and criteria considered for road traffic noise abatement and control

The fuzzy *TOPSIS* approach is utilized to select the best alternative among A1 to A7 satisfying the benefit criteria C1 to C8. Three experienced researchers having a long experience (>25 years) in noise monitoring, abatement, and control and had successfully completed many projects on noise abatement and control were requested to give their opinion (in terms of linguistic variables) as decision-makers D1, D2, and D3. A specific site as a case study was considered in Delhi city of India having a moderate road traffic density (4000–5000 vehicles per hour) and honking noise dominant.

The topological features, site characteristics including the vehicular traffic, road condition, nearby buildings, etc. for the site were provided to experts along with the noise monitoring database for 24×365 hours via the CPCB website, http://www.cpcbnoise.com [45]. This commercial site experienced the annual average day equivalent level (L_{day}) of 69.2 dB(A) and annual average night equivalent level (L_{night}) of 59.1 dB(A) for the year 2015, which was above the ambient noise limits of 65 dB(A) L_{day} and 55 dB(A) L_{night} recommended in ambient noise standards [22]. The annual average 24 h equivalent sound level, $L_{\mathrm{Aeq,24h}}$ experienced by the site was 67 dB(A), which is not acceptable considering the US Department of Housing and Urban Development (HUD) criteria [46] that recommends $49 < L_{\mathrm{Aeq}} \leq 62$ dB(A) as normally acceptable.

The experienced decision-makers used the linguistic weighting variables for ascertaining the importance of each criterion, as shown in Table 10.3. Then linguistic rating variables for evaluating the rating of alternatives with respect to each criterion are assigned by the three decision-makers as shown in Table 10.4. It can be observed that expert's opinion varied in assigning the weightings as a result of which it makes imperative to use such techniques (Fuzzy *TOPSIS*) for arriving at a definite solution to the problem. The linguistic variables are converted into triangular fuzzy numbers enlisted in Tables 10.1 and 10.2. Table 10.4 exhibits the ratings of seven alternatives by decision-makers under all criteria. The fuzzy decision matrix and fuzzy weights of their alternatives are determined as shown in Table 10.5. For the criteria "Expected noise abatement in dB(A)" and "Economic considerations," a relative comparison was done by the decision-makers based on the expected noise abatement provided by the seven alternatives in dB(A) and costs incurred in Indian currency per sq. m. (in Rupees/m^2). As these two factors are based on actual calculations, Table 10.5 incorporates the decision D1, D2, and D3 as same for these criterions. Thus, for the installation of low height noise barriers having the highest expected noise abatement, a linguistic variable of very good (VG) is assigned, which is transformed into equivalent triangular fuzzy numbers listed in Table 10.2. The improvement in road surface was assigned a fair rating (F). While comparing the costs, the installation of low height noise barriers and sound insulation of facades have the highest costs incurred in Rupees/m^2, and thus it is assigned very poor (VP) rating as minimal cost was the preferred alternative. However, prohibiting the horn noise has the minimum costs associated in Rupees/m^2 and thus it was assigned very good (VG) rating. Consequently, a fuzzy normalized decision matrix is developed as shown in Table 10.6. The weighted normalized fuzzy decision matrix is developed as shown in Table 10.7. Table 10.8 shows the *FPIS* and *FNIS* values

Table 10.5 The fuzzy decision matrix and fuzzy weights of their alternatives

	C1	C2	C3	C4	C5	C6	C7	C8
A1	(8.3; 9.7; 10)	(8.3; 9.7; 10)	(1.3; 3; 5)	(6.3; 8.3; 9.7)	(1.3; 3; 5)	(0; 1; 3)	(5.7; 7.7; 9.3)	(3; 5; 7)
A2	(5; 7; 8.7)	(6.3; 8.3; 9.7)	1.3; 3; 5)	(4.3; 6.3; 8.3)	(0.3; 1.7; 3.7)	(0; 0; 1)	(0.3; 1.7; 3.7)	(9; 10; 10)
A3	(8.3; 9.7; 10)	(9; 10; 10)	(6.3; 8.3; 9.7)	(9; 10; 10)	(5.7; 7.7; 9.3)	(5; 7; 9)	(6.3; 8.3; 9.7)	(7; 9; 10)
A4	(0.3; 1.3; 3)	(4.3; 6.3; 8.3)	(0.3; 1.7; 3.7)	(6.3; 8.3; 9.7)	(0; 0.7; 2.3)	(0; 0; 1)	(0; 0.7; 2.3)	(9; 10; 10)
A5	(5; 7; 8.7)	(7; 9; 10)	(7.7; 9.3; 10)	(7.7; 9.3; 10)	(7.7; 9.3; 10)	(5; 7; 9)	(4.3; 6.3; 8.3)	(7; 9; 10)
A6	(9; 10; 10)	(9; 10; 10)	(8.3; 9.7; 10)	(7; 9; 10)	(9; 10; 10)	(9; 10; 10)	(8.3; 9.7; 10)	(7; 9; 10)
A7	(7.7; 9.3; 10)	(7.7; 9.3; 10)	(5.7; 7.7; 9.3)	(6.3; 8.3; 9.7)	(7.7; 9.3; 10)	(1; 3; 5)	(6.3; 8.3; 9.7)	(7; 9; 10)
Weight	(0.9; 1; 1)	(0.7; 0.9; 1)	(0.2; 0.4; 0.6)	(0.4; 0.6; 0.8)	(0.4; 0.6; 0.8)	(0.7; 0.9; 1)	(0.6; 0.8; 1)	(0.8; 0.9; 1)

Table 10.6 The fuzzy normalized decision matrix

	C1	C2	C3	C4	C5	C6	C7	C8
A1	(0.83; 0.97; 1)	(0.83; 0.97; 1)	(0.13; 0.3; 0.5)	(0.63; 0.83; 0.97)	(0.13; 0.3; 0.5)	(0; 0.1; 0.3)	(0.57; 0.77; 0.93)	(0.3; 0.5; 0.7)
A2	(0.5; 0. 7; 0.87)	(0.63; 0.83; 0.97)	(0.13; 0.3; 0.5)	(0.43; 0.63; 0.83)	(0.03; 0.17; 0.37)	(0; 0; 0.1)	(0.03; 0.17; 0.37)	(0.9; 1; 1.0)
A3	(0.83; 0.97; 1)	(0.9; 1; 1)	(0.63; 0.83; 0.97)	(0.9; 1; 1)	(0.57; 0.77; 0.93)	(0.5; 0.7; 0.9)	(0.63; 0.83; 0.97)	(0.7; 0.9; 1)
A4	(0.03; 0.13; 0.3)	(0.43; 0.63; 0.83)	(0.03; 0.17; 0.37)	(0.63; 0.83; 0.97)	(0; 0.07; 0.23)	(0; 0; 0.1)	(0; 0.07; 0.23)	(0.9; 1; 1)
A5	(0.5; 0.7; 0.87)	(0.7; 0.9; 1.0)	(0.77; 0.93; 1.0)	(0.77; 0.93; 1.0)	(0.77; 0.93; 1.0)	(0.5; 0.7; 0.9)	(0.43; 0.63; 0.83)	(0.7; 0.9; 1.0)
A6	(0.9; 1.0; 1.0)	(0.9; 1.0; 1.0)	(0.83; 0.97; 1.0)	(0.7; 0.9; 1.0)	(0.9; 1.0; 1.0)	(0.9; 1.0; 1.0)	(0.83; 0.97; 1.0)	(0.7; 0.9; 1.0)
A7	(0.77; 0.93; 1.0)	(0.77; 0.93; 1.0)	(0.57; 0.77; 0.93)	(0.63; 0.83; 0.97)	(0.77; 0.93; 1.0)	(0.1; 0.3; 0.5)	(0.63; 0.83; 0.97)	(0.7; 0.9; 1)

Table 10.7 The fuzzy weighted normalized decision matrix

	C1	C2	C3	C4	C5	C6	C7	C8
A1	(0.75; 0.97; 1)	(0.58; 0.87; 1)	(0.03; 0.12; 0.3)	(0.25; 0.5; 0.78)	(0.05; 0.18; 0.4)	(0; 0.09; 0.3)	(0.34; 0.62; 0.93)	(0.24; 0.45; 0.7)
A2	(0.45; 0. 7; 0.87)	(0.44; 0.75; 0.97)	(0.03; 0.12; 0.3)	(0.17; 0.38; 0.66)	(0.01; 0.1; 0.3)	(0; 0; 0.1)	(0.02; 0.14; 0.37)	(0.72; 0.9; 1.0)
A3	(0.75; 0.97; 1)	(0.63; 0.9; 1)	(0.13; 0.33; 0.58)	(0.36; 0.6; 0.8)	(0.23; 0.46; 0.74)	(0.35; 0.63; 0.9)	(0.38; 0.66; 0.97)	(0.56; 0.81; 1)
A4	(0.03; 0.13; 0.3)	(0.3; 0.57; 0.83)	(0.01; 0.07; 0.22)	(0.25; 0.5; 0.78)	(0; 0.04; 0.18)	(0; 0; 0.1)	(0; 0.06; 0.23)	(0.72; 0.90; 1.0)
A5	(0.45; 0.7; 0.87)	(0.49; 0.81; 1.0)	(0.15; 0.37; 0.6)	(0.31; 0.56; 0.8)	(0.31; 0.56; 0.8)	(0.35; 0.63; 0.9)	(0.26; 0.5; 0.83)	(0.56; 0.81; 1.0)
A6	(0.81; 1.0; 1.0)	(0.63; 0.9; 1.0)	(0.17; 0.39; 0.6)	(0.28; 0.54; 0.8)	(0.36; 0.6; 0.8)	(0.63; 0.9; 1.0)	(0.57; 0.78; 1.0)	(0.56; 0.81; 1.0)
A7	(0.69; 0.93; 1.0)	(0.54; 0.84; 1.0)	(0.11; 0.31; 0.56)	(0.25; 0.5; 0.78)	(0.31; 0.56; 0.8)	(0.07; 0.27; 0.5)	(0.38; 0.66; 0.97)	(0.56; 0.81; 1)

Table 10.8 Determination of *FPIS* and *FNIS* values

	C1	C2	C3	C4	C5	C6	C7	C8
P	(0.81; 1.0; 1.0)	(0.63; 0.9; 1.0)	(0.17; 0.39; 0.6)	(0.36; 0.6; 0.8)	(0.36; 0.6; 0.8)	(0.63; 0.9; 1.0)	(0.57; 0.78; 1.0)	(0.72; 0.9; 1.0)
N	(0.03; 0.13; 0.3)	(0.3; 0.57; 0.83)	(0.01; 0.07; 0.22)	(0.17; 0.38; 0.66)	(0; 0.04; 0.18)	(0; 0; 0.1)	(0; 0.06; 0.23)	(0.24; 0.45; 0.7)

determined using step 6. The similarity degree of each alternative from *FPIS* is given in Table 10.9, while that from *FNIS* in Table 10.10. Finally, the closeness coefficient of each alternative is calculated and presented in Table 10.11.

It can be observed that the ranking order of seven alternatives is A6, A3, A5, A7, A1, A2, and A4. The best optimal alternative is A6 which is prohibiting horn noise, and this is followed by installation of green belt, traffic management, enhancement in sound insulation of window glazing, improvement in road surface, installation of low height barrier, and enhancing sound insulation of facades. It may be noted here that present order cannot be generalized to every site and situation, and thus, regional road traffic noise management strategies (*RTNMS* described in [20]) may differ from one site to another. Also, the experts opinion may differ on each and every aspect pertaining to site specific conditions and requirement of a suitable noise abatement solution. However, the present case study serves as an illustrative example that can

Table 10.9 The similarity degree of each alternative from *FPIS*

	C1	C2	C3	C4	C5	C6	C7	C8
A1	0.9665	0.9698	0.8002	0.9199	0.7200	0.5751	0.8768	0.7025
A2	0.7807	0.8784	0.8002	0.8377	0.6826	0.5444	0.6252	1.0000
A3	0.9665	1.0000	0.9559	1.0000	0.8929	0.8085	0.9077	0.9118
A4	0.5536	0.7725	0.7680	0.9199	0.6533	0.5444	0.5944	1.0000
A5	0.7807	0.9176	0.9839	0.9654	0.9654	0.8085	0.8037	0.9118
A6	1.0000	1.0000	1.0000	0.9475	1.0000	1.0000	1.0000	0.9118
A7	0.9313	0.9445	0.9382	0.9199	0.9654	0.6325	0.9077	0.9118

Table 10.10 The similarity degree of each alternative from *FNIS*

	C1	C2	C3	C4	C5	C6	C7	C8
A1	0.5631	0.7903	0.9497	0.9003	0.8726	0.8999	0.6451	1.0000
A2	0.6516	0.8618	0.9497	1.0000	0.9345	1.0000	0.9198	0.7025
A3	0.5631	0.7725	0.7940	0.8377	0.7049	0.6186	0.6292	0.7489
A4	1.0000	1.0000	1.0000	0.9003	1.0000	1.0000	1.0000	0.7025
A5	0.6516	0.8259	0.7760	0.8618	0.6663	0.6186	0.6916	0.7489
A6	0.5536	0.7725	0.7680	0.8748	0.6533	0.5444	0.5944	0.7489
A7	0.5744	0.8070	0.8063	0.9003	0.6663	0.7874	0.6292	0.7489

Table 10.11 Ranking of various alternatives for road traffic noise abatement

Alternatives	S+	S-	CC	Rank
A1	6.5308	6.6211	0.5034	**V**
A2	6.1491	7.0199	0.5331	**VI**
A3	7.4432	5.6689	0.4323	**II**
A4	5.8060	7.6028	0.5670	**VII**
A5	7.1369	5.8406	0.4501	**III**
A6	7.8593	5.5098	0.4121	**I**
A7	7.1514	5.9197	0.4529	**IV**

be easily applied with no computational complexities inviting the opinion of any number of experts (fuzzy numbers will change) to formulate an effective strategy for road traffic noise management considering all these alternatives with respect to the criteria's discussed [47].

The method is suitable for multiple-criteria decision-making (*MCDM*) and is flexible enough to consider both the subjective and objective weights. Thus, it overcomes the limitations of conventional weighted sum approach or the weighted product approach that suffers from the limitations of no solution in cases wherein equal weights of decision matrix are there and for multidimensional problems [32, 48]. Since alternatives of low noise pavements, traffic management, and building insulation of more than 5 dB alone do not suffice to achieve the necessary attenuation, they can be contemplated in combination with one another and with noise barriers [49].

The Fuzzy *TOPSIS* approach is flexible enough to utilize the exact values of the criterion for selection of optimal noise abatement strategy or noise action plans (NAPs). The method is applied in many problems pertaining to manufacturing, product design, transportation, selection of robot, facility location selection, solid waste management, water management, etc. as recommended by Shih et al. [38]. The linguistic variables assigned to each of road traffic noise abatement alternatives versus criteria considered cannot be generalized to every situation and thus may change depending upon the site-specific conditions. For instance, honking noise may not be dominant for traffic on a freeway in a less populated suburban area, and thus may not serve as the best optimal noise abatement alternative unlikely for a heavily congested area. Thus, the ranking of noise abatement alternatives could be varied depending upon site situation conditions and eight criteria discussed. However, the fuzzy *TOPSIS* approach can prove to be a powerful tool in selection of optimal strategies for regional road traffic noise management and control and for noise abatement and control of other noise sources as well. The use of this methodology provides sound arguments from the standpoint of objectivity and rigor, both for the authorities responsible for the approval and implementation of a noise action plan. The approach is based on technical support, and other subjective factors such as political influences are not accounted in the decision-making process. However, the methodology does not impose a sole or right solution; rather it proposes and sorts the best feasible alternatives on the basis of technical suitability [49].

10.4 Validation of the Fuzzy *TOPSIS* Method

The validation of the ranking order sought by this method is very important as different *MCDM* methods may arrive at different solutions. For instance, in one such study for prioritizing the noise action plans (NAPs), the *TOPSIS* method selected a combination of noise barrier and rubberized asphaltic road as the best alternative and a combination of rubberized asphalt and building sound insulation as the second, while the weighted sum method chose the combination of rubberized asphalt and building sound insulation as the best alternative and secondly selects the combination of noise barrier and rubberized asphaltic road [49]. The validation of the Fuzzy *TOPSIS* approach presented in this chapter was done with Li method [50] that proposed a Fuzzy model on the basis of α-level weighted, fuzzy preference relation. It recommended ordering the alternatives according to α-level weighted, fuzzy preference relation and selected the best alternative with the maximum ranking value as the best alternative. It was observed that the method proposed and method of Li (1999) had the same answer to the present problem, which validates the present solution arrived using fuzzy *TOPSIS* approach. The ranking of alternatives evaluated by Li method [50] is A6, A3, A5, A7, A1, A2, and A4, which is the same as that of the proposed method as described in Table 10.12.

It can be thus observed that Fuzzy *TOPSIS* approach is a simple, logical, and validated methodology that can be easily programmed and assigns a scalar value that

Table 10.12 Comparison of results sought from the present method with Li method [50] for ranking of the various alternatives for road traffic noise abatement

Alternatives	Present method		Li method [50]	
	CC	Rank	Index value, *r*	Rank
A1	0.5034	V	0.2840	V
A2	0.5331	VI	0.0850	VI
A3	0.4323	II	0.8510	II
A4	0.5670	VII	0.0090	VII
A5	0.4501	III	0.7300	III
A6	0.4121	I	0.9410	I
A7	0.4529	IV	0.7120	IV

accounts for both best and worst alternative simultaneously [33, 38]. Besides, it has fewest rank reversals [51] and is flexible enough to accommodate any number of alternatives when compared to the other methods [52–54]. Figure 10.4 shows an illustrative flow chart of noise abatement program utilizing the proposed Fuzzy *TOPSIS* approach for traffic noise control in metropolitan cities of India.

10.5 *MCDM* Tool for Alternatives in Combination

In many situations, alternatives of low noise pavements, traffic management, and building insulation of more than 5 dB alone do not suffice to achieve the necessary attenuation, and thus these are contemplated in combination with one another and with noise barriers. Ruiz-Padillo et al. [49] study showed the ranking of various alternatives in combinations as shown in Table 10.13. Utilizing the *TOPSIS* approach, using an acoustic screen in combination with rubberized asphaltic roads was ranked as the best alternative and placed rubberized asphaltic roads and enhancing building sound insulation as the second option. For instance, it was observed from Table 10.13 that the alternative of installing traffic light for speed control to 40 km/h in combination with building insulation and the installation of berm along

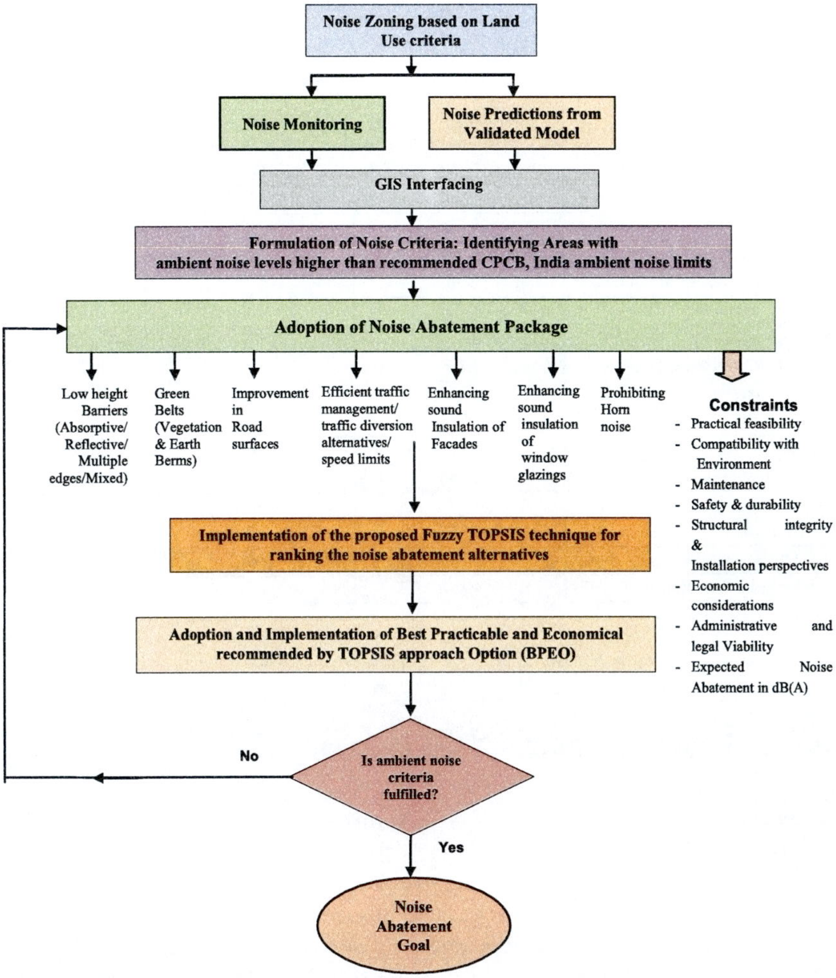

Fig. 10.4. Flow chart of a noise abatement program utilizing the proposed Fuzzy *TOPSIS* approach for selecting optimal strategy for road traffic noise abatement and control in metropolitan cities of India

with traffic light do not offer a favorable approach. It may be noted that the two alternatives in combination may be implemented in two clearly separate phases as they involve two independent and inclusive actions [49]. Thus, utilizing *MCDM* tool supports decision-making in prioritization of the noise control action plans for reducing noise pollution in urban agglomerations.

Table 10.13 Results of *MCDM* tool on various alternatives in combination (Ruiz-Padillo et al. [49])

Alternatives	Summation value weighted sum	Numbering of alternatives by which it is outranked (*ELECTRE*)	Similarity ratio value (*TOPSIS*)
Acoustic screen	5.211	9	0.406
Berm	5.398	9	0.428
Berm + screen	5.313	10	0.418
Screen + porous asphalt	7.967	2	0.786
Berm + porous asphalt	8.035	3	0.758
Screen + rubberized asphalt	8.845	0	0.859
Berm + rubberized asphalt	8.190	0	0.809
Screen + traffic light	4.763	8	0.306
Berm + traffic light	4.815	10	0.313
Screen + insulation	5.900	6	0.457
Berm + insulation	6.383	4	0.467
Insulation	7.176	6	0.480
Porous asphalt + insulation	7.933	2	0.809
Rubberized asphalt + insulation	9.214	0	0.849
Traffic light + insulation	4.856	7	0.330

10.6 Selection of Optimal Noise Barrier

Design considerations for erection of noise barriers of optimum dimensions providing a high insertion loss are essential and depend upon site and situation. Barrier thickness is insignificant, but diffraction over the top edge of a barrier is affected by its cross section. Transparent barriers, tilted, dispersive barriers, and multiple edge design barriers prove to be fruitful options for road traffic noise abatement. May and Osman [55] had observed higher noise reduction for T-profile (wide top) barriers and T-shape with absorptive top with cap widths of 0.6 m or more and of small cap thickness. Noise measurements with a 4 m highway noise barrier with a horizontal cap of 0.75 m (T-profile) in Toronto produced a noise reduction of 1–1.5 dB greater than that obtained for the same barrier without the cap [55]. In a review of the barrier studies involving T-, Y-, and arrow-profile noise barriers, numerical modeling, and boundary element approach, Hothersall et al. [56] concluded that for barriers with reflecting surfaces, the changes in insertion losses for various profiles can be described by using the concept of path difference. Further the introduction of absorptive upper surfaces produces a significant increase in insertion loss. In most conditions, the Y- and arrow-profiles perform less efficiently than the T-profile. The experimental results of the performance of barriers having different shapes and surface conditions under well-controlled environment showed that a 3 m high T-shaped barrier provided the same performance as a 10 m high plain barrier [57].

The application of the fuzzy *TOPSIS* approach can be thus utilized for the selection of optimal road traffic noise abatement strategy as reported earlier by Garg et al. (2015) [27]. The hierarchical structure of problem is shown in Fig. 10.5. The various benefit criteria considered for the problem are summarized as:

- Insertion Loss (C1)
- Structural Integrity (C2)
- Compatibility with Environment (C3)
- Maintenance (C4)
- Safety & Durability (C5)
- Installation (C6)
- Corrosion Resistance (C7)
- Economic Considerations (C8)
- Ventilation, Lighting, Drainage, Vandalism (C9)

The various alternatives, that is, types of noise barriers considered for road traffic noise abatement are:

- T-top
- Y-top
- Jagged top
- Mushroom top
- Cylindrical top
- Multiple edges
- Absorptive
- Reflective
- Transparent
- Concrete
- Polycarbonate sheet

It is imperative for the town planners to consider all these criteria before finally selecting and installing the noise barrier for a particular site for noise abatement. There is no simple solution to such a problem wherein among the variable alternative one has to find the optimal strategy considering the various benefit criteria. The application of *TOPSIS* approach in selection of optimal noise barrier for road traffic noise abatement using the fuzzy database of the linguistic variable assigned by three experienced decision-makers can provide an optimal solution for this problem with minimum numerical complexities and in minimum time. Arizona Department of Transportation (ADOT) study on scoring the barrier options recommends the T-top design with absorptive material over the conventional barrier of concrete or masonry construction [58]. Table 10.14 shows the strength, weaknesses, opportunities, and threats (*SWOT*) for various types of noise barrier options available, which can help a decision-maker to identify the pros and cons associated with each alternative while awarding the linguistic ratings.

It may be noted here that present order cannot be generalized to every site and situation, and thus, selection of optimal alternative may differ from one site to another. Also, the experts opinion may differ on each and every aspect pertaining

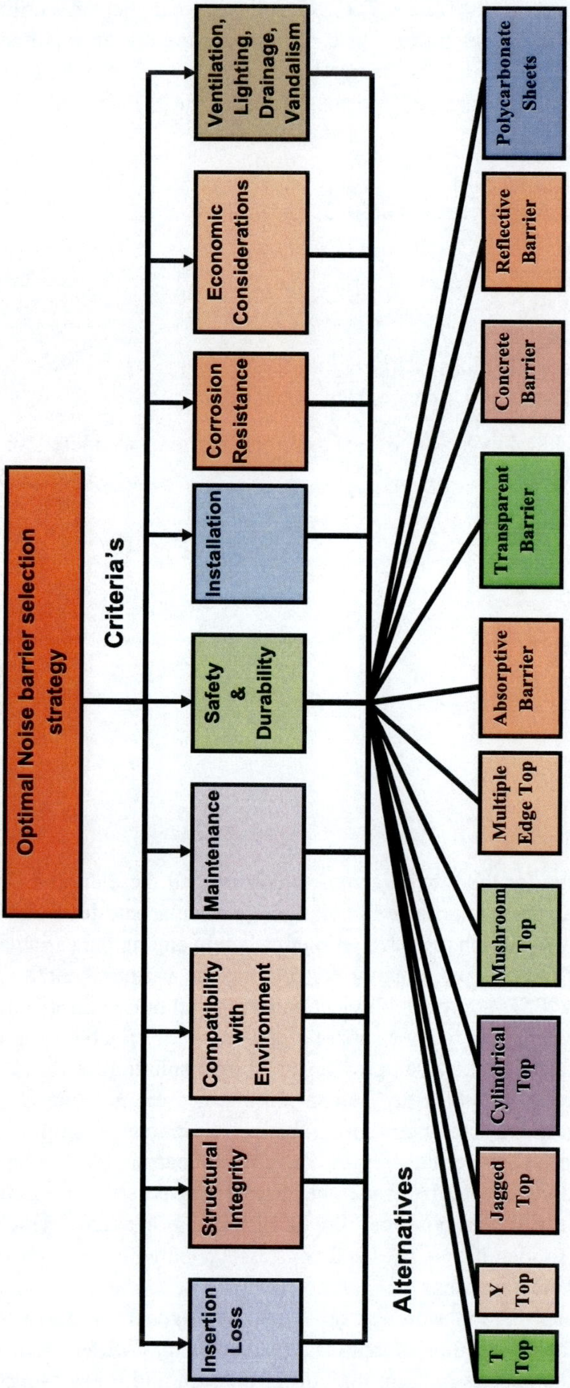

Fig. 10.5 Various alternatives and criteria considered for the selection of optimal barrier for road traffic noise abatement (Garg et al. [27])

Table 10.14 *SWOT* analysis of various noise barrier options (Garg et al. [27])

Barrier type	Strengths	Weaknesses	Opportunities	Threats
T-Top barrier	High insertion loss, reduced barrier height by 1 m on average	Maintenance costs associated with absorptive material	Instrumental for specific source screening	Debris collection, drainage issues
Y-Top/half Y	High insertion loss	Wind loading, maintenance issue	Instrumental for specific source screening	Debris collection, drainage issues
Jagged top	Fair insertion loss, high-frequency mitigation	--	Traffic noise abatement	Structural integrity
Mushroom top/ Multiple-edge top treatment/ cylindrical top	High insertion loss	Structural integrity, debris associated with use of absorptive material	Traffic noise abatement, specific source noise screening	Wind loading issue, degradation and maintenance associated
Absorptive barrier	High insertion loss	Debris associated with use of absorptive material	Traffic noise abatement, specific source noise screening	Debris collection, drainage issues; vulnerable to accretion of particles thrown by traffic and of water and ice due to natural precipitation
Transparent barrier	Fair insertion loss and economical solution	Noise reflecting surface may add up to cause a reverberant noise field	Traffic noise abatement; specific source noise screening	Vandalism
Concrete barrier/hollow concrete blocks	High insertion loss, minimal maintenance, high structural integrity, minimal maintenance	High costs and installation time	Hollow blocks for low-frequency noise control, absorptive GRC barriers	High costs of erection of barrier
Reflective barrier	Fair insertion loss	Noise reflecting surface may add up to cause a reverberant noise field	Traffic noise abatement	Reflective surface needs regular maintenance and "glare effects" associated
Metal/composite/ polycarbonate sheets/sintered metals/plastics	Metals, e.g., aluminum with high strength to width ratio	Non transparent materials, e.g., steel, aluminum, require greater efforts in surface treatment to soften visual impact	Perforated metal sheeting with absorptive material is good for low-frequency mitigation	Can produce glare effects at certain incident angles

(continued)

Table 10.14 (continued)

Barrier type	Strengths	Weaknesses	Opportunities	Threats
Phase reversal type barriers/ reactive surface barriers	Low-frequency noise control	Technical complications, practical difficulty	Specific source noise mitigation	High costs and technical complications
Active noise control barriers	Low-frequency noise control	Technical complications, practical difficult for moving noise sources	Specific source noise mitigation	High costs and technical complications

to the site-specific conditions as the effectiveness of noise barrier depends upon barrier geometry; acoustical characteristics such as sound reflection, diffusion of surface exposed to noise, airborne sound insulation; installation factors; deterioration with time; site geometry; ground profile; site acoustical characteristics; ground impedance; and meteorological conditions [59].

However, the present case study serves as an illustrative example that can be easily applied with minimal computational complexities inviting the opinion of any number of experts (fuzzy numbers will change) to formulate an effective strategy for selection and installation of noise barrier considering all these alternatives w.r.t. criteria's considered. It may be noted that in present case study, the multiple edges noise barrier is selected as least priority, which may not be applicable for every site or source noise mitigation. The application of multiple edges noise barriers or the so-called profiled noise barriers having better acoustical performance than conventional concrete structures can be very effective for specific applications wherein a single source has to be screened. The noise barrier designed for Delhi Metro Rail Corporation (DMRC), New Delhi, India, consisting of half Y-shaped structure that includes a 2 mm thick aluminum alloy as the backplate, 1 mm thick aluminum alloy sheet (30% open) as facing surface, 50 mm thick glasswool of density 40 kg/m^3 wrapped in tissue bags and polythene cover, and the whole assembly fixed in GI framework serves as an illustration of specific source noise mitigation [60].

10.7 Aircraft Noise Control Strategies

Reduction of noise at source involves use of new technologies for reducing the propulsion noise and air traffic management and adoption and implementation of noise certification standards. The noise propagation can be controlled by increasing the sound insulation of facades particularly the window glazing and sound insulated walls [61–63]. A recent study reported by Amudsen et al. (2011) [63] shows that the percentage highly annoyed dropped from 42% of respondents to 16% wherein an average of equivalent noise reduction inside dwelling of 7 dB was obtained from the facade insulation. Land-use planning and management around airports is one of the principal elements of the balanced approach. Noise zoning is the exercise of the legal powers of a national or local government that enables it to designate the uses that are permitted for each parcel of land depending on the noise exposure [64]. Zoning and preventing noise-sensitive land uses, for example, residential buildings, schools, and hospitals near to airports or flight path should be avoided as far as possible. Construction, administration, and operation of facilities that have little concerns of receiving disturbances from aircraft noise can be a good step. The Federal Interagency Committee on Urban Noise (1980) and Federal Interagency Committee on Noise (1992) recommended criterion of incompatibility for residential-land usage is that no more than 15% of the people should be highly or more annoyed by transportation noise [65]. The construction of a new runway with a different orientation will shift noise away from noise sensitive uses to less populated or compatible areas. However, the orientation of runway is dependent upon many factors including prevailing wind conditions, topography, obstacles, and other conditions. A new runway cannot be constructed if wind direction and topographic conditions are such that safety criteria cannot be met [66]. New runways may in fact improve the social welfare [67, 68]. The noise abatement operational procedures involves noise-preferential routes or runways, limited engine running while on ground, displaced landing thresholds, reducing power/drag, continuous descent approach, concentrating flights over less sensitive areas, avoiding the use of thrust reversals after landing in some cases with prior safety precautions, etc. [69], Three degree decelerating approach [70], and noise optimal departure and arrival trajectories [71]. Figure 10.6 shows the various aircraft noise abatement and mitigation strategies, and Table 10.15 shows the *SWOT* analysis of all these control measures based on review of various studies [67–74]. All these alternatives can be thus ranked subject to the criteria using fuzzy *TOPSIS* approach to find out the optimal solution for aircraft noise abatement and control.

The noise barriers must be structured and positioned accurately to provide any meaningful relief [64]. The operating restrictions as a last resort can be movement caps or curfews such as night landing curfew. However, imposition of night landing curfew in one region may contribute to an increased need for nighttime movements in other region. Imposing noise tax on noisier aircraft can also be proactive step. Combining various strategies, direct restrictions with fair financial impositions can be employed to accomplish noise control goals [72]. Figure 10.7 shows the various alternatives and criteria for aircraft noise control.

10.8 Contemporary Noise Control Measures and Balanced Approach

The International Civil Aviation Organization (ICAO) "balanced approach" suggested that noise policy should focus on combination of the solutions as the most appropriate option rather than single solutions for aircraft noise reduction and

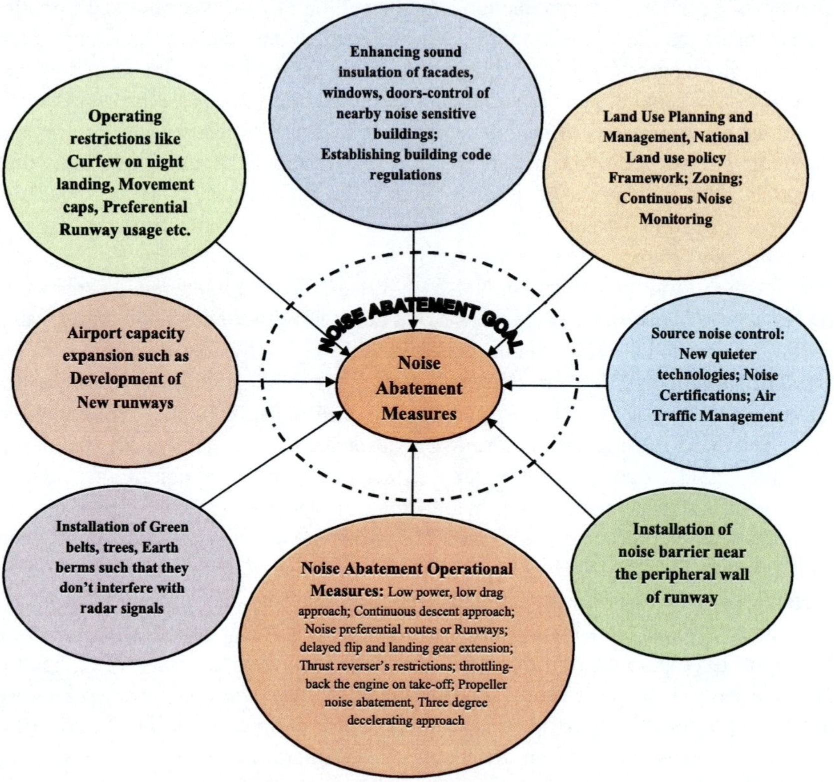

Fig. 10.6 Aircraft noise abatement and control measures

control [73–76]. The balanced approach recommends decreasing the noise at source; land use planning and management; operational procedures; and operating restrictions on air traffic [74]. Netjasov's [74] study analyzed 18 different contemporary noise reduction measures for noise reduction. Figure 10.8 shows the various noise abatement measures highlighted by Netjasov [74] categorized in balanced approach recommended by International Civil Aviation Organization (ICAO). The study shows that the noise abatement procedures and engine run-up restrictions and preferential runways are the three most commonly used options by large number of airports across the globe. The noise abatement procedures relate to the procedures pertaining to the arrival or departure trajectories and flying methods and techniques, etc. The engine run-up restrictions primarily focus on the engine working and some of the techniques used like "reversed thrust" while landing. Preferential runways refer to the allocated runways based on thorough planning for departure or arrival for airports having multiple runways. The measures pertaining to the "Land use planning and management" category were the mostly used worldwide [74]. However, the

Table 10.15 Strengths, weaknesses, opportunities, and threats (*SWOT*) analysis of various aircraft noise control measures

Noise control measures	Strengths	Weaknesses	Opportunities	Threats
Land-use planning and management	Lower financial impact for newer projects, reduces the aircraft noise annoyance	For existing projects, the cost is high for re-location of runways or re-location of areas around the airport for specific purposes	Lower implementation costs for newer projects and can be realized in a shorter time period	The financial burden of re-location of runways or relocating of areas around airport for specific purposes is very high
Source noise control	New quieter technologies shall be helpful in reducing the noise impact of aircrafts	Higher R&D costs for quieter technologies	Maximum permissible noise limits have to be formulated and legally enforced in some countries	The sound propagation path and receiver ends have to be considered for realizing the best noise control benefit
Installation of noise barrier near the peripheral wall of runway	Noise generated from ground level sources such as aircraft movement and engine reversed thrust can be controlled	Higher costs associated with maintenance considerations for barriers with absorptive materials and concrete barriers	Positioning of barrier from runway centerline and height of barrier are two major attributes	Cost versus benefit criteria is a major consideration prior implementation of noise barrier
Noise abatement operational measures	Lowers costs associated with noise abatement and control	Passenger and aircraft safety concerns and effect on airport capacity must be considered prior implementing these measures	Cost-effective, easy to implement and follow provided safety aspects are considered	Airlines may object to these procedures if these operating measures raise the operating costs by increasing the fuel consumption or flight time or increased emissions
Airport capacity expansion	Noise load on each runway will be reduced leading to reduction of annoyance	Higher costs incurred in airport capacity expansion	Long-term solution for reducing aircraft noise annoyance, although expansion may take a longer time duration	Airport capacity expansion requires longer time and other considerations favorable for flight take-off, etc.

(continued)

Table 10.15 (continued)

Noise control measures	Strengths	Weaknesses	Opportunities	Threats
Operating restrictions	Lower costs and time associated with implementation of these restrictions	May lead to significant burden on some runway	Can be instrumental for shorter time period only	Higher costs associated with night time curfew that includes operator's lost revenue
Enhancing the sound insulation of facades and building sound insulation requirements	Shall provide a permanent solution for acoustic comfort for community residing near airports	Higher costs and longer time period required	One time solution for aircraft noise abatement and control	Practical difficulties in implementation and enforcement of building sound insulation codes

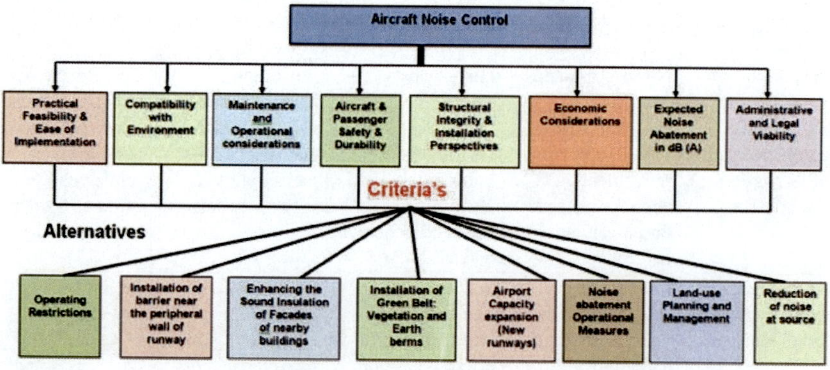

Fig. 10.7 Various alternatives and criteria considered for aircraft noise control

main hindrance in the execution of the noise abatement procedures lies in the inability of air traffic controllers to maintain manually the precise sequencing and spacing required for maximum take-off and landing particularly when the runway experiences heavy air traffic [77]. Reducing the first dB of noise is almost costless, but the cost function increases steeply, especially at noise reduction levels above 2.5 dB, wherein the only option left is to reduce the number of flights. The reduction in number of flights becomes quite expensive after a certain limit [78]. A study for Girona airport, Catalonia, Spain, employed a multiobjective optimization strategy for the design of noise abatement procedures and trajectory optimization algorithms for aircraft take-offs that lead to substantial reduction of noise annoyance [79]. The strengths, weaknesses, opportunities, and threats (*SWOT*) analysis of various aircraft noise control measures can be helpful to analyze the suitability of all the noise control measures for a specific situation or area.

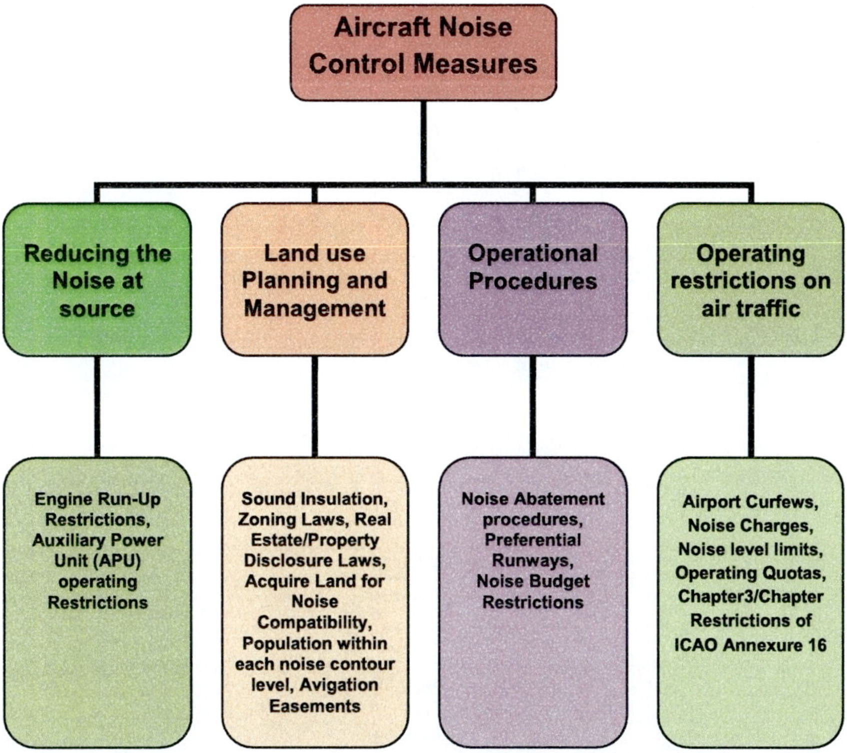

Fig. 10.8 Noise abatement measures categorized in balanced approach recommended by International Civil Aviation Organization (ICAO) (Based on Netjasov [74])

10.9 Railway Noise Control Strategies

The effectiveness of vehicle-related measures has the best cost-benefit ratio as recommended in European report [80]. So the introduction of composite brakes on freight wagons should be approached with the highest priority. Other measures can be done complementarily. Roughness of rails and wheels, especially corrugation in rails and out-of-round wheels, is a major cause of rail noise and needs to be monitored and controlled. The use of composite brake blocks rather than cast iron brake blocks will significantly improve the wheel running contact surface and reduce noise levels. Retrofitting existing wagons with composite brake blocks is possible, and the use of LL-blocks in particular can substantially reduce the noise emissions. Resilient wheels can reduce noise and improve ride quality, and can be very effective at reducing squeal noise in tight curves. Noise barriers have a large ongoing maintenance cost, have a high visual impact, and create problems for track access. Rail dampers can be tuned to the local needs of the railway and left in place for the life of the track; these can be an effective alternative to noise barriers. Gauge-face

Table 10.16 Measures, effects, and costs of railway noise (Reducing Railway Noise Pollution, EU report [80])

Measure	Avoided source of noise	Effect	Cost/unit
K-blocks	Rolling noise	Up to 8–10 dB(A)	4000–10,000 € per wagon
LL-blocks	Rolling noise	Up to 8–10 dB(A)	500–2000 € per wagon
General grinding of bad track	Rolling noise	10–12 dB(A) (up to 20 dB (A) at very bad tracks)	
Special acoustic grinding	Rolling noise	1–4 dB(A)	
Disc brakes	Rolling noise	10 dB(A)	
Wheel-tuned absorbers	Wheel noise	2–7 dB(A)	3000–8,000 € per wheel
Bogie Shrouds together with low height barriers	Wheel noise	8–10 dB(A)	
Rail dampers	Rail noise	3–7 dB(A)	300–400 € per meter
Slab tracks	Rail noise	5 dB(A)	
Rail Pads	Rail noise	3–4 dB(A)	
Different measures to lower squeal noise	Squeal noise	Up to 20 dB(A) depending on local conditions	
Shielding of pantographs	High speed trains	5–10 dB(A)	
Barriers 2 meter high	All source	10 dB(A)	1000 € per meter
Barriers 3–4 meter high	All sources	15 dB(A)	1700 € per meter for 4m high barrier
Insulated windows	All sources	10–30 dB(A)	3000–8000 € per house (4 windows)

lubrication is the traditional means for controlling wear of the high rail in narrow-radius curves, which has a secondary effect of reducing noise levels, including squeal noise in some cases. Top-of-rail friction modifiers are a relatively new extension of this technology, and are used to prevent corrugation of low rails and squeal noise in curves, as well as brake squeal in shunting yards. Aerodynamic designs like shielding or special materials like porous coating of pantographs can be used to reduce aerodynamic noise [80].

Table 10.16 summarizes all the measures, effects, and costs of railway noise [80]. Oertli [81] study provided a cost-benefit approach for railway noise abatement in Switzerland. The study proposed an optimal strategy of spending 65% of available finances on rolling stock improvement, 30% on the noise control barriers, and 5% on the insulated windows [81]. It can be observed that the range of noise mitigation measures is quite broad. Thus, in light of varied solutions for noise abatement and control with their pros and cons, *MCDM* approach can be very effective in arriving at an optimal solution for developing a Best Practicable and Economical Option for railway noise control.

10.10 Conclusions and Recommendations

This chapter presents an application of *TOPSIS* method with fuzzy data to determine the most preferable choice among all the possible alternatives for regional road traffic noise abatement and control. The fuzzy *TOPSIS* model proposed by Mahdavi et al. [37] is utilized in selecting the optimal road traffic noise abatement strategy. A case study of seven traffic noise abatement alternatives with respect to eight different criteria's is presented whereby three decision-makers have assigned the linguistic variables for rating these alternatives. Consequently, a fuzzy decision matrix is developed obtaining the *FPIS* and *FNIS* values. The closeness coefficient of each alternative is finally calculated so as to determine the ranking order of all alternatives. The proposed methodology is simple, validated, and easy to apply and thus finds its immense utility in various decision-making problems. It is envisaged that fuzzy *TOPSIS* approach can be efficiently utilized by town planners, architects, expert committees and development authorities in the selection of optimal regional road traffic noise abatement strategy. This approach can also be used for selection of optimal barriers, selection of optimal aircraft noise abatement and control strategies, and for noise abatement and control of railway and other noise sources as well. A recent study [82] utilized this approach in prioritizing the noise control solutions in a distillation unit of refinery (Abadan oil refinery, Khuzestan, Iran). This methodology finds an immense utility for arriving at an optimal strategy for aircraft noise abatement and control strategies especially for the constitutional committees set up for recommending noise control strategies. The proposed technique finds the best suitability and applicability especially when the opinion of different experts vary and no consensus can be reached even using a simple weighted or additive mean of the expert's opinion (assigned in terms of linguistic variables). However, the best optimized results and a noise abatement goal, can be in some cases realized by a balanced approach, whereby all the alternatives have to be considered for implementation in a coordinated manner.

The Fuzzy *TOPSIS* approach is flexible in handling a large number of alternatives for several criteria which makes it indispensable to be considered in development of regional road traffic noise management strategies. The alternatives may also be contemplated in combination with one another or with noise barriers [49] for accomplishing the desired noise level reductions. Also, other *MCDM* techniques such as *ELECTRE* and *AHP* may also be explored for their suitability in prioritizing the noise action plans as shown in this chapter. It may be noted that the linguistic variables assigned for each of road traffic noise abatement alternatives versus criteria considered and the weight of the criteria cannot be generalized to every situation and thus shall change according to the site specific conditions. However, the methodology presented can be widely applied to any site in decision-making process concerning noise action plans (NAPs) to reduce and control the transportation noise.

References

1. Garg N, Maji S (2014) A critical review of principal traffic noise models: Strategies and implications. Environ Impact Assess Rev 46:68–81
2. Steele C (2001) A critical review of some traffic noise prediction models. Appl Acoust 62:271–287
3. Garg N, Mangal SK, Saini PK, Dhiman P, Maji S (2015) Comparison of *ANN* and analytical models in traffic noise modeling and predictions. Acoust Australia 43:179–189
4. European Noise Directive (2002) Assessment and management of environmental noise. 2002/49/EU
5. Murphy E, King EA (2011) Scenario analysis and noise action planning: modelling the impact of mitigation measures on population exposure. Appl Acoust 72:487–494
6. Wilmink I, Vonk T (2015) Applying intelligent transport systems to manage noise impacts. Euronoise 2015, 31st May to 3rd June, Maastricht, pp 1053–1058
7. Sandberg U (2003) The multi-coincidence peak around 1000 Hz in tyre/road noise spectra. Euronoise, Naples
8. Buna S, Burgess M (1987) Methods of controlling traffic noise impact. In: Nelson PM (ed) Transportation noise reference book. Butterworths, London
9. Amundsen AH, Klæboe R (2005) A Nordic perspective on noise reduction at the source. TOI report 806/2005
10. Sandberg U, Kalman B (2004) The poroelastic road surface-results of an experiment in Stockholm. In Proceedings of Forum Acusticum, Budapest, Hungary
11. Garg N, Sharma O, Maji S (2011) Design consideration of building elements for traffic and aircraft noise abatement. Indian J Pure Appl Phys 49:437–450
12. Garg N, Sharma O, Maji S (2011) Experimental investigations on sound insulation through single, double and triple window glazing for traffic noise abatement. J Sci Indust Res 70:471–478
13. Rasmussen B (2010) Sound insulation between dwellings-requirements in building regulations in Europe. Appl Acoust 71:373–385
14. Garg N, Kumar A, Maji S (2013) Significance and implications of airborne sound insulation criteria in building elements for traffic noise abatement. Appl Acoust 74:1429–1435
15. Cohn L, Harris R (1996) Special treatments for highway noise barriers. Inter-noise 1996:747–750
16. Fujiwara K, Furuta N (1991) Sound shielding efficiency of a barrier with a cylinder at the edge. Noise Control Eng J 37:5–11
17. Mohanan V, Akhtar N (2011) An opinion survey of noise pollution in the residential premises: need for noise barriers. Indian J Air Pollut Contr XI:92–98
18. Santra SC, Chakrabarty D, Roy B (1998) Urban traffic noise abatement with vegetation barriers. J Acoust Soc India 26:1–10
19. Pathak V, Tripathi BD, Mishra VK (2008) Dynamics of traffic noise in a tropical city Varanasi and its abatement through vegetation. Environ Monit Assess 146:67–75
20. Naish D (2010) A method of developing regional road traffic noise management strategies. Appl Acoust 71:640–652
21. Dintrans A, Préndez M (2013) A method of assessing measures to reduce road traffic noise: A case study in Santiago, Chile. Appl Acoust 74:1486–1491
22. Noise Pollution (Regulation and Control) Rules (2000) Ministry of Environment and Forests, India
23. Garg N, Kumar A, Saini PK, Maji S (2015) A retrospective view of ambient noise standards in India: Status and proposed revision. Noise Contr Eng J 63(3):266–278
24. Tachibana H, Lang WW (2009) Final report of I-INCE TSG 3-survey of legislation, regulations and guidelines for control of community noise, I-INCE Publication, 09-1, July, 2009
25. U.S Department of Transportation (1995) Federal Highway Administration, Highway Traffic Noise Analysis and Abatement Policy and Guidance, Washington DC, June 1995

26. Hwang CL, Yoon K (1981) Multiple attribute decision making: Methods and applications. Springer, Berlin
27. Garg N, Vishesh MS (2015) Fuzzy TOPSIS approach in selection of optimal noise barrier for traffic noise abatement. J Achiev Acoust 40:453–467
28. Garg N, Sinha AK, Gandhi V, Bhardwaj RM, Akolkar AB (2016) A pilot study on the establishment of national ambient noise monitoring network across the major cities of India. Appl Acoust 103:20–29
29. Oyedepo SO (2012) Noise map: tool for abating noise pollution in urban areas. Open Access Sci Rep 1(3):1–7
30. Garg N, Kumar A, Maji S (2013) Parametric sensitivity analysis of factors affecting sound insulation of double glazing using Taguchi method. Appl Acoust 74:1406–1413
31. Jahanstaplas GR, Hosseinzadeh LI, Izadikhah M (2006) Extension of the TOPSIS method for decision making problems with fuzzy data. Appl Math Comput 181:1544–1551
32. Aruldoss M, Lakshmi TM, Venkatesan VP (2013) A survey on multicriteria decision making methods and its applications. Am J Inform Syst 1:31–43
33. Kim G, Park CS, Yoon KP (1997) Identifying investment opportunities for advanced manufacturing systems with comparative-integrated performance measurement. Intl J Product Econom 50:23–33
34. Chen SJ, Hwang CL (1992) Fuzzy multiple attribute decision making methods and application, Lecture notes in economics and mathematical sciences. Springer, New York
35. Chen TC (2000) Extensions of the TOPSIS for group decision making under fuzzy environment. Fuzzy Sets Syst 114:1–9
36. Prascevic Z, Prascevic N (2013) One modification of fuzzy TOPSIS method. J Model Manage 8:81–102
37. Mahdavi I, Mahdavi-Amiri N, Heidarzade A, Nourifar R (2008) Designing a model of fuzzy TOPSIS in multiple criteria decision making. Appl Math Comput 206:607–617
38. Shih HS, Shyur HJ, Lee ES (2007) An extension of TOPSIS for group decision making. Math Comput Model 45:801–813
39. Janic M (2003) Multicriteria evaluation of high-speed rail, transrapid maglev and air passenger transport in Europe. Transport Plan Technol 26:491–512
40. Kwang CK, Tam SM (2002) Case based reasoning approach to concurrent design of low power transformers. J Mater Process Technol 128:136–141
41. Milani AS, Shanian A, Madoliat R (2005) The effect of normalization norms in multiple attribute decision making models: a case study in gear material selection. Struct Multidisc Optim 29:312–318
42. Yang T, Chou P (2005) Solving a multi response simulation-optimization problem with discrete variables using a multi-attribute decision making method. Math Comput Simulat 68:9–21
43. Ding JF (2011) An integrated Fuzzy TOPSIS method for ranking alternatives and its application. J Marine Sci Technol 19:341–352
44. Jin Y (2003) Advanced fuzzy systems design and applications. Springer
45. Central Pollution Control Board India. Ambient Noise monitoring data display site, http://www.cpcbnoise.com
46. U.S Department of Housing and Urban Development (1984) Environmental criteria and standards, 24 CFR Part 51, V 12 July, 1979, amended by 49FR 880, 6 January, 1984
47. Garg N, Sharma O, Mohanan V, Maji S (2012) Passive noise control measures for traffic noise abatement in Delhi, India. J Sci Indust Res 71:226–234
48. Wang TC, Lee H-D (2009) Developing a fuzzy TOPSIS approach based on subjective weights and objective weights. Expert Syst Appl 36:8980–8985
49. Ruiz-Padillo A, Ruiz DP, Torija AJ, Ridao AR (2016) Selection of suitable alternatives to reduce the environmental impact of road traffic noise using a fuzzy multi-criteria decision model. Environ Impact Assess Rev 61:8–18
50. Li RJ (1999) Fuzzy method in Group decision making. Comput Math Appl 38:91–101

51. Zanakis SH, Solomon A, Wishart N, Dublish S (1998) Multi-attribute decision making: a simulation comparison of selection methods. Eur J Operat Res 107:507–529
52. Paxkan C, Wu ML (1997) On the equivalence of operational performance measurement and multiple attribute decision making. Intl J Prod Res 35:2963–2988
53. Triantaphyllou E, Lin CT (1996) Development and evaluation of five fuzzy multiattribute decision making methods. Intl J Approx Reason 14:281–310
54. Tavana M, Marbini A H- (2011) A group AHP-TOPSIS framework for human spaceflight mission planning at NASA. Expert Syst Appl 38:13588-13603.
55. May DN, Osman MM (1980) Performance of sound absorptive, reflective and T-profile noise barriers in Toronto. J Sound Vibrat 71(1):65–71
56. Hothersall DC, Crombie DH, Chandler-Wilde SN (1991) The performance of T-profile and associated noise barriers. Appl Acoust 32(4):269–287
57. Ishizuka T, Fujiwara K (2004) Performance of noise barriers with various edge shapes and acoustic conditions. Appl Acoust 65(2):125–141
58. Watson D (2006) Evaluation of benefits and opportunities for innovative noise barrier designs. Final report 572. Arizona Department of Transportation
59. Garai M, Guidorzi P (2000) European methodology for testing the airborne sound insulation characteristics of noise barriers in situ: Experimental verification and comparison with laboratory data. J Acoust Soc Am 108(3):1054–1067
60. Mohanan V, Sharma O (2001) Acoustical design of a noise barrier for an elevated rail corridor. J Acoust Soc India 28:95–100
61. Narang PP, Butler KR (1996) Reducing aircraft noise impact by sound insulation of houses. Acoust Australia 24:97–103
62. Garg N, Sharma O, Maji S (2011) Design consideration of building elements for traffic and aircraft noise abatement. Ind J Pure Appl Phys 49:437–450
63. Amundsen AH, Klæboe R, Aasvang GM (2011) The Norwegian Facade Insulation Study: the efficacy of façade insulation in reducing noise annoyance due to road traffic. J Acoust Soc Am 129(3):1381–1389
64. Guidance on the balanced Approach to Aircraft Noise Management (2008) International Civil Aviation Organization (ICAO), Doc. 9829, AN/451
65. Kryter KD (2007) Acoustical, sensory and psychological research data and procedures for their use in predicting effects of environmental noises. J Acoust Soc Am 122:2601–2614
66. Detroit Metropolitan Wayne County Airport (2003) Potential noise abatement measures. Part 150 Noise Compatibility Study Update
67. Girvin R (2009) Aircraft noise-abatement and mitigation strategies. J Air Transport Manage 15: 14–22
68. Nero G, Black JA (2000) A critical examination of an airport noise mitigation scheme and an aircraft noise charge: the case of capacity expansion and externalities at Sydney (Kingsford Smith) airport. Transport Res Part D Trans Environ 5:433–461
69. Civil Air Navigation Services Organisation and Airports Council International (2015) Managing the impacts of aviation noise, a guide for airport operators and air navigation service providers, September, 2015
70. Ren L, Clarke J-P, Ho NT (2003) Achieving low approach noise without sacrificing capacity. Digital avionics systems conference, AIAA, Indianapolis, Indiana, USA, 12th to 16th October, 2003
71. Visser HG, Wijner RAA (2001) Optimization of noise abatement departure trajectories. J Aircraft 38:620–627
72. Xie H, Heng L, Kang J (2014) The characteristics and control strategies of aircraft noise in China. Appl Acoust 84:47–57
73. ICAO (2001) Resolutions adopted at the 33rd session of the assembly provisional edition. International Civil Aviation Organization, Montreal
74. Netjasov F (2012) Contemporary measures for noise reduction in airport surroundings. Appl Acoust 73:1076–1085

75. ACI (2004) Airports and environment legislation. Airport Council International, Brussels
76. Celikel A, Hustache J, de Lepinay I, Martin K, Melrose A (2005) Environmental tradeoffs assessment around airports. Paper presented at 6th FAA/EUROCONTRL Air Traffic Management R&D Seminar, Baltimore, USA
77. Clarke J-P (2003) The role of advanced air traffic management in reducing the impact of aircraft noise and enabling aviation growth. J Air Trans Manage 9(3):161–165
78. Lijesen M, van der Straaten W, Dekkers J, van Elk R, Blokdijk J (2010) How much noise reduction at airports? Transport Res Part D 15(1):51–59
79. Prats X, Puig V, Quevedo J (2011) A multi-objective optimization strategy for designing aircraft noise abatement procedures. case study at Girona airport. Transport Res Part D 16(1): 31–41
80. European Report (2012) Reducing railway noise pollution. Directorate General for Internal Policies Policy Department B: Structural and Cohesion Policies, March 2012
81. Oertli J (2000) Cost-benefit analysis in railway noise control. J Sound Vibrat 231:505–509
82. Mousavi SM, Abbasi M, Yazdanirad S, Yazdanirad M, Khatooni E (2019) Fuzzy AHP-TOPSIS method as a technique for prioritizing noise control solutions. Noise Contr Eng J 67(6):415–421

Appendices

Appendix I

Glossary of Acoustical Terms

Acoustics The science of sound and hearing/listening.

Ambient noise All pervasive noise existing in an environment (area/locality/city/town).

Absorption coefficient It is defined as the ratio of energy absorbed by a material to the energy incident upon its surface. Sound absorption coefficient is dependent on the frequency of the sound wave.

Attenuate Reduction in the amplitude level of a signal.

A-weighted sound level A measure of sound pressure level designed to reflect the response of the human ear by reducing the effects of the low and high frequencies with respect to the mid-range frequencies.

Agglomeration Agglomeration shall mean part of a territory, delimited by the Member State, in EU having a population in excess of 100,000 persons and a population density such that the Member State considers it to be an urbanized area.

Background noise It is the environmental noise that exists at a particular site or location due to the various noise sources.

Barrier Noise barrier is an exterior structure erected to provide acoustic comfort from noise exposure levels.

Cycles per second The frequency of a signal or sound measured in Hertz (Hz).

Decibel A unit of measure of sound pressure level. One decibel is one-tenth of one bel.

N. Garg, *Environmental Noise Control*,
https://doi.org/10.1007/978-3-030-87828-3

Damping Damping is restraining of vibratorymotion, such as mechanical oscillations, noise by dissipation of energy.

Dead room Anechoic room whose surfaces are covered with highly absorptive materials.

Diffraction The bending of a sound wave around an obstacle or through an opening, such as slits.

Environmental noise Noise governed by environmental legislations, and usually enforced by local authorities.

Echo Echo is a reflection of sound that arrives at the listener with a delay after the direct sound.

Frequency Number of cycles per second for a periodic phenomena.

Free-field noise level The noise level measured away from any reflecting surfaces.

Facade effect The phenomenon of sound energy (noise) being reflected form the hard rigid, external surface of a building or structure. Where a facade is present, this effect adds approximately 2.5 or 3 dB(A) to the free-field noise level (at a distance of 1 meter from the facade).

Impulsive noise Short, transient, acoustical signal such as cracker burst and gun shot.

Inverse square law A condition in which the magnitude of a physical quantity follows an inverse relationship to the square of the distance. The sound pressure level decreases 6 dB for each doubling of the distance for a spherical point source.

Live room A reverberant room with many reflective surfaces.

Mass law The law that states that a material's ability to reduce the transmission of sound is proportional to its mass. According to the mass law, to increase a wall's transmission loss by 6 dB, it is necessary to double the thickness (mass) of the wall.

Microphone A transducer that converts the incident sound energy into equivalent voltage that can be measured by a sound level meter.

Noise Unwanted sound that is unpleasant or disturbing based on subjective perception.

Noise criteria Standard spectrum curves by which a given measured room's ambient noise may be described by a single NC number line NC, NCB, RNC, etc.

Noise reduction coefficient The arithmetic average of the sound absorption coefficients of a material at 250, 500, 1000, and 2000 Hz.

Normalized level Difference, D_n The sound insulation index measured under field conditions calculated as $D_n = D - 10 \log (A/A_0)$, $A_0 = 10\ m^2$.

Octave Interval between one musical pitch and another with double its frequency.

Pink noise Broadband noise whose energy content is inversely proportional to frequency (-3 dB per octave). This gives the noise equal energy per octave.

Reflection Bouncing of a sound wave from a surface.

Refraction The bending of waves when they enter a medium with different sound velocity.

Resonance Phenomenon that occurs when the frequency at which a force is periodically applied is equal or nearly equal to one of the natural frequencies of the system on which it acts causing the system to oscillate with larger amplitude.

Reverberation Reverberation is the persistence of sound after it is produced due to continuous reflections, until it reaches zero amplitude.

Sound Sound is a pressure wave that propagates through an elastic medium at a particular characteristic speed.

Sound pressure level The fundamental measure of sound pressure.

Sound transmission loss Ratio of sound energy emitted by an acoustical material or structure to the energy incident.

Sound transmission class A single number rating for describing sound transmission loss of a wall or partition. Higher the STC, higher the sound insulation that a material provides.

Speech intelligibility A measure of sound clarity, the degree to which one can understand the spoken language.

Sound reduction index Ten times the logarithm of the ratio of the sound power incident on the test specimen to the sound power transmitted through the specimen.

Standardized level difference, D_{nT} Standardized level difference (the level difference D is standardized to the 0.5 seconds reference value of the reverberation time in the receiving room. D is the difference in sound pressure level in the source and the receiving room.

Spectrum adaptation terms The spectrum adaptation terms introduced in ISO 717-1, viz., C and C_{tr} value, are calculated either from 100 Hz to 3150 Hz or from 50 Hz to 5 kHz. The spectrum adaptation term C pertains to living activities, children playing, railway traffic, highway road traffic, jet aircrafts, and factories emitting mainly medium and high frequency noise, while the spectrum adaptation term C_{tr} considers urban traffic noise, railway traffic at low speeds, aircraft, propeller driven, jet aircraft, disco music, etc.

Standardized impact sound pressure level, L'_{nT} The impact sound pressure level in a stated frequency band, corrected for the standardized reverberation time of 0.5 s.

$$L'_{nT} = L_i - 10 \log (T/To) \text{ dB}$$

where T is the measured reverberation time in seconds, T_0 for dwellings = 0.5 seconds, and L_i is the sound pressure level in the receiving room during impact noise testing.

Spectrum adaptation term C_I Value in dB to be added to the single number quantity to take account of the unweighted impact sound level, thereby representing the characteristics of typical walking noise spectra.

White noise Broadband noise having a constant energy per unit of frequency.

Weighting network An electronic filter in a sound level meter such as A-weighting and C-weighting.

Weighted sound reduction index, R_w It is the measured quantity that characterizes the sound insulating properties of a material or building element in a stated frequency band. It is a single figure rating descriptor whereby the normalized or standardized levels are compared to the reference curves published in ISO 717 for airborne sound insulation.

Weighted standardized level difference, $D_{nT,w}$ Field measurement of airborne sound insulation standardized to 0.5 seconds reverberation time and weighted to provide a single figure value. Weighted standardized level difference is the value of the reference curve at 500 Hz after shifting the reference curve.

Appendix II

Performance Evaluation of Sound Level Meter/Noise Monitoring Terminal (NMT) for Noise Monitoring

The current international standard that specifies sound level meter pattern evaluation tests is the IEC 61672-2. The performance evaluation is must for accurate and precise measurements associated with the sound level measurements done by a sound level meter or a noise monitoring terminal along with calibration of the microphone used. In Indian scenario, with the setting of NMTs in seven major cities and many studies reported in recent years on noise monitoring and mapping, performance/pattern evaluation is must and followed periodically.

A sound level meter in general is a combination of a microphone, an amplifier with controlled frequency weighting and a detector-indicator device with controlled time-weighting characteristics. The performance of a sound level meter (SLM) can be specified mainly by the following characteristics:

- Indication at the calibration check frequency
- Frequency-weighted free-field response level

- Level linearity
- Time weighting detector and indicator characteristics
- Tone burst response
- Pulse range and peak C sound level
- Overload, Long-term stability, self-generated noise with microphone installed and replaced by the electrical input signal device

Both electrical and acoustical tests are carried out to verify compliance. The output signal of the microphone is frequency weighted to produce one or more of the three characteristics designated as A-, C-, and Z-weighting. The electrical chain consists of a signal generator feeding electrical signals directly into the preamplifier portion of SLM, and the various characteristics can be studied by applying constant electrical input into the meter. The acoustical setup consist of a signal generator feeding into a power amplifier and then into a loudspeaker for generating the required acoustic field. The receiver side consists of a measurement microphone, preamplifier, and the measuring voltmeter/amplifier. Plane progressive sound waves are generated under anechoic conditions with the help of generator-amplifier-loudspeaker combination. The relative frequency-weighted free-field response level of the sound level meter (under test) can be studied by keeping the acoustic field constant and changing the generator frequency. Table A1 describes the performance evaluation tests in detail.

Table A1 Performance evaluation of sound level meter

Performance characteristics	Details of measurement process
Indication at the calibration check frequency	It can be measured using a sound calibrator and Pistonphone. The response of the sound level meter to a sinusoidal sound pressure signal at the calibrationcheck frequency and at the calibration level of the sound level meter is registered.
Frequency response	A, C, and Lin weighting characteristics can be measured in anechoic chamber using a signal generator feeding into a power amplifier and then into a loudspeaker for generating the required acoustic field. The relative frequency response in an acoustic coupler w.r.t 1 kHz is measured as Sound Pressure Level in the Acoustic Calibrator minus Free-Field-Correction
Level linearity	Linearity error can be measured in 1 dB and 10 dB steps for sound pressure level
Frequency and Time weightings at 1 kHz	Frequency and time weighting measured at 1 kHz as the deviation of indication (Fast A, C and Z -weighted and Slow A-weighted, L_{Aeq}) from the indication of A-weighted sound level with Fast time weighting
Peak C sound level	Peak-response to a 500 Hz and 8 kHz half-cycle sine measured in least-sensitive range.
R.M.S. detector response	The response of the sound level meter to sequences of repeated tonebursts when measuring the time-weighted sound level is registered for various tonebursts duration of the individual tonebursts. The peak level is held constant for all the signals and the repetition frequency is 40 Hz i.e. different crest factors.

(continued)

Table A1 (continued)

Performance characteristics	Details of measurement process
Free-field tests and Directional response	At each test frequency, the frequency weighting shall be calculated from the sound level indicated by the sound level meter minus the non-frequency-weighted sound -pressure level measured with the laboratory standard microphone.The directional response of a sound level meter shall be determined with plane progressive sinusoidal sound waves in a free-field test facility.
Toneburst response	Response to 4 kHz toneburst (200 ms, 2 ms, 0.25 ms burst) measured in reference range for time-weighting Fast, Slow and Leq
Level range control	The response of the sound level meter to a sinusoidal voltage signal is registered in each of the level ranges of the sound level meter
Long-term stability	Difference between the initial and final A-weighted sound level indicated in response to steady 1 kHz electrical signal
Overload indication	The difference between the levels of the positive and negative one-half-cycles input signals that first cause the displays of overload indication
Electrical inherent noise level (A, C, Lin)	Self-generated noise is measured in most sensitive range, with microphone and electrical substitution for microphone, according to manufacturer's specifications

Appendix III

Mining Noise Control in India

(a) Recommendations of 11th Conference on Safety in Mines held on 4th and 5th July 2013 at New Delhi, India, pertaining toNoise Mapping

Noise mapping should be made mandatory for various workplaces in the mine premises based on the various machines being used in concerned mines along with personal noise dosimetry of individual workmen exposed to noise level above 85 dB (A). Vibration studies of various mining machineries before their introduction in mining operation should be done as per the ISO standards.

(b) Suggested Recommendations

Most of the exposure to noise in mining comes from the need to use heavy machinery underground. Noise assessment in coal mines primarily requires identifying the noise sources in order to implement effectively the hierarchy of controls. Some of the recommendations suggested in mining environment in Indian scenario for noise control are as follows:

- Noise mapping should be made mandatory for coal mines. The critical noise sources and their control strategies should be devised based on the noise mapping and assessment. Table A2 provides some noise sources and exposures in mines

Table A2 Noise sources and exposures in mines and mining plants [1]

Noise source	Range (dB)
Cutting machines	83–93
Locomotives (electrical)	85–95
Haulage truck	90–100
Loaders	95–100
Long-wall shearers	96–101
Chain conveyors	97–100
Continuous miners	97–103
Loader-dumper	97–102
Fans	90–110
Pneumatic percussion tools	114–120

and mining plants. Thus, all these noise sources should be exclusively characterized in terms of amplitude and frequency spectrum in order to implement effectively the hierarchy of controls.
- The workers in coal mines should be provided with ear muffs so that they are protected from the high noise levels emitted.
- Audiometric monitoring is essential for the employees in such an environment.

A comprehensive noise controlprogram that addresses noise at the source, interrupts noise on its path, and shields receivers from the noise should be followed in Mines.

References

1. McBride DI (2004) Noise-induced hearing loss and hearing conservation in mining. Occupational Med 54(5):290–296
2. Seiler JP, Pon M (2008) Engineering and administrative noise controls for the mining industry. Sound vibration, July 2008, 8–9

Appendix IV

*Noise Impact Index (*NII*) and Its Suitability in Indian Perspectives*

The concept and methodology of a single number value or index for quantifying the noise impact assessment shall be very helpful in Indian scenario. A single number noise index such as Noise Impact Index (*NII*) developed either exclusively for the four different zones or irrespective of the zones and limits associated shall be instrumental in analyzing the severity and annoyance associated with the noise scenario at a particular site/area/zone in urban cities. There have been very few studies such as one reported by Gorai and Pal (2006) for an iron ore mining complex that intends to calculate noise impact index using the subjective response of the community. This method is based on a social survey which leads to the assessment of a quantity called sound level weighted population (*SLWP*), which measures the total number of people expected to be highly annoyed by the noise among the entire population [1]. This approach can be used in Indian scenario using the 24 hours equivalent sound pressure level, $L_{\text{Aeq,24h}}$ instead of L_{dn} as:

$$SLWP = \sum \left[P_i W \left(L_{Aeq,24h} \right)_i \right] \tag{A.1}$$

where P_{i} is the population distribution function and $W(L_{\text{Aeq,24h}})_i$ is the fractional impact or weighting function that characterizes the severity of the noise impact as a function of $L_{\text{Aeq,24h}}$. Noise impact index (*NII*) can be thus calculated as the sound level weighted population divided by the total population (P_{total}) that exists in the study zone.

$$NII = \frac{SWLP}{P_{total}} \tag{A.2}$$

The annoyance weighting factors have been described in Bies and Hansen (2009) corresponding to values of day-night average sound levels, L_{dn} ranging from 0.01 (35-40 L_{dn}) to 2.31 (85-90 L_{dn}) [2]. The benefit of using such descriptor is that they account the adverse reaction of the people to noise. Also, it can be used for noise impact assessment studies by comparing the existing index and indicative index calculated on the basis of the existing noise levels and the permissible noise levels, respectively. Such a descriptor can be analogous to the National Air Quality Index (*AQI*) used in Indian perspectives for describing the air pollution in the cities. The existing noise impact index (*NII*) can be calculated on the basis of existing noise level in the monitoring stations, whereas the indicative noise impact index (*NII*) may be calculated on the basis of permissible noise level in the monitoring stations. A noise stress in an area exists if the indicated *NII* is lower than the existing *NII*. For instance, a case study of a particular area with five different locations, namely A, B, C, D, and E, lying in different zones (Commercial: C, Residential: R, Industrial: I, and Silence: S) is presented in Table A3. The percentage highly annoyed can be calculated from a socio-acoustic survey of all the sites quantifying the number of people highly annoyed (% HA) with the existing ambient noise levels (in $L_{\text{Aeq,24h}}$).

Table A3 Sound level weighted population and noise impact index (*NII*)

Location	Category	$L_{\text{Aeq,24h}}$ in dB(A)	Fractional noise impact	No. of people in sub-neighborhood	$W(L_{Aeq,\ 24h})_i P_i]$	*NII*
A	R	68.2	0.14	900	126	**0.18**
B	C	70.5	0.11	500	55	
C	R	65.1	0.09	450	40.5	
D	C	74.4	0.21	560	117.6	
E	S	59.7	0.33	700	231	
				$P_{\text{total}} = 3110$	$\Sigma[P_i W(L_{Aeq,\ 24h})_i] = 570.1$	

Table A4 Indicative noise impact index (*NII*).

Location	Category	$L_{\mathrm{Aeq,24h}}$ in dB(A)	Fractional noise impact	No. of people in sub-neighborhood	$W(L_{Aeq,\ 24h})_i P_i]$	*NII*
A	R	53.5	0.04	900	36	**0.08**
B	C	63.5	0.09	500	45	
C	R	53.5	0.03	450	13.5	
D	C	63.5	0.09	560	50.4	
E	S	48.5	0.16	700	112	
				$P_{\mathrm{total}} = 3110$	$\sum[P_i W(L_{Aeq,\ 24h})_i] = 256.9$	

As a result, a regression equation can be developed between the % HA and $L_{\mathrm{Aeq,24h}}$ (in dB(A)) levels based on the social survey.

The existing ambient noise standards in India as given by the Central Pollution Control Board (CPCB) for the day time for Commercial area: 65 dB(A), Residential area: 55 dB (A), and Silence zone area:50 dB (A). In case of night time, the ambient noise standard for Commercial area: 55 dB(A), Residential area: 45 dB (A), and Silence zone area: 40 dB (A). Thus, the equivalent $L_{\mathrm{Aeq,24h}}$ limits can be assessed for Commercial area as: 63.5 dB(A), Residential area: 53.5 dB(A), and Silence zone area: 48.5 dB(A). Consequently, the % Highly annoyed will be less, which can be calculated from the developed formulation correlating the % HA with $L_{\mathrm{Aeq,24h}}$ levels. For instance, Table A4 calculates the Indicative Noise Impact Index (*NII*) based on assumption of less people annoyed than the earlier case with permissible limits noise level existing at that site. In this case, fractional noise impact is arbitrarily assumed to cite an example of calculating *NII*.

Thus, it can be observed that indicative *NII* is less than the existing *NII*, which shows that noise stress in that area is quite high. The existing noise impact index can be reduced to a value near to the indicative *NII*, if the existing$L_{\mathrm{Aeq,24h}}$ values for all the localities can be reduced. Thus, the concept of noise impact index can be used for evaluating the noise impact assessment at various localities. The best benefit is that it includes the subjective response of a part from the ambient sound level measurements in characterizing a particular locality. Such a descriptor can be analogous to the National Air Quality Index (*AQI*) used for describing the air pollution in the Indian cities and can be widely used for evaluating noise impact in Indian perspectives. As such in Indian perspectives, the use of National Noise Impact Index or Noise Quality Index is essentially required that prior requires a comprehensive socio-acoustic survey for the commercial, residential, silence, industrial, and mixed zones in each city for establishing an authorized relationship between % HA and $L_{\mathrm{Aeq,24h}}$ values. It may be also noted that the European models calculate the % HA which is the maximum value of annoyance reported for particular noise levels as the noise sensitivity in general is quite high in Indian community as compared to European community. Thus, the exposure-response relations (ERR) between noise levels of road traffic noise and the percentage of highly annoyed residents (% HA) as presented by Guski et al. (2017) [3] (discussed in Chap. 9) based on results of large

Table A5 Fractional noise impact values for the four zones considering the permissible limits (in case of unavailability of an authorized data of % HA correlating noise levels) for calculating the indicative noise impact index (*NII*)

Category	Permissible $L_{Aeq,24h}$ in dB(A)	Fractional noise impact using Guski et al. (2017) relationship for road traffic noise
Industrial	73.9	0.354
Commercial	63.5	0.189
Residential	53.5	0.101
Silence	48.5	0.082

socio-acoustic surveys may be used to calculate the indicative noise impact index values as shown in Table A5.

References

1. Gorai AK, Pal AK (2006) Noise impact assessment of an iron ore mining complex. Noise Control Eng J 54(6):352–359
2. Bies DA, Hansen CH (2009) Engineering noise control: theory and practice. Spon Press
3. Guski R, Schreckenberg D, Schuemer R (2017) WHO environmental noise guidelines for the european region: a systematic review on environmental noise and annoyance. Int J Environ Res Public Health 14:1539

Appendix V

Socio-Acoustic Survey on Annoyance of Transportation Noise Sources

The survey is intended for correlating the subjective responses with measured noise exposure levels. So, it is requested to kindly give your valuable comments only after reading the questions thoroughly.

Name : ____________________
Age : ____________________
Gender : ☐ Male / ☐Female
Area in which residing : ____________________
Occupation : ____________________
Annual Salary : ____________________

Education : ☐ High School / ☐ Graduate / ☐ Post Graduate / ☐ Doctorate

Social Status : ____________________(Please rate yourself based on your perception whether you belong to Lower Class/Middle class/Upper Middle Class /Higher Class)

Please Tick the only right option based on your judgment for the following Questions:

1. How **many persons** are there in your household?
☐ 1; ☐ 2; ☐ 3; ☐ 4; ☐ 5; ☐ More than 5

2. How **long** you had been staying in this household?
☐ One year; ☐ Less than one year; ☐ Two years; ☐ Between Two & Five years; ☐ More than Five years

3. a) Do you have **any fear** from any noise source (aircraft, any machine/vehicle etc)? ☐ Yes / ☐ No

b) How do you rate yourself as being **Noise Sensitive**?

☐ Highly sensitive; ☐ Medium Sensitive; ☐ Low sensitive; ☐ Not at all ☐ Can't Say
Please fill the questionnaire on judging your noise sensitivity attached in Annexure A.

c) Do you think that **sound insulation** of your house is adequate? ☐ Yes ☐ No ☐ Can't Say

4. Are you **Willing to Pay** for reducing the noise by sound proofing your house or installing a barrier
☐ Yes, but not much ☐ Yes, No issues on costs
☐ No, Satisfied with present status ☐ Not at all, State responsibility

5. How do you rate **metro train noise** in your near vicinity?
☐ Not at all annoyed
☐ Slightly Annoyed
☐ Moderately Annoyed ☐ Extremely Annoyed

6. How do you rate **traffic noise** in your near vicinity?
☐ Not at all annoyed
☐ Slightly Annoyed
☐ Moderately Annoyed
☐ Extremely Annoyed

7. How do you rate **Aircraft noise** in your area? **(Please leave blank if your locality has no aircrafts flyover).**
- ☐ Not at all annoyed
- ☐ Slightly Annoyed
- ☐ Moderately Annoyed
- ☐ Extremely Annoyed

8. How do you react to **Horn noise** of vehicles?
- ☐ Not at all annoyed
- ☐ Slightly Annoyed
- ☐ Moderately Annoyed
- ☐ Extremely Annoyed

9. How do you rate **industrial noise (or commercial noise)** in your near vicinity?
- ☐ Not at all annoyed
- ☐ Slightly Annoyed
- ☐ Moderately Annoyed
- ☐ Extremely Annoyed

10. When you pass through traffic/ exposed to aircraft flyover events /metro trains, what is your **response**?
- ☐ Stressed
- ☐ Irritated and angry
- ☐ Too worried to trip
- ☐ Other health disorders (e.g. Hypertension, Hearing loss etc)
- ☐ Mentally tired
- ☐ Headache
- ☐ No comments

11. Do you feel noise induced **sleep** disturbances?
- ☐ Not at all
- ☐ Difficulty falling asleep
- ☐ Disturbed sleep quality seldom
- ☐ Wakes up
- ☐ Can't say with confidence

12. How seriously are following activities disturbed by **transportation noise** you can hear while at home or at work place (noise disturbance)?

	Not at all	----------	----------	----------	**Very Much**
• Watching Television	☐1	☐2	☐3	☐4	☐5
• Concentrating	☐1	☐2	☐3	☐4	☐5
• Conversation	☐1	☐2	☐3	☐4	☐5

■ Do the other members in your **family** agree with your answers for the above questionnaire?
☐Yes / ☐ No /☐Can't say

(Signature)

Surveyor Observations/ Conclusions: ■Any interesting Observation: ____________________

■ Remarks about the **Noise Sensitivity** of subject (based on Weinstein Noise Sensitivity Scale):
☐ Low ☐ Medium ☐ High ☐ Couldn't Judge

■ **Noise Exposure** Characteristics of Area where subject is residing:

Please mail your comments at following address: ngarg@ nplindia.org

Scoring

The Weinstein Noise Sensitivity Scale (WNSS) was developed to measure individual sensitivity to perceived noise. Responses to each of the 21 items are scaled along 6 points as discussed in Luz (2005) study. However, in original Weinstein Noise Sensitivity Scale (WNSS) (Weinstein, 1978) responses to each of the 21 items are scaled along 6 points with 0 indicating strong disagreement and 5 indicating strong agreement. After reverse coding relevant items, the unweighted sum of scores from each of the items is tallied. Stronger agreement with the items results in a higher score (ranging from 0 to 105), indicating greater individual noise sensitivity (Weinstein, 1978).

Weinstein Noise Sensitivity Scale (WNSS)

Instructions – Circle the number corresponding to how well you agree or disagree. Don't be disturbed by the reversals of order from one line to another.

1. I wouldn't mind living on a noisy street if the apartment I had was nice.	AGREE 1 2 3 4 5 6 DISAGREE
2. I am more aware of noise than I used to be.	AGREE 6 5 4 3 2 1 DISAGREE
3. No one should mind much if someone turns up his stereo full blast once in a while.	AGREE 1 2 3 4 5 6 DISAGREE.
4. At movies, whispering and crinkling candy wrappers disturb me.	AGREE 6 5 4 3 2 1 DISAGREE
5. I am easily awakened by noise.	AGREE 6 5 4 3 2 1 DISAGREE
6. If it's noisy where I'm studying, I try to close the door or window or move someplace else.	AGREE 6 5 4 3 2 1 DISAGREE
7. I get annoyed when my neighbors are noisy.	AGREE 6 5 4 3 2 1 DISAGREE
8. I get used to most noises without much difficulty.	AGREE 1 2 3 4 5 6 DISAGREE
9. How much would it matter to you if an apartment you were interested in renting was located across from a Fire station.	A LOT 6 5 4 3 2 1 NOT MUCH
10. Sometimes noises get on my nerves and get me irritated.	AGREE 6 5 4 3 2 1 DISAGREE
11. Even music I normally like will bother me if I'm trying to concentrate.	AGREE 6 5 4 3 2 1 DISAGREE
12. It wouldn't bother me to hear the sounds of everyday living from my neighbors (footsteps, running water, etc).	AGREE 1 2 3 4 5 6 DISAGREE
13. When I want to be alone, it disturbs me to hear outside noises.	AGREE 6 5 4 3 2 1 DISAGREE
14. I'm good at concentrating no matter what is going on around me.	AGREE 1 2 3 4 5 6 DISAGREE
15. In a library, I don't mind if people carry on a conversation if they do it quietly.	AGREE 1 2 3 4 5 6 DISAGREE
16. There are often times when I want complete silence.	AGREE 6 5 4 3 2 1 DISAGREE
17. Motorcycles ought to be required to have bigger mufflers.	AGREE 6 5 4 3 2 1 DISAGREE
18. I find it hard to relax in a place that's noisy.	AGREE 6 5 4 3 2 1 DISAGREE
19. I get mad at people who make noise that keeps me from falling asleep or getting work done.	AGREE 6 5 4 3 2 1 DISAGREE
20. I wouldn't mind living in an apartment with thin walls.	AGREE 1 2 3 4 5 6 DISAGREE
21. I am sensitive to noise.	AGREE 6 5 4 3 2 1 DISAGREE

Add up the numbers for your score. **TOTAL SCORE** ___________

References

1. Weinstein ND (1978) Individual differences in reactions to noise: a longitudinal study in a college dormitory. J Appl Psychol 63: 458–466
2. Weinstein ND (1980) Individual differences in critical tendencies and noise annoyance. J Sound Vibration 68:2 41–248
3. Worthington DL, Bodie GD (2018) The sourcebook of listening research: methodology and measures, 1st Ed. Wiley. pp 475–481
4. Smith A (2003) The concept of noise sensitivity: implications for noise control. Noise Health 5:57–59
5. Stansfeld SA, Sharp DS, Gallacher J, Babisch W (1993) Road traffic noise, noise sensitivity and psychological disorder. Psychol Med 88: 167–188.
6. Luz George A (2005) Noise Sensitivity Rating of Individuals, Sound and Vibration, August, 2005, 14–17

Appendix VI

Performance Evaluation of Noise Reducing Devices

The European classification system for road traffic noise reducing devices (NRDs) is based on an advanced framework based on "product standard" EN 14388 that specifies the performance requirements for NRDs in terms of measurable characteristics and supporting standards covering acoustic, nonacoustic and long-term performance [1]. The intrinsic properties that are independent of the environment of noise reducing devices (NRDs) or noise barriers are the airborne sound insulation and sound absorption. However, the extrinsic properties depend upon the specific environment like terrain characteristics, meteorological parameters, etc. The insertion loss of noise barrier is an illustration of an extrinsic property as it may depend on the environment. The recent European project named as SOPRANOISE (Securing and Optimizing the Performance of Road trAffic noise barriers with New methOds and In-Situ Evaluation) focuses on the acoustic performances of noise barriers and developing the in situ inspection methods [2–4]. The EN 1793 standards [5–10] primarily deal with the intrinsic acoustic properties of the noise barriers as follows:

Diffuse Sound Field Method for Testing Sound Absorption (EN 1793-1)

The change of reverberation time inside a reverberation chamber after installing a test sample of the NRD at the room's surface is the basis for this procedure (as measured with ISO 354). A single-number rating of the sound absorption evaluation $DL_{\alpha,\,\mathrm{NRD}}$ is described by the following equation (in accordance with EN 1793-3) [2]:

$$DL_{\alpha,NRD} = -10\log_{10}\left[1 - \frac{\sum_{i=1}^{18}\alpha_{NRD,i}10^{0.1L_i}}{\sum_{i=1}^{18}10^{0.1L_i}}\right] \quad (A.3)$$

where α_{NRD} is the absorptance of the test sample and L_i is the relative A-weighted sound pressure levels (dB) of the normalized traffic noise spectrum, as defined in EN 1793-3. The classification of the single-number sound absorption evaluation rating is done on the basis of Table A6 placed in the older version of the standards, but not in the current one [11, 12].

Table A6 Categories of the absorptive performance [11]

Category	$DL_{\alpha,\ NRD}$
A0	Not determined
A1	<4
A2	4–7
A3	8–11
A4	12–15
A5	>15

The values of DL_α are limited to ≤ 0.99 (corresponding $DL_\alpha = 20$ dB), thus it can range from 0 to 20 dB [13]. Most highway projects would require A3 or higher for a reasonable amount of absorption. Close "corridors," high barriers (>3 m), and reverberant areas where higher degrees of absorption are justified should prefer using A4 [11]. Figure A1 represents the normalized traffic spectrum in one-third octave band according to EN 1793-3 standard. The spectrum is normalized to an overall value of 0 dB as its shape is more important rather than the absolute value [13].

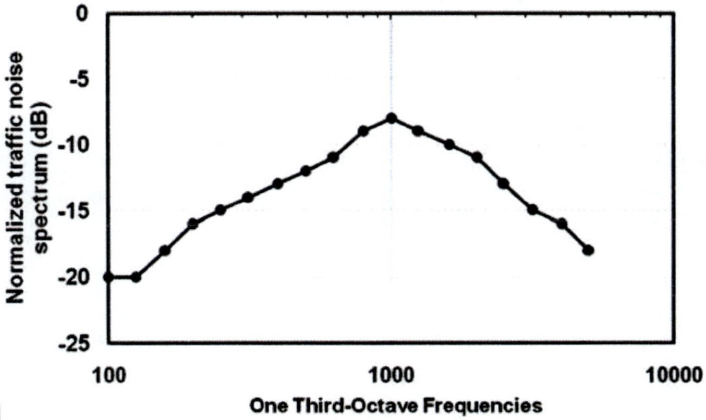

Fig. A1 Normalized traffic noise spectrum (in dB) at one-third octave frequency band according to EN 1793-3 standard

Diffuse Sound Field Method for Testing Airborne Sound Insulation (EN 1793-2)

This method requires two adjacent reverberation rooms, the "source room" and the "receiving room," with an opening between them in which the test specimen is inserted. The method is based on ISO 10140-2. A single number rating, DL_R, for summarizing the airborne sound insulation in third-octave bands can be given by [2]:

$$DL_R = -10\log_{10}\left[\frac{\sum_{i=1}^{18} 10^{0.1L_i}\, 10^{-0.1R_i}}{\sum_{i=1}^{18} 10^{0.1L_i}}\right] \tag{A.4}$$

where R is the airborne sound insulation of the test specimen. Table A7 represents the classification of the single-number airborne sound insulation evaluation rating placed in the older version of standards [11].

Table A7 Categories of the airborne sound insulation performance [11]

Category	DL_R
B0	Not determined
B1	≤15
B2	15–24
B3	≥24

B3 is required for the majority of highway developments. B2 may be acceptable for low barriers (less than 2 m); however, the rating level of at least 20 dB is required [11].

Direct Sound Field Method for Testing Sound Reflection (EN 1793-5)

This method in accordance with EN 1793-5 is based on a comparison between the sound energy reflected from the NRD and the sound energy emitted towards it. The third-octave band values are then averaged and weighted according to the standardized traffic noise spectrum L_i in EN 1793-3, to get a single number rating for the NRD as [2]:

$$DL_{RI} = -10\log_{10}\left[\frac{\sum_{i=1}^{18} RI_i 10^{0.1L_i}}{\sum_{i=1}^{18} 10^{0.1L_i}}\right] \tag{A.5}$$

where RI_i is the reflectance in i-th third-octave bands. For the calculated value of DL_{RI}, a higher value denotes a higher sound absorption, the property which is generally desired for an NRD. The basic test setup for EN 1793-5 measurements is schematically shown in the Fig. A2 that depicts the side view of the sound source (loudspeaker), the microphone grid, and the NRD under test. The sound is generated by a loudspeaker positioned 1.5 m in front of the NRD at half of its height. The corresponding microphones are positioned in between, in a 3 × 3 grid, with a distance of 0.25 m to the NRD and with a spacing of 0.4 m to neighboring ones. Multiple impulse response measurements are performed with this setup to average out noise [2]. The power spectra of the direct and the reflected components corrected to take into account the path length difference give the basis for calculating the reflection index [13].

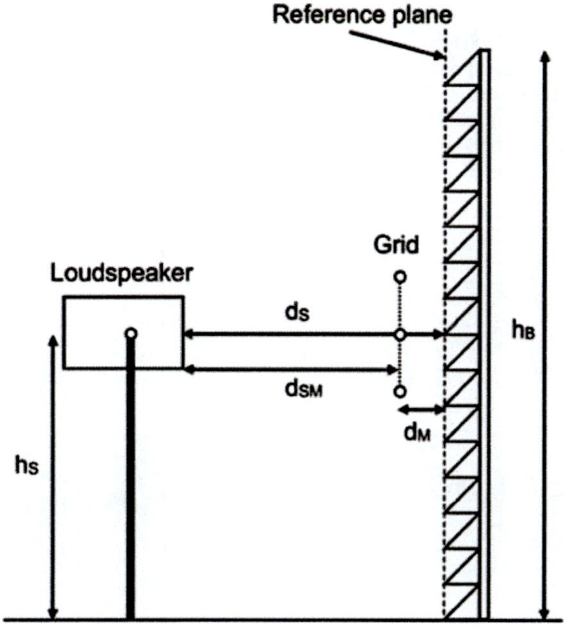

Fig. A2 Setup for direct sound field reflection measurement (D2.1-Review of the physical significance of EN 1793-1, EN 1793-2, EN 1793-5 and EN 1793-6, SOPRANOISE Project report, [2])

Direct Sound Field Method for Testing Airborne Sound Insulation (EN 1793-6)

This method, according to EN 1793-6, is based on a comparison between the sound energy arriving at the microphone positions with and without the NRD. The single-number rating by a weighted average with the standardized traffic noise spectrum L_i according to EN 1793-3 is described by [2]:

$$\mathrm{DL_{SI}} = -10\log_{10}\left[\frac{\sum_{i=1}^{18} 10^{0.1L_i}\, 10^{-0.1\mathrm{SI}_i}}{\sum_{i=1}^{18} 10^{0.1L_i}}\right] \quad \text{(A.6)}$$

The basic test arrangement for EN 1793-6 measurements is schematically shown in Fig. A3 depicting the side view of the sound source S, the microphone array M, and the NRD under test. A loudspeaker is mounted on the traffic side of the test barrier, while a microphone is mounted on the opposing side. The sound wave, emitted by the loudspeaker, is partly reflected, partly transmitted, and partly diffracted by the noise barrier. A signal is received by the microphone, which is postprocessed to produce an overall impulse response. The measurement in front of the barrier is reported at nine points placed on an ideal grid. The final sound insulation index is the logarithmic average of the results of the nine positions. The power spectra of the direct free-field component and the transmitted component corrected to take into account the path length difference of the two signals are used to calculate the sound transmission loss [14].

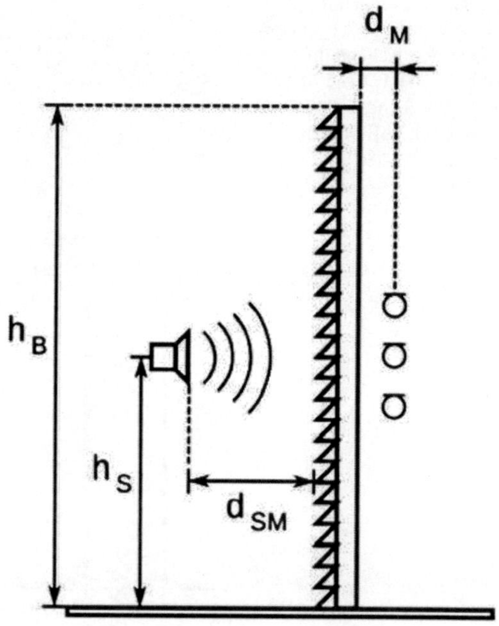

Fig. A3 Setup for direct sound field reflection measurement (D2.1-Review of the physical significance of EN 1793-1, EN 1793-2, EN 1793-5 and EN 1793-6, SOPRANOISE Project report, [2])

Thus, these single-number ratings can be used for the performance evaluation of noise barriers or noise reducing devices. The in situ method proved to be reliable and yield realistic values than the laboratory method and additionally facilitates the fault detection and ascertain the acoustic durability of the NRD [13]. The correlation between the values of the two single-number ratings observed form the laboratory measurements (100 Hz to 5 kHz), DL_R, specified in EN 1793-2 and R_w in ISO 717-1 is as [14]:

$$DL_R = 0.98\,R_w - 3.05\ (r = 0.995) \tag{A.7}$$

It has been observed that on average the EN single number rating DL_R is few decibels less than the most widely used sound reduction index, R_w is building acoustics. The correlation observed in the frequency range of 200 Hz to 5 kHz is [14]:

$$\mathrm{DL_R} = 0.93\, R_w + 0.37\ (r = 0.992) \tag{A.8}$$

The linear correlation between the single-number rating, $\mathrm{DL_R}$, obtained from the laboratory data and single-number rating, $\mathrm{DL_{SI}}$, obtained from the outdoor data measured close to the barrier and posts is as [14]:

$$\mathrm{DL_{SI}} = 1.18\ \mathrm{DL_R} - 0.94\ (r = 0.97)\ \text{for elements} \tag{A.9}$$

$$\mathrm{DL_{SI}} = 1.18\ \mathrm{DL_R} - 3.16\ (r = 0.93)\ \text{for posts} \tag{A.10}$$

The updated correlations between these single-number ratings can be referred in SOPRANOISE project report [15]. Table A8 shows an illustrative example of the acoustical requirements in tenders of Agency for Roads and Traffic of the Flemish Government [16].

Table A8 Recommended acoustical requirements by Agency for Roads and Traffic of the Flemish Government [16]

Standard	Parametric requirements
EN 1793-1	$\mathrm{DL_{\alpha,\,M}} \geq \mathrm{DL_{\alpha,\,min}}$ = 10 dB
EN 1793-2	$\mathrm{DL_{R,\,M}} \geq \mathrm{DL_{R,\,min}}$ = 25 dB
EN 1793-5	$\mathrm{DL_{RL,\,M}} \geq \mathrm{DL_{RI,\,min}}$ = 5 dB
EN 1793-6	At the element:$\mathrm{DL_{SI,\,E,\,M}} \geq \mathrm{DL_{SI,\,E,\,min}}$ = 28 dB
	At the post or junction:$\mathrm{DL_{SI,\,P,\,M}} \geq \mathrm{DL_{SI,\,P,\,min}}$ = 26 dB

References

1. Garai M, Clairbois JP (2003) The European Qualification system for road traffic Noise Reducing Devices. Euronoise 2003, Naples, Paper ID 117-OL
2. SOPRANOISE Project report, D2.1-Review of the physical significance of EN 1793-1, EN 1793-2, EN 1793-5 and EN 1793-6. June 2020, https://www.enbf.org/sopranoise/wp-content/uploads/2021/06/20200608_spnWP2_T2.1_D2.1.pdf
3. Conference of European Directors of Roads (CEDR) website, https://www.cedr.eu/peb-research-programme-2018-noise-and-nuisance
4. SOPRANOISE (Securing and Optimizing the Performance of Road trAffic noise barriers with New methOds and In-Situ Evaluation) project, https://www.enbf.org/sopranoise
5. EN 1793-1 (2017) Road traffic noise reducing devices – test method for determining the acoustic performance – part 1: intrinsic characteristics of sound absorption under diffuse sound field conditions

6. EN 1793-2 (2018) Road traffic noise reducing devices – part 2: intrinsic characteristics of airborne sound insulation under diffuse sound field conditions
7. EN 1793-3 (1998) Road traffic noise reducing devices – test method for determining the acoustic performance – Part 3: normalised traffic noise spectrum
8. EN 1793-4 (2015) Road traffic noise reducing devices – test method for determining the acoustic performance – Part 4: intrinsic characteristics – in situ values of sound diffraction
9. EN 1793-5 (2016) Road traffic noise reducing devices – test method for determining the acoustic performance. Intrinsic characteristics – in situ values of sound reflection under direct sound field conditions
10. EN 1793-6 (2018) Road traffic noise reducing devices – test method for determining the acoustic performance – part 6: Intrinsic characteristics – in situ values of airborne sound insulation under direct sound field conditions
11. Giles P (2006) Effective noise barrier design and specification. Proceedings of ACOUSTICS 2006, 20–22 November, 2006, New Zealand
12. Batko W, Pawlik P, Wszołek G (2017) Sensitivity analysis of the estimation of the single-number sound absorption evaluation index DL_{α}. Arch Acoust 42 (4):689–696
13. Garai M, Guidorzi P (2008) In situ measurements of the intrinsic characteristics of the acoustic barriers installed along a new high speed railway line. Noise Control Eng J 56(5):342–355
14. Garai M, Guidorzi P (2000) European methodology for testing the airborne sound insulation characteristics of noise barriers *in situ*: Experimental verification and comparison with laboratory data. J Acoust Soc Am 108(3):1054–1067
15. SOPRANOISE Project report (2021, April) D2.2-Final report on the main results of WP2-Acoustic assessment of the intrinsic performances of noise barriers. https://www.cedr.eu/docs/view/60ba4bae80acb-en.
16. Buytaert A, Vanhooreweder B, Clairbois J P, Houtave P, Glorieux C, Roozen NB (2017) In-situ measurements according EN 1793-5 and EN 1793-6-First results and impressions. Proceedings Internoise, 27th to 30th August, 2017, Hong Kong

Index

N. Garg, *Environmental Noise Control*,
https://doi.org/10.1007/978-3-030-87828-3

T

V